Heat Transfer and Thermal Engineering

Heat Transfer and Thermal Engineering

Edited by **Edgar Miller**

NY RESEARCH PRESS

New York

Published by NY Research Press,
23 West, 55th Street, Suite 816,
New York, NY 10019, USA
www.nyresearchpress.com

Heat Transfer and Thermal Engineering
Edited by Edgar Miller

International Standard Book Number: 978-1-63238-513-0 (Hardback)

Printed in the United States of America.

Contents

Preface

Heat transfer is the exchange of thermal energy between physical systems. Thermal engineering encompasses mass transfer, fluid mechanisms, heat transfer, and thermodynamics. There has been rapid progress in this field and its applications are finding their way across multiple industries. Solar heating, boiler design, thermal power plants, cooling systems, designing of combustion systems and cooling of computer chips are some of the important applications of these disciplines. This book is compiled in such a manner, that it will provide in-depth knowledge about the theory and practice of heat transfer and thermal energy. Such selected researches that redefine the concepts of this discipline have been presented in this text. With state-of-the-art inputs by acclaimed experts of this field, this book targets students and professionals alike and aims to contribute to the growth of the discipline.

The information shared in this book is based on empirical researches made by veterans in this field of study. The elaborative information provided in this book will help the readers further their scope of knowledge leading to advancements in this field.

Finally, I would like to thank my fellow researchers who gave constructive feedback and my family members who supported me at every step of my research.

Editor

1

FUNDAMENTALS AND APPLICATIONS OF NEAR-FIELD RADIATIVE ENERGY TRANSFER

Keunhan Park[a,*] and Zhuomin Zhang[b]

[a]*Department of Mechanical, Industrial and Systems Engineering, University of Rhode Island, Kingston, RI 02881, USA*
[b]*G.W. Woodruff School of Mechanical Engineering, Georgia Institute of Technology, Atlanta, GA 30332-0405, USA*

ABSTRACT

This article reviews the recent advances in near-field radiative energy transfer, particularly in its fundamentals and applications. When the geometrical features of radiating objects or their separating distances fall into the sub-wavelength range, near-field phenomena such as photon tunneling and surface polaritons begin to play a key role in energy transfer. The resulting heat transfer rate can greatly exceed the blackbody radiation limit by several orders magnitude. This astonishing feature cannot be conveyed by the conventional theory of thermal radiation, generating strong demands in fundamental research that can address thermal radiation in the near field. Important breakthroughs of near-field thermal radiation are presented here, covering from the essential physics that will help better understand the basics of near-field thermal radiation to the most recent theoretical as well as experimental findings that will further promote the fundamental understanding. Applications of near-field thermal radiation in various fields are also discussed, including the radiative property manipulation, near-field thermophotovoltaics, nanoinstrumentation and nanomanufacturing, and thermal rectification.

Keywords: *thermal radiation, micro/nanoscale, radiative properties, thermophotovoltaics, nanoinstrumentation, nanomanufacturing*

1. INTRODUCTION

Conventionally, the theory of thermal radiation is based on the concept of blackbody, cast by Gustav Kirchhoff in 1860. A blackbody absorbs all energy of the radiation rays reaching it geometrically. Among all objects at the same temperature with the same geometry, a blackbody emits the largest amount of energy when measured in the same angular and spectral ranges. As such, the Stefan-Boltzmann law and Planck's law provide descriptions of the total and spectral characteristics of blackbodies. Thermal emission from real materials can be described by comparison with that emitted by a blackbody at the same temperature using a property called emissivity (also called emittance). Although care should be taken with regards to the proper definition of emissivity (spectral, total, directional, individual polarization versus polarization averaged, etc.) (Howell *et al.*, 2010; Modest, 2003; Zhang, 2007), the emissivity should be always smaller than unity in conventional thermal radiation: that is, thermal emission from real materials is always smaller than that from the blackbody in the far-field regime.

Conventional radiative transfer approaches are often not applicable when the geometric features or distances are smaller than the characteristic wavelength of thermal radiation based on the Wien's displacement law (Zhang and Wang, 2012). Planck (1914) noted that the spectral distribution of blackbody radiation is derived based on the assumption that the geometric dimensions of the enclosure (also called a blackbody cavity) are much greater than the characteristic wavelength of thermal radiation. This condition makes Planck's law only applicable in the far field, i.e., away from the surface of any objects. In essence, thermal radiation can be understood as electromagnetic (EM) waves emitted due to the random fluctuation of charges in the material. When a material is in thermal equilibrium at temperature T, charges such as free electrons (for metals) or ions (for polar materials) experience a random thermal motion and radiate the fluctuating EM field. While the average of the fluctuating electric or magnetic field is zero due to its random nature, the energy density and Poynting vector (which characterizes the energy flux) are nonzero and can greatly exceed the blackbody radiation depending on the dielectric and magnetic properties of materials (Rytov *et al.*, 1987). In particular, evanescent EM fields near the interface, which do not carry energy alone and exponentially decay from the interface, are coupled to carry a significant portion of energy across the gap when two objects are placed closer than the characteristic wavelength of thermal radiation. This phenomenon is known as photon tunneling and is responsible for the enhanced energy transfer in the near field, along with other near-field effects such as interference and surface polaritons (Zhang, 2007; Fu and Zhang, 2006; Basu *et al.*, 2009). Such near-field effects can also alter the far-field properties of nanostructured surfaces or objects. In the far-field, no matter how complex the structure is, the emissivity and transmittance cannot exceed unity. However, unique spectral- and angular-dependent radiative properties can be achieved by engineering nano/microstructures (Zhang and Wang, 2011). Surface waves and photonic band structures are often utilized to enable unique optical properties of nano/microstructures

Corresponding author. Email: kpark@egr.uri.edu

(Fu and Zhang, 2009).

Near-field radiation holds promise for applications in energy systems, nanofabrication and near-field imaging. Rapidly depleting reserves of fossil fuels and the concern of the global warming have placed a great demand of alternative power generation technologies. One of such technology is a thermophotovoltaic (TPV) system, which operates on the principle similar to that of solar cells (but with a lower bandgap) to generate electricity from thermal emission. A possible method of improving the performance of TPV systems is to employ near-field thermal radiation for the energy conversion (Basu *et al.*, 2007). However, large near-field thermal radiation is not always favorable in some energy conversion systems: as revealed by Dillner (2008), near-field thermal radiation needs to be suppressed to increase the thermoelectric energy conversion efficiency of thermotunneling devices. Besides the energy conversion, near-field thermal radiation has also been used for imaging beyond the diffraction limit (De Wilde *et al.*, 2006; Kittel *et al.*, 2005). Furthermore, the concept of using near-field radiation as thermal rectifier has also been suggested (Otey *et al.*, 2010; Basu and Francoeur, 2011a). Limiting the magnitude of near-field radiation is critical for improving the performance of thermal tunneling devices (Dillner, 2008). Another important application of near-field radiation is in the field of nanomanufacturing. Enhanced transmission of metallic films perforated with subwavelength holes stirred the interest in studying light transmission through nanostructures. Nanolithography techniques based on the surface plasmon waves have been demonstrated for patterning structures of less than 50 nm (Liu *et al.*, 2005; Wang *et al.*, 2006). Furthermore, nanoscale direct writing has also been demonstrated using near-field optics coupled with femtosecond laser (Grigoropoulos *et al.*, 2007).

This review article provides a thorough review of near-field thermal radiation, covering the essential physics of fluctuational electromagnetism along with recent advances in fundamentals and applications of near-field thermal radiation. The remaining sections are organized as follows. Section 2 focuses on the essential physics of near-field thermal radiation by introducing the fluctuation-dissipation theorem, Dyadic Green's function, dielectric functions, and surface polaritons. Recent advances on the fundamental research of near-field thermal radiation are discussed in the following sections. Starting from near-field radiative heat transfer between two semi-infinite media, Section 3 discusses the upper limit of near-field radiative heat flux and the extremely small penetration depth of near-field thermal radiation, as well as energy streamlines – a novel way to elucidate the near-field energy propagation between semi-infinite media and multilayered structures. Near-field thermal radiation in other geometries, such as sphere-sphere and sphere-flat surface, and between emerging nanomaterials is covered in Section 4, followed by the discussion of experimental observations of near-field radiative heat transfer in Section 5. Section 6 looks into the applications of near-field thermal radiation, including the manipulation of radiative properties, near-field thermophotovoltaics, tip-based engineering, and thermal rectification. In conclusion, a brief summary on the recent progresses in fundamentally understanding and engineering near-field thermal radiation is provided along with remarks on future research opportunities and challenges of the field.

2. ESSENTIAL PHYSICS OF NEAR-FIELD THERMAL RADIATION

2.1. Fluctuation-Dissipation Theorem

Thermal radiation between solids is traditionally treated as a surface phenomenon with the concept of emissivity, reflectivity and absorptivity of the surfaces. Radiation heat transfer in a participating medium is thus analyzed using ray optics, leading to the development of the radiative transfer equation (RTE) that considers emission, absorption, and scattering of thermally emitted rays in the medium (Howell *et al.*, 2010). Although RTE can determine the radiation distribution within the medium by computing the intensity along the propagation of radiation, this phenomenological equation does not fully account for the fundamentals of thermal emission and breaks down when wave interference and diffraction become important. To speculate the origin of thermal radiation, Rytov and his co-workers (1987) combined the fluctuation-dissipation theorem and Maxwell's equations to establish the fluctuational electrodynamics. According to the fluctuation-dissipation theorem, thermal radiation is essentially EM waves emitted from the fluctuating currents due to the random thermal motion of charges, known as thermally induced dipoles, in a medium. Thus the propagation of thermal radiation and its interaction with matter can be fully described in the framework of the fluctuational electrodynamics, in both the far-field and near-field regimes.

When a material is in thermal equilibrium at temperature T, charges in the material – electrons in metals or ions in polar crystals – are subject to random thermal motions and generate fluctuating electric currents. The fluctuating electric density $\mathbf{j}(\mathbf{x}, t)$, or $\mathbf{j}(\mathbf{x}, \omega)$ in the frequency domain, can be implemented in Maxwell's equations as an external source to make the equations stochastic. The key issue in calculating thermally induced, fluctuating EM waves is then how to determine the statistical properties of these random sources. According to the fluctuation-dissipation theorem (FDT), while the fluctuating electric density is averaged to zero (i.e.,$\langle j_m(\mathbf{x}, \omega)\rangle = 0$) due to its random nature, the ensemble average of its cross-spectral spatial correlation function is nonzero and expressed as (Joulain *et al.*, 2005):

$$\langle j_m(\mathbf{x}', \omega) j_n^*(\mathbf{x}'', \omega')\rangle = \frac{4}{\pi}\omega\varepsilon_0 \mathrm{Im}(\varepsilon(\omega))\delta_{mn}\delta(\mathbf{x}' - \mathbf{x}'')\Theta(\omega, T)\delta(\omega - \omega') \quad (1)$$

where $\langle\ \rangle$ represents ensemble averaging, and $*$ denotes the complex conjugate. In Eq. (1), ε_0 is the electrical permittivity of the free space, j_m and j_n (m, n = 1, 2, or 3) stands for the x, y, or z component of $\mathbf{j}$, δ_{mn} is the Kronecker delta, and $\delta(\mathbf{x}' - \mathbf{x}'')$ and $\delta(\omega - \omega')$ are the Dirac delta function. $\Theta(\omega, T)$ is the mean energy of a Planck oscillator at the frequency ω in thermal equilibrium at temperature T and is given by $\Theta(\omega, T) = \hbar\omega/[exp(\hbar\omega/k_B T) - 1]$, where $\hbar$ is the Planck constant divided by 2π and k_B is the Boltzmann constant. Since only positive values of frequencies are considered here, a factor of 4 has been included in Eq. (1) to consistently use the conventional definitions of the spectral energy density and the Poynting vector (Fu and Zhang, 2006).

2.2. Dyadic Green's Function

For prescribed geometric conditions and temperature, Maxwell's equations need to be solved in order to obtain the electric and magnetic field distributions. This can be done with the help of the dyadic Green's function, which makes the formulations simple and compact. With the assistance of the dyadic Green's function $\overline{\overline{\mathbf{G}}}_e(\mathbf{x}, \mathbf{x}', \omega)$, the induced electric and magnetic fields due to the fluctuating current density can be expressed respectively in the frequency domain as volume integrations:

$$\mathbf{E}(\mathbf{x}, \omega) = i\omega\mu_0 \int_V \overline{\overline{\mathbf{G}}}_e(\mathbf{x}, \mathbf{x}', \omega) \cdot \mathbf{j}(\mathbf{x}', \omega) d\mathbf{x}' \quad (2)$$

$$\mathbf{H}(\mathbf{x}, \omega) = \int_V \overline{\overline{\mathbf{G}}}_h(\mathbf{x}, \mathbf{x}', \omega) \cdot \mathbf{j}(\mathbf{x}', \omega) d\mathbf{x}' \quad (3)$$

where $\overline{\overline{\mathbf{G}}}_h(\mathbf{x}, \mathbf{x}', \omega) = \nabla \times \overline{\overline{\mathbf{G}}}_e(\mathbf{x}, \mathbf{x}', \omega)$ is the magnetic dyadic Green's function, μ_0 is the magnetic permeability of vacuum, and the integral is over the region V that contains the fluctuating sources. The dyadic Green's function, $\overline{\overline{\mathbf{G}}}_e(\mathbf{x}, \mathbf{x}', \omega)$ is essentially a spatial transfer function between a point source at location $\mathbf{x}'$ and the resultant electric field $\mathbf{E}$ at $\mathbf{x}$ (Zhang, 2007). Based on the ergodic hypothesis, the spectral energy

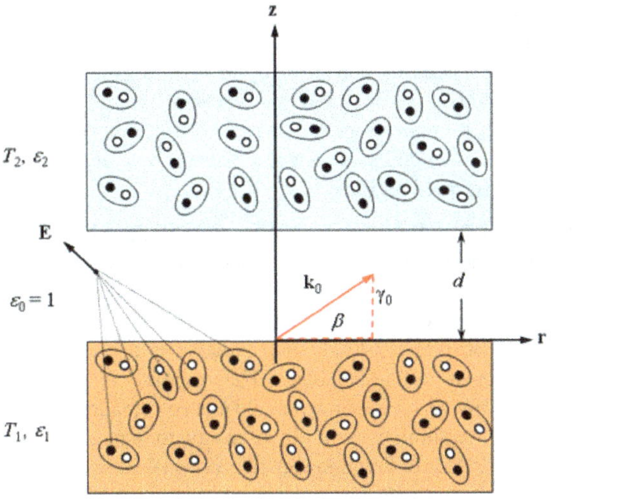

Fig. 1 Schematic of near-field radiative heat transfer between two closely spaced semi-infinite plates, at temperatures T_1 and T_2, separated by a vacuum gap d.

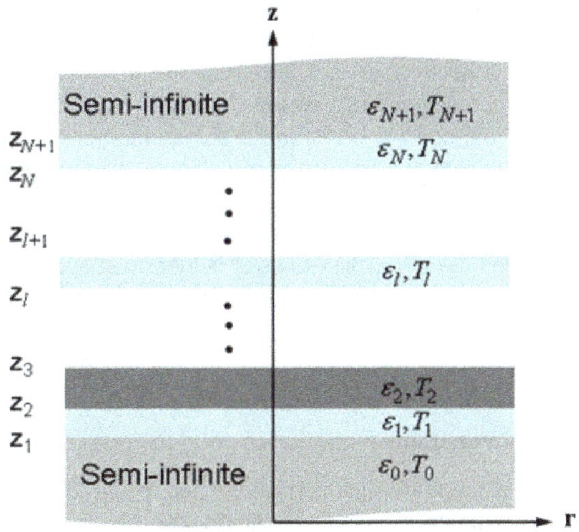

Fig. 2 Schematic of a multilayered thin-film structure for Green's function analysis. Each of the films may be at a different temperature.

flux is given by (Basu *et al.*, 2009)

$$\langle \mathbf{S}(\mathbf{x},\omega)\rangle = \int_0^\infty \frac{1}{2}\left\langle \mathrm{Re}[\mathbf{E}(\mathbf{x},\omega') \times \mathbf{H}^*(\mathbf{x},\omega')]\right\rangle d\omega' \tag{4}$$

where $\mathbf{S}$ is the spectral Poynting vector, ω (and ω') is the angular frequency. In order to compute the spectral Poynting vector at $\mathbf{x}$, we should compute the cross-spectral density of electric and magnetic field vectors $E_i(\mathbf{x},\omega)$ and $H_j(\mathbf{x},\omega')$. The cross-spectral density can be written as

$$\left\langle E_i(\mathbf{x},\omega)H_j^*(\mathbf{x},\omega')\right\rangle = i\omega\mu_0 \int_V d\mathbf{x}' \int_V d\mathbf{x}'' \\ \left\{G_{e,im}(\mathbf{x},\mathbf{x}',\omega)G_{h,jn}(\mathbf{x},\mathbf{x}'',\omega')\left\langle j_m(\mathbf{x}',\omega)j_n^*(\mathbf{x}'',\omega')\right\rangle\right\} \tag{5}$$

With the relationship between the fluctuating current densities and the temperature of the emitting medium being established through Eq. (1), the spectral radiative heat flux can be calculated using Eq. (5) once the dyadic electric Green's function, $\overline{\overline{\mathbf{G}}}_e(\mathbf{x},\mathbf{x}',\omega)$ is obtained. Since the dyadic Green's function depends on the geometry of the physical system, the following sections will briefly describe the dyadic Green's functions for two representative structures, i.e., for two semi-infinite media and multilayered media.

Two semi-infinite media: Let's consider near-field thermal radiation between two semi-infinite media separated by a vacuum gap (or a dielectric medium) of width d, when they are in thermal equilibrium at temperatures T_1 and T_2, respectively, where $T_1 > T_2$. Both media are nonmagnetic, isotropic, and homogeneous, and surfaces are parallel and smooth. As illustrated in Fig. 1, dipoles in the media are in random motions, radiating space-time dependent fluctuating electric field, $\mathbf{E}(\mathbf{x},t)$. Cylindrical coordinate system is used so that the space variable $\mathbf{x} = \mathbf{r} + \mathbf{z}$, with r-direction being parallel to the interface and z-direction perpendicular to the interface. β and γ_j refer to the r-component and z-component of the wavevector $\mathbf{k}_j$, respectively. Thus, $\mathbf{k}_j = \beta\hat{\mathbf{r}} + \gamma_j\hat{\mathbf{z}}$ and $k_j^2 = \beta^2 + \gamma_j^2$, for j = 0, 1, and 2. The magnitude of $\mathbf{k}_j$ is related to the dielectric function ε_j by $k_0 = \omega/c$, $k_1 = \sqrt{\varepsilon_1}\omega/c$, and $k_2 = \sqrt{\varepsilon_2}\omega/c$, with c being the speed of light in vacuum and ε_1 and ε_2 being the dielectric functions (or relative permittivity) of medium 1 and 2, respectively. For the described two semi-infinite media, the dyadic Green's function takes the following form (Fu and Zhang, 2006; Joulain *et al.*, 2005)

$$\overline{\overline{\mathbf{G}}}_e(\mathbf{x},\mathbf{x}',\omega) = \\ \int_0^\infty \frac{i}{4\pi\gamma_1}\left(\hat{\mathbf{s}}t_{12}^{\mathrm{s}}\hat{\mathbf{s}} + \hat{\mathbf{p}}_2 t_{12}^{\mathrm{p}}\hat{\mathbf{p}}_1\right)e^{i(\gamma_2 z - \gamma_1 z')}e^{i\beta(r-r')}\beta d\beta \tag{6}$$

where $\mathbf{x} = r\hat{\mathbf{r}} + z\hat{\mathbf{z}}$ and $\mathbf{x}' = r'\hat{\mathbf{r}} + z'\hat{\mathbf{z}}$. The unit vectors are $\hat{\mathbf{s}} = \hat{\mathbf{r}} \times \hat{\mathbf{z}}$ and $\hat{\mathbf{p}}_{1(2)} = \left(\beta\hat{\mathbf{z}} - \gamma_{1(2)}\hat{\mathbf{r}}\right)/k_1$. Note that t_{12}^{s} and t_{12}^{p} are the transmission coefficients from medium 1 to medium 2 for s- and p-polarizations, respectively, given by Airy's formula (Zhang, 2007). Provided that t_{12}^{s} and t_{12}^{p} take into account multiple reflections in the vacuum layer, the dyadic Green's function describes the transfer of the electromagnetic fields through propagating waves (i.e., $\beta < k_j$) and evanescent waves (i.e., $\beta > k_j$), from a point source at $\mathbf{x}'$ to a receiving point at $\mathbf{x}$.

Multilayered media: Dyadic Green's function for multilayered structures has been extensively used for calculating microwave thermal emission from layered media (Tsang *et al.*, 1974), thermal emission from 1-D photonic crystals (Narayanaswamy and Chen, 2005), power generation in near-field TPV systems (Park *et al.*, 2008), and near-field energy transfer between bodies with thin film coatings (Francoeur *et al.*, 2008; Fu and Tan, 2009). Figure 2 shows the schematic of a multilayered structure containing N thin films sandwiched between two semi-infinite half spaces. Properties of the layers are different and are assumed to vary only in the z-direction. The layers can be metallic, dielectric or even be a vacuum gap and can have a temperature gradient across them. The dyadic Green's function between any two layers s and l in Fig. 2 is given by (Park *et al.*, 2008)

$$\overline{\overline{\mathbf{G}}}_e(\mathbf{x},\mathbf{x}',\omega) = \frac{i}{4\pi}\int \frac{\beta d\beta}{\gamma_s}F(\beta)e^{i\beta(r-r')} \tag{7}$$

where

$$F(\beta) = Ae^{i(\gamma_l z - \gamma_s z')}\hat{\mathbf{e}}_l^+\hat{\mathbf{e}}_s^+ + Be^{i(-\gamma_l z - \gamma_s z')}\hat{\mathbf{e}}_l^-\hat{\mathbf{e}}_s^+ + \\ Ce^{i(\gamma_l z + \gamma_s z')}\hat{\mathbf{e}}_l^+\hat{\mathbf{e}}_s^- + De^{i(-\gamma_l z + \gamma_s z')}\hat{\mathbf{e}}_l^-\hat{\mathbf{e}}_s^- \tag{8}$$

Here, the subscript s denotes a source layer and l is the receiving layer. Note that $\hat{\mathbf{e}}^+$ and $\hat{\mathbf{e}}^-$ are two unit vectors, which are given by $\hat{\mathbf{e}}_l^+ = \hat{\mathbf{e}}_l^- = \hat{\mathbf{r}} \times \hat{\mathbf{z}}$ for s-polarization and $\hat{\mathbf{e}}_l^\pm = (\beta\hat{\mathbf{z}} \mp \gamma_l\hat{\mathbf{r}})/k_l$ for p-polarization, respectively. The coefficients A,B,C, and D can be determined using the transfer matrix formulation (Zhang, 2007; Park *et al.*, 2008; Francoeur *et al.*, 2009).

There are four terms in the expression of $F(\beta)$ because EM waves in each layer can be decomposed into upward and downward components due to multiple reflections at each interface. The first two terms account for the upward and downward waves in the l-th layer, respectively, which are induced by the upward waves in the source medium. Likewise, the last two terms denote the two waves in l-th layer due to the downward waves in the source medium (Tsang *et al.*, 2004). It should be pointed out that the terms having $\hat{\mathbf{e}}_s^-$ become zero if the source is in the bottom semi-infinite medium while the l-th layer is located above, and the terms having $\hat{\mathbf{e}}_l^-$ become zero if the l-th layer is the top semi-infinite medium or if there is free emission from multilayered structures. When both the source and receiver layers are semi-infinite, Eq. (7) will reduce to Eq. (6).

2.3. Dielectric Functions

Besides the fluctuation-dissipation theorem and dyadic Green's function, the dielectric function of materials should also be discussed to better understand near-field thermal radiation and its interactions with materials. If nonlinear optical effects are ignored, the polarization $\mathbf{P}$ is related to the electric field as $\mathbf{P}(\mathbf{x},\omega) = \varepsilon_0\chi_e(\mathbf{x},\omega)\mathbf{E}(\mathbf{x},\omega)$, where $\chi_e(\mathbf{x},\omega)$ is the electric susceptibility of the medium and ε_0 is the permittivity of vacuum (Griffiths, 2012). The electric susceptibility indicates the degree of polarization of a dielectric material in response to the incident electric field, depending on the microscopic structure of the medium. The electric displacement vector $\mathbf{D}$ can be expressed as

$$\mathbf{D}(\mathbf{x},\omega) = \varepsilon(\omega)\mathbf{E}(\mathbf{x},\omega) \tag{9}$$

where $\varepsilon(\omega)$ is the dielectric function or relative permittivity of the medium and is related with the electric susceptibility as $\varepsilon(\omega) = \varepsilon_0\,[1+\chi_e(\omega)]$. It should be noted that the spatial dependence term in the susceptibility and the relative permittivity drops out under the local assumption. This local assumption remains valid for near-field thermal radiation unless the vacuum gap is extremely small (less than 1 nm distance). In the extreme proximity, the dielectric function becomes *nonlocal* and its wavevector dependence must be considered (Joulain, 2008). Recently, Chapuis *et al.* (2008a) calculated the near-field heat transfer between two semi-infinite gold plates using non-local dielectric function models and compared their results with the heat flux calculated using the Drude model for gold. They found that the non-local dielectric function saturates the near-field thermal radiation as the vacuum gap approaches zero, whilst local dielectric function erroneously diverges the thermal radiation .

Equation (9) represents the displacement of charges inside the material upon the incidence of electric waves. Thus the dielectric function is the key property in understanding the light-matter interactions, and needs to be further discussed. Under the local assumption, the following sections will discuss two models of the dielectric function, the Drude model for metals (and semiconductors) and the Lorentz model for dielectrics.

Drude model for metals and semiconductors: The Drude model describes the frequency-dependent conductivity of metals and can also be extended to free-carriers in semiconductors. In a metal, electrons in the outermost orbits are "free" to move in accordance with the external electric field. The dielectric function of a metal can be modeled by considering the electron movement under the electric field and is related to the conductivity by (Zhang, 2007)

$$\varepsilon(\omega) = \varepsilon' + i\varepsilon'' = (n+i\kappa)^2 = \varepsilon_\infty - \frac{\sigma_0/\tau}{\varepsilon_0\,(\omega^2 + i\omega/\tau)} \tag{10}$$

where ε_∞ accounts for high-frequency contributions, τ is the relaxation time (inverse of scattering rate), σ_0 is the dc conductivity, and n and κ are the refractive index and extinction coefficient, respectively. Based on Eq. (10), the real and imaginary parts of the dielectric function can be expressed as $\varepsilon' = n^2 - \kappa^2$ and $\varepsilon'' = 2n\kappa$, respectively. The plasma frequency is defined as $\omega_p = \sqrt{\sigma_0/(\tau\varepsilon_0)}$, which is in the ultraviolet region for most metals. When $\omega < \omega_p$, n becomes smaller than κ and ε' becomes negative. At very low frequencies ($\omega\tau \ll 1$), the real part of the dielectric function is much smaller than the imaginary part, and therefore, $n \approx \kappa$. Generally speaking, metals become highly reflective in the visible and infrared regions.

Lorentz model for dielectrics: Unlike metals, the electrons in a dielectric are bound to molecules and cannot move freely. In contrast to free electrons, bound charges experience a restoring force given by the spring constant in addition to the damping force given by the scattering rate. There exist different kinds of oscillators in a real material, such as bound electrons or lattice ions. The response of a single-charge oscillator to a time-harmonic electric field can be extended to a collection of oscillators. Assuming N types of oscillators in a dielectric, the corresponding dielectric function can be given as (Zhang, 2007)

$$\varepsilon(\omega) = \varepsilon_\infty + \sum_j^N \frac{\omega_{p,j}^2}{\omega_{0,j}^2 - \omega^2 - i\omega/\tau_j} \tag{11}$$

where $\omega_{p,j}$, $\omega_{0,j}$, and τ_j may be viewed as the plasma frequency, resonance frequency and the relaxation time of the j-th oscillator, respectively. Since the parameters for the Lorentz model are more difficult to be modeled as compared to those for the Drude model, they are considered as adjustable parameters that are determined from fitting. It can be observed from Eq. (11) that for frequencies far greater or lower than the resonance frequency, the extinction coefficient becomes negligible and the dielectrics are completely transparent. Absorption is appreciable only when an interval (i.e., $1/\tau_j$) is around the resonance frequency. Therefore, the dielectric becomes highly reflective near the resonance frequency, and the radiation inside the material is rapidly attenuated or dissipated. The spectral region with a large imaginary part of the dielectric function is also called the region of resonance absorption.

2.4. Surface Plasmon (or Phonon) Polaritons

Another radiative phenomenon that is worthwhile to discuss here is the optical plasmon (or phonon) polariton. Plasmons are quasiparticles associated with oscillations of plasma, which is a collection of charged particles such as electrons in a metal or semiconductor (Raether, 1988). Plasmons are longitudinal excitations of electron charges that can occur either in the bulk or at the interface. The field associated with a plasmon is confined near the surface, while the amplitude decays away from the interface. Such a wave propagates along the surface, and it is called a surface electromagnetic wave. Surface plasmons can be excited by electromagnetic waves and are important for the study of optical properties of metallic materials, especially near the plasma frequency, which usually lies in the ultraviolet.

In addition to the requirement of evanescent waves on both sides of the interface, the polariton dispersion relations given below must be satisfied (Raether, 1988; Park *et al.*, 2005):

$$\frac{k_{1z}}{\varepsilon_1} + \frac{k_{2z}}{\varepsilon_2} = 0 \quad \text{for TM waves} \tag{12}$$

$$\frac{k_{1z}}{\mu_1} + \frac{k_{2z}}{\mu_2} = 0 \quad \text{for TE waves} \tag{13}$$

This means that the sign of permittivity must be opposite for media 1 and 2 in order to couple a surface polariton with a TM wave. A negative $\mathrm{Re}(\varepsilon)$ exists in the visible and near infrared for metals like Al, Ag, W, and Au. When Eq. (12) is satisfied, the excitation of surface plasmon polariton (SPP) interacts with the incoming radiation and causes strong absorption. Lattice vibration in some dielectric materials like SiC and SiO_2 can result in a negative $\mathrm{Re}(\varepsilon)$ in the mid-infrared. The associated surface electromagnetic wave is called a surface phonon polariton (SPhP). On the other hand, magnetic materials having negative permeability are necessary to excite a surface polariton for a TE wave. Some metamaterials can exhibit negative permeability in the optical frequencies, and negative index materials exhibit simultaneously negative permittivity and permeability in the same frequency region. Therefore, both TE and TM waves may excite SPPs with negative index materials (Park *et al.*, 2005) or with bilayer materials of alternating negative ε and μ, the so-called single negative materials (Fu *et al.*, 2005).

The condition for the excitation of surface polaritons is that the denominator of Fresnel's reflection coefficient be zero. A pole in the reflection coefficient is an indication of a resonance. Taking a TM wave for example, one can solve Eq. (12) to obtain (Zhang, 2007)

$$k_x = \frac{\omega}{c}\sqrt{\frac{\mu_1/\varepsilon_1 - \mu_2/\varepsilon_2}{1/\varepsilon_1^2 - 1/\varepsilon_2^2}} \tag{14}$$

This equation is called the polariton dispersion relation, which relates the frequency with the parallel component of the wavevector. For nonmagnetic materials, it becomes

$$k_x = \frac{\omega}{c}\sqrt{\frac{\varepsilon_1\varepsilon_2}{\varepsilon_1 + \varepsilon_2}} \tag{15}$$

One should bear in mind that the permittivities are in general functions of the frequency. For a metal with a negative real permittivity, the normal component of the wavevector is purely imaginary for any real k_x, because $(\mu\varepsilon\omega^2)/c^2 < 0$. Thus, evanescent waves exist in metals regardless of the angle of incidence.

The requirement of evanescent waves on both sides of the interface prohibits the coupling of propagating waves to the surface polaritons. Figure 3 qualitatively shows a dispersion curve of surface polaritons from Eq. (15) along with the dispersion line of the light propagating in a dielectric having a refractive index nd, suggesting that the propagating light cannot excite the surface polariton. In order to couple propagating light with surface polaritons, we critically need a coupler that can shift the dispersion line of the light to match the parallel (or in-plane) wavevector component to that of the surface polaritons (Raether, 1988). Two conventional surface polariton couplers being widely accepted are a metal-coated prism and a metallic grating structure, whose configurations and mechanisms of light-SP(h)P coupling are schematically illustrated in Figs. 3 and 4. For a metal-coated prism coupler (or Kretschmann coupler), the in-plane wavevector of the incident light becomes $k_x = n_p k_0 \sin\theta$, where n_p is the refractive index of the prism and $k_0 = \omega/c$ is the propagating wavevector in vacuum, when the light is incident on the metallic thin film of the prism side with the incidence angle θ. Owing to the large refractive index of the prism, the dispersion line of the incident light shifts to a greater k_x, or to the right in Fig. 3(b), to intersect with the dispersion curve for the surface polariton: A metal-coated prism can excite the surface polariton.

When the light is incident on the grating structure as shown in Fig. 4(a), the Bloch-Floquet condition becomes $k_{x,j} = k_x + 2\pi j/\Lambda$, where j is the diffraction order and Λ is the grating period. The in-plane wavevector of the diffracted light can increase by a factor of $2\pi j/\Lambda$ depending on the diffraction order, shifting the light dispersion to couple with the SPP. It should be noted that the grating coupler can excite multiple surface polaritons even at the normal incidence. Figure 4(b) shows the reduced dispersion relation for a binary grating made of Ag with $\Lambda = 1.7$ μm (Zhang, 2007). The dispersion curves (dash-dotted

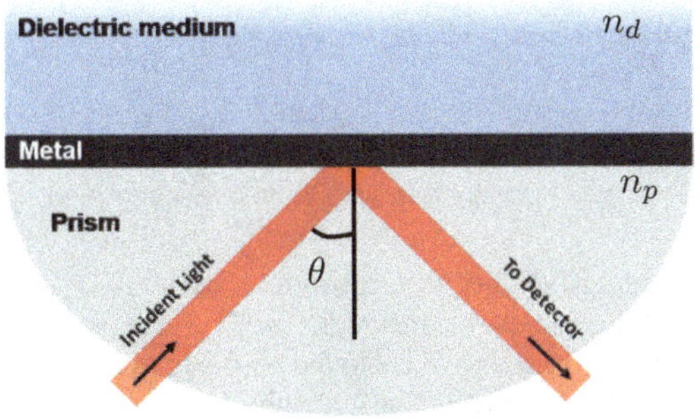

(a) A metal-coated prism coupler

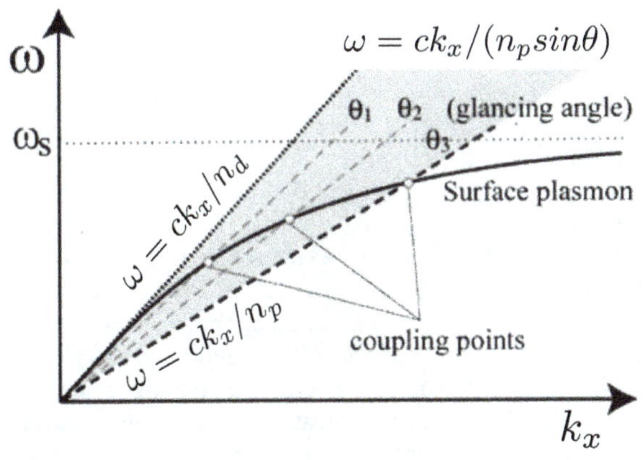

(b) Dispersion relation

Fig. 3 (a) Schematic diagrams of a metal-coated prism surface polariton coupler and (b) dispersion curves.

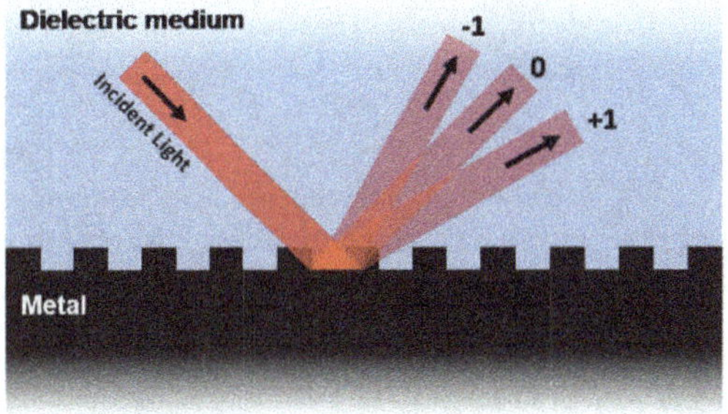

(a) A metal grating coupler

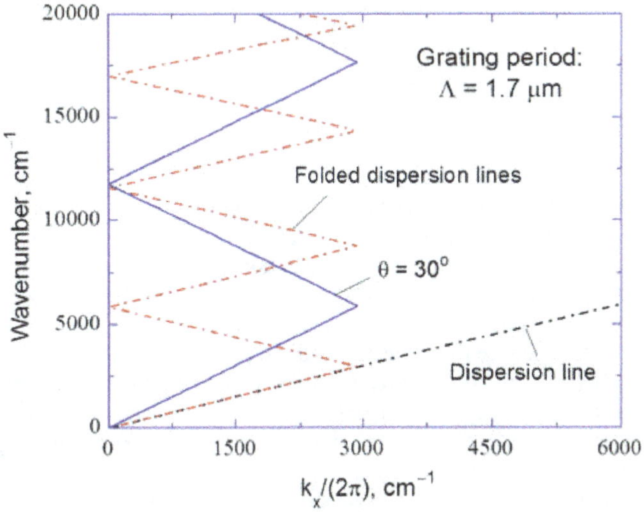

(b) Dispersion relation

Fig. 4 (a) Schematic diagrams of a metal grating surface polariton coupler and (b) Dispersion relation of SPP as manifested by an Ag grating (Zhang, 2007). Note that $k_x = (\omega/c)\sin\theta$.

lines) are folded. The solid lines correspond to an incidence angle of 30° and are also folded. The intersections identify the location where SPPs can be excited for a TM wave incidence, when the magnetic field is parallel to the grooves.

The excitation of surface polaritons makes a significant effect on near-field thermal radiation. As briefly mentioned in the introduction, the enhancement of heat transfer rate in near-field thermal radiation is because photon tunneling enables evanescent EM waves to carry radiative energy across the vacuum gap. Among the involved evanescent waves, those that match with the dispersion relation of the surface polariton will resonantly enhance the absorption of the evanescent EM fields. Thus near-field radiative heat transfer can be greatly enhanced with the surface polariton excitation. Moreover, surface polaritons play a crucial role in tailoring the spectral and directional radiative properties of materials. For example, coherent thermal emission can be realized by exciting surface polaritons in grating structures and truncated photonic crystals. Further discussion is deferred to later sections

3. NEAR-FIELD RADIATIVE ENERGY TRANSFER BETWEEN TWO SEMI-INFINITE MEDIA

3.1. Formulation of Near-Field Radiation

Consider the structure shown in Fig. 1, where both the emitter and receiver are n-doped silicon. The emitter and receiver are assumed to be at 400 and 300 K, respectively. The dielectric function model of doped Si may be modeled as a combination of Drude term and other contributions (Fu and Zhang, 2006) and the details are described in (Basu *et al.*, 2010a,b). The total heat transfer between two media can be expressed as (Basu *et al.*, 2009)

$$q''_{\text{net}} = \frac{1}{\pi^2}\int_0^\infty [\Theta(\omega, T_1) - \Theta(\omega, T_2)]\, X(\omega) d\omega \tag{16}$$

where $X(\omega) = \int_0^\infty s(\omega, \beta) d\beta$. Note that the integration of $s(\omega, \beta)$ over ω gives a weighted function to modify the Planck blackbody distribution function. Expression of $s(\omega, \beta)$ is different for propagating ($\beta < \omega/c$) and evanescent ($\beta < \omega/c$) waves,

$$s_{\text{prop}}(\omega, \beta) = \frac{\beta(1-\rho_{01}^{\text{s}})(1-\rho_{02}^{\text{s}})}{4|1-r_{01}^{\text{s}} r_{02}^{\text{s}} e^{i2\gamma_0 d}|^2} + \frac{\beta(1-\rho_{01}^{\text{p}})(1-\rho_{02}^{\text{p}})}{4|1-r_{01}^{\text{p}} r_{02}^{\text{p}} e^{i2\gamma_0 d}|^2} \tag{17}$$

and,

$$s_{\text{evan}}(\omega, \beta) = \frac{\text{Im}(r_{01}^{\text{s}})\text{Im}(r_{02}^{\text{s}})\beta e^{-2\text{Im}(\gamma_0)d}}{|1-r_{01}^{\text{s}} r_{02}^{\text{s}} e^{-2\text{Im}(\gamma_0)d}|^2} + \frac{\text{Im}(r_{01}^{\text{p}})\text{Im}(r_{02}^{\text{p}})\beta e^{-2\text{Im}(\gamma_0)d}}{|1-r_{01}^{\text{p}} r_{02}^{\text{p}} e^{-2\text{Im}(\gamma_0)d}|^2} \tag{18}$$

In Eqs. (17) and (18), the first term on the right-hand side refers to the contribution of s-polarization or TE wave, while the second term refers to the contribution of p-polarization or TM wave. Note that $r_{0j}^{\text{s}} = (\gamma_0 - \gamma_j)/(\gamma_0 + \gamma_j)$ and $r_{0j}^{\text{p}} = (\varepsilon_j\gamma_0 - \gamma_j)/(\varepsilon_j\gamma_0 + \gamma_j)$ are the Fresnel reflection coefficients for s- and p-polarization, respectively, at the interface between vacuum and medium j (1 or 2). On the other hand, $\rho_{0j} = |r_{0j}|^2$ is the far-field reflectivity at the interface between vacuum and medium j. When different doping levels are considered, the location of the peak in $s(\omega, \beta)$ shifts towards higher frequencies with increased doping level.

Notice that $s(\omega, \beta)$ is independent of temperature and contains all the information about the material properties as well as the geometry of the emitting media. The predicted radiative heat transfer between two doped Si plates is plotted in Fig. 5(a) as a function of the vacuum gap width (Basu *et al.*, 2010b). Both plates are maintained at the same doping level, which is varied from 10^{18} to 10^{21} cm^3. The dotted line with circles is the radiative heat flux between two blackbodies. The net heat flux at

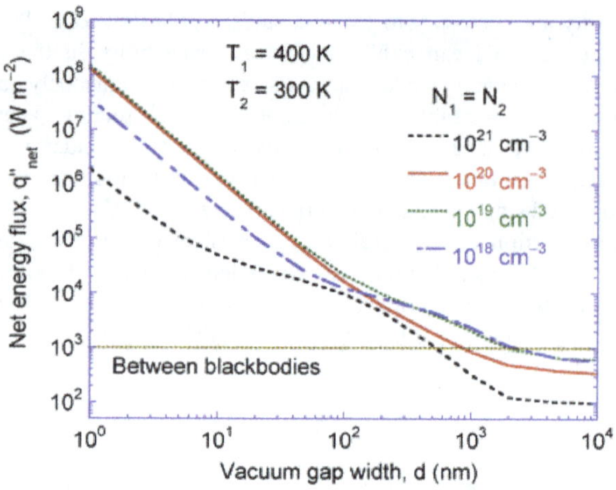

(a) Net energy flux vs. Vacuum gap width

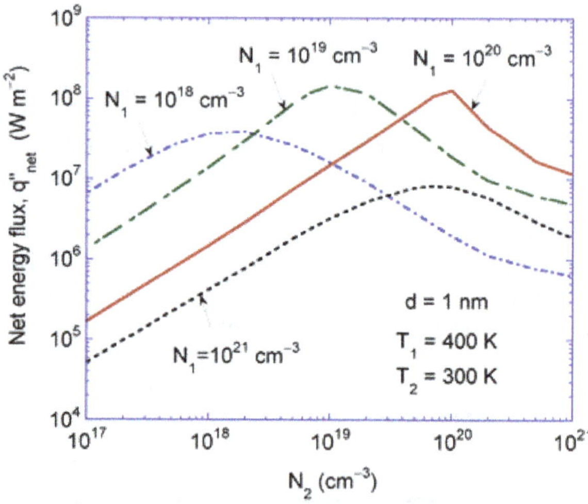

(b) Net energy flux vs. Doping concentration

Fig. 5 Net energy flux between medium 1 (at 400 K) and medium 2 (at 300 K) versus the gap width for Si at different doping levels. The dash-dotted line refers to the net energy transfer between two blackbodies maintained at 400 and 300 K, respectively; and (b) effect of doping on the net energy transfer between two doped Si plates separated by 1 nm vacuum gap (Basu *et al.*, 2010b).

$d = 1$ nm between 10^{19} or 10^{20} cm^3 doped Si plates can exceed that between two blackbodies by five orders of magnitude, because of photon tunneling and surface waves. Increase in the doping level of Si does not always enhance the energy transfer. In fact, the radiative heat transfer is the smallest for 10^{21} cm^3 doped Si plates as compared with other doping levels considered here. At $d > 200$ nm, doping concentrations between 10^{18} and 10^{19} cm^3 yield the largest radiative heat transfer, which is comparable to that between SiC and SiC. A detailed parametric study has been performed to determine the ideal Drude or Lorentz dielectric functions that yield the largest near-field enhancement (Wang *et al.*, 2009). Figure 5(b) illustrates the effect of doping concentration on nanoscale energy flux when the vacuum gap width is fixed at $d = 1$ nm. The doping level of medium 1 is represented as N_1 while that for medium 2 is represented by N_2. Generally speaking, surface waves are better coupled when the two media have similar dielectric functions. As a result there exist peaks when $N_1 \approx N_2$, at doping levels up to 10^{20} cm^3.

The enhancement of near-field radiation in metallic and polar materials can be well explained by surface polaritons. The coupling of SP(h)Ps allows a significant increase in the function given in Eq. (18) for evanescent waves. Furthermore, for magnetic materials, the enhancement can occur for both $s-$ and $p-$polarizations, resulting in multiple spectral peaks in near-field radiative transfer (Wang *et al.*, 2009; Joulain *et al.*, 2010; Zheng and Xuan, 2011). It should be noted that for intrinsic Si or dielectric materials without strong phonon absorption bands, the tunneling is limited and saturate at extremely small distances. Also, for good metals, the SPP excitation frequency is too high to significantly enhance thermal radiation unless the distance is less than 1 nm (Wang *et al.*, 2009; Basu and Francoeur, 2011b). Figure 6 illustrates the surface wave effects on the enhancement of near-field thermal radiation by comparing the contour plots of $s(\omega, \beta)/2\pi$ for SiC plates separated at 100 nm in Fig. 6(a) and for 10^{20} cm^3 $n-$doped Si plates separated at 10 nm in Fig. 6(b). Only TM waves are compared here since the contribution of TE waves is negligibly small. For simplicity, β is normalized with respect to ω/c. The brightest color represents the peak value at $\omega_{\mathrm{m}} = 1.79 \times 10^{14}$ rad/s and $\beta_{\mathrm{m}} = 50\omega/c$ for SiC, and at $\omega_{\mathrm{m}} = 2.67 \times 10^{14}$ rad/s and $\beta_{\mathrm{m}} = 62\omega/c$ for doped Si. The contribution of propagating waves ($\beta < \omega/c$) to the overall heat transfer is negligible. As mentioned earlier, the resonance energy transfer in the near field around ω_{m} is due to SPhP for SiC and SPP for doped Si, respectively.

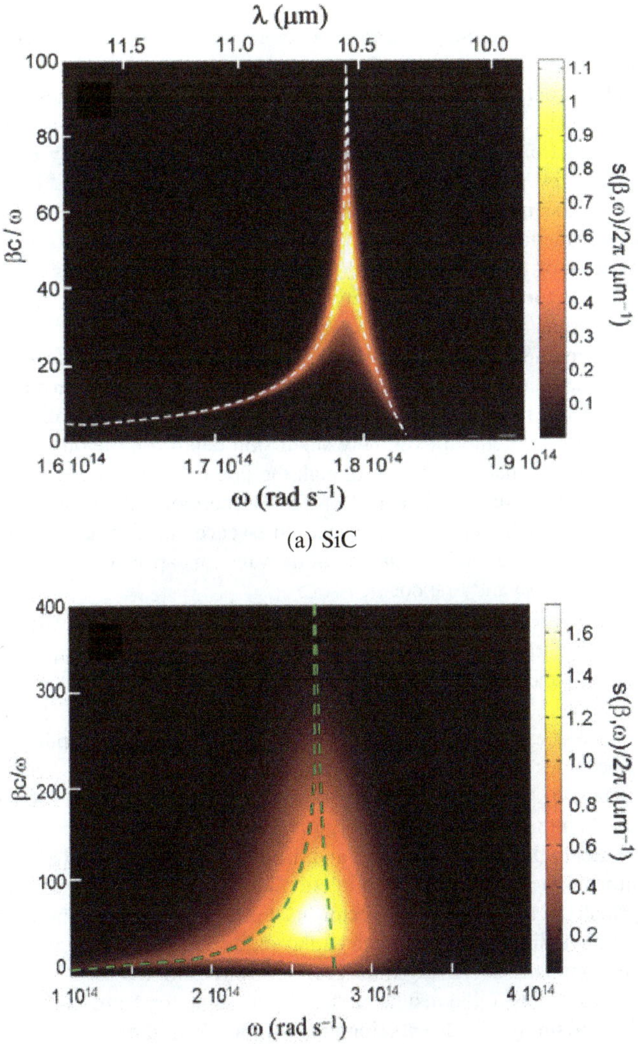

(a) SiC

(b) Doped Si with 10^{20} cm^{-3}

Fig. 6 Contour plot of $s(\omega, \beta)/2\pi$ for (a) SiC and (b) n-doped Si for doping concentration of 10^{20} cm^{-3}. Note that the parallel wavevector component is normalized to the frequency. The dashed curves represent the two branches of the surface-polariton dispersion (Lee and Zhang, 2008; Basu *et al.*, 2010b).

The calculated dispersion curves for surface polaritons between two SiC and doped Si plates are also plotted as dashed lines in Figs. 6(a) and 6(b), respectively. Due to the coupling of surface polaritons at vacuum-SiC and vacuum-doped Si interfaces, there exist two branches of dispersion curves for the p polarization as follows:

$$\text{Symmetric mode}: \quad \frac{\gamma_0}{\varepsilon_0} + \coth\left(-\frac{i\gamma_0 d}{2}\right) \cdot \frac{\gamma_1}{\varepsilon_1} = 0 \tag{19a}$$

$$\text{Asymmetric mode}: \quad \frac{\gamma_0}{\varepsilon_0} + \tanh\left(-\frac{i\gamma_0 d}{2}\right) \cdot \frac{\gamma_1}{\varepsilon_1} = 0 \tag{19b}$$

The lower-frequency branch corresponds to the symmetric mode, and the higher-frequency branch represents the asymmetric mode (Park *et al.*, 2005). Note that for both doped Si and SiC, the dispersion relation becomes almost flat at $\omega_{\max}$ implying that surface polaritons can be excited in a wide range of β, being responsible for the enhancement of thermal radiation through photon tunneling (Lee and Zhang, 2008).

3.2. Upper Limit of Near-Field Heat Flux

For nonmagnetic materials, when $\beta \gg \omega/c$ (evanescent waves), we have $\gamma_1 \approx \gamma_2 \approx \gamma_0 \approx i\beta$. As a result, for dielectrics, r_{01}^{s} and r_{01}^{s} are negligibly small, and the contribution of TE waves can be ignored. Furthermore, $r_{01}^{\mathrm{p}} \approx (\varepsilon_1 - 1)/(\varepsilon_1 + 1)$ and $r_{02}^{\mathrm{p}} \approx (\varepsilon_2 - 1)/(\varepsilon_2 + 1)$ are independent of β. Hence, Eq. (18) can be simplified as

$$s_{\mathrm{evan}}(\omega, \beta) \approx \frac{\mathrm{Im}(r_{01}^{\mathrm{p}})\mathrm{Im}(r_{02}^{\mathrm{p}})\beta e^{-2\beta d}}{\left|1 - r_{01}^{\mathrm{p}} r_{02}^{\mathrm{p}} e^{-2\beta d}\right|^2} \tag{20}$$

However, for metals, the contribution from TE waves is more significant when $\omega/c = \beta = \sqrt{|\varepsilon_1|}\omega/c$, whereas the contribution from TM waves is more important for $\beta \gg \sqrt{|\varepsilon_1|}\omega/c \gg \omega/c$ (Chapuis *et al.*, 2008a). As a result, for metals, heat transfer due to TM waves becomes dominant at very short distances.

Using the relation, $\mathrm{Im}\left[(\varepsilon - 1)/(\varepsilon + 1)\right] = (2\varepsilon'')/\left|\varepsilon + 1\right|^2$, and assuming identical permittivity for both media, the spectral heat flux from 1 to 2 in the limit $d \to \infty$ is given by (Basu and Zhang, 2009a)

$$q''_{\omega,1-2} \approx \frac{4\Theta(\omega, T_1)}{\pi^2 d^2} \int_{\xi_0}^{\infty} \frac{\varepsilon''^2 e^{-2\xi} \xi d\xi}{\left|(\varepsilon + 1)^2 - (\varepsilon - 1)^2 e^{-2\xi}\right|^2} \tag{21}$$

where $\xi = \beta d$, $\xi_0 = d\omega/c$, and ε'' is the imaginary part of the dielectric function. As observed from Eq. (21) the heat flux will be inversely proportional to d^2 in the proximity limit. This means that the heat flux will diverge as $d \to 0$ and its physical significance has been debated among researchers. It should be noted that the d^{-2} dependence is for contribution from the $p-$polarized electromagnetic waves only, since the contribution from the $s-$polarized waves will asymptotically reach a constant as $d \to 0$. As the vacuum gap decreases, the energy transfer shifts to large values of the parallel wavevector component. A cutoff in the order of the lattice constant is imposed as the minimum spatial wavelength, which subsequently sets a maximum wavevector component parallel to the interfaces (Volokitin and Persson, 2004). The imposed cutoff limits the number of modes for photon tunneling. Consequently, the radiative heat flux will experience a reduction as $d \to 0$.

In order to consider the upper limit of near-field radiative heat flux, $X(\omega)$ shown in Eq. (16) should be modified to $X(\omega) = \int_0^{\beta_c} s(\omega, \beta) d\beta$ to take into account the upper limit of the integration. Electrons in solids move in a periodic potential characterized by the Bloch wave, with a maximum wavevector of π/a at the edge of the first Brillouin zone. Here, a is the lattice constant, which is on the order of interatomic distance. This posts a limit on the smallest surface wavelength or cutoff wavevector parallel to the surface (Volokitin and Persson, 2004). Take a typical value of a as 0.5 nm and note that there exists a maximum of X, i.e., $X_{\max} = \beta_c^2/8$. There exists an upper limit of the near-field radiative heat flux given by (Basu and Zhang, 2009a; Volokitin and Persson, 2004)

$$q''_{\max} = X_{\max}\frac{k_B^2}{6\hbar}(T_1^2 - T_2^2) = \frac{k_B^2\beta_c^2}{48\hbar}(T_1^2 - T_2^2) \tag{22}$$

for nonmagnetic materials. Note that $q''_{\max}$ is the ultimate maximum heat flux and is only achievable when $d \to 0$. It is found that metals with a large imaginary part in the infrared can help reach such a limit at extremely small distances (Pendry, 1999). For distances greater than a few nanometers, however, the situation is different. Basu and Zhang (2009a) considered a case in which both the emitter and receiver are assumed to have frequency-independent permittivity in order to identify the expression of the complex dielectric constant that will result in maximum heat flux. It should be noted that such a constant dielectric function cannot exist in reality because of the violation of Causality. When X was plotted against ε' and ε'' in a 3D plot or a 2D contour for given d (say 10 nm), it was found that $X_{\max}$ corresponds to $\varepsilon' = -1$ at which surface waves exist.

Figure 7 shows the calculated radiative heat flux between the two media ($T_1 = 300$ K and $T_2 = 0$ K) as a function of the vacuum gap for different values of ε' and ε'' (Basu and Zhang, 2009a). In most cases, ε' is fixed at -1. For the sake of comparison, the energy transfer between two SiC plates is also shown in the figure using a frequency-dependent dielectric function. At 300 K, the upper limit of near-field heat flux is 1.4×10^{11} W/m^2, which is represented as the dashed horizontal line. The radiation flux between two blackbodies is 459 W/m^2, which is several orders of magnitude smaller than near-field radiative transfer. The cutoff in β sets an upper limit on the maximum energy transfer between the two media. Hence, for each of the dielectric functions, there exists an optimal vacuum gap width (d_m) that allows the maximum energy transfer. For $\varepsilon = -1 + i0.1$, it can be seen from Fig. 7 that $d_m = 0.6$ nm, which also

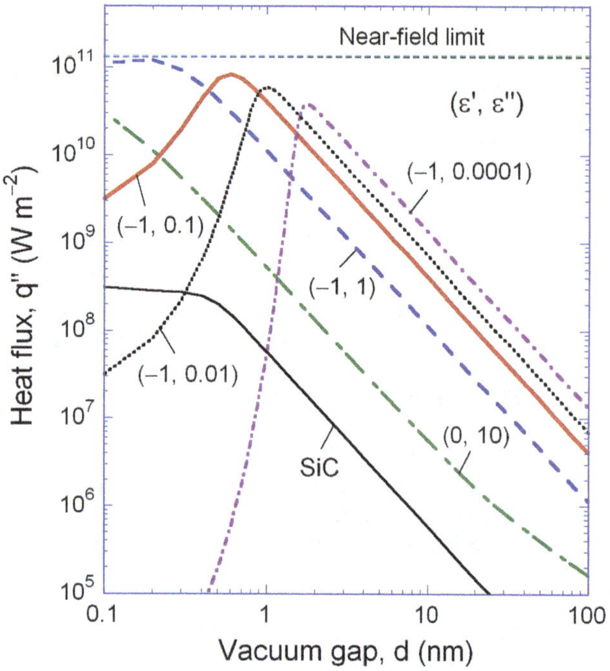

Fig. 7 Plot of radiative heat flux versus gap width for different dielectric functions (Basu and Zhang, 2009a). The temperatures of the two media are set to be 300 K and 0 K. The dielectric function is assumed to be frequency independent, except that for SiC for which the Lorentz model at 300 K is used.

maximizes X. The value of d_m decreases with increasing ε'', implying that the reduction in the energy transfer begins to take place at smaller vacuum gaps. Furthermore, the d^{-2} dependence in the energy transfer exists only when $d > d_m$. At $d > 2$ nm, increasing ε'' results in a decrease of the heat flux. When $\varepsilon = 0 + i10$, the radiative heat flux is generally much smaller than those with $\varepsilon' = -1$ but will keep increasing towards the upper limit as d unrealistically approaches zero. For the selected dielectric functions with $\varepsilon' = -1$ and $\varepsilon'' \ll 1$, the energy transfer can be orders of magnitude greater than that between SiC plates. This is due to the assumed frequency-independent dielectric function, which results in the excitement of surface waves at almost every frequency. While no such materials can exist, the results provide some hints of appropriate dielectric functions that will result in optimal heat flux at different vacuum gaps. By introducing the cutoff in β, even for SiC, the d^{-2} trend ceases to exist at $d < 0.6$ nm. Instead, the near-field radiative transfer reaches a plateau below $d = 0.5$ nm. Wang *et al.* (2009) performed a design optimization of the parameters in the Drude model and Lorentz model that can result in the highest near-field radiative flux at given distance and temperatures.

3.3. Penetration Depth in Nanoscale Thermal Radiation

Traditionally, radiation penetration depth in a solid, also called skin depth or photon mean free path, is defined as $\delta_\lambda = \lambda/(4\pi\kappa)$, where κ is the extinction coefficient as discussed earlier. A film whose thickness is six times the skin depth can be treated as opaque in most applications. In the optical spectrum, the penetration depth of noble metals is usually $10-20$ nm. For an evanescent wave, such as that induced under the total internal reflectance setup when light is incident from an optically denser medium to a rarer medium, the skin depth may be defined according to the $1/e$ attenuation of the field as $\delta = 1/\mathrm{Im}(\gamma)$, where γ (purely imaginary) is the wavevector component perpendicular to the interface in the optically rarer medium. The electric and magnetic fields will decay exponentially and become negligible at a distance greater than about one wavelength. Hence, the skin depth is expected to be several tenths of a wavelength in a dielectric medium. However, in near-field radiation especially when SP(h)Ps are excited, an extremely small skin depth (on the order of the vacuum gap d) may exist even though the dominant wavelengths are in the infrared (Basu and Zhang, 2009b). Furthermore, the skin depth is proportional to the separation distance. In essence, the skin depth in near-field thermal radiation is a function of the vacuum gap as well as material properties (Basu and Francoeur, 2011b).

For a very small gap, while ω_m (see Fig. 6) remains constant as d decreases, the energy transfer is shifted towards a larger β, leading to greater near-field enhancement. For $\beta \gg \sqrt{\varepsilon_j}\omega/c$, $\gamma_j \approx i\beta$ or $\mathrm{Im}(\gamma_j) \approx \beta$. There exists an evanescent wave (in medium 3) whose amplitude decays according to $e^{-\beta(z-d)}$. Hence, the skin depth of the field becomes $\delta_F \approx 1/\beta$ and the power penetration depth becomes $\delta_P \approx 1/(2\beta)$. Using the multilayer Green's function, the $z-$component of the Poynting vector, which is proportional to the heat flux, can be calculated both inside the emitter and the receiver. The spectral and total Poynting vector distributions are plotted in Fig. 8 for SiC. The ordinate is normalized to the Poynting vector inside the vacuum gap. The energy flux in the emitter increases towards the surface, remains constant in the vacuum gap, and decreases in the receiver away from the surface. When the abscissa is z/d, the results are nearly the same for 1 nm $< d <$ 100 nm. Surprisingly, the distributions are symmetric in the emitter and the receiver. The $1/e$ decay line is shown as the horizontal dashed line so that the penetration depth can be evaluated. Note that the calculated Poynting vector is integrated over all β values. As mentioned earlier, when SPP is excited, the energy transfer is pushed towards large β values; hence, the spectral penetration depth has a minimum near ω_m.

As shown in Fig. 8, the penetration depth is approximately $0.19d$ at 10.54 μm, where SPP is excited at the vacuum-SiC interfaces. The actual minimum penetration depth is located at 10.47 μm, corresponding to the

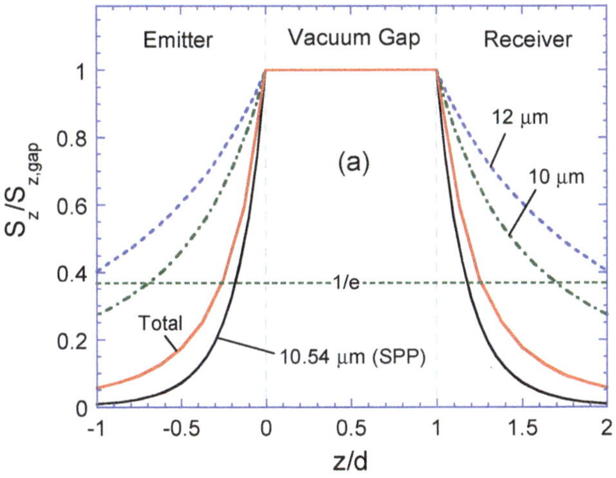

Fig. 8 The distributions of the spectral and total Poynting vector (z-component) near the surfaces of the emitter and receiver (refer to Fig. 1), both made of SiC, normalized to that in the vacuum (Basu and Zhang, 2009b).

maximum of $X(\omega)$. The penetration depth increases towards longer or shorter wavelengths, and the overall penetration depth based on the total energy flux is $0.25d$, which is about 30% greater than δ_P evaluated at ω_m and β_m. For a thin vacuum gap, the SPhP dispersion is shifted to large β values, resulting in a shorter penetration depth. Hence, a 10-nm coating of SiC can act as an optically thick medium when $d = 10$ nm as predicted in Refs. (Francoeur *et al.*, 2008; Fu and Tan, 2009). When $d < 1$ nm, the penetration depth is less than a monolayer, implying that the SiC emitter is completely a 2D solid. It should be mentioned that δ_P cannot be arbitrarily small. When d is comparable to or less than the interatomic distance, the radiative transfer cannot be explained by the local electromagnetic theory. Obviously, in such case, one cannot use its bulk dielectric function and also cannot set β_c as infinity. Note that with magnetic materials, surface waves can be excited by both TE and TM waves. The penetration depth in near-field radiation between metamaterials has also been examined (Basu and Francoeur, 2011b).

3.4. Energy Streamlines

The direction in near-field transfer cannot be determined by the wavevector as in the case of a propagating wave. From the wave point of view, phonon tunneling is through the coupling of evanescent waves since there exist a forward decaying and backward decaying waves in the vacuum gap, both with purely imaginary γ, the $z-$component of a wavevetor. In such case, the Poynting vector represents the direction of energy flow and the trace of Poynting vectors provides the energy streamlines (ESLs), which can be used to elucidate the energy propagation like fluid flow (Zhang and Lee, 2006). Due to the random fluctuation of charges, the Poynting vectors are decoupled for different values of β (Lee and Zhang, 2008; Lee *et al.*, 2007). The ESLs are laterally displaced as they leave the surface of the emitter and reach the surface of the receiver. This lateral displacement is called a *lateral shift* (Basu and Zhang, 2009a), which is different from the well-known Goos-Hänchen shift (Zhang and Lee, 2006), may be important to determining the lateral dimension of the real system which can be modeled as infinite plates in near-field radiation.

Figure 9 shows the ESL projected to the $x-z$ plane at $\lambda = 10.55\,\mu m$ and $d = 100$ nm for $\beta = 40\omega/c$ in all three media for $p-$polarized waves (Lee *et al.*, 2007). The magnitude of magnetic field is overlaid as depicted by the colored contours (i.e., the brighter color indicates the greater value). To calculate the magnetic field, thin-film optics is employed with an assumption that a plane wave is incident from medium 1. The emission originated deeper from the surface than the radiation penetration depth could not reach the SiC-vacuum interface. Hence, the field distribution is

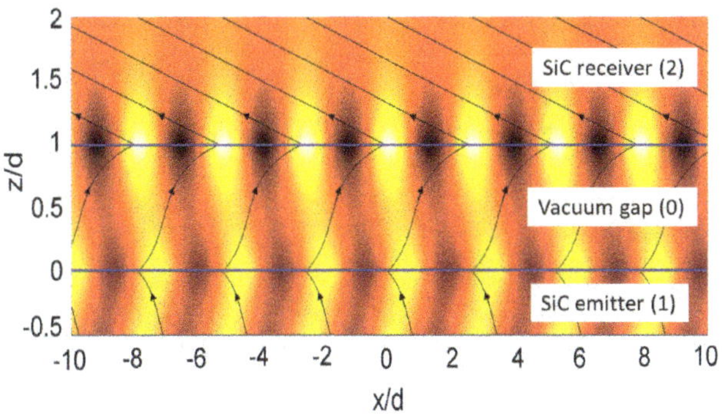

Fig. 9 Energy streamlines for TM waves in SiC-vacuum-SiC for $d = 100$ nm. The magnitude of the magnetic field is denoted by colored contours and plotted along with the ESLs (Lee *et al.*, 2007).

plotted in the vicinity of the vacuum gap. It can be seen from Fig. 9 that negative refraction of energy path occurs at the interfaces between SiC and vacuum due to the opposite sign of their dielectric functions. The energy streamlines are curved except for medium 3 where no backward waves exist. The magnetic field oscillates in the lateral direction as a result of the excitation of SPhPs.

Basu *et al.* (2011) applied fluctuational electrodynamics in multilayered structures to directly trace the energy streamlines not only in the gap and receiver but also in the emitter. It was found that when surface waves are excited, there is a larger lateral shift inside the emitter. Figure 10 shows the ESLs for combined TE and TM waves at $d = 10$ nm with different β values. Note that for propagating waves, $\beta^* = \beta c/\omega < 1$ and the shape of ESLs is independent of d in the proximity limit. For evanescent waves, the contribution of TM waves to the near-field radiation is dominant. At the SPhP frequency, ESLs for the same βd value are essentially the same. The resonance conditions may be denoted by $\omega_{\rm m}$ and βm. When d is very small, $\omega_{\rm m}$ depends little on d, whereas βm is inversely proportional to d as mentioned previously. The value corresponds to $\beta_{\rm m}^* = \beta_{\rm m} c/\omega_{\rm m} = 450$ when $d = 10$ nm. For propagating waves, all ESLs are located inside the conical surfaces bounded by the ESL at $\beta = \omega/c$ (Basu and Zhang, 2009a). The ESLs inside the emitter and the vacuum gap are curved much more for evanescent waves than for propagating waves, as it is assumed to be semi-infinite and no backward waves exist. Because the receiver is treated as non-emitting (i.e., at zero absolute temperature), the streamlines in the receiver are straight lines. Figure 10(b) suggests that the largest lateral shift occurs inside the emitter and the lateral shift increases with β. Hence, it is important to take the lateral shift inside the emitter into consideration when determining the minimum area needed for the emitter and receiver to be approximated as infinitely extended plates. In the receiver, the lateral shift can be written as $\theta(z, \omega, \beta) = \tan^{-1}(\varepsilon''/\varepsilon')$ when $\beta^* \gg 1$ (or $\beta \gg \omega/c$). Hence inside the receiver, ESLs for evanescent waves are parallel as seen in Fig. 10(b). But this is not so for propagating waves when ESLs can intercept each other. The results obtained from this study will facilitate the design of experiments for measuring nanoscale thermal radiation. The method discussed above can be extended to the study of energy flux and streamlines between layered structures and materials with coatings.

4. NEAR-FIELD RADIATIVE ENERGY TRANSFER IN VARIOUS MEDIA

The previous section has made intensive discussions on the near-field radiative heat transfer between two half-spaces separated with a vacuum gap, including when they have thin coatings. We now discuss radiative heat transfer between different geometries, in particular between two spheres in Sec. 4.1, between a sphere and a half-space in Sec. 4.2, and between other geometries such as cylinders and gratings in Sec. 4.3, when they are held at different temperatures and separated in vacuum. In

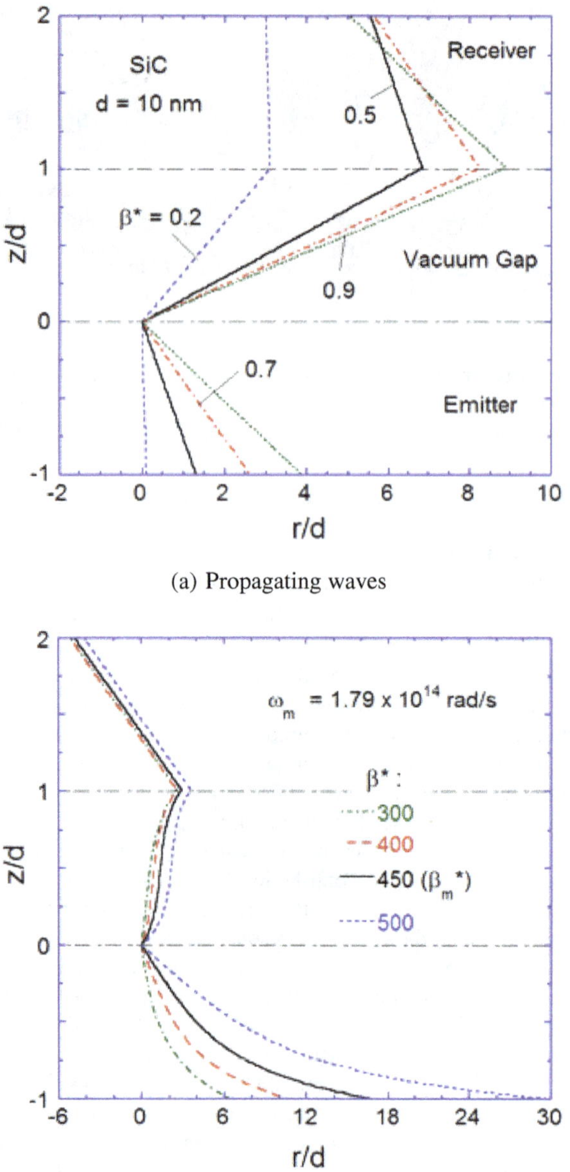

(a) Propagating waves

(b) Evanescent waves

Fig. 10 ESLs for combined TE and TM waves at the SPhP frequency for SiC with d = 10 nm at different $\beta^* = \beta c/\omega$ values: (a) propagating waves; (b) evanescent waves (Basu *et al.*, 2011).

addition, near-field radiative energy transport in emerging materials will also be discussed in Sec. 4.4.

4.1. Near-Field Radiative Heat Transfer between Two Spheres

Near-field radiative heat transfer between two spherical nanoparticles has been theoretically predicted by modeling the nanoparticles as fluctuating dipoles (Volokitin and Persson, 2001; Dorofeyev, 2008; Chapuis *et al.*, 2008b). When there are two spherical nanoparticles whose dielectric constants are ε_1 and ε_2, the spectral power dissipated in particle 2 by the electromagnetic field induced by particle 1 can be written using the dipolar approximation as

$$Q_{1\rightarrow 2}(\omega) = \varepsilon_0 \frac{\omega}{2} \mathrm{Im}(\alpha_2) |\mathbf{E}_{\mathrm{inc}}(\mathbf{x}_2, \omega)|^2 \quad (23)$$

where $\mathbf{x}_2$ is the position of the particle 2 and $\alpha_2 = 4\pi R^3(\varepsilon_2 - 1)/(\varepsilon_2 + 2)$ is the polarizability of a sphere of radius R having the relative permittivity of ε_2. The electric field incident on the particle 2, $\mathbf{E}_{\mathrm{inc}}(\mathbf{x}, \omega)$, is created by the thermal fluctuating dipole of particle 1 at temperature T_1: $\mathbf{E}_{\mathrm{inc}}(\mathbf{x}_2, \omega) = \mu_0 \omega^2 \overline{\overline{\mathbf{G}}}_e(\mathbf{x}_2, \mathbf{x}_1, \omega) \cdot \mathbf{p}$. Here, $\overline{\overline{\mathbf{G}}}_e(\mathbf{x}_2, \mathbf{x}_1, \omega)$ is the electric dyadic Green's function between two dipoles in vacuum and expressed as (Domingues *et al.*, 2005)

$$\overline{\overline{\mathbf{G}}}_e(\mathbf{x}_2, \mathbf{x}_1, \omega) = \frac{ke^{ikd}}{4\pi} \left[\left(\frac{1}{kd} + \frac{i}{(kd)^2} + \frac{1}{(kd)^3} \right) \overline{\overline{\mathbf{I}}} + (\hat{\mathbf{u}}_r \hat{\mathbf{u}}_r) \left(\frac{3}{(kd)^3} - \frac{3i}{(kd)^2} - \frac{1}{kd} \right) \right] \quad (24)$$

where $d = |\mathbf{x}_2 - \mathbf{x}_1|$ is the distance between the sphere centers, $\overline{\overline{\mathbf{I}}}$ is the identity tensor, and $\hat{\mathbf{u}}_r \hat{\mathbf{u}}_r$ is the dyadic notation of unit vectors. Because of thermal fluctuations, particle 1 has a random electric dipole that yields the correlation function of the dipole:

$$\langle p_m(\omega) p_n^*(\omega') \rangle = \frac{4\varepsilon_0}{\pi\omega} \mathrm{Im}[\alpha_1(\omega)] \Theta(\omega, T_1) \delta_{mn} \delta(\omega - \omega') \quad (25)$$

Equation (25) is in fact the primitive form of Eq. (1). By combining the above equations, we finally obtain the thermal conductance between two dipoles due to near-field radiative heat transfer that can be expressed as (Domingues *et al.*, 2005):

$$G_{12}(T) = \frac{3}{4\pi^3 d^6} \int_0^\infty \frac{d\Theta(\omega, T)}{dT} \mathrm{Im}[\alpha_1(\omega)] \mathrm{Im}[\alpha_2(\omega)] d\omega \quad (26)$$

It should be noted that radiative heat transfer between two spheres has the d^{-6} spatial dependence, which is typical of the dipole-dipole interactions. The thermally fluctuating dipole at one nanoparticle induces electromagnetic field on the other nanoparticle to cause the second dipole to fluctuate. Equation (26) suggests that near-field radiative thermal conductance between two nanoparticles has a resonant behavior when the polarizability α has a resonance, or the dielectric constant approaches -2 in $\alpha_{1(2)} = 4\pi R_{1(2)}^3(\varepsilon_{1(2)} - 1)/(\varepsilon_{1(2)} + 2)$. Provided that the surface polariton resonance occurs when the dielectric constant approaches -1 in case the material is interfaced with the vacuum (Raether, 1988), this resonant behavior is not directly related with the surface polariton resonance: instead, is named as the localized surface polariton resonance – grouped oscillations of the charge density confined to nanostructures (Hutter and Fendler, 2004). The localized surface polariton resonance appears in the visible range for metals and in the infrared for polar materials.

While the dipole approximation elucidates the d^{-6} dependence of the near-field radiative heat transfer between two spheres, this dependence is valid only when $R \ll \lambda_T$ and $d \gg R_1 + R_2$, where λ_T is the characteristic wavelength defined from Wien's displacement law (Zhang and Wang, 2012), and R_1 and R_2 are radii of nanoparticles. However, near contact, the near-field radiative thermal conductance deviates drastically from the dipole approximation. This deviation results from the fact that when particles become very close, the charge distributions become nonsymmetric and cannot be described merely as two interacting dipoles (Pérez-Madrid *et al.*, 2008). For cases when the dipole approximation is not valid, calculation of near-field thermal radiation between two spheres becomes computationally challenging, mainly due to the difficulty in determining the dyadic Green's function. More realistic Green's function for the two-sphere configuration has been suggested by approximating nanoparticles as fluctuating multipoles (Pérez-Madrid *et al.*, 2008) and by implementing the vector spherical wave expansion method (Narayanaswamy and Chen, 2008). Narayanaswamy and Chen (2008) investigated the scattering between two spheres by expanding the electromagnetic field in terms of the vector spherical waves at each sphere and re-expanding the vector spherical waves of one sphere with the vector spherical waves of the second sphere to satisfy the boundary conditions. Recurrence relations for vector spherical waves were used to reduce the computational demands in determining translation coefficients of each spherical wave function term. Due to complexities in the formulations, the equations of the dyadic Green's function for two spheres are not included here but can be found from Narayanaswamy and Chen (2008) for a detailed derivation and Sasihithlu and Narayanaswamy (2011) for the convergence limit of the vector spherical wave expansion approach.

Domingues *et al.* (2005) attempted to overcome the limitation of the dipole-approximation by using the molecular dynamics (MD) scheme. After computing all the atomic positions and velocities as function of time using Newton's second law, $\sum_j \mathbf{f}_{ij} = m_i \ddot{\mathbf{x}}_i$, where m_i and $\ddot{\mathbf{x}}_i$ are the atomic mass and acceleration and $\mathbf{f}_{ij}$ is the interatomic force exerted by atom j on atom i, the power exchange between two nanoparticles (NP_1 and NP_2) is computed as the net work done by a particle on the ions of the other particle:

$$Q_{1\leftrightarrow 2} = \sum_{\substack{i\in NP_1 \\ j\in NP_2}} \mathbf{f}_{ij}\cdot\mathbf{v}_j - \sum_{\substack{i\in NP_1 \\ j\in NP_2}} \mathbf{f}_{ji}\cdot\mathbf{v}_i \qquad (27)$$

The interatomic force $\mathbf{f}_{ij}$ is derived from the van Beest, Karmer, and van Santen (BKS) interaction potential (van Beest *et al.*, 1990), in which a Coulomb and a Buckingham potentials are included.

When the interparitlce distance is larger than the radii of spheres, i.e., $d \geq 4R$ for identical spheres with the radius R, the aforementioned methods (Domingues *et al.*, 2005; Pérez-Madrid *et al.*, 2008; Narayanaswamy and Chen, 2008) have a good agreement with the dipole approximation (e.g., Volokitin and Persson, 2001): See Fig. 11. However, at smaller interparticle distances, they show different trends. Figure 11(a) shows that the thermal conductance predicted in Refs. (Domingues *et al.*, 2005) and (Pérez-Madrid *et al.*, 2008) has a higher gap dependence than d^{-6}, resulting in four orders of magnitude higher than the dipole approximation in the intermediate distance range, i.e., $2R < d < 4R$. This enhanced heat transfer appears to be due to the contribution of multipolar Coulomb interactions (Pérez-Madrid *et al.*, 2008). However, as shown in Fig. 11(b), the thermal conductance calculated by Narayanaswamy and Chen (2008) asymptotically approaches a d^{-1} slope when the interparticle gap distance is much smaller than the particle radius. This slope change is consistent with the result of the proximity approximation or the Derjaguin approximation (Derjaguin *et al.*, 1956). In the Derjaguin approximation, the radiative heat flux between curved surfaces is approximated as the summation of heat fluxes between flat plates that integrate to form the profile of curved surfaces. The consequent thermal conductance is simplified as $G_{12} \approx \pi R \cdot d \cdot h_r(d,T)$, where h_r is the radiative heat transfer coefficient between flat surfaces. From the previous studies (Mulet *et al.*, 2002; Fu and Zhang, 2006; Basu and Zhang, 2009a), it was found that h_r has d^{-2} dependence in the near-field regime. Therefore, the proximity approximation predicts the d^{-1} dependence in the near-field thermal conductance, which is in contradiction to the $> d^{-6}$ dependence predicted in (Domingues *et al.*, 2005; Pérez-Madrid *et al.*, 2008). While this discrepancy is likely due to the difference in the considered nanoparticle sizes, i.e., R = 0.72 nm – 1.79 nm (Domingues *et al.*, 2005; Pérez-Madrid *et al.*, 2008) versus $R \geq 20$ nm (Narayanaswamy and Chen, 2008), further theoretical and experimental investigations are required to resolve this contradiction in near-field thermal radiation between two spheres.

Another unresolved issue regarding the near-field thermal radiation between spheres is the drastic decrease of thermal conductance when they become in contact. MD simulation (Domingues *et al.*, 2005) predicts that the contact thermal conductance would be 2-3 orders of magnitude lower than the conductance just before contact: see Fig. 11(a). This drastic reduction is still an open question that cannot be explained with the fluctuation-dissipation theorem. At such sub-nanometric distance, nanoparticles cannot be treated as a thermodynamic system at local equilibrium: fluctuation-dissipation theorem is not valid (Pérez-Madrid *et al.*, 2003). The knowledge gap on the thermal conductance change upon contact was explored by implementing the mesoscopic nonequilibrum thermodynamics for the calculation of the random current density, based on the assumption of the validity of the second law in the phase space (Pérez-Madrid *et al.*, 2009). The computed near-field radiative thermal conductance has a strong enhancement as d decreases below around double the radii of spheres, $d < 4R$, due to multipolar Coulomb interactions (Pérez-Madrid *et al.*, 2008). However, it sharply falls to the contact thermal conductance when both nanoparticles are in contact. This sharp reduction is attributed to an intricate conglomerate of energy barriers inherent to the amorphous character of nanoparticles, which is generated by the strong interaction (Pérez-Madrid *et al.*, 2009).

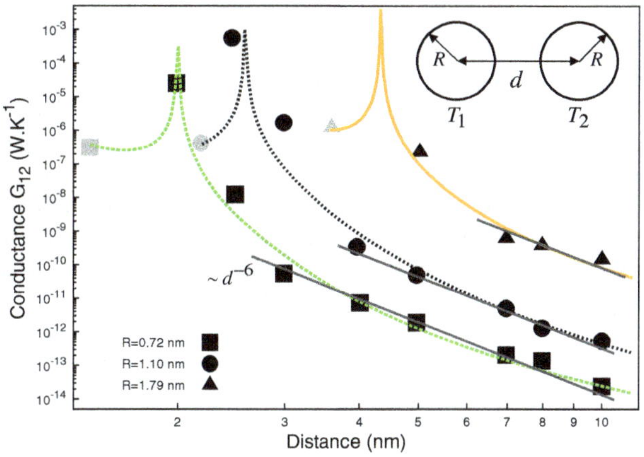

(a) Pérez-Madrid *et al.* (2009)

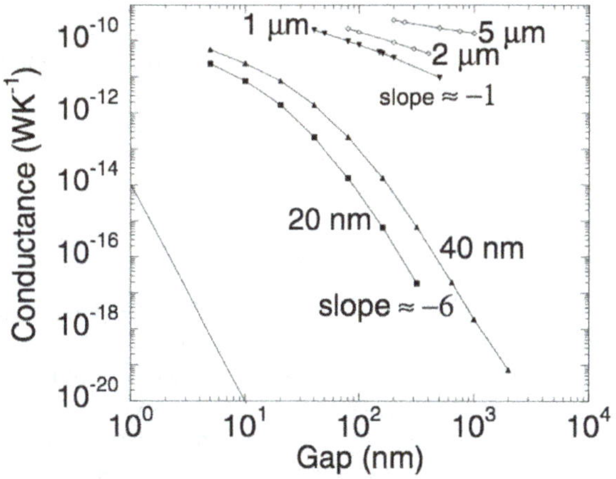

(b) Narayanaswamy and Chen (2008)

Fig. 11 Thermal conductance between two identical micro/nanospheres computed (a) in the framework of mesoscopic nonequilibirum thermodynamics (Pérez-Madrid *et al.*, 2009) and (b) using the vector spherical wave expansion method (Narayanaswamy and Chen, 2008). The inset illustrates the sphere-sphere configuration for the identical sphere case. In (a), the marks represent molecular dynamics data obtained by Domingues *et al.* (2005), where particles with effective radius $R = 0.72$, 1.10 and 1.79 nm were considered. The grey points represent the conductance when the particles are in contact. When bigger spheres are considered in (b), the gap dependence asymptotically approaches d^{-1} as the gap distance further decreases below the dipole approximation limit. (Figures reprinted with permission of the American Physical Society.)

4.2. Near-Field Radiative Heat Transfer between a Sphere and a Half Space

In this part, we discuss near-field thermal radiation between a small spherical particle and a semi-infinite medium, which can be considered as a simplified model of the scanning tunneling microscopy or scanning thermal microscopy (Volokitin and Persson, 2001). Similarly to the previous section, the small particle can be approximated as a dipole of radius R with dielectric constant $\varepsilon_{\mathrm{P}}(\omega)$ and temperature T_{P}. The semi-infinite surface is maintained at temperature T_{B} and has the dielectric constant $\varepsilon_{\mathrm{B}}(\omega)$. The center of the particle is at a distance d above the interface. Then, the spectral mean power radiated by the half space and absorbed by the particle can be written as (Mulet *et al.*, 2001)

$$P_{\mathrm{B}\to\mathrm{P}}(\omega) = \frac{2\omega^4}{\pi c^4}\mathrm{Im}[\varepsilon_{\mathrm{B}}(\omega)]\mathrm{Im}[\alpha_{\mathrm{P}}(\omega)]\Theta(\omega, T_{\mathrm{B}}) \sum_{n,m=x,y,z}\int_{\mathrm{B}}\left|G_{nm}(\mathbf{x}_{\mathrm{p}},\mathbf{x}',\omega)\right|^2 d^3\mathbf{x}' \quad (28)$$

When the fluctuating currents inside the particles radiate into the half space and dissipate, the locally dissipated power per unit volume at a point $\mathbf{x}$ inside the space can be written as

$$P_{\mathrm{P}\to\mathrm{B}}(\mathbf{x},\omega) = \frac{2\omega^4}{\pi c^4}\mathrm{Im}[\varepsilon_{\mathrm{B}}(\omega)]\mathrm{Im}[\alpha_{\mathrm{P}}(\omega)]\Theta(\omega, T_{\mathrm{B}}) \sum_{n,m=x,y,z}\left|G_{nm}(\mathbf{x},\mathbf{x}_{\mathrm{p}},\omega)\right|^2 \quad (29)$$

where $\alpha_{\mathrm{P}}(\omega) = 4\pi R^3[\varepsilon_{\mathrm{P}}(\omega) - 1]/[\varepsilon_{\mathrm{P}}(\omega) + 2]$ is the polarizability of the dipole and $\mathbf{G}_{nm}(\mathbf{x},\mathbf{x}',\omega)$ is (n, m) component of the dyadic Green's function at $\mathbf{x}$ due to a point source at $\mathbf{x}'$ in a system constituted by two semi-infinite media whose dielectric constants are either 1 if $z \geq 0$ or $\varepsilon_{\mathrm{B}}(\omega)$ if $z < 0$. It should be noted that the dipole polarizability needs to be corrected to take into account the interaction between the dipole and the interface when d is comparable to R (Dorofeyev, 1998).

When $P_{\mathrm{B}\to\mathrm{P}}(\omega)$ is calculated for a SiC spherical particle at temperature $T_{\mathrm{P}} = 300$ K of radius $R = 5$ nm at different distances above the SiC surface (Mulet *et al.*, 2001), two peaks are observed at frequency $\omega_1 \approx 1.756 \times 10^{14}$ rad/s and $\omega_2 \approx 1.787 \times 10^{14}$ rad/s, each of which corresponds to the localized surface polariton resonance of the SiC particle (i.e., $\mathrm{Re}\,[\varepsilon_{\mathrm{P}}] = -2$) and the surface polariton resonance on the SiC surface (i.e., $\mathrm{Re}\,[\varepsilon_{\mathrm{B}}] = -1$), respectively. Moreover, near-field radiative heat transfer increases showing the d^{-3} dependence as d decreases, enhanced by several orders of magnitude than the far-field thermal radiation. Another example is shown in Fig. 12, which plots $P_{\mathrm{P}\to\mathrm{B}}(\omega)$ emitted from a SiC particle of radius $R = 25$ nm heated at $T_{\mathrm{P}} = 1000$ K to a gold surface right below the particle. There is a sharp peak at $\sim$930 cm^{-1}, corresponding to $\omega \approx 1.756 \times 10^{14}$ rad/s or 10.6 μm when the tip is near the surface, suggesting that the heated SiC particle emits a quasi-monochromatic thermal radiation at around 10.6 μm in the near-field regime.

It should be noted again that the dipole approximation for a sphere is valid when $R \ll \lambda_T$ and $d \geq 2R$. In the other extreme case, i.e., $d \ll R \sim \lambda_T$, the dipole approximation may not be valid, and instead we should use the proximity approximation (or Derjaguin approximation). Applying the proximity approximation for the sphere-flat surface case, the near-field thermal conductance has the d^{-1} dependence, instead of d^{-3}. This result was experimentally validated as will be discussed in the following section. A further research was conducted to seek the shape dependence of near-field radiative heat transfer when a spheroidal metallic nanoparticle is in proximity to a planar metallic sample (Huth *et al.*, 2010). By changing the aspect ratio of a spheroidal particle from 1/5 (long axis perpendicular to the surface) to 5 (long axis parallel to the surface) while fixing its volume, they predicted that the total radiative heat transfer between a gold spheroid at 100 K and a gold planar surface at 300 K at the gap width of 100 nm changes from about half to two times of that between sphere and planar surface.

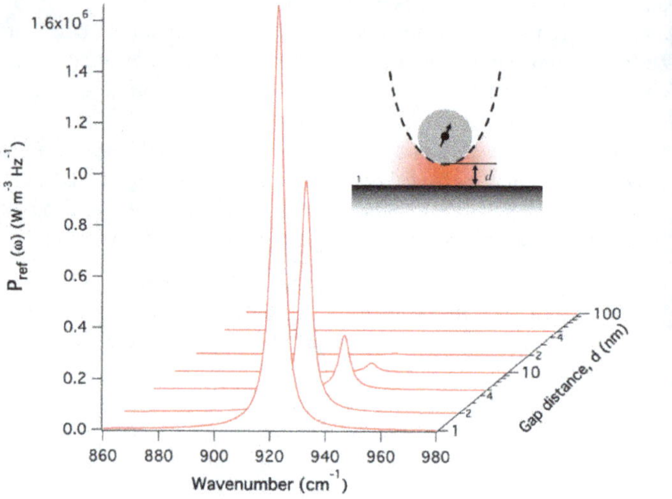

Fig. 12 Mean spectral power radiated from the SiC particle (of radius $R =$ 25 nm and held at $T_{\mathrm{P}} = 1000$ K.) to the gold surface right below the particle.

4.3. Near-Field Radiative Heat Transfer in Cylindrical Objects

While various geometries of nanostructures are being developed as potential candidates for efficient heat transfer sources, it is not an easy task to theoretically investigate radiative heat transfer from or between such nanostructures. The thermal radiation of long cylindrical objects whose thickness is in the range of the thermal wavelength is a good example, as it can be very different from the classical blackbody radiation described by Planck's law (Öhman, 1961). Recent experimental studies on thermal radiation of an antenna-like platinum microheater (Ingvarsson *et al.*, 2007; Au *et al.*, 2008), a single SiC whisker antenna (Schuller *et al.*, 2009), thin tungsten wires (Bimonte *et al.*, 2009) and individual carbon nanotubes (Fan *et al.*, 2009; Singer *et al.*, 2011) demonstrated that thermal radiation from sub-wavelength cylindrical emitters has coherent features and thus can be polarized depending on the orientation of the emitters. Polarization effects were also observed from the thermal radiation of carbon nanotube bundles (Li *et al.*, 2003), which could be used to probe the degree of alignment inside the bundle. Some works (Schuller *et al.*, 2009; Singer *et al.*, 2011) further studied the polarized thermal radiation by developing a simple theoretical model based on the Mie theory and comparing with the measurement, revealing that the degree of polarization is affected by the optical conductance along with the geometry of the cylindrical thermal emitter. However, the Mie theory could not correctly deliver the frequency-dependence of the degree of polarization, requiring a more comprehensive theoretical model. The Kadar group at MIT has developed a general formalism to compute the heat radiation of arbitrary objects in terms of their classical scattering properties based on the fluctuation-dissipation theorem (Krüger *et al.*, 2011; Golyk *et al.*, 2012). They predicted that the degree of polarization of the emitted radiation depends on the cylinder radius: if the radius is much smaller than the thermal wavelength, the radiation is polarized parallel to the cylindrical axis and becomes perpendicular when the radius is comparable to the thermal wavelength.

Recently, near-field radiative heat transfer between cylinder-cylinder and cylinder-plate has been formulated. Carillo and Bayazitoglu (Carillo and Bayazitoglu, 2011, 2012) modified the aforementioned vector spherical wave expansion approach (Narayanaswamy and Chen, 2008) for near-field sphere-to-sphere radiative exchange to make it applicable to cylindrical geometry of nanorods. The computation results on silica nanorods clearly show strong near-field effect on radiative heat transfer for the cylinder-cylinder case, but uniquely from the spherical case, this near-field enhancement is due to the larger geometric surface area that enables more photon tunneling of evanescent waves. They also asymptotically computed a nanorod-to-plate geometric configuration, where a nanorod lies in parallel to the plane, by increasing the radius of a nanorod while the other nanorod radius is held constant until the corresponding radiative heat transfer rate converges. A cylinder-to-plate configuration where the cylinder axis is perpendicular to the plate was investigated and compared with a sphere-to-plate and a sharp tip-to-plate cases (McCauley, 2012). The near-field cylinder-to-plate heat transfer rate has a $\sim d^{-2}$ gap dependence while the sphere-to-plate case shows a d^{-1} dependence, consistently with the proximity approximation. The tip-to-plate configuration has the least gap dependence in the near-field heat transfer rate and, interestingly, exhibits a local minimum heat flux directly below the tip that becomes deeper as the tip becomes sharper. The authors believe that this dip in the local heat flux is attributed to the more dipole-like radiative behaviors as the tip becomes sharper. While further experimental verifications need to be followed, the accumulated results manifest that a general numerical method to calculate near-field radiative heat transfer between arbitrary objects, including all the aforementioned geometries, is at hand.

4.4. Near-Field Radiative Heat Transfer in Emerging Materials

Most near-field thermal radiation studies to date have focused on naturally occurring materials, such as dielectrics, semiconductors, metals and polar materials. However, the advent of nanotechnology has enabled the integration of emerging materials, such as graphenes, photonic crystals, and metamaterials, in the near-field radiative heat transfer research. This section is thus devoted to provide up-to-date review of near-field radiation research where novel materials are involved.

Graphene: Graphene has recently received a keen attention due to its unique electronic (Novoselov *et al.*, 2004, 2005), mechanical (Frank *et al.*, 2007), and thermal (Lee *et al.*, 2011) properties, thus has been actively explored for next-generation technologies (Geim and Novoselov, 2007). Among many graphene-based electronic and sensor applications, graphene field-effect transistors (FETs) are considered as a promising candidate that would resolve current challenges of Si-based FETs (Schwierz, 2010). However, heat generation and dissipation in the graphene FET must be fundamentally understood prior to its reliable operation in commercial integrated circuits.

Freitag *et al.* (2009) measured the temperature distribution of a biased single-layer graphene FET and found that 77% of the heat dissipation occurs at the gate stack (300 nm SiO_2 film on silicon) directly below the active graphene channel, having an effective graphene-SiO_2 interface heat transfer coefficient of $h_{Gr-SiO_2} \approx 2.4 \times 10^7$ W/m^2-K. Theoretical studies have followed to fundamentally understand the graphene-SiO_2 interfacial thermal interactions. Persson and Ueba (2010a,b) studied heat transfer mechanisms between graphene and amorphous SiO_2 in contact by including the heat transfer from the area of real contact as well as between the surfaces in the non-contact region, which may occur due to rough surfaces. Although it was concluded that most of the heat flows through the area of real contact, they were the first that considered near-field thermal radiation in graphene-involved configurations. However, their near-field radiation calculation was based on the assumption that the free carriers in graphene had vanishing drift velocity, which may cause significant discrpancies from actual behaviors of graphene. The effect of drift velocity on the graphene-SiO_2 near-field thermal radiation was explored by considering quantum fluctuations of free carriers along with thermal fluctuations in graphene (Volokitin and Persson, 2011). For nonsuspended graphene on SiO_2 (having a subnanometer separation), near-field radiation gives a significant contribution to the heat transfer in addition to the contribution of phononic coupling: the near-field radiative heat transfer coefficient can reach $\sim 10^8$ W/m^2-K at $d = 0.35$ nm when the biased electric field is in the low and intermediate range. On the other hand, suspended graphene with separation in the order of 1 nm has the near-field radiative heat transfer coefficient of $\sim 10^4$ W/m^2-K, which is significantly less than for the nonsuspended graphene case but still ~ 3 orders of magnitude larger than the blackbody radiation limit ($h_{bb} \approx 5$ W/m^2-K).

Recently, two papers were concurrently published that analyzed the contributions of plasmons to near-field radiative heat transfer in graphene (Svetovoy *et al.*, 2012; Ilic *et al.*, 2012a). Both studies were motivated by the tunability of plasmon frequencies in graphene from terahertz to the near infrared by changing the electron density (Ju *et al.*, 2011). Near-field radiative heat transfer between graphene-coated dielectrics can be larger than the best known materials for the radiative heat transfer – around two times the near-field radiative heat transfer coefficient between SiO_2 ($h_{SiO_2-SiO_2} \approx 300$ W/m^2-K at room temperature when the gap distance is ~ 100 nm), and can be reduced 100 times or so (Svetovoy *et al.*, 2012). In addition, near-field radiative heat transfer between two graphene sheets can be enhanced up to three orders of magnitude the blackbody radiation limit as the gap distance is reduced to 10 nm (Ilic *et al.*, 2012a). These unique radiative features of graphene may offer a potential for a novel, hybrid thermophotovoltaic/thermoelectric solid-state energy conversion platform, as reported in Ref. (Ilic *et al.*, 2012b) that proposed the application of graphene as a thermal emitter in a near-field thermophotovoltaic system.

Periodic Structures: Near-field radiative heat transfer in periodic structures, such as *nanoporous media* (Biehs *et al.*, 2011a), *gratings* (Biehs *et al.*, 2011b), and *photonic crystals* (Ben-Abdallah *et al.*, 2010; Rodriguez *et al.*, 2011), has become a very interesting topic with promising future applications. By combining the fluctuation-dissipation theorem and the Maxwell-Garnett effective medium description for effective media, Biehs and his colleagues (2011a) studied radiative heat transfer between two semi-infinite nanoporous media, made of SiC having cylindrical inclusions oriented in the perpendicular direction to the surface. The obtained results reveal that for the small distance regime ($d < 100$ nm), the heat flux between the nanoporous media can be significantly larger than that between two homogeneous SiC plates in the same thermal conditions (e.g., $\sim$50% larger radiative heat flux when nanoporous media have a 0.5 filling factor and are separated by $d = 10$ nm). This increase is seemingly due to additional surface waves arising at the uniaxial material-vacuum interface. The same numerical scheme was applied to near-field heat transfer between two misaligned 1-D gratings (Biehs *et al.*, 2011b). As the twisting angle changes from the parallel grating configuration ($\phi = 0°$), the near-field radiative heat flux between gratings is modulated significantly, up to 90% reduction for the perpendicular ($\phi = 90°$) configuration when gold gratings with a filling factor of 0.3 are taken into account. This allows the manipulation of the heat flux at nanoscale.

Photonic crystals are another example of periodic structures. By periodically repeating dielectric or metallo-dielectric layers of high and low dielectric constants, the propagation of electromagnetic waves can be controlled to have a photonic band gap, i.e., wavelength band of disallowed photon propagation, giving rise to distinct optical phenomena such as inhibition of spontaneous emission, high-reflecting mirrors, and low-loss-waveguide (Yablonovitch, 1987; John, 1987). Due to such unique radiative properties, photonic crystals have been applied to manipulate thermal radiation, such as wavelength-selective thermal emission (Pralle *et al.*, 2002; Lin *et al.*, 2003; Narayanaswamy and Chen, 2005, 2004; O'Sullivan *et al.*, 2005) and spectrally and directionally coherent thermal emission (Lee *et al.*, 2005; Lee and Zhang, 2006a,b; Lee *et al.*, 2008a). Near-field thermal radiation between photonic crystals has also been addressed with different approaches, including the use of dyadic Green's function along with the scattering matrix method (Francoeur *et al.*, 2009), the expansion of Green tensors in terms of the intracavity fields (Ben-Abdallah *et al.*, 2010), and the finite-difference time-domain method (Rodriguez *et al.*, 2011). These theoretical investigations report that the strong coupling of surface Bloch states supported by photonic crystals makes near-field heat transfer several folds larger than that between two homogeneous plates at the same separation distance (Ben-Abdallah *et al.*, 2010). Moreover, even frequency-selective near-field radiative heat transfer is possible with careful design of photonic crystals (Rodriguez *et al.*, 2011).

Metamaterials: Metamaterials are broadly defined as any artificial material engineered to achieve material properties that may not be found in nature, but narrowly referred to as materials with negative refractive index (Pendry, 2000; Shelby *et al.*, 2001; Smith *et al.*, 2002). Metamaterials has emerged as a new frontier of optical and thermal radiation research, as it may realize innovative technologies such as superlenses, invisibility cloaks, and manipulation of radiative properties (Fu and Zhang, 2009; Liu and Zhang, 2011). Near-field radiative heat transfer between metamaterials is thus an attractive research topic that may pave the way to the development of novel thermal management or energy harvesting. Earlier studies demonstrated that surface polaritons could be excited for both p- and s-polarizations when the refractive index becomes negative (Ruppin, 2000, 2001; Park *et al.*, 2005). Later, radiative and nonradiative heat exchanges between metamaterials were conducted by considering fluctuations of electric and magnetic currents density in semi-infinite metamaterials (Joulain *et al.*, 2010). They showed that the excitation of magnetic polariton resonance (when the magnetic permeability becomes -1) and the ferromagnetic behavior of materials under a strong magnetization (when the magnetic permeability becomes large) yield novel channels for energy transfer enhancement. The enhancement factor of near-field radiation, normalized to the blackbody radiation, becomes in the order of 10^4 in the extreme near-field regime (i.e., $5.31 \times 10^{-4}\lambda_p$, where λ_p is the plasma wavelength), where heat transfer is dominated by polariton-like waves. Similar enhancement mechanisms of radiative heat transfer in near field were also found in different types of metamaterials, such as chiral metamaterials (Cui *et al.*, 2012) and SiC sphere-embedded potassium bromide (Francoeur *et al.*, 2011a), confirming that the presence of negative magnetic permeability in metamaterials is beneficial in enhancing near-field thermal radiation. More recently, hyperbolic metamaterials were also studied for potentially enhanced near-field radiation (Biehs *et al.*, 2012).

5. EXPERIMENTAL OBSERVATIONS OF NEAR-FIELD RADIATIVE HEAT TRANSFER

Due to difficulties in maintaining the nanoscale gap distance between the emitter and the receiver, experimental investigations of near-field thermal radiation have been rather limited. Cravalho *et al.* (1966), Domoto *et al.* (1969), and Hargreaves (1969) were among the first to measure the radiative flux of two parallel plates at cryogenic temperatures. Domoto *et al.* (1969) measured the radiative heat transfer at cryogenic temperatures between two copper plates at gaps from 1 to 10 μm. While the near-field heat transfer was 2.3 times greater than that of the far field, the measured heat flux was only 3% of the energy transfer between blackbodies. Hargreaves (1969) measured the near-field heat transfer between two chromium plates separated by vacuum gaps from 6 to 1.5 μm. At 1.5 μm vacuum gap, the near-field heat transfer at room temperature was five times greater than that in the far field. However, the measured heat flux was still only 40% of that between two blackbodies. In 1994, Xu *et al.* (1994) tried to measure near-field radiative heat transfer through a sub-micrometer vacuum gap by using an indium needle of 100 μm in diameter and a thin-film thermocouple on a glass substrate, but could not observe a substantial increase of radiative heat transfer. On the other hand, Muller-Hirsch *et al.* (1999) found that near-field radiation plays an important role in heat transfer between a STM thermocouple probe and a substrate. However, due to the limit of their experimental setup, they were not able to determine the absolute value of near-field thermal radiation. This limit was overcome in their following work (Kittel *et al.*, 2005) by successfully calibrating the STM thermocouple probe, demonstrating the d^{-3} dependence in the near-field thermal radiation from the surface to the thermocouple tip of $R = 60$ nm when the gap is larger than 10 nm. However, for gaps less than 10 nm, the measured heat flux saturates and differs from the divergent behavior of the predicted results. The authors attributed this difference to the spatial dependence of the dielectric function of materials.

Continuous efforts have been made to experimentally demonstrate the near-field enhancement of energy transfer for other relatively simple geometries, such as parallel plates separated by micro-particle spacers (Hu *et al.*, 2008) and microsphere-plate geometry (Narayanaswamy *et al.*, 2008; Shen *et al.*, 2009; Rousseau *et al.*, 2009; Shen *et al.*, 2012). The thermal conductance of near-field radiation was successfully measured for a gap distance as small as 30 nm by using a vertically aligned bimetallic AFM cantilever having a silica or gold-coated silica microsphere at the free end. The plate was heated to produce a temperature difference ΔT between the sphere and the plate, typically on the order of 10–20 K, leading to the near-field radiative heat flux of the order of nanowatts. In order to measure such small heat flux, the measurement was conducted in a vacuum condition ($\sim 10^{-6}$ mbar). Near-field thermal radiation was measured by monitoring the deflection of the bimetallic cantilever, which has a minimum measurable temperature of 10^{-4}–10^{-5} K and a minimum detectable power of 5×10^{-10} W (Narayanaswamy *et al.*, 2008). Comparison of their measurement with the Derjaguin approximation confirms that the near-field thermal radiation between the microsphere and flat surface is more than two orders of magnitude larger than that of blackbody radiation and has d^{-1} dependence. At a 30 ± 5 nm gap, the heat transfer coefficient was measured to be $\sim$400 W/m^2-K for Au-Au, which is around 4 times smaller than that for SiO_2-SiO_2 and much greater than the blackbody radiation limit of $\sim$5 W/m^2-K (Shen *et al.*, 2012). Although the enhancement of near-field radiative heat transfer between SiO_2 surfaces can be explained with strong coupling of surface phonon polaritons (Shen *et al.*, 2009), the radiative heat transfer enhancement for the Au-Au case is somewhat counterintuitive since metals are highly reflective for infrared lights. However, the theoretical study of near-field radiation between metals reveals that although metals are highly reflective, thermal radiation emitted from the hot surface experiences multiple reflections in a nanoscale gap until eventually absorbed by the cold surfaces (Chapuis *et al.*, 2008b).

6. APPLICATIONS OF NEAR-FIELD RADIATIVE TRANSFER

6.1. Manipulation of Radiative Properties

Controlling the radiative properties has important applications in photonic and energy conversion systems such as solar cells and solar absorbers, thermophotovoltaic (TPV) devices, radiation filters, selective emitters, photodetectors, semiconductor processing, and optoelectronics (Zhang and Wang, 2011; Basu *et al.*, 2007; Zhang *et al.*, 2003; Zhu *et al.*, 2009). The performance of various devices can be greatly enhanced by the modification of the reflection, transmission, absorption and emission spectra using one-, two-, or three-dimensional micro/nanostructures. Surface microstructures can also strongly affect the directional behavior of absorption and emission due to multiple reflections and diffraction, allowing the radiative properties to be tailored. Because of the important applications to energy transport and conversion, the study of engineered surfaces with desired thermal radiative characteristics has become an active research area.

As briefly introduced in the previous section, photonic crystals have been studied to control the thermal emission to be wavelength-selective (Pralle *et al.*, 2002; Lin *et al.*, 2003; Narayanaswamy and Chen, 2005, 2004; O'Sullivan *et al.*, 2005) or to be spectrally and directionally coherent (Lee *et al.*, 2005; Lee and Zhang, 2006a,b; Lee *et al.*, 2008a).

For the application as wavelength-selective diffuse TPV emitters and infrared detectors, Chen and Zhang (2007, 2008) proposed the concept of complex gratings whose surface profile is superposed by two or more 1D grating profiles. The complex grating may improve simple 1D gratings by reducing the sharpness in the spectral peak and the directional sensitivity. Moreover, heavily-doped silicon complex gratings exhibit a broad band absorptance peak that is insensitive to the angle of incidence by properly choosing the carrier concentration and geometry. The peak wavelength can be engineered by changing either the height of the ridges or the period. Such a type of absorptance peak comes from the SPP excitation and is dominated by the first evanescent diffraction order.

Recent studies on periodic gratings have revealed that localized magnetic polaritons (MPs), which are responsible for extraordinary transmission in metamaterials (Liu *et al.*, 2006, 2009), can also occur in periodic gratings to significantly alter the thermal radiative properties of the structure (Lee *et al.*, 2008c). Figure 13 illustrates the effect of MP for a deep grating structure (Wang and Zhang, 2009). The induced current flow, shown as red arrows, in the 1-D grating can be modeled by an equivalent LC circuit model shown in Fig. 13(a) and the included inset. The contour plot of $1 - R$, or the sum of the transmittance T and absorptance α, as a function of ω and k_x is shown in Fig. 13(b). The radiative properties of considered structure are calculated with the rigorous coupled-wave analysis (RCWA) (Lee *et al.*, 2008b) and the predicted resonance frequency from the LC model for the fundamental mode (MP1) is illustrated as triangles. Excellent agreement between the LC model and the RCWA results further confirms the mechanism of magnetic resonance. The bright bands indicate usually a strong transmission, but can also be associated with a strong absorption, due to the resonance behavior of SPPs or MPs. The inclined line close to the light line, which is then folded due to the Bloch-Floquet condition in the gratings, is associated with the excitation of SPP at the Ag-vacuum interface. Several relatively flat dispersion curves correspond to the fundamental, second, and third modes of MPs and are marked as MP1, MP2 and MP3 in the figure. The flatness of MP dispersion curves indicates their unique feature as directional independence. The directional independence of MPs can be understood by the diamagnetic response, as the oscillating magnetic field is always along the y-direction no matter what incident angles is for TM waves. It should be noted that, the cavity-like resonance or coupled SPPs were previously proposed to explain the resonance phenomenon in simple gratings, but MPs seem to more quantitatively account for the geometric effects on the resonance conditions (Wang and Zhang, 2009).

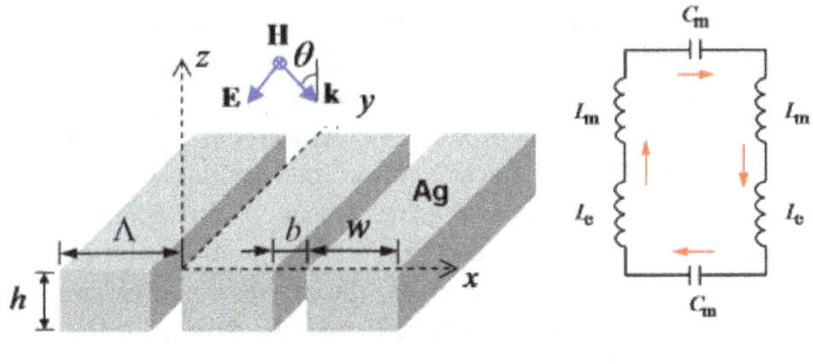

(a) Schematics of a deep grating

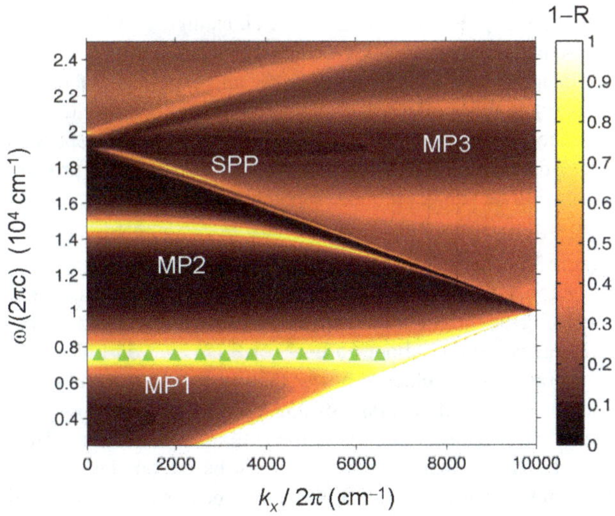

(b) MP and SPP dispersion contour

Fig. 13 Effect of magnetic polaritons (MPs) on the radiative properties of a single grating (slit array): (a) Schematic of a deep grating with th inset that depicts the equivalent LC circuit model; (b) Contour plots of the sum of absorptance and transmittance (i.e., $1 - R$) for a Ag grating with period $\Lambda = 500$ nm, $h = 400$ nm, and $b = 50$ nm. Triangle marks indicate the frequency of the fundamental mode predicted by the LC circuit model (Wang and Zhang, 2009).

Another potential application of the MPs is the construction of coherent thermal emission sources. It has been demonstrated that a nanostructure consisting of a periodic metallic strips separated by a thin dielectric layer over an opaque metal film exhibits coherent emission characteristics (Lee *et al.*, 2008c; Zhang *et al.*, 2011). The coupling of the metallic strips and the film induces a magnetic response that is characterized by a negative permeability and positive permittivity. On the other hand, the metallic film intrinsically exhibits a negative permittivity and positive permeability in the near infrared. This artificial structure is equivalent to a pair of single-negative materials. By exciting surface MPs, large emissivity peaks can be achieved at the resonance frequencies, which are almost independent of the emission angle. This spectrally selective, diffuse thermal emission could be beneficially used for thermophotovoltaics if MPs are excited in the infrared range (Wang and Zhang, 2012). To this end, phonon-assisted MPs were also explored by designing deep SiC grating structures, observing similar features as metallic MP couplers (Wang and Zhang, 2011).

6.2. Near-Field Thermophotovoltaic Energy Conversion

It has been recently reported that the current global energy demand is approximately 14 TW and is expected to double to 25-30 TW by 2050 (Baxter *et al.*, 2009). When considering the serious energy dependence (i.e., more than 80% of the current energy consumption) on fossil fuels including oil, coal, and natural gas, raising prices of these energy sources and carbon-dioxide-driven global warming will pose a grave threat to the global economy and environment. Thus it is imperative to develop carbon-free, high-efficiency and low-cost renewable energy harvesting and recycling technologies.

Thermophotovoltaic (TPV) energy conversion is an energy harvesting technology that directly generates electric power from thermal sources emitting IR radiation. A TPV system consists of a thermal emitter and a TPV cell that is a p-n junction semiconductor converting radiative energy to electric power (Basu *et al.*, 2007). Since wasted heat from many industrial processes (e.g., glass manufacturing or power plants) can be used as an IR emission source, TPV systems are considered as one of the promising techniques for the wasted energy recovery and recycling. Moreover, having no moving parts allows quiet and reliable operations in harsh environments, making TPV ideal in military or space applications (Nelson, 2003). When compared with other solid-state technologies, the energy conversion efficiency of TPV ($\sim$25%, Lin *et al.*, 2003) is higher than those of thermoelectric ($<\sim$15%, Chen, 2006) and thermionic ($<\sim$13%, Lee *et al.*, 2009) devices. However, the TPV efficiency is still low and, more seriously, its low power throughput is a big challenge in applying the TPV for effective energy recycling. Thermal radiation at low working temperatures is not a compelling energy source due to the T^4

dependence of the radiative power: for example, the blackbody emissive power at 600 K is only ~7 kW/m^2, which is too small for thermal energy harvesting.

One solution for improving the power throughput and conversion efficiency of the TPV system may be to utilize near-field thermal radiation. The feasibility of the near-field TPV system has been investigated by several research groups. Pan *et al.* (2000) were the first to analyze the performance of near-field TPV systems. However, they used the same dielectric material for both the emitter and TPV cell to calculate the near-field energy enhancement, which is not only overly simplified but also impractical. Whale and Cravalho (2002) considered a more realistic system by using a fictitious Drude material with a low conductivity and InGaAs for the emitter and the TPV cell, respectively. Narayanaswamy and Chen (2003) theoretically demonstrated the effect of surface polaritons in improving the performance of near-field TPV systems. However, their work focused only on the thermal radiation enhancement, leaving questions on the near-field effect on the photocurrent generation. Laroche *et al.* (2006) provided an analysis on the performance and efficiency of near-field TPV systems based on the assumption of 100 % quantum efficiency in calculating the photocurrent generation; this may result in an overestimation of the TPV system performance. Park *et al.* (2008) performed a more realistic analysis of the power generation in a near-field TPV system by calculating the photocurrent generation in different regions of the TPV cell. Francoeur *et al.* (2011b) developed a coupled near-field thermal radiation and the charge and heat transfer model within the cell and stressed the thermal management issues in near-field TPV systems.

A near-field TPV energy conversion system consists of a TPV cell and the thermal source that are separated with a very small vacuum gap. Figure 14 illustrates a conceptual near-field TPV device used in (Park *et al.*, 2008), where a tungsten thermal emitter is placed in proximity to a TPV cell within a subwavelength gap distance. In order to demonstrate the operation concept of TPV device, the thermal source is set to be made of tungsten and maintained at $T_H = 2000$ K, so that the characteristic wavelength of thermal emission is around 1.5 μm. As for the TPV cell, $In_{0.18}Ga_{0.82}Sb$ is chosen because its energy bandgap of 0.56 eV is sufficiently low for the thermophotovoltaic energy conversion. Doping concentration of the p-layer is set to 10^{19} cm^{-3} whilst the tellurium-doped n-layer has a doping concentration of 10^{17} cm^{-3} (González-Cuevas *et al.*, 2006). The concentration gradient of the majority carriers across the p-n junction diffuses electrons from the n- to p-region, and vice versa for the holes. As a result, the depletion region having only ionized dopants is formed, and its width is estimated to be 0.1 μm from the given doping concentrations (González-Cuevas *et al.*, 2006). When thermal radiation is incident on the TPV cell, photons whose energy is greater than the bandgap energy E_g will generate electron-hole pairs inside the TPV cell. However, some electron-hole pairs generated outside the depletion region will be recombined as they diffuse toward the edge of the depletion region. In order to predict the near-field thermophotovoltaic power generation, Park *et al.* (2008) first calculated the distribution of radiative energy absorption in the TPV cell due to near-field thermal radiation using the fluctuation dissipation theorem and dyadic Green's function for a multilayered structure, and calculated photocurrent generation in each region to have $J_\lambda(\lambda) = J_e(\lambda) + J_h(\lambda) + J_{dp}(\lambda)$, where $J_{e(h)}(\lambda)$ is the photocurrent generated in the $p(n)$-doped region and $J_{dp}(\lambda)$ is a drift current generated in the depletion region.

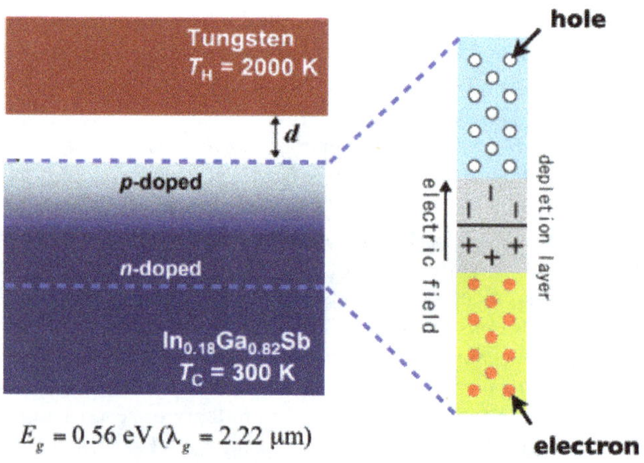

Fig. 14 Schematic of a near-field TPV system, where $In_{0.18}Ga_{0.82}Sb$ is used as the TPV cell material and plain tungsten is used as the emitter. Both the emitter and the cell material are modeled as semi-infinite media.

The performance of a TPV system can be evaluated through two efficiencies: the quantum efficiency η_q and the conversion efficiency η. The quantum efficiency is the ratio of the number of electron-hole pairs used for the photocurrent generation to the number of photons absorbed. On the other hand, the conversion efficiency (or thermal efficiency) is the ratio of the electric power generated from a TPV cell to the absorbed radiative power. Figure 15(a) shows the total photocurrents integrated over all wavelengths as a function of the vacuum gap distance. In general, J_h is greater than J_e due to the large thickness of the n-region. However, when the vacuum gap is very small, i.e., $d < 4$ nm, J_e becomes greater than J_e because a significant amount of the near-field thermal radiation is absorbed very close to the surface. The sum of the three photocurrents is used to calculate the electrical power generated by the TPV cell. Apparently, the near-field enhancement occurs in both the thermal radiation and the electric power generation: see Fig. 15(b). From the calculated near-field thermal radiation and photocurrent, the conversion efficiency of the near-field TPV system can be obtained as shown in Fig. 15(c). For comparison, the conversion efficiency for the ideal case with 100% quantum efficiency is also plotted. If the quantum efficiency is 100%, all photogenerated electron-hole pairs contribute to the power generation without being recombined during the diffusion to the depletion region. For such case, the conversion efficiency increases as the vacuum gap decreases and can reach as high as 35% when $d = 5$ nm. The conversion efficiency calculated by considering recombination is lower than the ideal case by 5% to 10%, and experiences a decrease as the vacuum gap further decreases below 10 nm. This efficiency decrease is due to the extremely small penetration depth of evanescent waves on the order of a nanometer: electron-hole pairs generated in proximity to the surface will be subject to more recombination while they move to the depletion region (Basu and Zhang, 2009b).

Figure 15 clearly shows that the near-field TPV can greatly enhance the power generation with approximately 20% conversion efficiency. The power density at the vacuum gap of 100 nm is predicted to be ~20 W/cm^2, suggesting that about 65 cm^2 (or 8 cm × 8 cm) of a TPV cell could generate enough electric power that meets the demand of one US household, i.e., monthly average of 958 kWh in 2010 as reported by the US Energy Information Administration (http://www.eia.gov). However, it should be noted that this prediction is based on the semi-infinite tungsten emitter maintained at 2000 K, assumed as a reservoir, and thus energy required to maintain 2000 K was not taken into consideration. Another issue is that temperature increase of the TPV cell due to thermalization of high-energy charge carriers was not considered although it will adversely affect the performance of the TPV cell. Together with the promising results, these limiting factors strongly suggest the near-field effect on the TPV energy conversion should be experimentally and fundamentally investigated to validate the theoretical prediction of the device performance. However, it remains extremely challenging to design and fabricate a near-field TPV system, particularly in keeping a large area within a sub-100-nm vacuum gap with good parallelism. Although

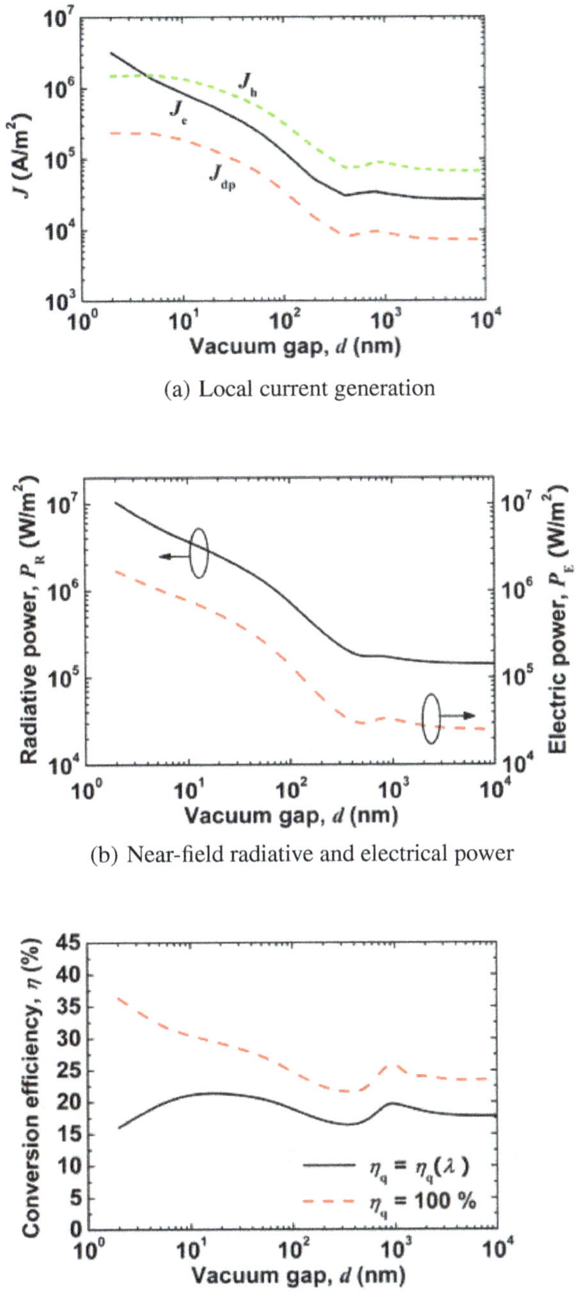

(a) Local current generation

(b) Near-field radiative and electrical power

(c) Conversion efficiency

Fig. 15 The effect of vacuum gap width d on (a) the local current generation, (b) the absorbed radiative power and electrical power generation, and (c) the conversion efficiency. The conversion efficiency for η_q = 100 % is also plotted for comparison (Park *et al.*, 2008).

near-field radiation has been experimentally verified, as discussed in the previous section, only near-field thermal radiation between a sphere and flat substrate has been measured near room temperature (Kittel *et al.*, 2005; Narayanaswamy *et al.*, 2008; Shen *et al.*, 2009, 2012; Rousseau *et al.*, 2009). Nanoscale thermal radiation between two flat surfaces has not been experimentally demonstrated despite several attempts in the past (Domoto *et al.*, 1969; Hargreaves, 1969; Xu *et al.*, 1994). Considering that the sphere-to-substrate case is not the adequate geometry for a near-field TPV system, parallelism between two flat surfaces with a small gap distance must be realized for the development of the near-field TPV system. Recently, Ottens *et al.* (2011) demonstrated near-field effect on the radiative heat flux between flat sapphire plates separated at a distance as small as 1-2 μm. DiMatteo *et al.* (2003) suggested using tubular spacers between the emitter and TPV cell to realize a sub-micron vacuum gap and also to prevent parasitic conduction heat transfer between the thermal emitter and the TPV cell. Based on previous research results, their concept is in the phase of commercialization (http://www.mtpvcorp.com): the first generation module that generates the current density of 1 W/cm^2 has been developed, being anticipated to advance it to 40 - 50 W/cm^2 within two years.

6.3. Tip-Based Applications Using Near-Field Thermal Radiation

Highly enhanced near-field thermal radiation between a tip and substrate has been used to develop novel scanning probe microscopies and spectroscopies. De Wilde *et al.* (2006) has developed thermal radiation scanning tunneling microscopy (TRSTM), which is a scattering-type near-field scanning optical microscopy (NSOM) in the infrared spectrum (Tersoff and Hamann, 1985), but without any external illumination. When a gold-coated tip scans over a heated SiC samples with gold patterns, thermally excited surface waves in the infrared, i.e., SPPs on gold and SPhPs on SiC, are near-field interacted and scattered by the tip. By measuring the scattered thermal emission with a HgCdTe detector, they could achieve the near-field image of the sample along with the AFM topographic image. In fact, TRSTM can measure the electromagnetic local density of states (LDOS) at a frequency that can be defined by a suitable filter: this is analogous to the scanning tunneling microscopy that probes the electronic LDOS (Tersoff and Hamann, 1985). Recently, Jones and Raschke (2012) reported another exciting tip-based metrology utilizing near-field thermal radiation. By combining scattering-type NSOM with Fourier-transform spectroscopy and using a heated atomic force microscope tip as both a local thermal source and scattering probe, they obtained the mid-infrared spectrum of thermal near-field scattered from the tip with the spatial resolution of ~50 nm. The developed thermal infrared near-field optical spectroscopy can measure a highly localized spectral near-field energy density change associated with vibrational, phonon, and phonon-polariton modes of a substrate, enabling broadband chemical nanospectroscopy without the need for an external excitation source.

In addition to nanoimaging and nanospectroscopy instrumentations, tip-induced near-field radiation can be beneficially used for laser-based processing and structuring of materials at the nanoscale, in the order of ~50 nm or smaller. Upon illuminating a silicon tip or a metal-coated tip with a femtosecond laser, electromagnetic fields will be highly concentrated at the tip apex due to the optical antenna effect (Au *et al.*, 2008; Schuller *et al.*, 2009) and the excitation of localized surface polaritons (Chimmalgi *et al.*, 2003; Milner *et al.*, 2008). This EM field concentration may cause surface modification either through a hot tip interaction with a surface, leading to the melting/evaporation of the material (Kirsanov *et al.*, 2003), or EM field enhancement under tip triggering the material ablation (Chimmalgi *et al.*, 2003; Milner *et al.*, 2008). When compared with other tip-based nanomanufacturing technologies, such as dip-pen nanolithography (e.g., Piner *et al.*, 1999), thermal tip-based processing (e.g., Pires *et al.*, 2010; Lee *et al.*, 2010; Wei *et al.*, 2010), and chemomechanical nanoscale patterning (e.g., Wacaser *et al.*, 2003; Liu *et al.*, 2004), the laser-based nanoscale material processing has a compelling advantage in manufacturable materials: its high energy concentration enables the nanoscale ablation and deposition of high melting-point metals, such as Au and FeCr (Chimmalgi *et al.*, 2003; Kirsanov *et al.*, 2003; Milner *et al.*, 2008; Grigoropoulos *et al.*, 2007). Moreover, the tip-induced scattering of laser beam could be collected to enable nanoscale optical imaging of the surface under fabrication, which will help the post-processing examination. Slow speed and low throughput still remain as challenging issues to be overcome for further advances of tip-based nanomanufacturing. However, various schemes are being proposed to operate multiple probes in parallel (Minne *et al.*, 1998; Vettiger *et al.*, 2002) at high speeds (Ando, 2012), which will enhance throughput by more than two orders of magnitude.

6.4. Radiation-Based Thermal Rectification

Thermal rectification has recently attracted great attention since it could allow heat to flow in a preferred direction, and may have promising applications in thermal management and energy systems. Solid-state thermal rectifiers can be realized by asymmetric geometric or interface arrangements, dissimilar materials with different temperature-dependent thermal conductivity, and quantum structures (Stevenson *et al.*, 1990; Li *et al.*, 2004; Chang *et al.*, 2006; Dames, 2008; Hu *et al.*, 2009; Wu and Segal, 2009; Roberts and Walker, 2011). While most solid-state thermal rectifiers are based on the nonlinear phononic, electronic or mechanical properties of materials near the interfaces, a photonic device may be advantageous for obtaining large rectification factors over a broad temperature range. Near-field radiation has been theoretically demonstrated for potential application as thermal rectifiers between planar structures (Otey *et al.*, 2010; Iizuka and Fan, 2012; Basu and Francoeur, 2011a; Wang and Zhang, 2013). The basic concept is based on the different temperature dependences of the dielectric functions ε_1 and ε_2 of the two materials as shown in Fig. 1. The forward heat flux $q''_{\text{forward}} = q''_{12,\text{net}}$ refers to the situation when medium 1 is at a higher temperature $T_1 = T_{\text{H}}$ and medium 2 is at a lower temperature $T_2 = T_{\text{L}}$. The reverse-bias heat flux $q''_{\text{forward}} = q''_{21,\text{net}}$ refers to the situation when $T_1 = T_{\text{L}} < T_{\text{H}} = T_2$. Thermal rectification factor or simply thermal rectification is defined as (Dames, 2008; Wang and Zhang, 2013)

$$R = \frac{q''_{\text{forward}}}{q''_{\text{backward}}} - 1 \tag{30}$$

which depends on the materialâĂŹs choice as well as both T_{H} and T_{L}.

Otey *et al.* (2010) theoretically obtained a rectification factor R = 0.41 by considering two SiC plates of different phases, an isotropic 3C-SiC plate and a uniaxial 6H-SiC plate, at $T_{\text{H}} = 600$ K and $T_{\text{L}} = 300$ K with a separation distance less than 100. The dielectric functions of SiC with different crystalline structures exhibit different temperature dependence, particularly due to the fact that the resonance frequencies of the two structures of SiC varies with temperature in opposite directions. This allows photon tunneling to be enhanced when the two resonance frequencies of SPhPs at each interface become closer. Iizuka and Fan (2012) studied the thermal rectification between coated and uncoated SiC plates, and optimized the permittivity and thickness of the coating to achieve a maximal rectification factor of 0.44 when the high and low temperatures are 500 K and 300 K, respectively. Basu and Francoeur (2011a) considered a thin Si film and a semi-infinite Si medium with different doping levels and obtained R = 0.51 at $d = 10$ nm with only 100 K temperature difference between the two media. Photon-mediated thermal rectifiers may be applicable to a large temperature range.

Wang and Zhang (2013) investigate the thermal rectification effect enabled by near-field radiative heat transfer between intrinsic silicon and several dissimilar materials including doped Si, SiO_2, and Au. The temperature-dependent properties of doped Si were taken from (Fu and Zhang, 2006). For Au, the Drude model was used and scattering rate is treated as proportional to temperature due to electron-phonon scattering. At elevated temperatures, free-carrier absorption becomes important in intrinsic Si due to the thermally excited charge carriers. When the intrinsic Si is at a temperature of $T_{\text{H}} = 1000$ K and the doped Si is at a temperature of $T_{\text{L}} = 300$ K, the dielectric functions of both media are dominated by free carriers, resulting in greatly enhanced heat flux, particularly as the vacuum gap is at the nanoscales. In the reverse-bias scenario, the intrinsic Si at 300 K behaves like a non-absorbing medium at wavelengths longer than 1.1 μm, except for some weak phonon absorptions. As shown in Fig. 16(a), the reverse heat flux is much lower at nanometer scales. Rectification factors $R = 0.71$, 2.7, and 67 were predicted at $d = 10$, 5, and 1 nm. The strong enhanced near-field

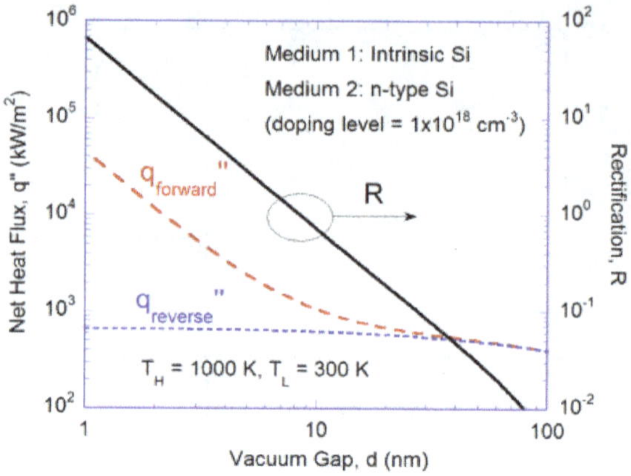

(a) Intrinsic Si vs. heavily doped Si

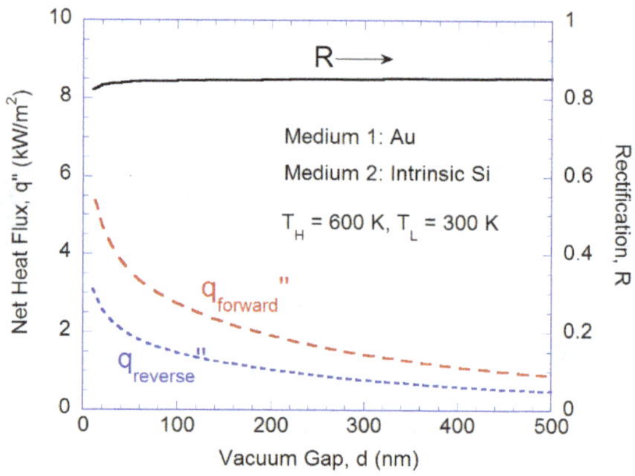

(b) Vacuum gap dependence (Au - intrinsic Si)

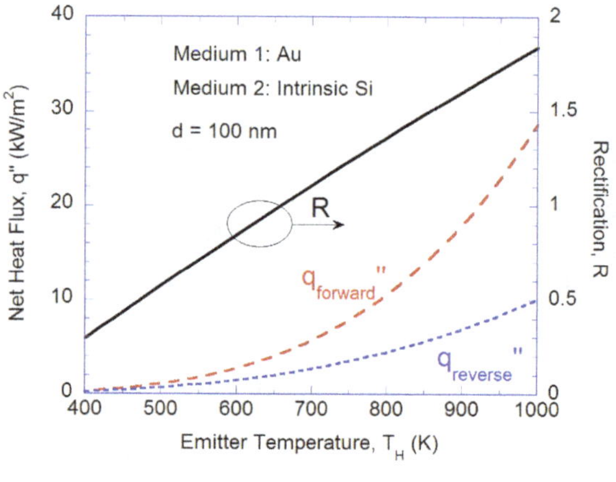

(c) T_{H} dependence (Au - intrinsic Si)

Fig. 16 Heat fluxes and thermal rectification between different materials (Wang and Zhang, 2013). (a) intrinsic Si versus heavily doped Si (n-type with a doping concentration of 10^{18} cm^3) at temperatures of T_{H} = 1000 K and T_{L} = 300 K. (b) Au (as medium 1) and intrinsic Si (as medium 2) at T_{H} = 600 K and T_{L} = 300 K for varying vacuum gap d. (c) Au versus intrinsic Si at d = 100 nm and T_{L} = 300 K for varying T_{H}.

radiation for forward bias is attributed to coupled SPPs (Wang and Zhang, 2013; Basu *et al.*, 2010b).

The calculated heat fluxes and rectification between Au and the intrinsic Si are shown in Fig. 16(b) with $T_{\text{H}} = 600$ K and $T_{\text{L}} = 300$ K. Since the emitter temperature is much lower (600 K compared to 1000

K) and the coupling of non-resonant evanescent waves between Au and intrinsic Si is weaker, the heat fluxes are several orders lower than that between intrinsic Si and doped Si. Interestingly, the rectification factor R between Au and intrinsic Si is nearly the same for 10 nm $< d <$ 500 nm, between 0.82 and 0.85. As the emitter temperature T_H increases, as shown in Fig. 16(c) for d = 100 nm, the rectification factor can be further enhanced, e.g., R = 1.84 for T_H = 1000 K, indicating almost twice as much heat can be transferred from Au (1000 K) to intrinsic Si (300 K) than vice versa. The thermal rectification between Au and intrinsic Si may facilitate thermal management and heat control at intermediate temperatures with relatively large vacuum gaps.

7. SUMMARY AND OUTLOOK

This article reviews recent achievements of near-field radiative heat transfer research from fundamental to application perspectives. Contrary to far-field thermal radiation carried by propagating EM waves, radiative heat transfer in the near field is dominated by evanescent EM waves (or surface waves) and photon tunneling. Based on the flucutational electrodynamics, the contributions of such near-field phenomena to thermal radiation are discussed for semi-infinite planar structures. Near-field thermal radiation is influenced by the vacuum gap and radiative properties of materials. For example, when doped Si plates are placed in proximity, the free carriers caused by dopants in Si give rise to fluctuating currents that result in significant augments in the net radiative energy flux. The excitation of surface polaritons plays a key role in the enhancement of near-field thermal radiation. In ideal but not realistic materials, the maximum near-field thermal radiation, would occur when surface polaritons were excited at all wavelengths within extremely small gap distances on the order of one nanometer. Surface waves are also responsible for the extremely small penetration depth and the laterally shifted energy flow of near-field thermal radiation between planar structures.

Besides the semi-infinite media, near-field radiative heat transfer between various geometrics has been theoretically investigated. Mathematical and computational challenges in solving stochastic Maxwell's equations for objects with arbitrary geometries have been addressed by implementing certain approximations, such as dipole, multipole, and proximity approximations, or by applying the molecular dynamics numerical scheme. Various geometrical configurations, including sphere-sphere, sphere-plane, cylinder-cylinder, and cylinder-plane cases have been investigated to reveal the geometry-dependence of near-field radiative transfer. However, there are discrepancies in the obtained gap dependence of near-field thermal radiation between different numerical schemes; this is an issue that demands further research. Near-field radiative heat transfer in emerging materials, such as graphenes, photonic crystals, and negative-index metamaterials, has become a very interesting research topic. It has been theoretically shown that unique radiative properties of such materials, e.g., the tunability of surface polariton excitation frequencies in graphenes, the photonic band gap in photonic crystals, and the negative refractive index in metamaterials, could further enhance near-field thermal radiation.

Despite significant progress in the theory of near-field thermal radiation, quantitative measurements have remained a challenge at nanometer distances. During the past few years, some meaningful measurements have been made in the sphere-plane configuration to validate the theoretical predictions. However, measuring near-field heat transfer between two flat surfaces at the nanometer distances is extremely challenging due to surface roughness and nonparallelism of the plates. Several suggestions have been made to achieve good parallelism between two planar surfaces by placing micro/nanospacers between plates or by feedback-controlling the emitter position, but experiments were successful only in the micrometer range. Since the experimental investigation on near-field thermal radiation between flat plates is crucial for the development of near-field TPV devices, innovative approaches that can overcome these barriers need to be designed and developed. Experimental investigations using emerging materials have not been scrutinized yet, although the material selection is critically important for manipulating near-field thermal radiation.

Among many potential applications utilizing near-field thermal radiation, including the manipulation of the radiative properties, nanoscale imaging and analysis, and thermal rectification, near-field TPV holds great promise as a novel renewable energy harvesting technology. Theoretical studies have revealed that the power throughput of the near-field TPV can be enhanced by 1-2 orders of magnitude due to near-field effects with approximately 20 % conversion efficiency. However, there remain issues that must be addressed. The cost of TPV systems should be reduced by developing inexpensive alloys for TPV cells and achieving a cost-effective vacuum sealing with good parallelism. Effective cooling systems need to be developed in order to prevent overheating of the TPV cells. Efforts are also needed to recycle the unusable photons back to the emitter. More importantly, a second-law thermodynamic analysis of TPV systems is required in order to establish the fundamental achievable efficiencies to guide future TPV development. Since the radiation entropy in the near-field regime has not been clarified yet, a satisfactory thermodynamic second-law interpretation of near-field thermal radiation does not exist. Non-equilibrium entropy needs to be employed to develop thermodynamic relations for near-field radiation and to provide a second-law analysis of photon tunneling and surface polariton phenomena.

Improvement in computational resources, ever-advancing micro/nanofabrication and nano-instrumentation techniques, and the unprecedented growth in materials science have created compelling opportunities in the fundamental research and applications of near-field radiative heat transfer. The field of nanoscale radiation keeps growing with excitement, providing a deeper understanding of the interplay among optical, thermal, mechanical, and electrical properties of materials at the nanoscale. The understanding of near-field interactions can also help design systems for far-field applications. The authors strongly believe that near-field thermal radiation will be beneficially used for novel applications in biological sensing, materials processing and manufacturing, and energy systems in the near future.

ACKNOWLEDGEMENTS

This work was supported by the National Science Foundation under grant CBET-1236239 for KP and grant CBET-1235975 for ZMZ.

NOMENCLATURE

c	speed of light in vacuum (2.998×10^8 m s^{-1})
$\mathbf{D}$	electric displacement vector (C m^{-2})
d	vacuum gap distance (m)
$\mathbf{E}$	electric field vector (V m^{-1})
$\overline{\overline{\mathbf{G}}}$	dyadic Green's function (m^{-1})
$\mathbf{H}$	magnetic field vector (A m^{-1})
h	heat transfer coefficient (W m^{-2} K^{-1})
$\hbar$	Planck constant divided by 2π (1.055×10^{-34} J s)
J	photocurrent density (A m^{-2})
$\mathbf{j}$	fluctuating current density (A m^{-2})
$\mathbf{k}$	wavevector (m^{-1})
k_B	Boltzmann constant (1.381×10^{-23} J K^{-1})
N	doping concentration (cm^{-3})
n	refractive index
q''	heat flux (W m^{-2})
R	radius (m)
$\mathbf{r}$	vector in the radial direction (m)
r	Fresnel reflection coefficient
$\mathbf{S}$	spectral Poynting vector (W m^{-2} s rad^{-1})
T	temperature (K)
t	Fresnel transmission coefficient

x	position vector (m)
z	vector in the normal direction to surfaces (m)

Greek Symbols

α	polarizability (m^3)
β	parallel wavevector component (m^{-1})
γ	wavevector component in the z-direction (m^{-1})
δ	penetration depth or skin depth (m)
ε	relative permittivity (i.e., dielectric function)
ε_0	electrical permittivity of vacuum (8.854×10^{-12} F m^{-1})
η	conversion efficiency
η_q	quantum efficiency
Θ	mean energy of the Planck oscillator (J)
κ	extinction coefficient
Λ	period of a grating structure (m)
λ	wavelength in vacuum (m)
μ	relative permeability
μ_0	magnetic permeability of vacuum ($4\pi \times 10^{-7}$ H m^{-1})
σ	Stefan-Boltzmann constant (5.67×10^{-8} W m^{-2} K^{-4})
σ_0	dc electrical conductivity (S m^{-1})
τ	relaxation time (s)
ω	angular frequency (rad s^{-1})

Superscripts

p	p-polarization or TM wave
s	s-polarization or TE wave

Subscripts

0,1,2	medium index
dp	depletion region
e	electron
evan	evanescent wave
h	hole
m	magnetic
max	maximum
prop	propagating wave
λ, ω	spectral

REFERENCES

Ando, T., 2012, "High-Speed Atomic Force Microscopy Coming of Age," *Nanotechnology*, **23**(6), 062001. http://dx.doi.org/10.1088/0957-4484/23/6/062001.

Au, Y.Y., Skulason, H.S., Ingvarsson, S., Klein, L.J., and Hamann, H.F., 2008, "Thermal radiation spectra of individual subwavelength microheaters," *Physical Review B*, **78**(8), 085402. http://dx.doi.org/10.1103/PhysRevB.78.085402.

Basu, S., Chen, Y.B., and Zhang, Z.M., 2007, "Microscale Radiation in Thermophotovoltaic Devices - a Review," *International Journal of Energy Research*, **31**(6-7), 689–716. http://dx.doi.org/10.1002/er.1286.

Basu, S., and Francoeur, M., 2011a, "Near-Field Radiative Transfer Based Thermal Rectification Using Doped Silicon," *Applied Physics Letters*, **98**(11), 113106. http://dx.doi.org/10.1063/1.3567026.

Basu, S., and Francoeur, M., 2011b, "Penetration Depth in Near-Field Radiative Heat Transfer Between Metamaterials," *Applied Physics Letters*, **99**(14), 143107. http://dx.doi.org/10.1063/1.3646466.

Basu, S., Lee, B.J., and Zhang, Z.M., 2010a, "Infrared Radiative Properties of Heavily Doped Silicon at Room Temperature," *Journal of Heat Transfer*, **132**(2), 023301. http://dx.doi.org/10.1115/1.4000171.

Basu, S., Lee, B.J., and Zhang, Z.M., 2010b, "Near-Field Radiation Calculated with an Improved Dielectric Function Model for Doped Silicon," *Journal of Heat Transfer*, **132**(2), 023302. http://dx.doi.org/10.1115/1.4000179.

Basu, S., Wang, L.P., and Zhang, Z.M., 2011, "Direct Calculation of Energy Streamlines in Near-Field Thermal Radiation," *Journal of Quantitative Spectroscopy and Radiative Transfer*, **112**(7), 1149–1155. http://dx.doi.org/10.1016/j.jqsrt.2010.08.027.

Basu, S., and Zhang, Z.M., 2009a, "Maximum Energy Transfer in Near-Field Thermal Radiation at Nanometer Distances," *Journal of Applied Physics*, **105**(9), 093535. http://dx.doi.org/10.1063/1.3125453.

Basu, S., and Zhang, Z.M., 2009b, "Ultrasmall Penetration Depth in Nanoscale Thermal Radiation," *Applied Physics Letters*, **95**(13), 133104. http://dx.doi.org/10.1063/1.3238315.

Basu, S., Zhang, Z.M., and Fu, C.J., 2009, "Review of Near-Field Thermal Radiation and Its Application to Energy Conversion," *International Journal of Energy Research*, **33**(13), 1203–1232. http://dx.doi.org/10.1002/er.1607.

Baxter, J., Bian, Z., Chen, G., Danielson, D., Dresselhaus, M.S., Fedorov, A.G., Fisher, T.S., Jones, C.W., Maginn, E., and Kortshagen, U., 2009, "Nanoscale Design to Enable the Revolution in Renewable Energy," *Energy & Environmental Science*, **2**(6), 559–588. http://dx.doi.org/10.1039/b821698c.

Ben-Abdallah, P., Joulain, K., and Pryamikov, A., 2010, "Surface Bloch Waves Mediated Heat Transfer Between Two Photonic Crystals," *Applied Physics Letters*, **96**(14), 143117. http://dx.doi.org/10.1063/1.3385156.

Biehs, S.A., Ben-Abdallah, P., Rosa, F.S.S., Joulain, K., and Greffet, J.J., 2011a, "Nanoscale Heat Flux Between Nanoporous Materials," *Optics Express*, **19**(S5), A1088–A1103. http://dx.doi.org/10.1364/OE.19.0A1088.

Biehs, S.A., Rosa, F.S.S., and Ben-Abdallah, P., 2011b, "Modulation of Near-Field Heat Transfer Between Two Gratings," *Applied Physics Letters*, **98**(24), 243102. http://dx.doi.org/10.1063/1.3596707.

Biehs, S.A., Tschikin, M., and Ben-Abdallah, P., 2012, "Hyperbolic Metamaterials as an Analog of a Blackbody in the Near Field," *Physical Review Letters*, **109**(10), 104301. http://dx.doi.org/10.1103/PhysRevLett.109.104301.

Bimonte, G., Cappellin, L., Carugno, G., Ruoso, G., and Saadeh, D., 2009, "Polarized Thermal Emission by Thin Metal Wires," *New Journal of Physics*, **11**(3), 033014. http://dx.doi.org/10.1088/1367-2630/11/3/033014.

Carrillo, L.Y., and Bayazitoglu, Y., 2011, "Nanorod near-field radiative heat exchange analysis," *Journal of Quantitative Spectroscopy and Radiative Transfer*, **112**(3), 412–419. http://dx.doi.org/10.1016/j.jqsrt.2010.10.011.

Carrillo, L.Y., and Bayazitoglu, Y., 2012, "Sphere Approximation for Nanorod Near-Field Radiative Heat Exchange Analysis," *Nanoscale and Microscale Thermophysical Engineering*, **15**(3), 195–208. http://dx.doi.org/10.1080/15567265.2011.597493.

Chang, C.W., Okawa, D., Majumdar, A., and Zettl, A., 2006, "Solid-State Thermal Rectifier," *Science*, **314**(5802), 1121–1124.
http://dx.doi.org/10.1126/science.1132898.

Chapuis, P.O., Laroche, M., Volz, S., and Greffet, J.J., 2008a, "Radiative heat transfer between metallic nanoparticles," *Applied Physics Letters*, **92**(20), 201906.
http://dx.doi.org/10.1016/j.physleta.2007.09.050.

Chapuis, P.O., Volz, S., Henkel, C., Joulain, K., and Greffet, J.J., 2008b, "Effects of Spatial Dispersion in Near-Field Radiative Heat Transfer Between Two Parallel Metallic Surfaces," *Physical Review B*, **77**(3), 035431.
http://dx.doi.org/10.1103/PhysRevB.77.035431.

Chen, G., 2006, "Nanoscale Heat Transfer and Nanostructured Thermoelectrics," *Components and Packaging Technologies, IEEE Transactions on*, **29**(2), 238–246.
http://dx.doi.org/10.1109/TCAPT.2006.875895.

Chen, Y.B., and Zhang, Z.M., 2007, "Design of Tungsten Complex Gratings for Thermophotovoltaic Radiators," *Optics communications*, **269**(2), 411–417.
http://dx.doi.org/10.1016/j.optcom.2006.08.040.

Chen, Y.B., and Zhang, Z.M., 2008, "Heavily Doped Silicon Complex Gratings as Wavelength-Selective Absorbing Surfaces," *Journal of Physics D: Applied Physics*, **41**(9), 095406.
http://dx.doi.org/10.1088/0022-3727/41/9/095406.

Chimmalgi, A., Choi, T.Y., Grigoropoulos, C.P., and Komvopoulos, K., 2003, "Femtosecond Laser Aperturless Near-Field Nanomachining of Metals Assisted by Scanning Probe Microscopy," *Applied Physics Letters*, **82**(8), 1146.
http://dx.doi.org/10.1063/1.1555693.

Cravalho, E.G., Tien, C.L., and Caren, R.P., 1966, "Effect of Small Spacings on Radiative Transfer Between Two Dielectrics," *Journal of Heat Transfer*, **89**(4), 351–358.
http://dx.doi.org/10.1115/1.3614396.

Cui, L., Huang, Y., and Wang, J., 2012, "Near-Field Radiative Heat Transfer Between Chiral Metamaterials," *Journal of Applied Physics*, **112**(8), 084309.
http://dx.doi.org/10.1063/1.4759055.

Dames, C., 2008, "Solid-State Thermal Rectification With Existing Bulk Materials," *Journal of Heat Transfer*, **131**(6), 061301.
http://dx.doi.org/10.1115/1.3089552.

De Wilde, Y., Formanek, F., Carminati, R., Gralak, B., Lemoine, P.A., Joulain, K., Mulet, J.P., Chen, Y., and Greffet, J.J., 2006, "Thermal Radiation Scanning Tunnelling Microscopy," *Nature*, **444**(7120), 740–743.
http://dx.doi.org/10.1038/nature05265.

Derjaguin, B.V., Abrikosova, I.I., and Lifshitz, E.M., 1956, "Direct Measurement of Molecular Attraction Between Solids Separated by a Narrow Gap," *Quarterly Reviews of the Chemical Society*, **10**(3), 295–329.
http://dx.doi.org/10.1039/QR9561000295.

Dillner, U., 2008, "The Effect of Thermotunneling on the Thermoelectric Figure of Merit," *Energy Conversion and Management*, **49**(12), 3409–3416.
http://dx.doi.org/10.1063/1.3238315.

DiMatteo, R.S., Greiff, P., Finberg, S.L., Young-Waithe, K.A., Choy, H.K.H., Masaki, M.M., and Fonstad, Jr., C.G., 2003, "Microngap ThermoPhotoVoltaics (MTPV)," *AIP Conference Proceedings*, **653**, 232–240.
http://dx.doi.org/10.1063/1.1539379.

Domingues, G., Volz, S., Joulain, K., and Greffet, J.J., 2005, "Heat Transfer Between Two Nanoparticles Through Near Field Interaction," *Physical Review Letters*, **94**, 085901.
http://dx.doi.org/10.1103/PhysRevLett.94.085901.

Domoto, G.A., Boehm, R.F., and Tien, C.L., 1969, "Experimental Investigation of Radiative Transfer Between Metallic Surfaces at Cryogenic Temperatures," *Journal of Heat Transfer*, **92**(3), 412–416.
http://dx.doi.org/10.1115/1.3449677.

Dorofeyev, I.A., 1998, "Energy dissipation rate of a sample-induced thermal fluctuating field in the tip of a probe microscope," *Journal of Physics D: Applied Physics*, **31**(6), 600.
http://dx.doi.org/10.1088/0022-3727/31/6/004.

Dorofeyev, I., 2008, "Rate of Heat Transfer Between a Probing Body and a Sample Due to Electromagnetic Fluctuations," *Physics Letters A*, **372**(9), 1341-1347.
http://dx.doi.org/10.1016/j.physleta.2007.09.050.

Fan, Y., Singer, S.B., Bergstrom, R., and Regan, B.C., 2009, "Probing Planck's Law with Incandescent Light Emission From a Single Carbon Nanotube," *Physical Review Letters*, **102**(18), 187402.
http://dx.doi.org/10.1103/PhysRevLett.102.187402.

Francoeur, M., Basu, S., and Petersen, S.J., 2011a, "Electric and Magnetic Surface Polariton Mediated Near-Field Radiative Heat Transfer Between Metamaterials Made of Silicon Carbide Particles," *Optics Express*, **19**(20), 18774–18788.
http://dx.doi.org/10.1364/OE.19.018774.

Francoeur, M., Mengüç, M.P., and Vaillon, R., 2008, "Near-Field Radiative Heat Transfer Enhancement via Surface Phonon Polaritons Coupling in Thin Films," *Applied Physics Letters*, **93**(4), 043109.
http://dx.doi.org/10.1063/1.2963195.

Francoeur, M., Mengüç, M.P., and Vaillon, R., 2009, "Solution of Near-Field Thermal Radiation in One-Dimensional Layered Media Using Dyadic Green's Functions and the Scattering Matrix Method," *Journal of Quantitative Spectroscopy and Radiative Transfer*, **110**(18), 2002–2018.
http://dx.doi.org/10.1016/j.jqsrt.2009.05.010.

Francoeur, M., Vaillon, R., and Mengüç, M.P., 2011b, "Thermal Impacts on the Performance of Nanoscale-Gap Thermophotovoltaic Power Generators," *IEEE Transactions on Energy Conversion*, **26**(2), 686–698.
http://dx.doi.org/10.1109/TEC.2011.2118212.

Frank, I.W., Tanenbaum, D.M., van der Zande, A.M., and McEuen, P.L., 2007, "Mechanical properties of suspended graphene sheets," *Journal of Vacuum Science & Technology B*, **25**(6), 2558–2561.
http://dx.doi.org/10.1116/1.2789446.

Freitag, M., Steiner, M., Martin, Y., Perebeinos, V., Chen, Z., Tsang, J.C., and Avouris, P., 2009, "Energy Dissipation in Graphene Field-Effect Transistors," *Nano Letters*, **9**(5), 1883–1888.
http://dx.doi.org/10.1021/nl803883h.

Fu, C.J., and Tan, W.C., 2009, "Near-Field Radiative Heat Transfer Between Two Plane Surfaces with One Having a Dielectric Coating," *Journal of Quantitative Spectroscopy and Radiative Transfer*, **110**(12), 1027–1036.
http://dx.doi.org/10.1016/j.jqsrt.2009.02.007.

Fu, C.J., and Zhang, Z.M., 2006, "Nanoscale Radiation Heat Transfer for Silicon at Different Doping Levels," *International Journal of Heat and Mass Transfer*, **49**(9–10), 1703–1718.
http://dx.doi.org/10.1016/j.ijheatmasstransfer.2005.09.037.

Fu, C.J., and Zhang, Z.M., 2009, "Thermal Radiative Properties of Metamaterials and Other Nanostructured Materials: a Review," *Frontiers of Energy and Power Engineering in China*, **3**(1), 11–26.
http://dx.doi.org/10.1007/s11708-009-0009-x.

Fu, C.J., Zhang, Z.M., and Tanner, D.B., 2005, "Planar Heterogeneous Structures for Coherent Emission of Radiation," *Optics Letters*, **30**(14), 1873–1875.
http://dx.doi.org/10.1364/OL.30.001873.

Geim, A.K., and Novoselov, K.S., 2007, "The rise of graphene," *Nature Materials*, **6**(3), 183–191.
http://dx.doi.org/10.1038/nmat1849.

Golyk, V.A., Krüger, M., and Kardar, M., 2012, "Heat Radiation From Long Cylindrical Objects," *Physical Review E*, **85**(4), 046603.
http://dx.doi.org/10.1103/PhysRevE.85.046603.

González-Cuevas, J.A., Refaat, T.F., Abedin, M.N., and Elsayed-Ali, H.E., 2006, "Modeling of the Temperature-Dependent Spectral Response of $In_{1-x}Ga_xSb$ Infrared Photodetectors," *Optical Engineering*, **45**(4), 044001.
http://dx.doi.org/10.1117/1.2192772.

Griffiths, D.J., 2012, *Introduction to Electrodynamics*, 4th ed., Addison Wesley.

Grigoropoulos, C.P., Hwang, D.J., and Chimmalgi, A., 2007, "Nanometer-Scale Laser Direct-Write Using Near-Field Optics," *MRS Bulletin*, **32**(1), 16–22.
http://dx.doi.org/10.1557/mrs2007.10.

Hargreaves, C.M., 1969, "Anomalous Radiative Transfer Between Closely-Spaced Bodies," *Physics Letters A*, **30A**(9), 491–492.
http://dx.doi.org/10.1016/0375-9601(69)90264-3.

Howell, J.R., Siegel, R., and Mengüç, M.P., 2010, *Thermal Radiation Heat Transfer*, 5th ed., CRC Press.

Hu, L., Narayanaswamy, A., Chen, X., and Chen, G., 2008, "Near-Field Thermal Radiation Between Two Closely Spaced Glass Plates Exceeding Planck's Blackbody Radiation Law," *Applied Physics Letters*, **92**(13), 133106.
http://dx.doi.org/10.1063/1.2905286.

Hu, M., Goicochea, J.V., Michel, B., and Poulikakos, D., 2009, "Thermal Rectification at Water/Functionalized Silica Interfaces," *Applied Physics Letters*, **95**(15), 151903.
http://dx.doi.org/10.1063/1.3247882.

Huth, O., Rüting, F., Biehs, S.A., and Holthaus, M., 2010, "Shape-Dependence of Near-Field Heat Transfer Between a Spheroidal Nanoparticle and a Flat Surface," *The European Physical Journal Applied Physics*, **50**(1), 10603.
http://dx.doi.org/10.1051/epjap/2010027.

Hutter, E., and Fendler, J.H., 2004, "Exploitation of Localized Surface Plasmon Resonance," *Advanced Materials*, **16**(19), 1685–1706.
http://dx.doi.org/10.1002/adma.200400271.

Iizuka, H., and Fan, S., 2012, "Rectification of Evanescent Heat Transfer Between Dielectric-Coated and Uncoated Silicon Carbide Plates," *Journal of Applied Physics*, **112**(2), 024304.
http://dx.doi.org/10.1063/1.4737465.

Ilic, O., Jablan, M., Joannopoulos, J.D., Celanovic, I., Buljan, H., and Soljačić, M., 2012a, "Near-Field Thermal Radiation Transfer Controlled by Plasmons in Graphene," *Physical Review B*, **85**(15), 155422.
http://dx.doi.org/10.1103/PhysRevB.85.155422.

Ilic, O., Jablan, M., Joannopoulos, J.D., Celanovic, I., and Soljačić, M., 2012b, "Overcoming the Black Body Limit in Plasmonic and Graphene Near-Field Thermophotovoltaic Systems," *Optics Express*, **20**(S3), A366–A384.
http://dx.doi.org/10.1364/OE.20.00A366.

Ingvarsson, S., Klein, L., Au, Y.Y., Lacey, J.A., and Hamann, H.F., 2007, "Enhanced thermal emission from individual antenna-like nanoheaters," *Optics Express*, **15**(18), 11249–11254.
http://dx.doi.org/10.1364/OE.15.011249.

John, S., 1987, "Strong Localization of Photons in Certain Disordered Dielectric Superlattices," *Physical Review Letters*, **58**(23), 2486–2489.
http://dx.doi.org/10.1103/PhysRevLett.58.2486.

Jones, A.C., and Raschke, M.B., 2012, "Thermal Infrared Near-Field Spectroscopy," *Nano Letters*, 120301010022003.
http://dx.doi.org/10.1021/nl204201g.

Joulain, K., 2008, "Near-Field Heat Transfer: a Radiative Interpretation of Thermal Conduction," *Journal of Quantitative Spectroscopy and Radiative Transfer*, **109**(2), 294–304.
http://dx.doi.org/10.1016/j.jqsrt.2007.08.028.

Joulain, K., Drevillon, J., and Ben-Abdallah, P., 2010, "Noncontact Heat Transfer Between Two Metamaterials," *Physical Review B*, **81**(16), 165119.
http://dx.doi.org/10.1103/PhysRevB.81.165119.

Joulain, K., Mulet, J.P., Marquier, F., Carminati, R., and Greffet, J.J., 2005, "Surface Electromagnetic Waves Thermally Excited: Radiative Heat Transfer, Coherence Properties and Casimir Forces Revisited in the Near Field," *Surface Science Reports*, **57**(3–4), 59–112.
http://dx.doi.org/10.1016/j.surfrep.2004.12.002.

Ju, L., Geng, B., Horng, J., Girit, C., Martin, M., Hao, Z., Bechtel, H.A., Liang, X., Zettl, A., Shen, Y.R., and Wang, F., 2011, "Graphene Plasmonics for Tunable Terahertz Metamaterials," *Nature Nanotechnology*, **6**(10), 630–634.
http://dx.doi.org/10.1038/nnano.2011.146.

Kirsanov, A., Kiselev, A., Stepanov, A., and Polushkin, N., 2003, "Femtosecond Laser-Induced Nanofabrication in the Near-Field of Atomic Force Microscope Tip," *Journal of Applied Physics*, **94**(10), 6822.
http://dx.doi.org/10.1063/1.1621722.

Kittel, A., Muller-Hirsch, W., Parisi, J., Biehs, S.A., Reddig, D., and Holthaus, M., 2005, "Near-Field Heat Transfer in a Scanning Thermal Microscope," *Physical Review Letters*, **95**(22), 224301.
http://dx.doi.org/10.1103/PhysRevLett.95.224301.

Krüger, M., Emig, T., and Kardar, M., 2011, "Nonequilibrium electromagnetic fluctuations: Heat transfer and interactions," *Physical Review Letters*, **106**(21), 210404.
http://dx.doi.org/10.1103/PhysRevLett.106.210404.

Laroche, M., Carminati, R., and Greffet, J.J., 2006, "Near-Field Thermophotovoltaic Energy Conversion," *Journal of Applied Physics*, **100**(6), 063704.
http://dx.doi.org/10.1063/1.2234560.

Lee, B.J., Chen, Y.B., and Zhang, Z.M., 2008a, "Surface Waves Between Metallic Films and Truncated Photonic Crystals Observed with Reflectance Spectroscopy," *Optics Letters*, **33**(3), 204–206. http://dx.doi.org/10.1364/OL.33.000204.

Lee, B.J., Chen, Y.B., and Zhang, Z.M., 2008b, "Transmission Enhancement Through Nanoscale Metallic Slit Arrays From the Visible to Mid-Infrared," *Journal of Computational and Theoretical Nanoscience*, **5**, 201–213. http://dx.doi.org/10.1166/jctn.2008.008.

Lee, B.J., Fu, C.J., and Zhang, Z.M., 2005, "Coherent Thermal Emission From One-Dimensional Photonic Crystals ," *Applied Physics Letters*, **87**(7), 071904. http://dx.doi.org/10.1063/1.2010613.

Lee, B.J., Park, K., and Zhang, Z.M., 2007, "Energy Pathways in Nanoscale Thermal Radiation," *Applied Physics Letters*, **91**(15), 153101. http://dx.doi.org/10.1063/1.2793688.

Lee, B.J., Wang, L.P., and Zhang, Z.M., 2008c, "Coherent Thermal Emission by Excitation of Magnetic Polaritons Between Periodic Strips and a Metallic Film," *Optics Express*, **16**(15), 11328–11336. http://dx.doi.org/10.1364/OE.16.011328.

Lee, B.J., and Zhang, Z.M., 2006a, "Coherent Thermal Emission From Modified Periodic Multilayer Structures," *Journal of Heat Transfer*, **129**(1), 17–26. http://dx.doi.org/10.1115/1.2401194.

Lee, B.J., and Zhang, Z.M., 2006b, "Design and Fabrication of Planar Multilayer Structures with Coherent Thermal Emission Characteristics ," *Journal of Applied Physics*, **100**(6), 063529. http://dx.doi.org/10.1063/1.2349472.

Lee, B.J., and Zhang, Z.M., 2008, "Lateral Shifts in Near-Field Thermal Radiation with Surface Phonon Polaritons," *Nanoscale and Microscale Thermophysical Engineering*, **12**(3), 238–250. http://dx.doi.org/10.1080/15567260802247505.

Lee, J.U., Yoon, D., Kim, H., Lee, S.W., and Cheong, H., 2011, "Thermal conductivity of suspended pristine graphene measured by Raman spectroscopy," *Physical Review B*, **83**(8), 081419. http://dx.doi.org/10.1103/PhysRevB.83.081419.

Lee, J.I., Jeong, Y.H., No, H.C., Hannebauer, R., and Yoo, S.K., 2009, "Size Effect of Nanometer Vacuum Gap Thermionic Power Conversion Device with CsI Coated Graphite Electrodes," *Applied Physics Letters*, **95**(22), 223107. http://dx.doi.org/10.1063/1.3266921.

Lee, W.K., Dai, Z., King, W.P., and Sheehan, P.E., 2010, "Maskless Nanoscale Writing of NanoparticleâĹŠPolymer Composites and Nanoparticle Assemblies using Thermal Nanoprobes," *Nano Letters*, **10**(1), 129–133. http://dx.doi.org/10.1021/nl9030456.

Li, B., Wang, L., and Casati, G., 2004, "Thermal Diode: Rectification of Heat Flux," *Physical Review Letters*, **93**(18), 184301. http://dx.doi.org/10.1103/PhysRevLett.93.184301.

Li, P., Jiang, K., Liu, M., Li, Q., Fan, S., and Sun, J., 2003, "Polarized Incandescent Light Emission From Carbon Nanotubes," *Applied Physics Letters*, **82**(11), 1763–1765. http://dx.doi.org/10.1063/1.1558900.

Lin, S.Y., Moreno, J., and Fleming, J.G., 2003, "Three-Dimensional Photonic-Crystal Emitter for Thermal Photovoltaic Power Generation," *Applied Physics Letters*, **83**(2), 380–382. http://dx.doi.org/10.1063/1.1592614.

Liu, H., Genov, D.A., Wu, D.M., Liu, Y.M., Steele, J.M., Sun, C., Zhu, S.N., and Zhang, X., 2006, "Magnetic Plasmon Propagation Along a Chain of Connected Subwavelength Resonators at Infrared Frequencies," *Physical Review Letters*, **97**(24), 243902. http://dx.doi.org/10.1103/PhysRevLett.97.243902.

Liu, H., Li, T., Wang, Q.J., Zhu, Z.H., Wang, S.M., Li, J.Q., Zhu, S.N., Zhu, Y.Y., and Zhang, X., 2009, "Extraordinary Optical Transmission Induced by Excitation of a Magnetic Plasmon Propagation Mode in a Diatomic Chain of Slit-Hole Resonators," *Physical Review B*, **79**(2), 024304. http://dx.doi.org/10.1103/PhysRevB.79.024304.

Liu, J.F., Von Ehr, J.R., Baur, C., Stallcup, R., Randall, J., and Bray, K., 2004, "Fabrication of High-Density Nanostructures with an Atomic Force Microscope," *Applied Physics Letters*, **84**(8), 1359–1361. http://dx.doi.org/10.1063/1.1647281.

Liu, Y., and Zhang, X., 2011, "Metamaterials: a New Frontier of Science and Technology," *Chemical Society Reviews*, **40**(5), 2494–2507. http://dx.doi.org/10.1039/C0CS00184H.

Liu, Z.W., Wei, Q.H., and Zhang, X., 2005, "Surface Plasmon Interference Nanolithography," *Nano Letters*, **5**(5), 957–961. http://dx.doi.org/10.1021/nl0506094.

McCauley, A.P., 2012, "Modeling near-field radiative heat transfer from sharp objects using a general three-dimensional numerical scattering technique," *Physical Review B*, **85**(16), 165104. http://dx.doi.org/10.1103/PhysRevB.85.165104.

Milner, A.A., Zhang, K., and Prior, Y., 2008, "Floating Tip Nanolithography," *Nano Letters*, **8**(7), 2017–2022. http://dx.doi.org/10.1021/nl801203c.

Minne, S.C., Yaralioglu, G., Manalis, S.R., Adams, J.D., Zesch, J., Atalar, A., and Quate, C.F., 1998, "Automated Parallel High-Speed Atomic Force Microscopy," *Applied Physics Letters*, **72**(18), 2340. http://dx.doi.org/10.1063/1.121353.

Modest, M.F., 2003, *Radiative Heat Transfer*, 2nd ed., Elsevier Science, San Diego.

Mulet, J.P., Joulain, K., Carminati, R., and Greffet, J.J., 2001, "Nanoscale Radiative Heat Transfer Between a Small Particle and a Plane Surface," *Applied Physics Letters*, **78**(19), 2931–2933. http://dx.doi.org/10.1063/1.1370118.

Mulet, J.P., Joulain, K., Carminati, R., and Greffet, J.J., 2002, "Enhanced Radiative Heat Transfer at Nanometric Distances," *Microscale Thermophysical Engineering*, **6**(3), 209–222. http://dx.doi.org/10.1080/10893950290053321.

Muller-Hirsch, W., Kraft, A., Hirsch, M.T., Parisi, J., and Kittel, A., 1999, "Heat Transfer in Ultrahigh Vacuum Scanning Thermal Microscopy," *Journal of Vacuum Science & Technology A*, **17**(4), 1205–1210. http://dx.doi.org/10.1116/1.581796.

Narayanaswamy, A., and Chen, G., 2003, "Surface Modes for Near Field Thermophotovoltaics," *Applied Physics Letters*, **82**(20), 3544–3546. http://dx.doi.org/10.1063/1.1575936.

Narayanaswamy, A., and Chen, G., 2004, "Thermal Emission Control with One-Dimensional Metallodielectric Photonic Crystals," *Physical Review B*, **70**(12).
http://dx.doi.org/10.1103/PhysRevB.70.125101.

Narayanaswamy, A., and Chen, G., 2005, "Thermal Radiation in 1D Photonic Crystals ," *Journal of Quantitative Spectroscopy and Radiative Transfer*, **93**(1–3), 175–183.
http://dx.doi.org/10.1016/j.jqsrt.2004.08.020.

Narayanaswamy, A., and Chen, G., 2008, "Thermal Near-Field Radiative Transfer Between Two Spheres," *Physical Review B*, **77**(7), 075125.
http://dx.doi.org/10.1103/PhysRevB.77.075125.

Narayanaswamy, A., Shen, S., and Chen, G., 2008, "Near-Field Radiative Heat Transfer Between a Sphere and a Substrate," *Physical Review B*, **78**(11), 115303.
http://dx.doi.org/10.1103/PhysRevB.78.115303.

Nelson, R.E., 2003, "A Brief History of Thermophotovoltaic Development," *Semiconductor Science and Technology*, **18**(5), S141–S143.
http://dx.doi.org/10.1088/0268-1242/18/5/301.

Novoselov, K.S., Geim, A.K., Morozov, S.V., Jiang, D., Katsnelson, M.I., Grigorieva, I.V., Dubonos, S.V., and Firsov, A.A., 2005, "Two-dimensional gas of massless Dirac fermions in graphene," *Nature*, **438**(7065), 197–200.
http://dx.doi.org/10.1038/nature04233.

Novoselov, K.S., Geim, A.K., Morozov, S.V., Jiang, D., Zhang, Y., Dubonos, S.V., Grigorieva, I.V., and Firsov, A.A., 2004, "Electric Field Effect in Atomically Thin Carbon Films," *Science*, **306**(5696), 666–669.
http://dx.doi.org/10.1126/science.1102896.

Öhman, Y., 1961, "Polarized Thermal Emission from Narrow Tungsten Filaments," *Nature*, **192**(4799), 254.
http://dx.doi.org/10.1038/192254a0.

O'Sullivan, F., Celanovic, I., Jovanovic, N., Kassakian, J., Akiyama, S., and Wada, K., 2005, "Optical Characteristics of One-Dimensional Si/SiO2 Photonic Crystals for Thermophotovoltaic Applications," *Journal of Applied Physics*, **97**(3), 033529.
http://dx.doi.org/10.1063/1.1849437.

Otey, C.R., Lau, W.T., and Fan, S., 2010, "Thermal Rectification Through Vacuum," *Physical Review Letters*, **104**(15), 154301.
http://dx.doi.org/10.1103/PhysRevLett.104.154301.

Ottens, R., Quetschke, V., Wise, S., Alemi, A.A., Lundock, R., Mueller, G., Reitze, D.H., Tanner, D.B., and Whiting, B.F., 2011, "Near-Field Radiative Heat Transfer Between Macroscopic Planar Surfaces," *Physical Review Letters*, **107**(1), 014301.
http://dx.doi.org/10.1103/PhysRevLett.107.014301.

Pan, J.L., Choy, H.K.H., and Fonstad, Jr., C.G., 2000, "Very Large Radiative Transfer Over Small Distances From a Black Body for Thermophotovoltaic Applications," *IEEE Transactions on Electron Devices*, **47**(1), 241–249.
http://dx.doi.org/10.1109/16.817591.

Park, K., Basu, S., King, W.P., and Zhang, Z.M., 2008, "Performance Analysis of Near-Field Thermophotovoltaic Devices Considering Absorption Distribution," *Journal of Quantitative Spectroscopy and Radiative Transfer*, **109**(2), 305–316.
http://dx.doi.org/10.1016/j.jqsrt.2007.08.022.

Park, K., Lee, B.J., Fu, C.J., and Zhang, Z.M., 2005, "Study of the Surface and Bulk Polaritons with a Negative Index Metamaterial," *Journal of the Optical Society of America B*, **22**(5), 1016–1023.
http://dx.doi.org/10.1364/JOSAB.22.001016.

Pendry, J.B., 1999, "Radiative Exchange of Heat Between Nanostructures," *Journal of Physics: Condensed Matter*, **11**(35), 6621–6633.
http://dx.doi.org/10.1088/0953-8984/11/35/301.

Pendry, J.B., 2000, "Negative Refraction Makes a Perfect Lens," *Physical Review Letters*, **85**(18), 3966–3969.
http://dx.doi.org/10.1103/PhysRevLett.85.3966.

Pérez-Madrid, A., Lapas, L.C., and Rubí, J.M., 2009, "Heat Exchange Between Two Interacting Nanoparticles Beyond the Fluctuation-Dissipation Regime," *Physical Review Letters*, **103**(4), 048301.
http://dx.doi.org/10.1103/PhysRevLett.103.048301.

Pérez-Madrid, A., Reguera, D., and Rubí, J.M., 2003, "Origin of the Violation of the Fluctuation–Dissipation Theorem in Systems with Activated Dynamics," *Physica A: Statistical Mechanics and its Applications*, **329**(3), 357–364.
http://dx.doi.org/10.1016/S0378-4371(03)00634-4.

Pérez-Madrid, A., Rubí, J.M., and Lapas, L.C., 2008, "Heat transfer between nanoparticles: Thermal conductance for near-field interactions," *Physical Review B*, **77**(15), 155417.
http://dx.doi.org/10.1103/PhysRevB.77.155417.

Persson, B.N.J., and Ueba, H., 2010a, "Heat Transfer Between Graphene and Amorphous SiO_2," *Journal of Physics: Condensed Matter*, **22**(46), 462201.
http://dx.doi.org/10.1088/0953-8984/22/46/462201.

Persson, B.N.J., and Ueba, H., 2010b, "Heat Transfer Between Weakly Coupled Systems: Graphene on a-SiO_2," *Europhysics Letters*, **91**(5), 56001.
http://dx.doi.org/10.1209/0295-5075/91/56001.

Piner, R., Zhu, J., Xu, F., Hong, S.H., and Mirkin, C.A., 1999, "" Dip-Pen" Nanolithography," *Science*, **283**(5402), 661–663.
http://dx.doi.org/10.1126/science.283.5402.661.

Pires, D., Hedrick, J.L., De Silva, A., Frommer, J., Gotsmann, B., Wolf, H., Despont, M., Duerig, U., and Knoll, A.W., 2010, "Nanoscale Three-Dimensional Patterning of Molecular Resists by Scanning Probes," *Science*, **328**(5979), 732–735.
http://dx.doi.org/10.1126/science.1187851.

Planck, M., 1914, *The Theory of Heat Radiation*, Blakiston's Son & Co, Philadelphia.

Pralle, M.U., Moelders, N., McNeal, M.P., Puscasu, I., Greenwald, A.C., Daly, J.T., Johnson, E.A., George, T., Choi, D.S., El-Kady, I., and Biswas, R., 2002, "Photonic Crystal Enhanced Narrow-Band Infrared Emitters," *Applied Physics Letters*, **81**(25), 4685–4687.
http://dx.doi.org/10.1063/1.1526919.

Raether, H., 1988, *Surface Plasmons on Smooth and Rough Surfaces and on Gratings*, Springer-Verlag, Berlin.

Roberts, N.A., and Walker, D.G., 2011, "A Review of Thermal Rectification Observations and Models in Solid Materials," *International Journal of Thermal Sciences*, **50**(5), 648–662.
http://dx.doi.org/10.1016/j.ijthermalsci.2010.12.004.

Rodriguez, A., Ilic, O., Bermel, P., Celanovic, I., Joannopoulos, J.D., Soljačić, M., and Johnson, S., 2011, "Frequency-Selective Near-Field Radiative Heat Transfer between Photonic Crystal Slabs: A Computational Approach for Arbitrary Geometries and Materials," *Physical Review Letters*, **107**(11), 114302.
http://dx.doi.org/10.1103/PhysRevLett.107.114302.

Rousseau, E., Siria, A., Jourdan, G., Volz, S., Comin, F., Chevrier, J., and Greffet, J.J., 2009, "Radiative Heat Transfer at the Nanoscale," *Nature Photonics*, **3**(9), 514–517.
http://dx.doi.org/10.1038/nphoton.2009.144.

Ruppin, R., 2000, "Surface Polaritons of a Left-Handed Medium," *Physics Letters A*, **277**(1), 61–64.
http://dx.doi.org/10.1016/S0375-9601(00)00694-0.

Ruppin, R., 2001, "Surface Polaritons of a Left-Handed Material Slab," *Journal of Physics: Condensed Matter*, **13**(9), 1811–1818.
http://dx.doi.org/10.1088/0953-8984/13/9/304.

Rytov, S.M., Kravtsov, Y.A., and Tatarskii, V.I., 1987, *Principles of Statistical Radiophysics*, vol. 3, Springer-Verlag, New York.

Sasihithlu, K., and Narayanaswamy, A., 2011, "Convergence of Vector Spherical Wave Expansion Method Applied to Near-Field Radiative Transfer," *Optics Express*, **19**(S4), A772–A785.
http://dx.doi.org/10.1364/OE.19.00A772.

Schuller, J.A., Taubner, T., and Brongersma, M.L., 2009, "Optical Antenna Thermal Emitters," *Nature Photonics*, **3**(11), 658–661.
http://dx.doi.org/10.1038/nphoton.2009.188.

Schwierz, F., 2010, "Graphene Transistors," *Nature Nanotechnology*, **5**(7), 487–496.
http://dx.doi.org/10.1038/nnano.2010.89.

Shelby, R.A., Smith, D.R., and Schultz, S., 2001, "Experimental Verification of a Negative Index of Refraction," *Science*, **292**(5514), 77–79.
http://dx.doi.org/10.1126/science.1058847.

Shen, S., Mavrokefalos, A., Sambegoro, P., and Chen, G., 2012, "Nanoscale Thermal Radiation Between Two Gold Surfaces," *Applied Physics Letters*, **100**(23), 233114.
http://dx.doi.org/10.1063/1.4723713.

Shen, S., Narayanaswamy, A., and Chen, G., 2009, "Surface Phonon Polaritons Mediated Energy Transfer Between Nanoscale Gaps," *Nano Letters*, **9**(8), 2909–2913.
http://dx.doi.org/10.1021/nl901208v.

Singer, S.B., Mecklenburg, M., White, E.R., and Regan, B.C., 2011, "Polarized Light Emission From Individual Incandescent Carbon Nanotubes," *Physical Review B*, **83**(23), 233404.
http://dx.doi.org/10.1103/PhysRevB.83.233404.

Smith, D.R., Schurig, D., and Pendry, J.B., 2002, "Negative Refraction of Modulated Electromagnetic Waves," *Applied Physics Letters*, **81**(15), 2713–2715.
http://dx.doi.org/10.1063/1.1512828.

Stevenson, P.F., Peterson, G.P., and Fletcher, L.S., 1990, "Thermal Rectification in Similar and Dissimilar Metal Contacts," *Journal of Heat Transfer*, **113**(1), 30–36.
http://dx.doi.org/10.1115/1.2910547.

Svetovoy, V., van Zwol, P., and Chevrier, J., 2012, "Plasmon Enhanced Near-Field Radiative Heat Transfer for Graphene Covered Dielectrics," *Physical Review B*, **85**(15), 155418.
http://dx.doi.org/10.1103/PhysRevB.85.155418.

Tersoff, J., and Hamann, D.R., 1985, "Theory of The Scanning Tunneling Microscope," *Physical Review B*, **31**(2), 805–813.
http://dx.doi.org/10.1103/PhysRevB.31.805.

Tsang, L., Kong, J.A., and Ding, K.H., 2004, *Scattering of Electromagnetic Waves, Theories and Applications*, Wiley, New York.

Tsang, L., Njoku, E., and Kong, J.A., 1974, "Microwave Thermal Emission From a Stratified Medium with Nonuniform Temperature Distribution," *Journal of Applied Physics*, **46**(12), 5127–5133.
http://dx.doi.org/10.1063/1.321571.

van Beest, B.W.H., Kramer, G.J., and van Santen, R.A., 1990, "Force Fields for Silicas and Aluminophosphates Based on Ab Initio Calculations," *Physical Review Letters*, **64**(16), 1955–1958.
http://dx.doi.org/10.1103/PhysRevLett.64.1955.

Vettiger, P., Cross, G., Despont, M., Drechsler, U., Dürig, U., Gotsmann, B., Haberle, W., Lantz, M.A., Rothuizen, H.E., Stutz, R., and Binnig, G.K., 2002, "The "Millipede" - Nanotechnology Entering Data Storage," *IEEE Transactions on Nanotechnology*, **1**(1), 39–55.
http://dx.doi.org/10.1109/TNANO.2002.1005425.

Volokitin, A.I., and Persson, B.N.J., 2001, "Radiative Heat Transfer Between Nanostructures," *Physical Review B*, **63**(20), 205404.
http://dx.doi.org/10.1103/PhysRevB.63.205404.

Volokitin, A.I., and Persson, B.N.J., 2004, "Resonant Photon Tunneling Enhancement of the Radiative Heat Transfer," *Physical Review B*, **69**(4), 045417.
http://dx.doi.org/10.1103/PhysRevB.69.045417.

Volokitin, A.I., and Persson, B.N.J., 2011, "Near-Field Radiative Heat Transfer Between Closely Spaced Graphene and Amorphous SiO_2," *Physical Review B*, **83**(24), 241407.
http://dx.doi.org/10.1103/PhysRevB.83.241407.

Wacaser, B.A., Maughan, M.J., Mowat, I.A., Niederhauser, T.L., Linford, M.R., and Davis, R.C., 2003, "Chemomechanical Surface Patterning and Functionalization of Silicon Surfaces Using an Atomic Force Microscope," *Applied Physics Letters*, **82**(5), 808–810.
http://dx.doi.org/10.1063/1.1535267.

Wang, L.P., and Zhang, Z.M., 2009, "Resonance Transmission or Absorption in Deep Gratings Explained by Magnetic Polaritons," *Applied Physics Letters*, **95**(11), 111904.
http://dx.doi.org/10.1063/1.3226661.

Wang, L.P., and Zhang, Z.M., 2011, "Phonon-Mediated Magnetic Polaritons in the Infrared Region," *Optics Express*, **19**(S2), A126–A135.
http://dx.doi.org/10.1364/OE.19.00A126.

Wang, L.P., and Zhang, Z.M., 2012, "Wavelength-Selective and Diffuse Emitter Enhanced by Magnetic Polaritons for Thermophotovoltaics," *Applied Physics Letters*, **100**(6), 063902.
http://dx.doi.org/10.1063/1.3684874.

Wang, L.P., and Zhang, Z.M., 2013, "Thermal Rectification Enabled by Near-field Radiative Heat Transfer Between Intrinsic Silicon and a Dissimilar Material," *Nanoscale and Microscale Thermophysical Engineering*, accepted for publication.

Wang, L., Uppuluri, S.M., Jin, E.X., and Xu, X., 2006, "Nanolithography Using High Transmission Nanoscale Bowtie Apertures," *Nano Letters*, **6**(3), 361–364.
http://dx.doi.org/10.1021/nl052371p.

Wang, X.J., Basu, S., and Zhang, Z.M., 2009, "Parametric Optimization of Dielectric Functions for Maximizing Nanoscale Radiative Transfer," *Journal of Physics D: Applied Physics*, **42**(24), 245403.
http://dx.doi.org/10.1088/0022-3727/42/24/245403.

Wei, Z., Wang, D., Kim, S., Kim, S.Y., Hu, Y., Yakes, M.K., Laracuente, A.R., Dai, Z., Marder, S.R., Berger, C., King, W.P., de Heer, W.A., Sheehan, P.E., and Riedo, E., 2010, "Nanoscale Tunable Reduction of Graphene Oxide for Graphene Electronics," *Science*, **328**(5984), 1373–1376.
http://dx.doi.org/10.1126/science.1188119.

Whale, M.D., and Cravalho, E.G., 2002, "Modeling and Performance of Microscale Thermophotovoltaic Energy Conversion Devices," *IEEE Transactions on Energy Conversion*, **17**(1), 130–142.
http://dx.doi.org/10.1109/60.986450.

Wu, L.A., and Segal, D., 2009, "Sufficient Conditions for Thermal Rectification in Hybrid Quantum Structures," *Physical Review Letters*, **102**(9), 095503.
http://dx.doi.org/10.1103/PhysRevLett.102.095503.

Xu, J.B., Lauger, K., Moller, R., Dransfeld, K., and Wilson, I.H., 1994, "Heat Transfer Between Two Metallic Surfaces at Small Distances," *Journal of Applied Physics*, **76**(11), 7209–7216.
http://dx.doi.org/10.1063/1.358001.

Yablonovitch, E., 1987, "Inhibited Spontaneous Emission in Solid-State Physics and Electronics," *Physical Review Letters*, **58**(20), 2059–2062.
http://dx.doi.org/10.1103/PhysRevLett.58.2059.

Zhang, Z.M., 2007, *Nano/Microscale Heat Transfer*, McGraw-Hill.

Zhang, Z.M., Fu, C.J., and Zhu, Q.Z., 2003, "Optical and Thermal Radiative Properties of Semiconductors Related to Micro/Nanotechnology," *Advances in Heat Transfer*, vol. 37, 179–296, Elsevier.
http://dx.doi.org/10.1016/S0065-2717(03)37003-0.

Zhang, Z.M., and Lee, B.J., 2006, "Lateral Shift in Photon Tunneling Studied by the Energy Streamline Method," *Optics Express*, **14**(21), 9963–9970.
http://dx.doi.org/10.1364/OE.14.009963.

Zhang, Z.M., and Wang, L.P., 2011, "Measurements and Modeling of the Spectral and Directional Radiative Properties of Micro/Nanostructured Materials," *International Journal of Thermophysics*, (available online).
http://dx.doi.org/10.1007/s10765-011-1036-5.

Zhang, Z.M., and Wang, X.J., 2012, "Unified Wien's Displacement Law in Terms of Logarithmic Frequency or Wavelength Scale," *Journal of Thermophysics and Heat Transfer*, **24**(1), 222–224.
http://dx.doi.org/10.2514/1.45992.

Zhang, Z., Park, K., and Lee, B.J., 2011, "Surface and Magnetic Polaritons on Two-Dimensional Nanoslab-Aligned Multilayer Structure," *Optics Express*, **19**(17), 16375–16389.
http://dx.doi.org/10.1364/OE.19.016375.

Zheng, Z., and Xuan, Y., 2011, "Theory of Near-Field Radiative Heat Transfer for Stratified Magnetic Media," *International Journal of Heat and Mass Transfer*, **54**(5-6), 1101–1110.
http://dx.doi.org/10.1016/j.ijheatmasstransfer.2010.11.012.

Zhu, Q., Lee, H.J., and Zhang, Z.M., 2009, "Radiative Properties of Materials with Surface Scattering or Volume Scattering: a Review," *Frontiers of Energy and Power Engineering in China*, **3**(1), 60–79.
http://dx.doi.org/10.1007/s11708-009-0011-3.

2

EFFECT OF MAGNETIC FIELD ON INDIRECT NATURAL CONVECTION FLOW ABOVE A HORIZONTAL HOT FLAT PLATE

Tapas Ray Mahapatra[a], Sumanta Sidui[b], Samir Kumar Nandy[c*]

[a] *Department of Mathematics, Visva-Bharati, Santiniketan - 731 235, India*
[b] *Department of Mathematics, Ajhapur High School, Burdwan - 713 401, India*
[c] *Department of Mathematics, A.K.P.C Mahavidyalaya, Hooghly – 712 611, India*

Abstract

The effect of variable transverse magnetic field on steady two-dimensional indirect natural convection flow of an incompressible viscous fluid over a horizontal hot flat plate is theoretically studied. The governing partial differential equations are transformed into ordinary ones by similarity transformation and solved numerically using fourth order Runge-Kutta method with shooting technique. The results are obtained for the skin friction coefficient and the local Nusselt number as well as the dimensionless velocities, temperature for some values of the magnetic parameter (M) subject to either prescribed (constant or variable) surface temperature or prescribed (variable) heat flux. It is seen that the skin friction coefficient decreases with increase in M for all the cases.
Keywords: Steady flow; Magnetic field; Indirect natural convection; Variable Heat Flux.

1. INTRODUCTION

Flow past a hot vertical plate adjacent to a viscous fluid at a lower temperature is a simple example of natural convection flow. In this case the convection takes place in boundary layer originating at the lower edge of the plate. Heat transferred from the plate to the fluid leads to an increase in temperature of the fluid near the wall causing decrease in density there and gaining buoyancy fluid moves upwards along the plate. The free convection of heat from a heated vertical plate in a fluid has been extensible studied for many years. Squire (1953) gave a review of the work. Subsequently Ostrach (1953) studied numerical solution for the free convection flow around a heated vertical plate for a wide range of values of Prandtl number.

However if the plate is horizontal the buoyancy has no component along its length and if the boundary layer exist it must be of a different kind. Natural convection flow of this different character can occur over a horizontal semi-infinite plate facing upward with temperature (T_W) higher than that (T_∞) of the surrounding fluid (see Fig.1). In front of the plate, the temperature of the fluid is T_∞ everywhere so that in this static field, there is a pressure distribution p, satisfying,$\partial p/\partial y = \rho_\infty g$ where ρ_∞ is the density of the fluid in this region. As heat is transferred from the plate to the fluid, the fluid temperature is larger than T_∞ above the plate and so the density (ρ) is less than ρ_∞.The reduced pressure gradient $|\partial p/\partial y| = \rho g < \rho_\infty g$ gives rise to reduced pressure close to the plate. Thus there is a pressure drop along the plate which is taken as the x-direction. Due to this induced pressure gradient in the x-direction fluid flows parallel to the plate. This flow has a boundary layer character at large Grashof number and is known as indirect natural convection and was first studied by Stewartson (1958). The same problem was revisited by Gill et al. (1965). He pointed out that Stewartson's conclusion that boundary layer solution exist only when the heated plate faces downward is erroneous. They showed that boundary layer solution exists only when the heated plate faces upward and gave the correct solution to the problem. On the other hand, Wickern (see in Schlichting (2000)) investigated natural convection flow over a horizontal plate subject to uniform heat flux.

There are many studies about natural convection flow caused by immersing a hot surface in a fluid saturated porous medium at constant ambient temperature. Chamkha (2003) analyzed the heat and mass transfer laminar boundary layer flow in the presence of heat generation/absorption. Later, Chen (2004) investigated the heat and mass transfer effects of an electrically conducting fluid in magneto hydrodynamic natural convection adjacent to vertical surface. Ahmed (2010) studied the effects of chemical reaction and viscous dissipation on unsteady heat and mass transfer along an infinite vertical porous plate in the presence of magnetic field. On the other hand, Kim (2000) studied MHD natural convection flow past a moving vertical plate embedded in a porous medium. Siddiqa et al. (2010) investigated laminar natural convection flow of a viscous fluid over a semi-infinite flat plate inclined at a small angle to the horizontal. The natural convection boundary layer flow on a vertical surface in a porous medium with prescribed constant surface heat flux was considered by Merkin (2012). Very recently, a boundary layer analysis was performed for the steady laminar natural convection of a electrically conducting viscous incompressible fluid above a horizontal plate in the presence of a transverse magnetic field by Samanta and Guha (2014).

In this paper we analyze the effect of variable transverse magnetic field on the steady two-dimensional indirect natural convection flow of an incompressible viscous fluid over a hot horizontal plate which is subject to either prescribed (constant or variable) surface temperature or prescribed (variable) heat flux.

2 FLOW ANALYSIS

A sketch of the physical problem already described in the above introduction is given in Fig. 1, where the x−axis is taken along the plate with the origin at the front end of the plate and y−axis is perpendicular

* *Corresponding author. Email: nandysamir@yahoo.com*

to the plate upward. Using boundary layer approximations, the governing Magneto-Hydro-Dynamics (MHD) equations for indirect natural convection flow are as follows:

The equation of continuity is

$$\frac{\partial u}{\partial x}+\frac{\partial v}{\partial y}=0 \tag{1}$$

where (u, v) are the velocity components in x and y directions, respectively.

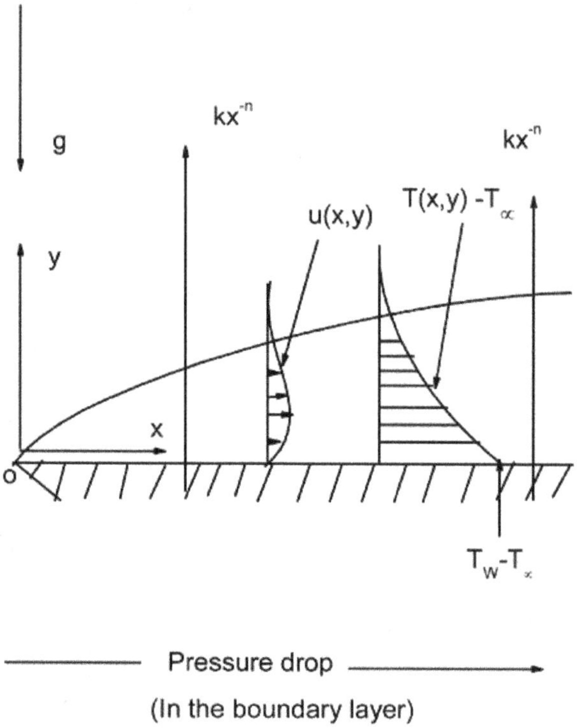

Fig 1. Indirect natural convection. Formation of a pressure gradient $\partial p/\partial x$ in the boundary layer via reduced static pressure above the horizontal hot plate.

The steady two-dimensional $u-$momentum and $v-$momentum equations for the flow in the presence of variable transverse magnetic field are given by

$$u\frac{\partial u}{\partial x}+v\frac{\partial u}{\partial y}=-\frac{1}{\rho_\infty}\frac{\partial p}{\partial x}+\upsilon_\infty\frac{\partial^2 u}{\partial y^2}+\frac{(\boldsymbol{J}\times\boldsymbol{B})_x}{\rho_\infty} \tag{2}$$

$$0=-\frac{1}{\rho_\infty}\frac{\partial p}{\partial y}+g\beta_\infty(T-T_\infty) \tag{3}$$

where T denotes the temperature, p is the pressure, ρ_∞ is the fluid density at the temperature T_∞, $\beta_\infty$is the coefficient of thermal expansion at temperature $T_\infty$and $\upsilon_\infty(=\mu/\rho_\infty)$ is the kinematic viscosity coefficient of the fluid. In writing (3), Boussinesq approximation is made so that density variation is taken into account in the buoyancy force term only. The last term in equation (2) stands for the x-component of the Lorentz force per unit mass where $\boldsymbol{J}$ is the electric current density vector and $\boldsymbol{B}$(=(0,$H(x)$,0)) is the imposed magnetic field. Here $H(x)(=kx^Q)$ denotes the variable applied transverse magnetic field where k and Q are constants. The value of Q is determined later. In equation (2) it is assumed that the induced magnetic field is negligible in comparison to the applied magnetic field. This is a valid assumption for flow at small magnetic Reynolds number. This assumption is justified for flow of electrically conducting fluids such as liquid metals e.g. mercury, liquid sodium (see Shercliff (1995), Mahapatra et al. (2011)). It is also assumed that the external electric field is zero and the electric field due to the polarization of charges is negligible. Here the electric current flows parallel to the z axis which is normal to x-y plane. So by Ohm's law

$$J_x=0,\ J_y=0,\ J_z=\sigma uH(x) \tag{4}$$

where σ is electrical conductivity of the fluid, which is assumed constant.

Using equation (4), equation (2) becomes

$$u\frac{\partial u}{\partial x}+v\frac{\partial u}{\partial y}=-\frac{1}{\rho_\infty}\frac{\partial p}{\partial x}+\upsilon_\infty\frac{\partial^2 u}{\partial y^2}-\frac{\sigma H^2(x)}{\rho_\infty}u \tag{5}$$

The energy equation, neglecting viscous and Ohmic dissipation, is given by

$$u\frac{\partial T}{\partial x}+v\frac{\partial T}{\partial y}=a_\infty\frac{\partial^2 T}{\partial y^2} \tag{6}$$

where a_∞denotes the thermal diffusivity of the fluid. We now introduce the dimensionless quantities (see Schlichting and Gersten (2000))

$$x^*=\frac{x}{l},\ \bar{y}=\frac{y}{l}(Gr)^{1/5},\ \bar{u}=\frac{u}{V_{IN}},\ \bar{v}=\frac{v}{V_{IN}}(Gr)^{1/5} \tag{7}$$

$$\bar{p}=\frac{p}{\rho_\infty V_{IN}^2},\ Gr=\frac{gl^3\beta_\infty B}{\upsilon_\infty^2},\ V_{IN}=(gl^{1/2}\upsilon_\infty^{1/2}\beta_\infty B)^{2/5}$$

Here l is a characteristic length scale, V_{IN} is the characteristic velocity for indirect natural convection, Gr is the Grashof number, B is a positive constant. In CST (Constant Surface Temperature) case $B=\Delta T(=T_W-T_\infty)$ when T_W is greater than T_∞. In PST (Prescribed Surface Temperature) case $B=A$; a positive constant which will be discussed in the section 3.2 and also in PHF (Prescribed Heat Flux) case $B=Dc_2/k_\infty$, another positive constant and c_2 is arbitrary and positive constant and k_∞ is the coefficient of thermal conductivity and this case will be discussed in the section 3.3.

Using non-dimensional quantities (7) in equations (1),(5), (3) and (6), the boundary layer equations for indirect natural convection are

$$\frac{\partial\bar{u}}{\partial x^*}+\frac{\partial\bar{v}}{\partial\bar{y}}=0 \tag{8}$$

$$\bar{u}\frac{\partial\bar{u}}{\partial x^*}+\bar{v}\frac{\partial\bar{u}}{\partial\bar{y}}=-\frac{\partial\bar{p}}{\partial x^*}+\frac{\partial^2\bar{u}}{\partial\bar{y}^2}-(x^*)^{2Q}M\bar{u} \tag{9}$$

$$\frac{\partial\bar{p}}{\partial\bar{y}}=\frac{T-T_\infty}{B} \tag{10}$$

$$\bar{u}\frac{\partial T}{\partial x^*}+\bar{v}\frac{\partial T}{\partial\bar{y}}=\frac{1}{\Pr}\frac{\partial^2 T}{\partial\bar{y}^2} \tag{11}$$

where $M(=\sigma k^2 l^{1-2n}/\rho_\infty V_{IN})$ is the Hartmann number and $\Pr(=\upsilon_\infty/a_\infty)$ is the Prandtl number. Note that the powers in equation (7) are chosen in such a way that the following remain the same after the transformation in the limit $Gr\to\infty$: the continuity equation, a viscous term in the $x-$momentum equation as well as the pressure and buoyancy terms in the $y-$momentum equation.

Introduce the dimensionless temperature θ as

$$\theta(\eta)=\frac{T-T_\infty}{T_w-T_\infty},\quad \text{for CST case} \tag{12}$$

$$\theta(\eta)=\frac{T-T_\infty}{Bx^{*s}},\quad \text{for PST and PHF cases,} \tag{13}$$

where s is the wall temperature parameter. In the CST case $s=0$ and in other two cases $s\neq 0$.

Introducing the dimensionless stream function and the similarity transformations

$$\bar{\psi}=c_1x^{*m}f(\eta),\bar{y}=c_2x^{*n}\eta,p=c_3x^{*(n+s)}g(\eta) \tag{14}$$

where m and n are constants, we find that (8) is identically satisfied and from the equations (9) - (11), we obtain

$$\frac{c_1}{c_2^3}f'''+\frac{c_1^2}{c_2^2}[mff''-(m-n)f'^2]$$
$$=c_3[(n+s)g-n\eta g']+\frac{c_1}{c_2}Mf' \tag{15}$$

$$g'=\frac{c_2}{c_3}\theta \tag{16}$$

$$\frac{1}{c_2^2}\theta'' + \frac{c_1}{c_2}\Pr[mf\theta' - s\theta f'] = 0 \quad (17)$$

It is observed that for the existence of similarity solutions, we must have $m-4n = 2m-3n-1 = s-1$ and $Q = -n$. Here a prime denotes differentiation with respect to η. It is clear from above that the velocity and temperature distributions depend on the dimensionless parameters Hartmann number M and Prandtl number Pr. Equations (15), (16) and (17) subject to the suitable boundary conditions are solved numerically by an efficient shooting method for different values of the parameter M. We first, eliminate θ between (16) and (17), the resulting equation and equation (15) are written as a system of six first order ordinary differential equations, which are solved by means of a standard fourth-order Runge-Kutta integration technique. Then a Newton iteration procedure is employed to assure quadratic convergence of the iterations required to satisfy outer boundary conditions. The constants c_1 and c_2 are arbitrary and positive. In Stewartson's (1958) notation $c_1 = c_2 = c_3 = 1$ and in the notation of Gill et al. (1965), $c_2 = c_3$. In this paper all computations are based on $c_1 = c_2 = c_3 = 1$ except when we compare our results (see Table 1) with the corresponding results computed by Stewartson (1958) and Gill et al. (1965).

The Nusselt number at the wall is given by

$$Nu(x) = \frac{q_w(x)l}{\lambda_\infty(T_w - T_\infty)} = -\frac{1}{c_2}\theta'(0)(x^*)^{-n}(Gr)^{1/5}, \text{ for CST case} \quad (18)$$

$$= -\frac{1}{c_2}\theta'(0)(x^*)^{-n}(Gr)^{1/5}, \text{ for PST case} \quad (19)$$

$$= \frac{1}{c_2}(x^*)^{-n}(Gr)^{1/5}, \text{ for PHF case} \quad (20)$$

where λ_∞ is the thermal conductivity and $n = (2-s)/5$. For the CST case $s = 0$ and for the PHF case $\theta'(0) = -1$.

3 Results and discussion

Three different cases are considered.

3.1 Constant surface temperature (CST case)

In this case, the boundary conditions are

$$u = v = 0,\ T = T_w \ \ at\ \ y = 0$$
$$u = u_\infty = 0,\ p = 0,\ T \to T_\infty \ as\ y \to \infty \quad (21)$$

where T_w and T_∞ are constants with $T_w > T_\infty$. The boundary conditions for f and g are derived from (7), (12), (14), (16) and (21) as

$$f(0) = 0, f'(0) = 0, g'(0) = \frac{c_2}{c_3} \quad (22)$$

$$f'(\infty) = 0, g(\infty) = 0, g'(\infty) = 0 \quad (23)$$

The equations of f, g and θ are obtained from (15)-(17) by putting s=0 as

$$\frac{c_1}{c_2^3}f''' + \frac{c_1^2}{c_2^2}[mff'' - (m-n)f'^2] = c_3 n[g - \eta g'] + \frac{c_1}{c_2}Mf' \quad (24)$$

$$g' = \frac{c_2}{c_3}\theta \quad (25)$$

$$\frac{1}{c_2^2}\theta'' + \frac{c_1}{c_2}\Pr mf\theta' = 0 \quad (26)$$

with the condition $m-4n = 2m-3n-1 = -1$.

Equation (24) – (26) are solved numerically subject to the boundary condition (22) and (23) using the technique as stated in earlier section.

Table 1 gives the comparison of the values of $f''(0), -\theta'(0), -g(0)$ and $f(\infty)$ calculated in the present study (A) with those of Stewartson (1958) (B) and Gill et al. (1965) (C) when $c_1 = c_2 = c_3 = 1$, $M = 0$ and $\Pr = 0.72$. A good agreement in the values is observed.

Table 1: Comparison of the values of $f''(0), -\theta'(0), -g(0)$ and $f(\infty)$ of (A) with those of (B) and (C) where M=0, Pr=0.72

	$f''(0)$	$-\theta'(0)$	$-g(0)$	$f(\infty)$
A	0.9784	0.3574	1.7349	2.3301
B	0.9710	0.3580	1.7300	2.3000
C	0.9787	0.3574	----	-----

Figure 2 shows the variation of horizontal velocity for different values of M where Pr = 0.72. It is observed that the velocity parallel to the plate decreases with increase in M up to a certain distance from the plate but beyond that distance opposite trend is observed. Variation of wall temperature for several values of M are presented in the Fig. 3 for Pr = 0.72. It shows that the temperature at a point increases with increase in M. Fig. 4 represents the variation of normal velocity to the plate for several values of M when Pr = 0.72. It is observed that up to a small distance from the plate the normal velocity increases with increase in M but beyond that distance opposite trend is observed.

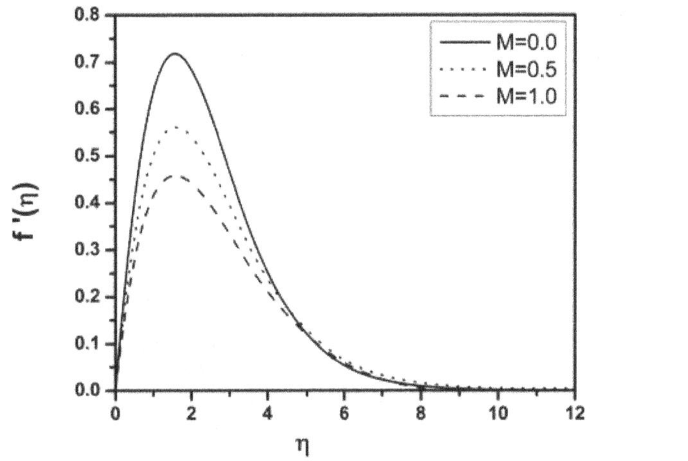

Fig. 2 Variation of $f'(\eta)$ for several values of M for CST case with Pr = 0.72.

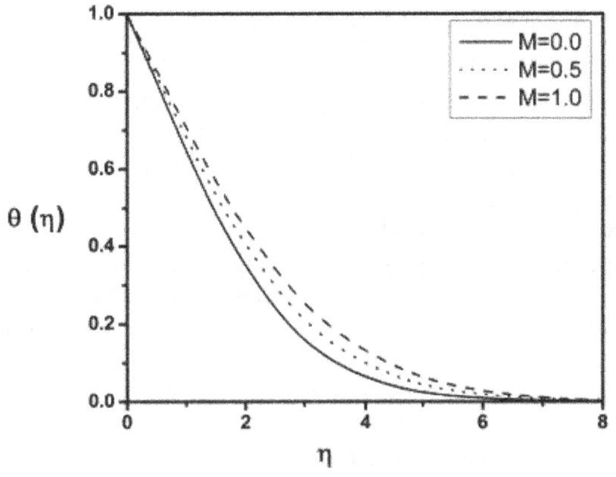

Fig. 3 Variation of temperature profile $\theta(\eta)$ for several values of M for CST case with Pr = 0.72.

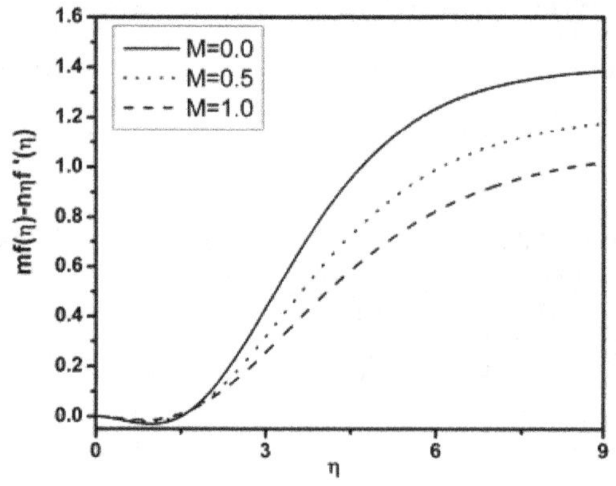

Fig. 4 Variation of $mf(\eta) - n\eta f'(\eta)$ velocity normal to the plate for several values of M for the CST case with Pr = 0.72.

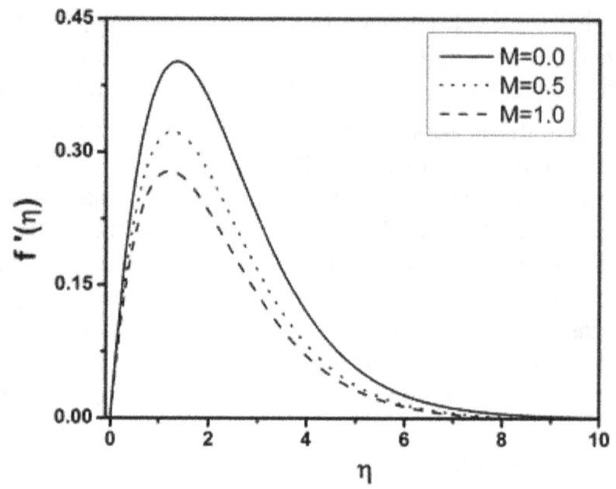

Fig. 5 Variation of $f'(\eta)$ for several values of M for PST case with Pr = 2.0 and s = 0.3.

Table 2: Values of $f''(0), -\theta'(0)$ for several values of M in CST case when Pr=0.72

M	0.0	0.5	1.0
$f''(0)$	0.9784	0.8321	0.7487
$-\theta'(0)$	0.3574	0.3212	0.2957

Table 2 gives the values of wall shear stress $(f''(0))$ and surface heat flux $(-\theta'(0))$ for several values of M, when Pr=0.72 for CST case. It can be seen from Table 2 that the surface shear stress decreases with increase in M. It is seen that the surface heat flux $-\theta'(0)$ also decreases with increase in M.

3.2 Prescribed surface temperature (PST case)

Here the boundary conditions are

$$u = v = 0,\ T = T_w(= T_\infty + Ax^{*s})\ \ at\ \ y = 0 \tag{27}$$

$$u = u_\infty = 0,\ p = 0,\ T \to T_\infty\ as\ y \to \infty$$

where $T_W > T_\infty$. Note that the temperature parameter s is introduced earlier is the index of power law variation of surface temperature shown above. Further the constant B in equation (7) that remained unspecified so far is now A, which is a positive constant. The boundary conditions for f and g are derived from (7), (13), (14), (16) and (27) as

$$f(0) = 0, f'(0) = 0, g'(0) = \frac{c_2}{c_3} \tag{28}$$

$$f'(\infty) = 0, g(\infty) = 0, g'(\infty) = 0 \tag{29}$$

In this case the equations of f, g and θ are (15), (16) and (17) with the condition $m-4n = 2m - 3n - 1 = s - 1$. It is clear from equations (15), (16) and (17) that the velocity and temperature distribution depend on the dimensionless parameter Hartmann number M, Prandtl number Pr and wall temperature parameter s. Equations (15), (16) and (17) subject to the boundary conditions (28) and (29) are solved numerically by using the same method which is used in CST case.

Figure 5 represents the variation of horizontal velocity for different values of M where Pr =0.72 and s = 0.3. It shows that the velocity parallel to the plate decreases with increase in M for fixed values of Pr and s. Variation of temperature are presented in Fig.6 for several values of M where Pr = 0.72 and s = 0.3. It is observed that for fixed value of Pr and s, temperature at a point increases with increase in M. Fig.7 shows the variation of normal velocity with M where Pr = 0.72 and s = 0.3. It shows that this velocity increases with increase in M up to a small distance from the plate but beyond that opposite trend is observed.

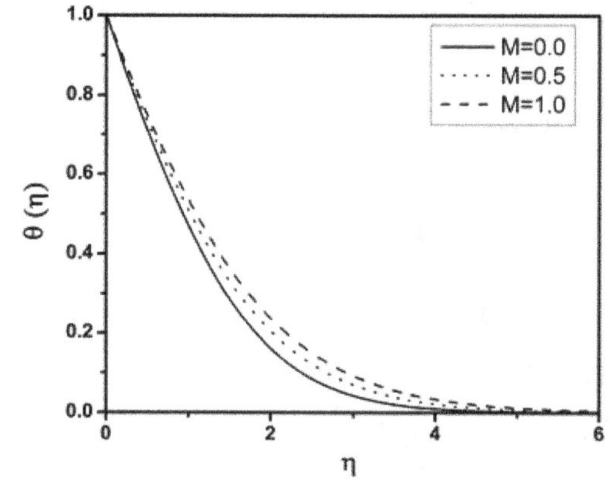

Fig. 6 Variation of temperature profile $\theta(\eta)$ for several values of M for PST case with Pr = 2., s=0.3.

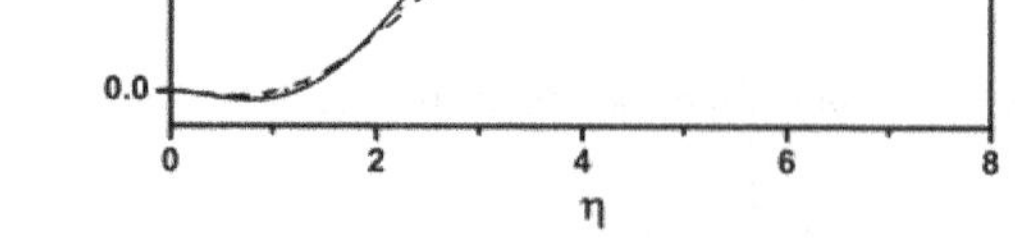

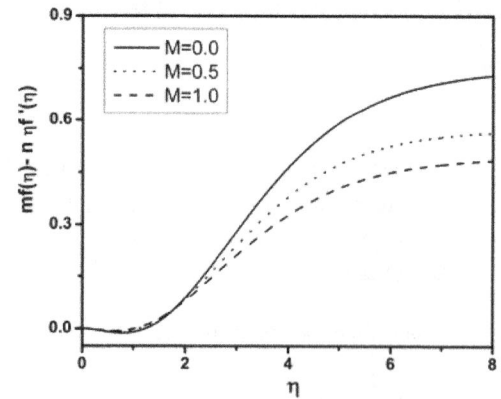

Fig. 7 Variation of $mf(\eta) - n\eta f'(\eta)$ velocity normal to the plate for several values of M for the CST case with Pr =2.0 and s=0.03.

Table 3: Values of $f''(0), -\theta'(0)$ for several values of M in PST case when Pr=2.0 and s=0.3

M	0.0	0.5	1.0
$f''(0)$	0.6924	0.6294	0.5925
$-\theta'(0)$	0.5832	0.5374	0.5066

Table 3 gives the values of $f''(0)$ and $-\theta'(0)$ for several values of M when Pr (= 2.0) and s (= 0.3) are kept fixed. It can be seen from Table 3 that the surface shear stress decreases with increase in M for fixed values of s and Pr. It can also be seen from Table 3 that the surface heat flux $-\theta'(0)$ decreases with increase in M for any fixed value of s(= 0.3) and Pr(= 2.0).

When s=-m, i.e., when s= -0.5, the energy equation (17) becomes

$$\theta'' + c_1 c_2 m\Pr(f\theta' + \theta f') = 0 \qquad (30)$$

Integrating (30) once w.r.t. η and using the boundary condition $\theta(\infty) = 0$, we obtain

$$\theta'(\eta) = -c_1 c_2 m\Pr f(\eta)\theta(\eta). \qquad (31)$$

As $f(0) = 0$, equation (31) shows that $\theta'(0) = 0$ when $s = -0.5$. It is interesting to note that wall heat flux vanishes when the surface temperature varies as $x^{-1/2}$. It may be noted that Gill et al. (1965) considered only a particular case of wall temperature variation viz., $T_W - T_\infty = Bx^{*-1/2}$.

3.3 Prescribed power law heat flux (PHF case)

In PHF case the dimensionless temperature variable $\theta(\eta)$ is defined as

$$\theta(\eta) = \frac{T(x,y) - T_\infty}{Dx^{*s}c_2 / k_\infty} \qquad (32)$$

Thus the constant B which remained so far unspecified, for this case is Dc_2 / k_∞, where k_∞ is the thermal conductivity of the fluid and D is a positive constant. In this case the corresponding boundary conditions are

$$u = v = 0,\ q_w = -k_\infty \left(\frac{\partial T}{\partial y}\right)_w = D(x^*)^{s-n} \quad at \ \ y = 0 \qquad (33)$$

$$u = u_\infty = 0,\ p = 0,\ T \to T_\infty \ as \ y \to \infty$$

The boundary conditions for f and g are derived from (7), (14), (16), (32) and (33) as

$$f(0) = 0, f'(0) = 0, g''(0) = -\frac{c_2}{c_3} \qquad (34)$$

$$f'(\infty) = 0, g(\infty) = 0, g'(\infty) = 0 \qquad (35)$$

In this case the equations of f, g and θ are (15), (16) and (17) with the condition $m-4n = 2m - 3n - 1 = s - 1$. Note that for the case of uniform heat flux, $s = n$ (see equation (33)). Then the above equation for m, n and s give $s = 1/3$ which agrees with the result of Wickern (2000).

Here the velocity and temperature distributions depend on three dimensionless parameters Hartmann number M, Prandtl number Pr and wall heat flux parameter s. Equations (15), (16) and (17) subject to the boundary conditions (34) and (35) are solved numerically by using the same method as described earlier. Here, wall temperature will be different for different physical situation contrast to the cases of PST where it attains the constant value.

This behavior of the temperature profile is the consequences of prescribed boundary conditions. The qualitative behaviors of velocity, temperature and normal velocity distributions due to variation of the parameter M are similar to those of the PST case.

Figure 8 shows the variation of horizontal velocity for several values of M. This velocity decreases with increase in M. Fig.9 represents the variation of temperature for several values of M for Pr = 0.1 and s = 0.1. The wall temperature $\theta(0)$ increases with increase in M. Variation of normal velocity for several values of M when Pr = 0.1 and s = 0.1 are presented in the Fig.10. It shows that this velocity almost remains constant with increase in M up to a small distance from the plate but beyond that it decreases with increase in M.

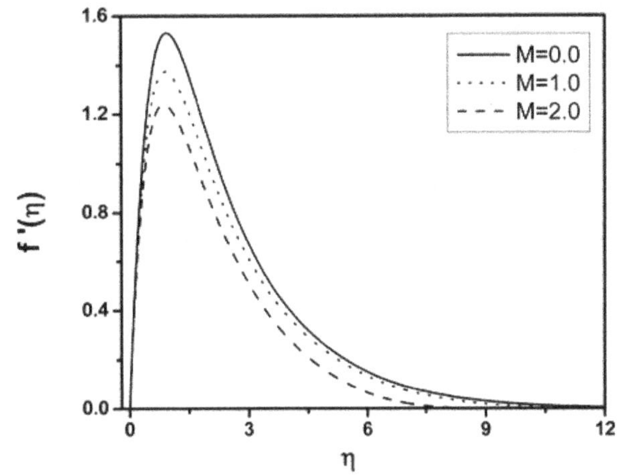

Fig. 8: Variation of $f'(\eta)$ for several values of M for PHF case with Pr = 0.1 and s = 2.0

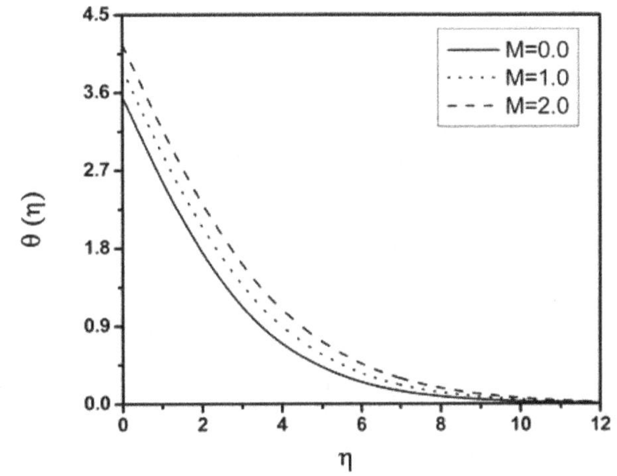

Fig. 9: Variation of temperature profile $\theta(\eta)$ for several values of M for PHF case with Pr =0.1, s=0.1.

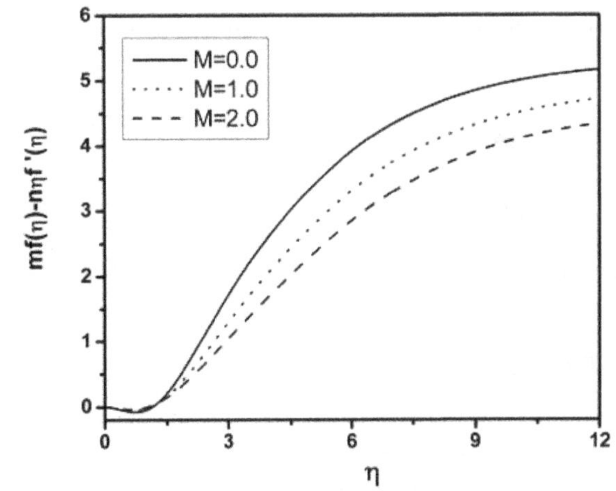

Fig. 10: Variation of $mf(\eta) - n\eta f'(\eta)$ velocity normal to the plate for several values of M for the PHF case with Pr =0.1 and s=0.1.

Table 4: Values of $f''(0)$ and $\theta(0)$ for several values of M in PHF case when Pr=1.0 and s=2.0

M	0.0	0.5	1.0
$f''(0)$	1.1989	1.1885	1.1820
$\theta(0)$	1.1884	1.2224	1.2509

Table 4 gives the values of $f''(0)$ and $\theta(0)$ for several values of M where Pr(= 1.0) and s(= 2.0) are kept fixed. It can be seen from Table 4 that the surface shear stress decreases with increase in M for any fixed value of s and Pr. It can also be seen from Table 4 that the temperature $\theta(0)$ increases with increase in M.

4 Concluding Remarks

We have obtained an exact similarity solution of the Navier-Stokes equations which represents steady two-dimensional indirect natural convection flow of an incompressible viscous fluid above a horizontal hot flat plate in the presence of a variable transverse magnetic field. Here the velocity field is dependent on the temperature field. Different boundary conditions are considered for temperature field and three different cases are examined. For all the three cases, surface shear stress decreases with increase in M for fixed values of Pr and s, where s is the parameter characterizing the wall temperature variation for PST and CST cases but wall heat flux parameter for PHF cases, Pr is the Prandtl number and M is the magnetic parameter.

Case 1. CST (Constant Surface Temperature):
It is found that at first the velocity at a point parallel to the plate decreases with increase in M but after a certain distance perpendicular to the sheet, opposite trend is observed. Temperature at a point increases with increase in M. The surface heat flux decreases with increase in M.

Case 2. PST (Prescribed Surface Temperature):
Similarity solutions for velocity and temperature distributions are obtained in this case. It is found that the velocity at a point parallel to the plate decreases with increase in M. The surface heat flux decreases with increase in M. For fixed values of Pr and s temperature at a point increases with increase in the parameter M.

Case 3. PHF (Prescribed Heat Flux):
Qualitative behaviours of temperature distribution in this case follow the similar pattern as those for the case of PST. It is seen that for the fixed values of Pr and s the wall temperature $\theta(0)$ increases with increase in M. It is also found that the velocity at a point parallel to the plate decreases with increase in M.

Acknowledgements

The work of one of the authors (T.R.M.) is supported under SAP (DRS Phase-III) under UGC, New Delhi, India. The authors thank the referees for their critical comments which enabled an improved presentation of the paper.

References

Squire, H.B., 1953, "Modern Developments in Fluid Dynamics: High Speed Flow," ed. L. Howarth; Oxford, Chap. 10. http://dx.doi.org/10.1002/zamm.19540341215

Ostrach, S., 1953, "An Analysis of Laminar Free-Convection Flow and Heat Transfer about a Flat Plate Parallel to the Direction of the Generating Body Force," N.A.C.A. Rep. No.1111. http://naca.central.cranfield.ac.uk/reports/1953/naca-report-1111.pdf

Stewartson, K., 1958, "On the Free Convection from a Horizontal Plate," *Z. Angew. Math. Phys. (ZAMP)*, pp. 276 – 282. http://dx.doi.org/10.1007/BF02033031

Gill, W.N., Zeh, D.W., and Casal E.D., 1965, "Free Convection on a Horizontal Plate," *Z. Angew. Math. Phys. (ZAMP)*, **16**, pp. 539 – 541. http://dx.doi.org/10.1007/BF01593934

Schlichting, H., 2000, *Boundary-Layer Theory*, 8th revised and enlarged ed., Springer-Verlag, Berlin/Heidelberg. http://dx.doi.org/10.1007/978-3-540-95998-4_2

Chamkha, A.J., 2003, "MHD Flow of a Uniform Stretched Vertical Permeable Surface In The Presence Of Heat Generation/absorption And a Chemical Reaction," *Int. Commu. Heat Mass Trans.*, **30**, pp. 413-422. http://dx.doi.org/10.1016/S0735-1933(03)00059-9

Chen, C.H., 2004, "Heat and Mass Transfer In MHD Flow by Natural Convection from a Permeable, Inclined Surface with Variable Wall Temperature and Concentration," *Acta Mech.*, **172**, pp. 219-235. http://dx.doi.org/10.1007/s00707-004-0155-5

Ahmed, S., 2010, "Effect of Viscous Dissipation and Chemical Reaction on Transient Free Convective Flow over a Vertical Porous Plate," *Journal of Energy Heat Mass Trans.*, **32**, pp. 311-332.

Kim, Y.J, 2000, "Unsteady MHD Convective Heat Transfer Past a Semi-Infinite Vertical Porous Moving Plate with Variable Suction," *Int. J. Eng. Sc.*, **38**, pp. 833-845. http://dx.doi.org/10.1016/S0020-7225(99)00063-4

Siddiqa, S., Asghar S. and Hossain M.A., 2010, "Natural Convection Flow over Inclined Flat Plate with Internal Heat Generation and Variable Viscosity," *Mathematical and Computer Modelling,*" **52**, pp. 1739-1761. http://dx.doi.org/10.1016/j.mcm.2010.07.001

Merkin, J.H., 2012, "Natural Convective Boundary Layer Flow in a Heat Generating Porous Medium with a Constant Surface Heat Flux," *European Journal of Mechanics- B/Fluids*, **36**, pp. 75-81. http://dx.doi.org/10.1016/j.euromechflu.2012.04.004

Samanta, S., Guha A., 2014, "Analysis of Heat Transfer and Stability of Magnetohydrodynamic Natural Convection above a Horizontal Plate with Heat Flux Boundary Condition," *Int. J. Heat Mass Trans.*, **70**, pp. 793-802. http://dx.doi.org/10.1016/j.ijheatmasstransfer.2013.10.049

Mahapatra, T.R., Nandi, S.K., Gupta, A.S., 2011, "Momentum and Heat Transfer in MHD Stagnation-point Flow over a Shrinking Sheet," *Journal of Applied Mechanics*, **78**, pp. 021015:1-8 http://dx.doi.org/10.1115/1.4002577

Shercliff, J.A., 1965, *A Text Book of Magneto-Hydro-Dynamics*, Pergamon, Oxford.

3

ENHANCEMENT OF THERMOELECTRIC DEVICE PERFORMANCE THROUGH INTEGRATED FLOW CHANNELS

B. V. K. Reddy, Matthew Barry, John Li, Minking K. Chyu*

Department of Mechanical Engineering and Materials Science, University of Pittsburgh, Pittsburgh, PA 15261, USA

ABSTRACT

In this study, the thermoelectric performance of an integrated thermoelectric device (iTED) with rectangular, round end slots, and circular flow channel designs applied to waste heat recovery for several hot stream flow rates has been investigated using numerical methods. An iTED is constructed with p- and n-type semiconductor materials bonded to the surfaces of an interconnector with flow channels drilled through it. This interconnector acts as an internal heat exchanger directing waste heat from the hot stream to thermoelectric elements. The quantity of heat extracted from the waste heat source and the subsequent amount of electrical power generated P_O from the iTED is increased significantly for the circular flow channels, followed by round end slots and rectangular flow channels, respectively. At Re = 100, the round end slots and the circular flow channels showed nearly 2.6 and 2.9 times increment in P_O, and 1.5 and 1.65 times in η when compared to the rectangular flow channels values. Conversely, when Re is increased from 100 to 500, the iTED with rectangular flow channels showed 2.67- and 1.6-fold improvement in P_O and η, respectively. However, the circular configurations showed 2.27- and 1.41-fold increases in P_O and η values, respectively. Within the Re range studied, the inclusion of flow channels' pumping power in η calculations showed negligible effect. For an iTED with circular flow channels, an increase in a cold side convective heat transfer coefficient h_c resulted in an enhancement in P_O and η values. Besides a h_c effect, the heat loss to the ambient via convective and radiation heat transfer exhibited an increase in P_O and decrease in η.

Keywords: *Circular, numerical model, performance, rectangular, slots, thermoelectrics, waste-heat recovery.*

1. INTRODUCTION

Over the last two decades, the increase in human population and demand of a higher quality of life has increased the energy requirements enormously. Thus, the carbon foot-print and waste heat released into the atmosphere are proliferated from both energy consuming and producing equipment. Based on the DOE report by Smith and Thornton (2009), it is approximated that two-thirds of the supplied input energy to these systems is rejected as a waste heat to the atmosphere. Scavenging even a small percentage of waste energy (5-10%) can reduce negative environmental impacts and make current equipment more economical. Hence, we are compelled to explore environmentally-benign technologies that can replace and/or enhance the performance of current energy conversion systems while mitigating the emission of greenhouse gases. Thermoelectric devices (TEDs) are one of the prominent technologies to recover the waste heat and convert it into useful electricity. TEDs work as electric power generators when the junction of two different electrically and thermally conductive materials are exposed to a temperature differential (Rowe, 2006). TEDs have no moving parts, are noise-free, scalable, reliable, and compact. The broad range of applications of these devices include temperature measurement, remote radio and satellite power generation stations, refrigeration cooling and waste heat recovery from exhaust streams and other low-grade heat sources, pocket electronics, biothermal batteries to power pacemakers, localized cooling in electronic components and automobile seats.

The maximum conversion efficiency of a TED depends on both materials' intrinsic properties (figure of merit ZT, expressed as $(\sigma\alpha^2 T)/\kappa$), here σ, α, T, and κ are the electrical conductivity, Seebeck coefficient, absolute temperature, and total thermal conductivity, respectively) and the hot and cold junction temperatures (T_h and T_c) (Rowe, 2006). Therefore, the TED efficiency can be enhanced either through material strategies (via methods of nano-structuring and fabrication (Sootsman *et al.*, 2009; Tritt, 2011; Biswas *et al.*, 2012) and use of new bulk materials (Rowe, 2006; Poudel *et al.*, 2008)) and/or through the system modeling, design, and optimisation (Caillat *et al.*, 1999; Kaibe *et al.*, 2005; Hodes, 2010; Crane *et al.*, 2012). Caillat *et al.* (1999) achieved a TED efficiency of 15% using segmented thermoelectric legs. When cascading TEDs are subjected to $T_h = 550\ ^\circ C$ and $T_c = 30\ ^\circ C$, Kaibe *et al.* (2005) reported an efficiency of 12.1%. Hodes (2010) investigated the optimum leg geometries for maximizing either thermoelectric efficiency or power output in the absence and then presence of interface contact resistances. Recently, Crane *et al.* (2012) built a full-scale cylindrical-shaped thermoelectric generator using segmented and high-power density elements and obtained a maximum power output of 608 W. They also demonstrated the validity of steady-state and transient TED models by constructing them on cylindrical gas/liquid heat exchangers. Further, Yazawa and Shakouri

*Leighton Orr Chair Professor and Chairman and Corresponding author, Email: mkchyu@pitt.edu, Ph: 412-624-9720 Fax: 412-624-4846.

(2012) studied the co-optimisation of TEDs by considering the TE leg shape, heat sink, load and contact resistances, and heat losses.

It has been demonstrated that the three-dimensional 3D thermoelectric modeling substantially influences the overall performance of TED (Harris *et al.*, 2006; Hu, 2009; Ziolkowski *et al.*, 2010; Chen *et al.*, 2011). Harris *et al.* (2006) studied the temperature and electrical field distributions in TEDs using a finite volume numerical model by taking into account both temperature-dependent material properties and contact resistances at the interfaces of the thermoelectric material and conducting shoes. Hu (2009) analysed the characteristics of two compact gas-phase heat exchangers placed on the hot and cold sides of the TED to enhance heat transfer and fluid flow while minimizing thermal stresses. Ziolkowski *et al.* (2010) investigated the performance of TEDs including the parasitic effects such as convection, radiation and conduction heat bypass. Further, accounting all temperature-dependent material properties and non-linear fluid-thermal-electric coupled mechanisms, Chen *et al.* (2011) developed a 3D numerical model for TEDs in FLUENT-UDS environment. Recently, using 3D numerical methods Reddy *et al.* (2012, 2013a) studied the enhanced heat transfer characteristics and performance of composite and integrated TEDs applied to waste heat recovery systems.

It is observed from the literature that the conventional TED designs, due to their geometrical structure, induce large amounts of thermal stresses. Furthermore, a thermal resistance present at the interfaces of the ceramic plates and the heat sinks. Keeping in mind the minimisation of contact resistances and thermal stresses, recently Reddy *et al.* (2012, 2013b) proposed an integrated thermoelectric device (iTED) where each leg is made of semiconductor slices bonded onto a highly conducting inter-connector material with flow channels which act as a heat exchanger between the flowing fluid and thermoelectric materials. It was demonstrated that such a design would increase the performance of iTED with given thermoelectric materials and geometric conditions. Furthermore, the iTED's inter-connector flow channel configuration type will play a significant role in achieving higher power output and efficiencies. Therefore, in this study, the theoretical performance of an iTED applied to waste heat recovery with various integrated flow channel designs (rectangular, round end slots and circular) and several hot fluid flow rates have been investigated. Further, the influence of cold wall convective heat transfer conditions and the heat loss to the ambient via the convective and radiation heat transfer on iTED's performance are studied.

2. GEOMETRY, GOVERNING EQUATIONS, AND SOLUTION PROCEDURE

Thermoelectric legs made of semiconductor slices bonded onto a highly conducting inter-connector material with a flow channel configuration can be treated as integrated Thermoelectric Devices (iTED). The schematic of such an iTED with a rectangular flow channel configuration is shown in Fig. 1a. It comprises of a vertical leg constructed with p- and n-type semiconductor materials and copper as a inter-connector. Leg with square cross-section W×D and height H is considered. Further, the thickness of n- and p-type semiconductor slices equal in size d is assumed. As shown in Fig. 1a, the inter-connector with rectangular flow channel grooved through it acts as a heat exchanger between n- and p-type semiconductor slices. The two dimensional view (in the direction of flow) of various flow channel configurations such as rectangular, slots with round end and circular channels are also shown in Fig. 1b. The two copper connectors placed one at the top and one at the bottom of the leg constitute two junctions. The connectors also act as terminals when the device is connected to the load circuitry of resistance R_L.

As shown in Fig. 1a, a hot fluid with constant inlet temperature T_{in} and uniform velocity U enters the main flow channel of cross-sectional area $D \times (L - 2d)$, flows through inter-connector's flow channels, and leaves at the exit of the fluid domain. The inner walls and left and right side surfaces of inter-connector are solely accountable for the heat transfer from the hot fluid to the thermoelectric (TE) elements. The top surface of the upper connector and the lower surface of the bottom connector are exposed to either constant cold temperature T_c or convective heat transfer conditions. The remaining surfaces of the iTED are subjected to either adiabatic or convective and radiative heat transfer boundary conditions while keeping the main flow channel walls adiabatic. An upstream buffer length $5W$ and a downstream buffer zone $20W$ have been considered to eliminate thermal back diffusion at the inlet and to meet an outflow boundary condition at the outlet of the flow domain, respectively. The iTED connected to a load resistance R_L delivers electrical power via the Seebeck voltage produced through a temperature difference $T_h - T_c$ between the inter-connector channel walls and the cold surface.

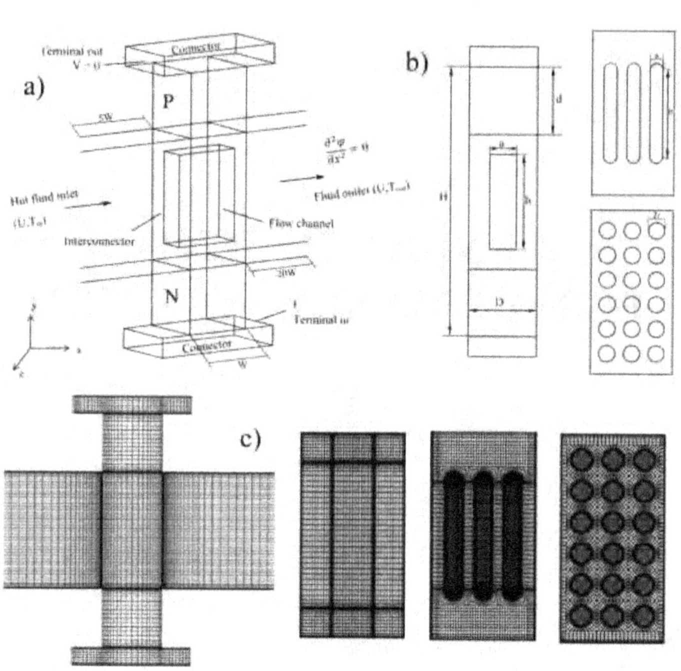

Fig. 1 Schematic of a) an integrated thermoelectric device with b) various flow channel configurations and c) their computational mesh.

The mass, momentum, and heat transport in the fluid regions and the electric current flow and heat transport in the TE materials are governed by the three-dimensional conservation equations subjected to the following assumptions: the thermophysical properties of the fluid are kept constant; the fluid flow is steady, laminar and incompressible; the materials are heterogeneous, isotropic, and the thermoelectric properties are temperature-dependent.

The governing equations for the incompressible fluid flow and heat transport in the fluid region and the thermo-electric coupling effects in the conductor and semiconductor materials under steady-state conditions are written as:

- Continuity equations:

$$\nabla \cdot \mathbf{v} = 0 \quad \text{fluid} \tag{1}$$

$$\nabla \cdot \mathbf{J} = 0 \quad \text{conductor and semiconductor} \tag{2}$$

- Momentum equation:

$$\rho_f(\mathbf{v} \cdot \nabla \mathbf{v}) = -\nabla P + \mu \nabla^2 \mathbf{v} \tag{3}$$

- Energy equations:

$$(\rho c_P)_f(\mathbf{v} \cdot \nabla T) = \nabla \cdot (k_f \nabla T) \quad \text{fluid} \tag{4}$$

$$\nabla \cdot (k \nabla T) + \rho \mathbf{J}^2 = 0 \quad \text{conductor} \tag{5}$$

$$\nabla \cdot (k\nabla T) + \rho \mathbf{J}^2 - T\mathbf{J} \cdot \left[\left(\frac{\partial \alpha}{\partial T} \right) \nabla T + (\nabla \alpha)_T \right] = 0 \quad \text{semiconductor} \quad (6)$$

In the left-hand side of Eq. (6), the second, third, and fourth terms represent the Joule heating, Thomson effect, and Peltier cooling respectively.

The total electric potential calculated using the non-ohmic current-voltage (Domenicali, 1954) relationship is written as:

$$\nabla V = \nabla V_O + \nabla V_S = -\rho \mathbf{J} - \alpha \nabla T \quad (7)$$

The first and second terms in Eq. (7), denotes the electrostatic and the Seebeck potential distributions, respectively. Here, the electrostatic potential is due to current flowing in the device whereas the Seebeck potential is due to the temperature differential created in the thermoelectric material.

The associated thermal and electrical boundary conditions for Eqs. (1) to (7) with respect to the geometry shown in Fig. 1a are, at terminal 'in': $J = \frac{V_0}{A_\xi (R_i + R_L)}$, $\frac{\partial T}{\partial \xi} = 0$; at terminal 'out': $V = 0$, $\frac{\partial T}{\partial \xi} = 0$; at the main flow channel walls: $\frac{\partial T}{\partial \xi} = 0$; at fluid flow inlet: $u = U$, $v = w = 0$, $T = T_{in}$ and at fluid flow outlet: $\frac{\partial^2 u}{\partial \xi^2} = \frac{\partial^2 T}{\partial \xi^2} = v = w = 0$. At the interface between the semiconductor and connector or inter-connector materials, the continuity of temperature, current density and the heat flux conditions are imposed (Reddy et al., 2012).

In the above paragraph, R_i is the total internal resistance due to n- and p-type slices, connectors and inter-connector materials, and R_L is the external load resistance. V_0 is the total built-in open circuit voltage of iTED at no-load condition and it is calculated as the summation of the Seebeck potential at the interfaces of the semiconductor and conductor materials. The Seebeck voltages are evaluated with reference to the hot surface temperature at the interface of the semiconductor and inter-connector materials. The total R_i and V_0 of the device are evaluated as:

$$R_i = \sum_{j=n,p,c,ic} \frac{L_j}{A_j} \left[\frac{1}{V_j} \int_{V_j} \rho_j dv_j \right] \quad \text{and}$$

$$V_0 = \sum_{j=n,p} \frac{1}{As_j} \int_{As_j} |\alpha_j| \frac{dT}{d\xi} dAs_j. \quad (8)$$

The boundary conditions at the top surface of the upper connector and the bottom surfaces of the lower connectors:

$$T = T_c \quad \text{or} \quad q'' = h_c(T - T_\infty), \quad \text{and} \quad \frac{\partial V}{\partial \xi} = 0, \quad (9)$$

and at all other iTED surfaces exposed to surroundings:

$$q'' = \frac{\partial T}{\partial \xi} = 0 \text{ or } q'' = h_s(T - T_\infty) + \epsilon \sigma(T^4 - T_{sur}^4) \text{ and } \frac{\partial V}{\partial \xi} = 0. \quad (10)$$

Here, ξ denotes the direction normal to the corresponding surface.

The pressure drop $\Delta P/L$, power output P_0, heat input Q_h at the inter-connector walls exposed to hot fluid and the thermoelectric conversion efficiency of iTED are evaluated as:

$$\frac{\Delta P}{L} = \frac{1}{25W} \left[\int_0^{H-2d} \int_0^D P dy dz |_{x=0} - \int_0^{H-2d} \int_0^D P dy dz |_{x=25W} \right],$$
$$(11)$$

$$P_0 = I^2 R_L, \quad (12)$$

$$Q_h = -\frac{1}{As} \int_{As} k_f \frac{\partial T}{\partial \xi} dAs, \quad (13)$$

and

$$\eta = \frac{P_0}{Q_h}. \quad (14)$$

Using the finite volume formulation of Eqs. (1) to (6) and the constitutive relation (Eq. 7) along with the above mentioned associated boundary conditions, the numerical simulations are performed in the FLUENT-UDS (User Defined Scalar) software. The coupling between velocity and

Table 1 Grid independence study for iTED with various flow channel configurations (grid size in bold face is chosen for further simulations) [$T_{in} = 550$ K, $T_c = 300$ K, d = 5 mm, and H = 20 mm].

Cells	P_0, W	% error in P_0	Q_h, W	% error in Q_h	η, %	% error in η
\multicolumn{7}{c}{Rectangular duct [$Re = 400$, $R_L = 4.69 \times 10^{-3}$ Ω]}
300124	0.0739		1.6977		4.3513	
633600	0.0724	1.998	1.6821	0.929	4.3057	1.060
1139152	0.0714	1.445	1.6714	0.642	4.2716	0.797
2000922	0.0720	0.859	1.6803	0.534	4.2857	0.327
\multicolumn{7}{c}{Multiple slots with rounded end [$Re = 400$, $R_L = 5.12 \times 10^{-3}$ Ω]}
598200	0.1593		2.6052		6.1140	
1175504	0.1590	0.151	2.6027	0.097	6.1107	0.054
2272752	0.1595	0.288	2.6137	0.420	6.1026	0.133
3307802	0.1583	0.739	2.6029	0.415	6.0830	0.323
\multicolumn{7}{c}{Circular ducts [$Re = 500$, $R_L = 5.29 \times 10^{-3}$ Ω]}
948960	0.2008		3.0114		6.6675	
1822976	0.1983	1.259	2.9919	0.653	6.6276	0.602
3723840	0.1978	0.230	2.9901	0.058	6.6162	0.172

pressure is handled by using the SIMPLE algorithm. The convective and diffusive terms are discretised with a power law scheme and the pressure term is handled with a standard scheme. The geometric models and mesh are generated in Gambit 2.4. The Seebeck and Ohmic potential distributions (Eq. (7)) and continuity of current density (Eq. (2)) are evaluated using UDS fields. The Ohmic heating, Peltier and Thomson effects are modeled as source terms in the energy equation (Eq. (6)). The electric current is calculated based on the open-circuit Seebeck voltage produced at a given load resistance R_L. Additional details on implementation of numerical model for TED in the FLUENT-UDS environment has been given in the articles (Chen et al., 2011; Reddy et al., 2012). The convergence criteria for mass and momentum, energy, and current density, Seebeck and Ohmic electric potentials are set as 10^{-5}, 10^{-15}, and 10^{-10}, respectively. After performing a grid independence study, suitable grid sizes were chosen for various integrated flow channel configurations and these are depicted in Table 1. The orthogonal, nonuniform grid is used, and for brevity, the computational mesh for an iTED along the front and side views is shown in Fig. 1c. Furthermore, the thermoelectric model has been validated with published results and the details are given in our earlier works (Reddy et al., 2012, 2013a).

3. RESULTS AND DISCUSSION

The performance of an integrated thermoelectric device (iTED) for various flow channel configurations in terms of power output P_0, heat input Q_h, conversion efficiency η, electric current I, Seebeck and Ohmic voltages V, and pressure drop $\Delta P/L$ with several hot fluid flow rates ($50 \le Re \le 500$) has been investigated using numerical methods. The rectangular, round end slots, and circular flow channels within the interconnector acting as an internal heat exchanger for the hot fluid flow are considered in the present study. Further, the effects of cold surface boundary condition (constant temperature and convective heat transfer conditions) on the performance of iTED with circular flow channel configuration alone has been studied when the remaining surfaces, except the flow channel walls, exposed to either adiabatic or convective and radiation conditions. The iTED made of n- (75% Bi_2Te 25% Bi_2Se) and p-type (25% Bi_2Te_3 75% Sb_2Te_3 (1.75% excess Se)) semiconductor materials, copper as connector and interconnector, and the air as hot fluid have been considered. The thermo-electrical properties of the semiconductor and conductor materials are varied with temperature (Reddy et al., 2012), and the thermo-physical properties of air are kept constant. Fur-

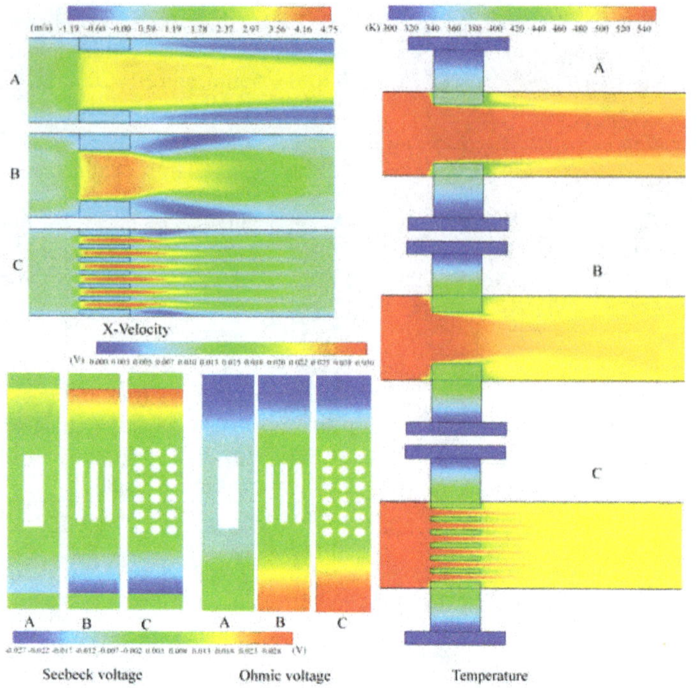

Fig. 2 The distributions of x-velocity, temperature, Seebeck and Ohmic voltage for various integrated flow channel configurations A) rectangular B) slots with round end and C) circular.

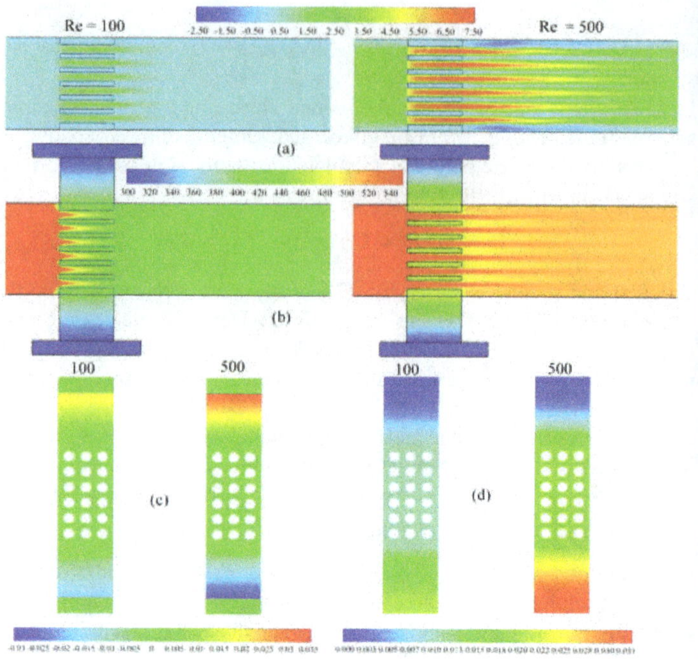

Fig. 3 The contours of a) x-velocity b) temperature c) Seebeck and d) Ohmic voltages for integrated circular flow channels at $Re = 100$ and 500 values.

thermore, to achieve maximum P_0 from an iTED, the load resistance R_L equaling to the total internal resistance R_i of the device (Eq. (8)) is used in the numerical simulations.

The geometrical dimensions: height $H = 20$ mm, depth $D = 5$ mm, connector thickness $t = 1.5$ mm and length $L = 10$ mm, the semiconductor and inter-connector width $W = 5$ mm, and semiconductor slice thickness $d = 5$ mm are maintained constant in the reported results. The hot fluid inlet temperature $T_{in} = 550K$ and cold wall temperature $T_c = 300K$ are invariant. Further, the ratio of flow cross-sectional areas of an integrated fluid flow channels and the main flow channel is kept fixed and it is expressed as

$$\phi_{FA} = \frac{A_c}{A} \tag{15}$$

Here, A and A_c are the cross-sectional areas of main flow channel $D \times (H - 2 \times d)$ and the integrated flow channels, respectively. For various flow channel configurations as shown in Fig. 1b, the A_c is evaluated as

$$A_c = \begin{cases} a \times b & \text{Rectangular} \\ n_s\left[a_s \times b_s + \frac{1}{4}\pi a_s^2\right] & \text{Slots with round end} \\ n_c \pi r^2 & \text{Circular.} \end{cases} \tag{16}$$

In Eq. (16), n_s, n_c, and r denote the number of slots, number of circles, and the circle radius, respectively. In the present study, $n_s = 3$ and $n_c = 6 \times 3$ are used. While maintaining $\phi_{FA} = 0.283$ invariant, the dimensions of the various flow channel designs are calculated using Eq. (16).

X-velocity, temperature, and Seebeck and Ohmic voltage distributions for integrated rectangular, round end slots, and circular flow channel configurations with hot fluid flow rate $Re = 300$ are shown in Fig. 2, whereas the same variable contours for circular flow channels alone with two Re values are depicted in Fig. 3. The x-velocity and temperature distributions are depicted on z-plane sectioned at 2.5 mm while the Seebeck and Ohmic voltage contours are shown on x-plane at 27.5 mm. Further, to show velocity and temperature contours in Figs. 2 and 3, an up-stream and down-stream lengths of W and $5W$, respectively have been taken. At $\phi_{FA} = 0.283$ and a given Re value, it is observed that the flow channel type has a substantial effect on fluid flow, heat transfer, and electric characteristics of iTED. As seen from Fig. 2, in the integrated flow channels, the fluid's average velocity increases and the bulk temperature decreases as the channel configuration varies from rectangular to circular and this subsequently resulted in enhanced heat transfer to the thermoelectric elements. The reason for enhanced heat transfer is due to the increase in the flow channel's surface area with decrease in size of the flow channels. For instance, for the flow configurations studied here, the surface area of round end slots and circular flow channels are 2.16 and 3.12 times higher when compared to the rectangular flow channels, respectively. Furthermore, as shown in Fig. 3 for the circular flow channels, an increment in temperature gradients near the flow channel walls is resulted in an increase in Re from 100 to 500. This increase in temperature gradients promotes higher convective heat transfer rates and thus in turn results in a reduction of iTED's total thermal resistance value. The recirculation zones and the flow patterns formed in the down stream of integrated flow channels have negligible influence on the thermo-electric performance of iTED, however they exhibit considerable effect on the total pressure drop experienced in their respective channels.

It is seen that an increase in heat transfer area of the flow channels either through round end slots or circular channels as well as the increase in Re resulted in larger temperature gradients in the thermoelectric (TE) elements. As shown in Figs. 2 and 3, this increase in temperature gradients generates higher Seebeck voltages in the TE elements. In each TE element, the Seebeck voltage drop is calculated with respect to the copper material. However, the Ohmic voltage distribution can be obtained through the produced electric current I when the iTED terminals are connected to load circuitry of resistance R_L. At $Re = 300$, the 2.16 and 3.12 times increase in heat transfer area of slots with round end and circular flow channels, respectively, results in 1.6 and 1.74 times increase in Seebeck voltage values when compared to the rectangular flow channel's value. Furthermore, as the flow channel configuration changes from rectangular to circular, the Ohmic voltage drop increased from 0.0168 to 0.0299 V; on the other hand, as Re increased from 100 to 500 for the circular flow channel, the Ohmic voltage drop raised from 0.0206 to 0.0333. It is also noticed that with $R_L = R_i$ condition, the Seebeck co-

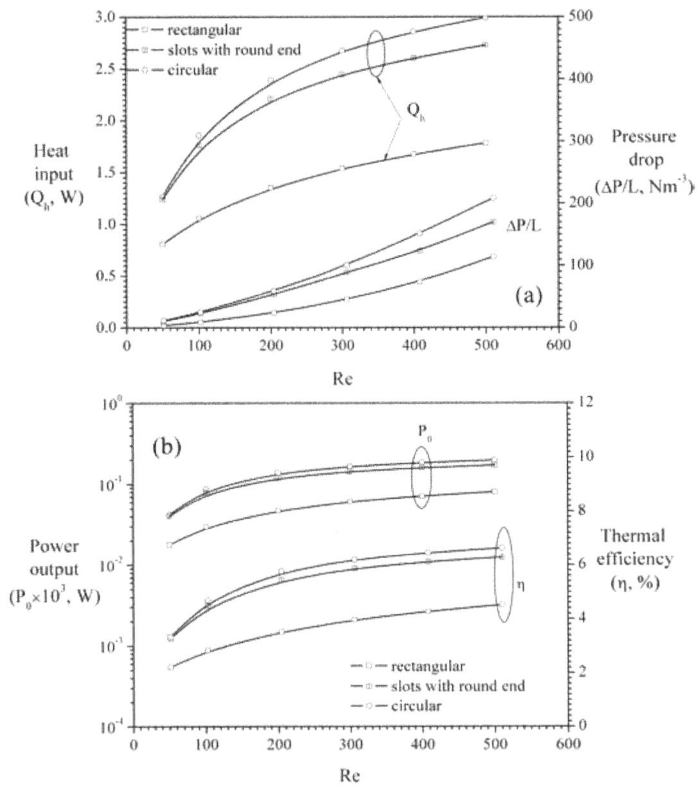

Fig. 4 Effects of Re on a) heat input and pressure drop and b) power output and thermal efficiency for flow channel configurations.

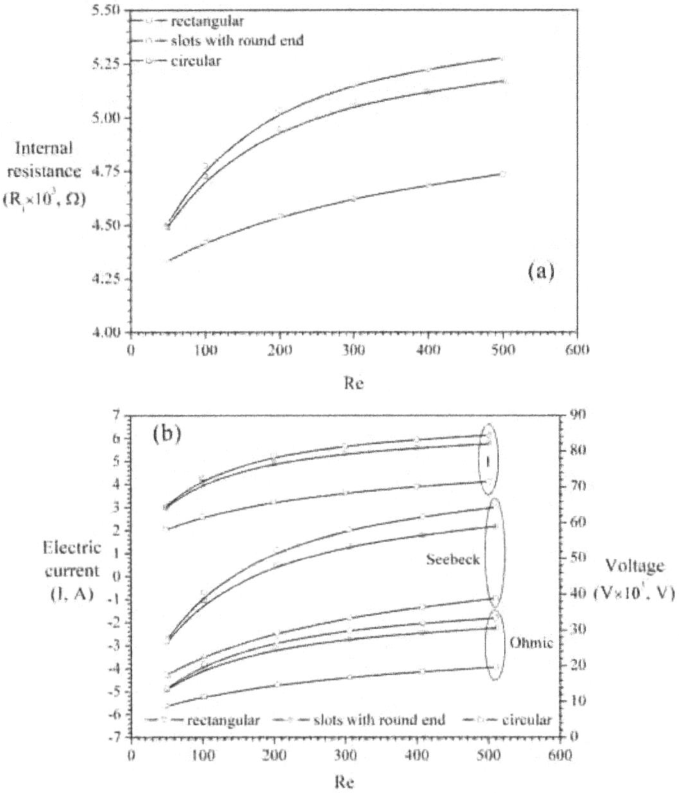

Fig. 5 Effects of Re on a) internal resistance and b) electric current, Seebeck and Ohmic voltages for flow channel configurations.

efficients and temperature gradients are only responsible for the Ohmic voltage drop variations and it is independent of R_i values.

The response of total internal resistance R_i and produced electric current I and Seebeck and Ohmic voltages V with Re for various integrated flow channel configurations is presented in Figs. 4a and b, respectively. The rise in temperature gradients in thermoelectric elements with an increase in Re produces larger Seebeck voltages and thus Ohmic voltages (as seen in Fig. 4a). Irrespective of Re value, the circular flow channels showed higher V-I values, followed by round end slots and rectangular flow channels, respectively. It is noticed that I and V exhibited similar trend with Re. At a given Re, the Ohmic voltage drop is higher than the Seebeck voltage drop value. Furthermore, from Fig. 4b, the Re in the specified range $50 \leq Re \leq 500$ has minimal effect on R_i predictions and this is attributed to variation of electrical resistivity behaviour with temperature. However, at a given Re value, a minor increment in R_i is noticed when the iTED's flow channels are varied from rectangular to circular flow configurations.

Figures 5a and b show the results of heat input Q_h and pressure drop $\Delta P/L$, and power output P_O and thermal efficiency η, respectively with Re for different integrated flow channel configurations. Irrespective of the flow channel design type, an increase in Re shows an enhancement in Q_h, $\Delta P/L$, P_O and η predictions. The increase in Re promotes higher heat transfer rate from the hot fluid to TE elements Q_h and this subsequently assists in achieving larger Seebeck voltages (Fig. 4b) and thus results in higher P_O and η (as seen in Fig. 5b) values. For instance, at $Re = 500$ the rectangular flow channel configuration showed 1.68-, 2.6-, and 1.6 times increment in Q_h, P_O and η, respectively when compared to the values at $Re = 100$. Conversely, the round end slots and circular flow channel configurations produced 1.55-, 2.17-, and 1.4 fold, and 1.61-, 2.27-, and 1.41 times increase Q_h, P_O and η, respectively at $Re = 500$ in comparison to values at $Re = 100$. For a given integrated flow channel configuration, the pressure drop $\Delta P/L$ measured using the Eq. (11) is shown a non-linear behaviour with Re and it is due to the inclusion of additional pressure drop anticipated from the recirculation zones and flow patterns formation in the down-stream and up-stream of integrated flow channels. At given Re value, the circular flow channel configuration displayed higher $\Delta P/L$, and the rectangular flow channel configuration showed the lower $\Delta P/L$ value. It is observed that when Re is less than 200, it has significant effect on device characteristics. It is due to the mutual interplay of temperature gradients and the thermo-electrical property dependency with temperature. It can also noticed that within the Re range, the consideration of pumping power in the η calculations has negligible effect.

It is inferred that the iTED with circular flow channel configuration always performs better when compared to other two configurations, i.e. round end slots and rectangular channels. Further, at given Re values, the discrepancy between the performance parameters of circular and round end slots flow channels is less when compared to the discrepancy between the circular and rectangular flow channel configurations values. However, this discrepancy increases minimally with an increase in Re. It is also surmised that the number of times increase in integrated heat transfer surface area of channels via round end slots and circular flow channels is not proportional to the increase in the performance parameters.

Figure 6 depicts the effects of cold side convective heat transfer coefficient h_c ($10 \leq h_c \leq 10^3$) on Q_h, R_i, P_O, η, I and V for an iTED with circular flow channel configuration. In Fig. 6, the data with filled bullets represents when the iTED walls (except the cold walls and the main flow channel walls) are subjected to adiabatic condition ($h_s = 0$). For given hot fluid flow rate Re = 500 and cold fluid temperature T_∞ = 300 K, A rise in h_c shows an improvement in Q_h, P_O, η, I and V values. However, the R_i drops as h_c increases. It is noticed that as h_c increases the cooling heat transfer rate also increases. Therefore, to achieve energy conservation, the heat transfer from hot fluid to iTED also increases

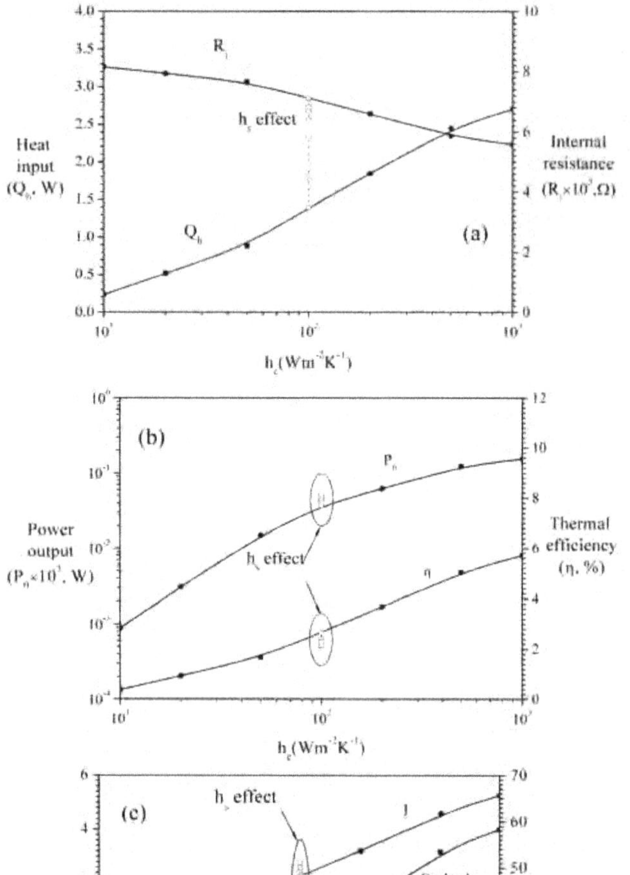

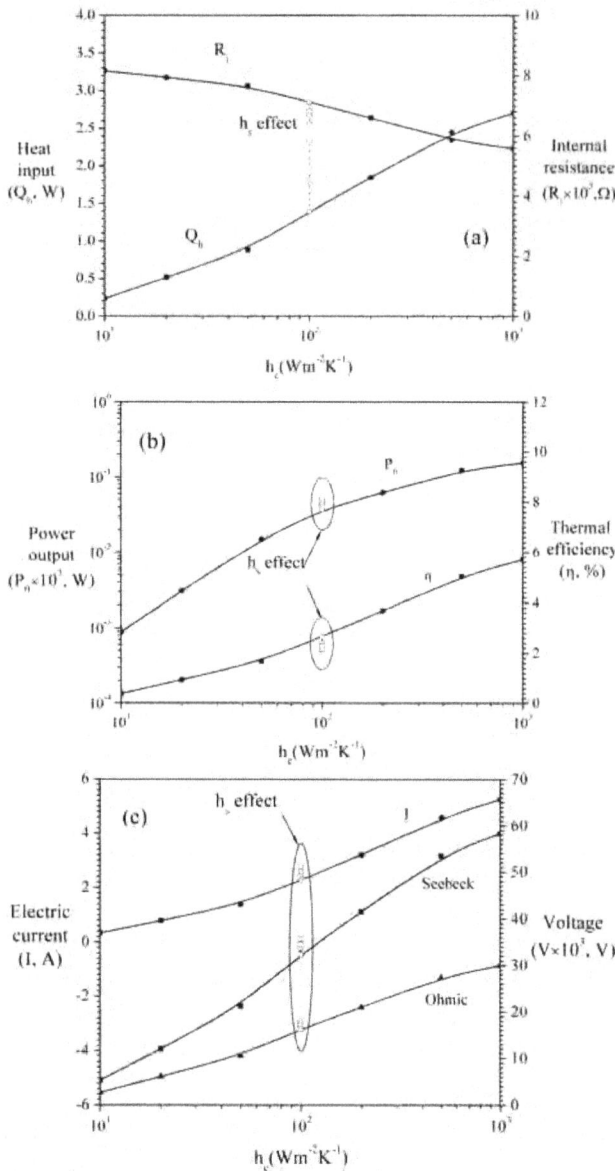

Fig. 6 The influence of cold side convective heat transfer coefficient h_c on a) heat input and internal resistance b) power output and thermal efficiency and c) electric current, Seebeck, and Ohmic voltages.

and this results in higher temperature gradients in TE elements. Further, this increase in temperature gradients showed an enhancement in iTED performance parameters. It is also shown in Fig. 6 the influence of convective and radiation heat transfer h_s (0, 5, 10, and 20 $Wm^{-2}K^{-1}$) on iTED performance at h_c = 10 $Wm^{-2}K^{-1}$ (the data with empty bullets). Here, the surrounding temperature T_{sur} = 300 K is fixed and the emissivities for the semiconductor and the highly polished copper conductor are taken as 0.45 and 0.03, respectively. It is observed that the rise in P_O and Q_h and decrease in η with an increase in h_s values. The heat loss from the walls exposed to ambient resulted in larger Q_h predictions; however, the decrease in R_i and larger temperature gradients showed enhancement in P_O values.

4. CONCLUSIONS

The performance of an integrated thermoelectric device (iTED) with various flow channel (rectangular, round end slots, and circular) configurations acting as internal heat exchangers between hot fluid and thermoelectric elements for different flow flow rates Re has been investigated using numerical simulations. Further, the influence of cold wall convective heat transfer conditions and the heat loss to the ambient via the convective and radiation heat transfer on iTED's performance with circular flow channel configuration has been studied.

It is observed that the integrated flow channel configuration type and Re have a substantial effect on iTED's performance. At a given flow channel cross-sectional areas ratio $\phi_{FA} = 0.283$, irrespective of Re value, an iTED with circular flow channel configuration showed higher power output P_O, efficiency η, and pressure drop values, followed by round end slots and then rectangular configurations, respectively. This behaviour is attributed to increase in flow channels' surface area with decrease in channel size.

The 2.16 and 3.12 fold increase in heat transfer surface area of the round end slots and the circular flow channels, respectively resulted in 1.6 and 1.74 times increase in Seebeck voltages, 2.36 and 2.72 times in P_O, and 1.48 and 1.56 times in η values when compared to the rectangular flow channel configuration values at $Re = 300$ with inlet temperature $T_{in} = 550K$ and cold surface temperature $T_c = 300K$ conditions.

When Re increased from 100 to 500, the rectangular flow channel configuration showed 2.67- and 1.6-times improvement in P_O and η, respectively. However, the round end slots and circular flow channels depicted 2.17- and 1.4 fold, and 2.27- and 1.41-fold increase in P_O and η values. For given Re and flow channel configuration, the produced current and Seebeck and Ohmic voltages exhibited similar behaviour as that of P_O and η. Due to mutual interplay of temperature gradients in thermoelectric material and the temperature dependency of the thermo-physical properties, the discrepancy in the performance of circular and slots with round end flow channels is less when compared to the discrepancy values between the circular and rectangular flow channel configurations.

For an iTED with circular flow channel configuration when the walls except the cold walls and the main flow channel walls are exposed to adiabatic conditions (h_s = 0), an increase in h_c resulted in enhancement in heat output Q_h, P_O, and η values. In addition to h_c effect, the rise in h_s (heat loss to ambient) showed an increase in P_O and Q_h and decrease in η values.

It has shown that under given set of operating conditions, the iTED with circular flow configuration performed better when compared to round end slots and rectangular flow channel configurations. Furthermore, within the Re range studied, the consideration of flow channel pumping power in η evaluations has showed negligible effect.

NOMENCLATURE

A	cross-sectional area (mm^2)
A_s	surface area (mm^2)
c_P	specific heat of fluid ($J/kg \cdot K$)
d	size of semiconductor material (mm)
D	depth of the thermoelectric leg (mm)
D_h	hydraulic diameter of main flow channel $\left(\frac{4(L-2d)\times D}{2(L-2d+D)}\right)$ (mm)
h	convective heat transfer coefficient ($W/m^2 \cdot K$)
H	height of the leg (mm)
I	electric current (A)
J	electric current density (A/m^2)
$\mathbf{J}$	current density vector
k	thermal conductivity ($W/m \cdot K$)
L	length (mm)
N	number of thermoelectric modules
P	pressure (N/m^2)
P_O	power output $I^2 R_L$ (W)
q''	heat flux (W/m^2)
Q	heat transfer (W)
R	electric resistance (Ω)
Re	Reynolds number $\rho U D_h/\mu$

T	temperature (K)
U	inlet velocity (m/s)
$\mathbf{v}$	velocity vector
V	voltage (V)
u, v, w	velocities in x,y,z directions (m/s)
W	width (mm)
x, y, z	coordinates (mm)
Greek Symbols	
α	Seebeck coefficient (V/K)
μ	dynamic viscosity ($N \cdot s/m^2$)
ρ	electrical resistivity ($\Omega \cdot m$)
ρ_f	density of fluid (kg/m^3)
η	conversion efficiency
Subscripts	
c	cold wall/conductor
in	inlet
f	fluid
h	hot wall
i	internal/integrated
ic	inter-connector
L	load
n	n-type semiconductor
O	Ohmic potential
p	p-type semiconductor
S	Seebeck potential
surr	surroundings
∞	ambient

REFERENCES

Biswas, K., He, J., Blum, I.D., Wu, C., Hogan, T.P., Seidman, D.N., Dravid, V.P., and Kanatzidis, M.G., 2012, "High-Performance Bulk Thermoelectrics with All-Scale Hierarchical Architectures," *Nature*, **489**, 414–418.
http://dx.doi.org/10.1038/nature11439.

Caillat, T., Fleurial, J.P., Snyder, G.J., Zoltan, A., Zoltan, D., and Borshchevsky, A., 1999, "Development of a High Efficiency Thermoelectric Unicouple for Power Generation Applications," *Proceedings of the XVIII International Conference on Thermoelectrics*, Baltimore, USA.
http://trs-new.jpl.nasa.gov/dspace/bitstream/2014/18364/1/99-1841.pdf.

Chen, M., Rosendahl, L.A., and Condra, T., 2011, "A Three-Dimensional Numerical Model of Thermoelectric Generators in Fluid Power Systems," *International Journal of Heat and Mass Transfer*, **54**, 345–355.
http://dx.doi.org/10.1016/j.ijheatmasstransfer.2010.08.024.

Crane, D.T., Koripella, C.R., and Jovovic, V., 2012, "Validating Steady-State and Transient Modeling Tools for High-Power-Density Thermoelectric Generators," *Journal of E*, **41**(6), 1524–1534.
http://dx.doi.org/10.1007/s11664-012-1955-3.

Domenicali, C.A., 1954, "Stationary Temperature Distribution in an Electrically Heated Conductor," *Journal of Applied Physics*, **25**, 1310–1311.
http://dx.doi.org/10.1063/1.1721551.

Harris, R., Hogan, T., Schock, H.J., and Shih, T.I.P., 2006, "Heat Transfer and Electric Current Flow in a Thermoelectric Couple," *44th Aerospace Sciences Meeting and Exhibition*, 575, AIAA, Reno, Nevada.
http://dx.doi.org/10.2514/6.2006-575.

Hodes, M., 2010, "Optimal Pellet Geometries for Thermoelectric Power Generation," *IEEE Trans Components and Packaging Technologies*, **33**, 307–318.
http://dx.doi.org/10.1109/TCAPT.2009.2039934
.

Hu, K.S.Y., 2009, *Heat Transfer Enhancement in Thermoelectric Power Generation*, M.s. thesis, Iowa State University.
http://lib.dr.iastate.edu/etd/12196.

Kaibe, H., Aoyama, I., Mukoujima, M., Kanda, T., Fujimoto, S., Kurosawa, T., Ishimabushi, H., Ishida, K., Rauscher, L., Hata, Y., and Seijirou, 2005, "Development of Thermoelectric Generating Stacked Modules Aiming for 15% of Conversion Efficiency," *Int. Conference on Thermoelectrics*, 227–232, IEEE.
http://dx.doi.org/10.1109/ICT.2005.1519929.

Poudel, B., Hao, Q., Ma, Y., Lan, Y., Minnich, A., Yu, B., Yan, X., Wang, D., Muto, A., Vashaee, D., Chen, X., Liu, J., Dresselhaus, M.S., Chen, G., and Ren, Z., 2008, "High-Thermoelectric Performance of Nanostructured Bismuth Antimony Telluride Bulk Alloys," *Science*, **320**(5876), 634–638.
http://dx.doi.org/10.1126/science.1156446.

Reddy, B.V.K., Barry, M., Li, J., and Chyu, M.K., 2012, "Three-Dimensional Multiphysics Coupled Field Analysis of an Integrated Thermoelectric Device," *Numerical Heat Transfer Part A*, **62**, 933–947.
http://dx.doi.org/10.1080/10407782.2012.715988.

Reddy, B.V.K., Barry, M., Li, J., and Chyu, M.K., 2013a, "Mathematical Modeling and Numerical Characterisation of Composite Thermoelectric Devices," *International Journal of Thermal Sciences*, **67**, 53–63.
http://dx.doi.org/10.1016/j.ijthermalsci.2012.11.004.

Reddy, B.V.K., Barry, M., Li, J., and Chyu, M.K., 2013b, "Thermoelectric Performance of Novel Composite and Integrated Devices Applied to Waste Heat Recovery," *ASME J Heat Transfer*, **135**(3), 031706 (1–11).
http://dx.doi.org/10.1115/1.4007892.

Rowe, D.M., editor, 2006, *Thermoelectrics Handbook Macro to Nano*, CRC Press, Taylor & Francis Group, Boca Raton.

Smith, K., and Thornton, M., 2009, "Feasibility of Thermoelectrics for Waste Heat Recovery in Conventional Vehicles," Tech. rep., U.S. Department of Energy.
www.nrel.gov/docs/fy09osti/44247.pdf.

Sootsman, J.R., Chung, D.K., and Kanatzidis, M.G., 2009, "New and Old Concepts in Thermoelectric Materials," *Angew Chem Int Ed*, **48**, 8616–8639.
http://dx.doi.org/10.1002/anie.200900598.

Tritt, T.M., 2011, "Thermoelectric Phenomena, Materials, and Applications," *The Annual Review of Materials Research*, **41**, 433–448.
http://dx.doi.org/10.1146/annurev-matsci-062910-100453.

Yazawa, K., and Shakouri, A., 2012, "Cost-Effective Waste Heat Recovery Using Thermoelectric Systems," *J Mater Res*, **27**(9), 1211–1217.
http://dx.doi.org/10.1557/jmr.2012.79.

Ziolkowski, P., Poinas, P., Leszczynski, J., Karpinski, G., and Muller, E., 2010, "Estimation of Thermoelectric Generator Performance by Finite Element Modeling," *Journal of Electronic Materails*, **39**(9), 1934–1943.
http://dx.doi.org/10.1007/s11664-009-1048-0.

4

ENHANCEMENT OF FRESH WATER PRODUCTION ON TRIANGULAR PYRAMID SOLAR STILL USING PHASE CHANGE MATERIAL AS STORAGE MATERIAL

Ravishankar Sathyamurthy[a,*], P.K. Nagarajan[b], Hyacinth J. Kennady[a], T.S. Ravikumar[a], V. Paulson[c], Amimul Ahsan[d]

[a] *Department of Mechanical Engineering, Hindustan Institute of Technology and Science, Chennai, Tamil Nadu, 603103, India*
[b] *Department of Mechanical Engineering, S.A. Engineering College, Chennai, Tamil Nadu, India*
[c] *Department of Aeronautical Engineering, Hindustan Institute of Technology and Science, Chennai, Tamil Nadu, 603103, India*
[d] *Department of Civil Engineering, Green Engineering & Sustainable Technology Lab, Institute of Advanced Technology, University Putra Malaysia, Selangor, Malaysia*

ABSTRACT

This paper presents the method of improvement of enhancing the performance of triangular pyramid solar still with and without latent heat energy storage. For comparing the productivity of solar still with and without LHTESS a solar still is designed, fabricated. Experiments are conducted in hot and humid climate of Chennai, India. Paraffin wax is used as LHTESS due to its feasible general and economic properties. The hourly productivity is slightly higher in case of solar still without LHTESS during sunny days. There is an increase of about 35% in production of fresh water with LHTESS than that of solar still without LHTESS. Also it was found that during the off shine period the fresh water produced from the still is higher. The solar still with and without LHTESS were found to be 4.5 L/m^{2}day and 3.5 L/m^{2}day during summer and in the winter the productivity was found to be 3.4 L/m^{2}day and 2.3 L/m^{2}day for the still with and without LHTESS respectively.

Keywords: *Enhancement, Phase change material, hourly variation, Efficiency.*

1. INTRODUCTION

Solar energy is the earliest source of energy, inexhaustible and non-pollutant in nature, solar distillation can provide an alternative source to generate clean water. Solar distillation exhibits a considerable economic advantage over other salt water distillation processes because of its use of free energy and its insignificant operating costs. This process removes salt impurities. Solar stills suffer from low efficiencies due to loss of heat of condensation to the surroundings from the glass cover. In order to improve the productivity, it is planned to incorporate partial thermal energy storage (Farell et al., 2006; Tabrizi et al., 2010).

The thermal energy storage system has become an important issue in the global energetic scene and has been widely used to increase energetic efficiency of different applications (El-Sebaii et al., 2009). These systems may be either sensible or latent heat systems. This method utilizes the heat dissipated from the bottom of the still. The latent heat storage system has many advantages over sensible heat storage systems including a large energy storage capacity per unit volume and almost constant temperature for charging and discharging. Here, phase change materials will act as a thermal energy storage medium. It is a substance with a high heat of fusion which, melting and solidifying at a certain temperature, is capable of storing and releasing large amount of energy. It changes its phase by absorbing latent heat during sun shine hours and it discharges the stored energy which is suitable for distillation purpose during off sunshine hours (Jinjia Wei at al, 2005) (Naim, 2002). With a thin layer of PCM under the basin liner of a solar still, a considerable amount of heat will be stored within PCM during sunshine hours instead of wasting it to surroundings. During freezing of PCM, the stored heat discharges to keep the basin water at a temperature enough to produce fresh water during night even thin layers of basin water. This causes enhancement of still productivity especially during night period (Demirbas, 2006).

Tabrizi et al. (2010) investigated the effect of built-in latent heat thermal energy storage on a weir-type cascade solar still. The hourly production of fresh water of solar still with PCM is relatively low when compared to solar still without PCM on summer. It has been reported that the solar still without PCM is well suited for sunny days and still with PCM was suited for cloudy conditions. (Tabrizi et al., 2010) experimentally investigated the effect of water flow on internal heat and mass transfer on a weir cascade solar still. It has been reported that the increase in mass flow on the basin of weir there will be a decrease in fresh water production. Results show that an accumulated yield of 7.4 and 4.3 kg/m^2 were achieved for minimum and maximum flow rates respectively. A triangular pyramid solar still with built-in latent heat thermal energy storage system was fabricated to improve the still productivity by (RaviShankar et al., 2013).

It was been found that the solar still with PCM and without PCM were found to be 4.2 kg/m^2.Another still with the same characteristics without PCM was also constructed for investigation of the internal convective heat transfer coefficient. In this paper an attempt has been made to find the use of paraffin wax as a heat storage material. Paraffin wax acts as a reliable, less expensive and non-corrosive heat storage medium (Naim, 2002).

* *Corresponding author. Email: raviannauniv23@gmail.com*

2. EXPERIMENTATION

Fig. 1 shows the experimental setup of triangular pyramid solar distiller. Experiments were carried out from 7 am- 6 pm. The plastic storage tank of capacity 50 l was used in order to avoid corrosion. Water from the storage tank enters the still through flexible hoses and a valve V, to maintain constant water level in the still. The valve 'V' can control the mass flow rate. Poly vinyl chloride (PVC) hoses were used for greater flexibility. The black painted still basin was placed inside the wooden box at a predetermined height. The area below the basin was filled with saw dust for insulation purpose. A small glass piece obstruction was fixed on the inside surface of the glass cover, to facilitate the deflection of the condensate return in to the collection channel, which in turn affixed with the still. The gliding water from the channel was transferred in to the measuring jar through the flexible piping. A heat reservoir is integrated with the still and filled by a phase change materials (PCM) that acts as a latent heat storage subsystem (LHTESS). Paraffin wax was selected as a LHTESS due to its thermal storage, safety, reliability and low cost. During the sunshine, when the absorber temperature is higher than the temperature of PCM, the heat is transferred to PCM and charging process is started to store solar energy as a sensible heat till PCM reaches its melting temperature. Additional charging heat is stored as the latent heat during the melting process. When the absorber temperature is lower than PCM (after sunset), reverse process is occurred (discharging process) till the PCM layer is fully solidified. Table 1 physical property of paraffin wax which is used as heat recovery. Calibrated NiCr-Ni thermocouples connected to a FLUKE 73 digital multimeter with accuracy 0.5°C were used to measure the temperatures of the still elements, e.g. basin liner T_b, basin water T_w and the inner T_{gi} and outer T_{go} surfaces of the still cover every half an hour starting at 8:00 am until sunset. The experiments were continued until 8:00 am of the next day. The yield from the still was collected and measured every 1 h. The productivity of the still during the night period was also collected and measured using a measuring jar. The ambient temperature had been also measured. The horizontal global solar radiation was measured using an Epply EPSP pyranometer coupled to an Epply instantaneous solar radiation meter with sensitivity of 8.79×10^{-6} (V/Wm^2) and accuracy better than 5% in the range from 0 to 2000 W/m^2. The wind speed V was also measured using the Environmental Products Anemometer.

The thermocouples are used to measure the temperature of the glass plate, absorber plate, and water and phase change materials temperatures. Also the quantity of distilled water was measured. This experimental setup was designed, installed and tested at Hindustan Institute of Technology and Science, Chennai, Tamil Nadu, India.

Table 1. Thermo physical properties of paraffin wax (Haji-Sheikh, 1982)

Property	Value
Melting temperature (°C)	40-60
Specific heat of solid/liquid (kJ/kg °C)	2.95/2.51
Density of solid/liquid (kg/m^3)	818/760
Thermal conductivity of solid/liquid (W/m °C)	0.24/0.24
Heat of fusion (kJ/kg)	226

Table 2. Relevant parameters for calculation (Wong, 1977) and (Duffie, 1991)

Parameter	Value	Parameter	Value
A_b (m^2)	1.0	α_g	0.05
τ_g	0.90	k_s(W/m^{2}K)	0.059
τ_w	0.95	x_s (m)	0.05
α_b	0.90	C_w (J/kgK)	4190
α_w	0.05	k_g (W/m^{2}K)	1.05
σ (W/m^{2}K^4)	5.66×10^{-8}	x_g (m)	0.003
V m/s	0-20	k_w(W/m^{2}K)	0.628

Fig. 1 Pyramid type distiller

3. ENERGY BALANCE

The energy received by the saline water in the still I (t) solar radiation and Q_{cb-w} convective heat transfer between basin and water are equal to the summation of energy lost by Q_{cw-g} convective heat transfer between water and glass, Q_{rw-g} radiative heat transfer between water and glass, Q_{ew-g} evaporative heat transfer between water and glass and energy gained by the saline water: The remaining is by evaporation, due to partial vapour pressure difference between the water surface and lower surface of the glass cover. Water evaporated condenses at the distillate collector through the glass cover. A small part of heat is lost to atmosphere through basin bottom and side wall by conduction and convection. For shallow basin still, the basin bottom surface and water are assumed as single element (Prakash, 1986) and the temperature is taken as constant for basin and water. Raw water is continuously supplied to the basin to keep the water mass in the basin always constant. This compensating water mass takes sensible heat to attain equilibrium with basin water.

4. RESULTS AND DISCUSSIONS

Fig. 2 and 3 show the variation of solar intensity from which it is observed to be maximum at the mid-noon. The observation shows that the solar radiation profile seems to be the same on the experimental days. Fig. 4 shows the hourly variation of experimental and theoretical temperatures of basin of a pyramid type solar still. The theoretical and experimental values are found to be very close during the morning and reach the maximum of 82°C. The RMS error for basin temperature is found to be 12% between theoretical and experimental values and these errors are minimized by proper insulation and higher wind velocity over the still which reduces the heat loss from water to ambient conditions.

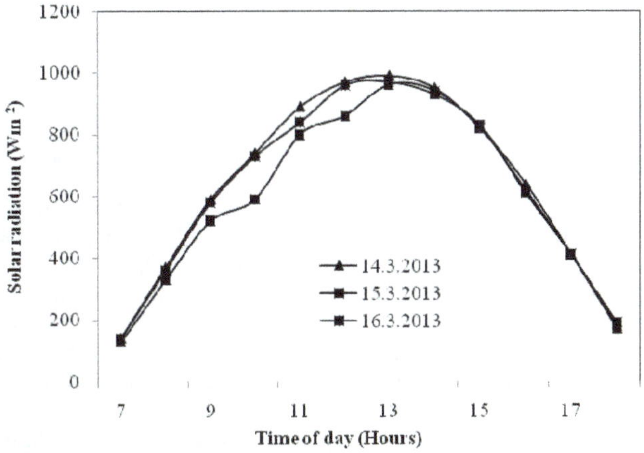

Fig. 2 Variation of solar radiation on summer day on Chennai climatic condition

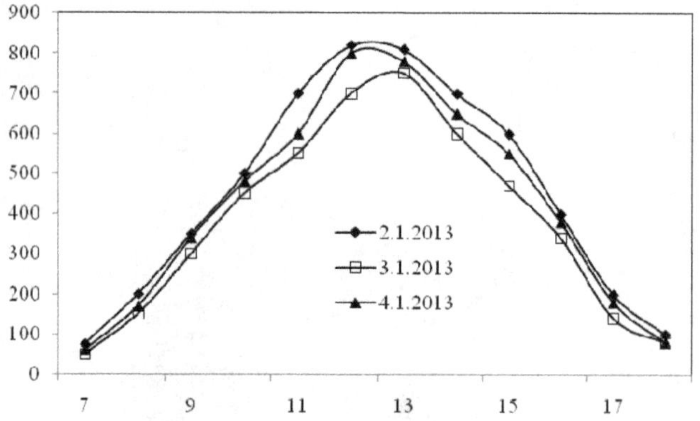

Fig. 3 Variation of solar radiation on winter day on Chennai climatic condition

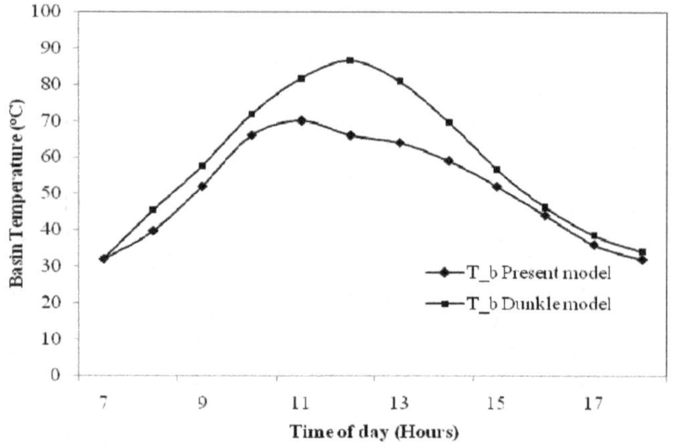

Fig. 4 Theoritical and experimental absorber plate temperature with time

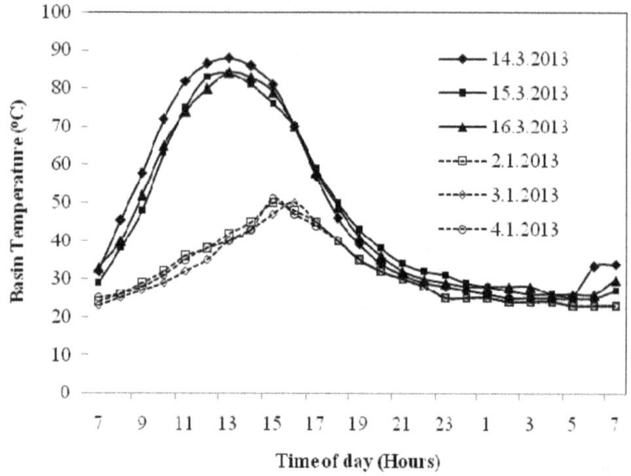

Fig. 5 Variations of absorber plate temperature without PCM

Fig. 5 and Fig. 6 show the variation of absorber plate with respect to time of solar still with and without use of LHTESS. The absorber plate temperature increases gradually with increase in solar intensity and have peak around 1 pm. The maximum obtained values for plate are 90 °C for still without LHTESS. Similarly for solar still with LHTESS, maximum obtained values for T_b is found to be 65°C. Temperature of plate is higher for still with LHTESS, due to the fact that some of energy was observed as phase change energy. It is cleared that in the early hours of the day the plate temperature is slightly higher than water temperature because in that period's glass is directly faces the solar radiation and its temperature rises faster in comparison with water temperature. Then, the increase in water temperature is faster in comparison with glass plate temp due to higher heat losses from the glass plate to the ambient.

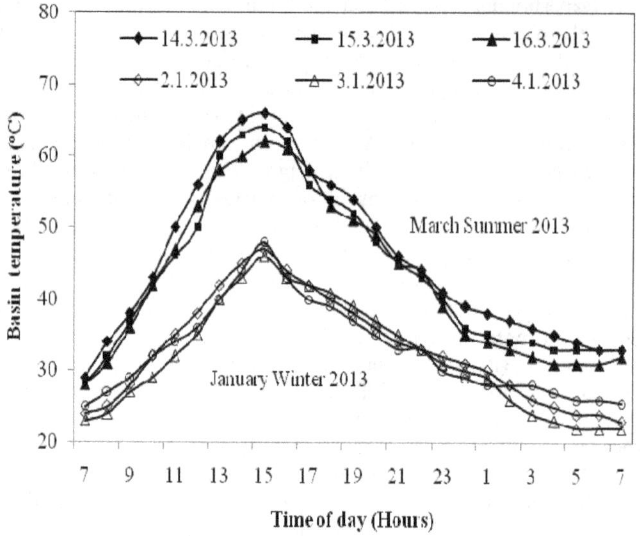

Fig. 6 Variations of absorber plate temperature with PCM

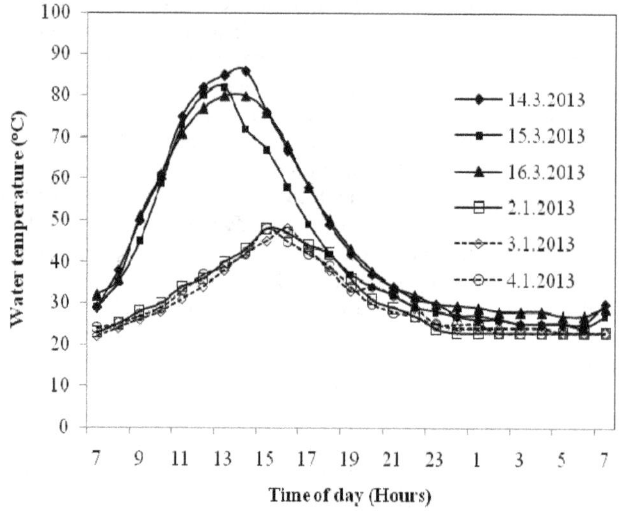

Fig. 7 Variations of water temperature without PCM

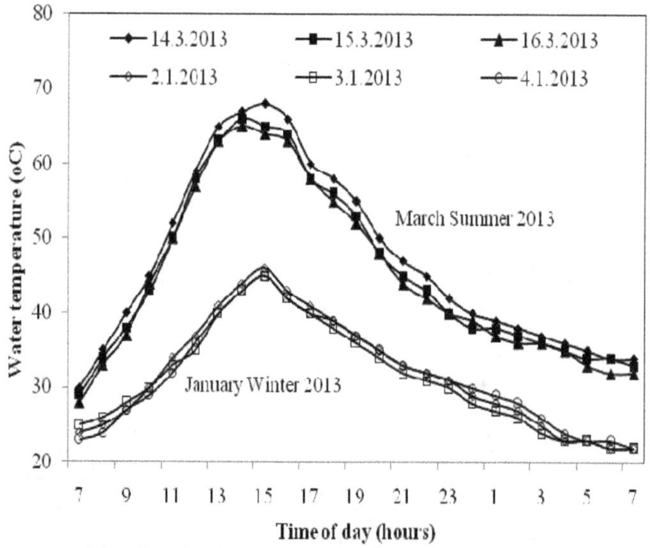

Fig. 8 Variations of water temperature with PCM

Fig. 7 and Fig. 8 show the hourly variation of water temperature with and without PCM. From the theoritical analysis the maximum acheivable temperature of water is about 75°C. In the present model the temperatuer of water in the still with and without PCM achieved is about 70 and 85°C respectively. During the winter days, for the same solar still temperatue of water inside the still with and without PCM found to be 45°C and 48°C respectively.

Fig. 9 andFig. 10 show the diurnal variation of glass temperatures. It is observed that the driving force between water and glass acts as the prime source for fresh water production. Due to the forcing of wind over the surface reduces the temperature of glass for better enhancement.

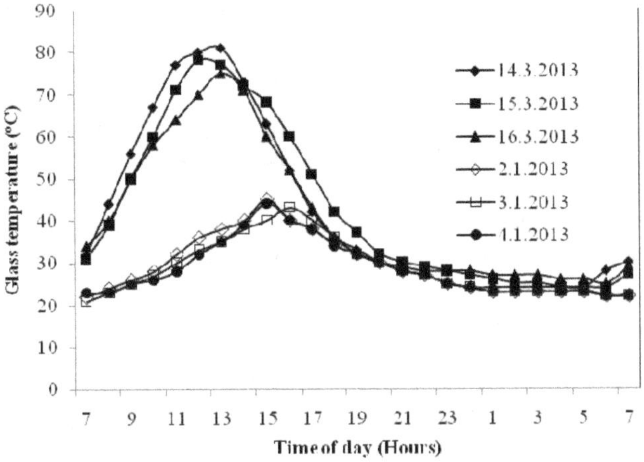

Fig. 9 Variations of glass temperature without PCM

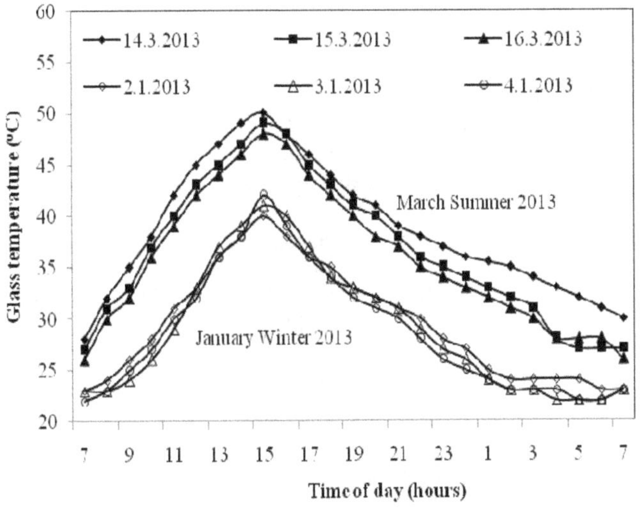

Fig. 10 Variations of glass temperature with PCM

Fig. 11 and Fig. 12 show the diurnal variation of fresh water producureion from the solar still with and without phase change material. It is clear obvious that the fresh water production from the solar still depends on solar intensity during the sunshine hours. And the results shows that the solar intensity is directly proportional to yield from the solar still. The yield obtained from the still without PCM shows a higher output compared to still with PCM. It is due to the amount of partial recovery of heat from the water to PCM in the basin. During the off shine period the heat which is recovered during the charging mode, discharges its heat to water.

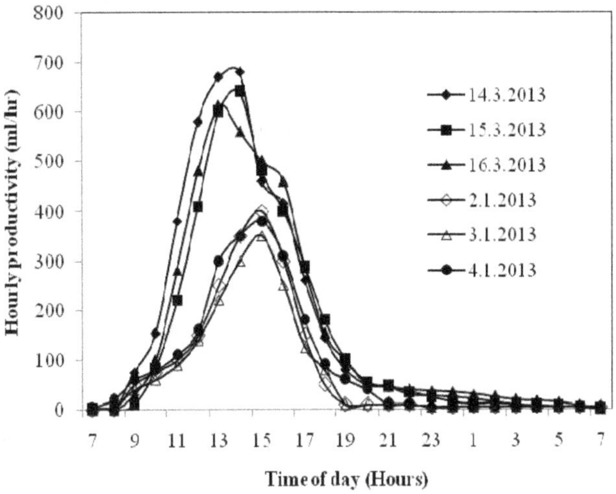

Fig. 11 Variations of distillate output without PCM

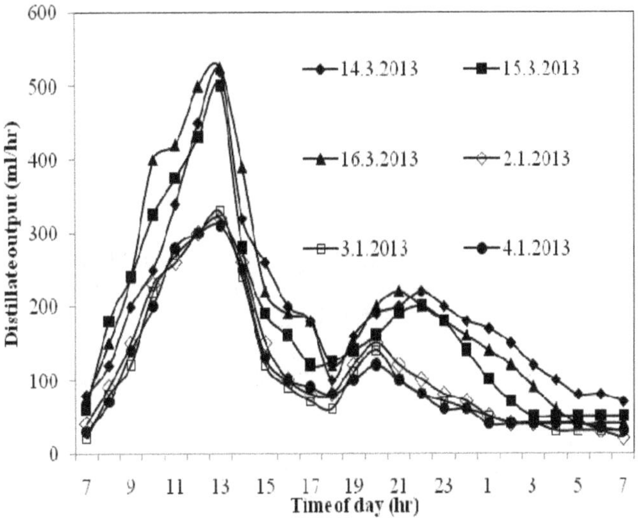

Fig. 12 Variations of absorber plate temperature with PCM

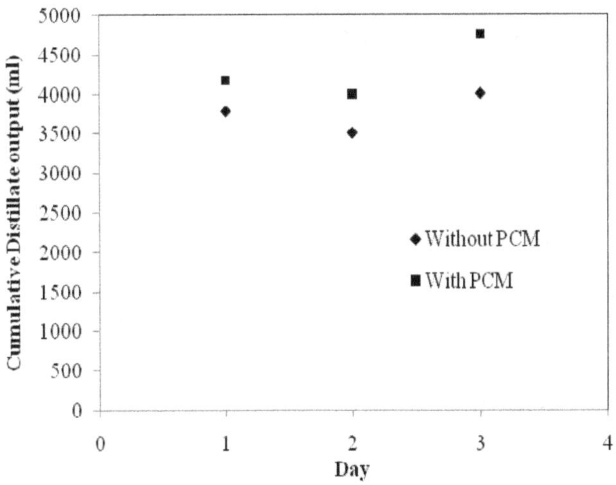

Fig. 13 Variations of cumulative yield on different experimental days in summer

Fig. 13 compares the accumulated yield of solar still with and without latent heat energy storage material. A triangular pyramid solar still with built-in latent heat thermal energy storage system was fabricated to improve the still productivity. Another still with the same characteristics without LHTESS was also constructed for investigation on hourly production of fresh water. Using a $1m^2$ area of solar still the average production of fresh water about 3.4 liters/m^2 without LHTESS effect is possible. The effect of introducing PCM in the above setup shows an increase of 20% in the production of fresh water.

5. CONCLUSIONS

The daily efficiency was found to be 53% with LHTESS and 45% without LHTESS. Solar radiation on the test days reveals that the maximum intensity occurred during the mid-day, and the productivity shows that the solar radiation and production rate are directly proportional. The temperature difference between water and glass varies from 10-15.5°C during the off-shine period. Experimental results concludes that with PCM the production of fresh water improved to about 4.3 liters/day for a 24 hour operation, its due to the higher specific heat capacity, better latent heat of fusion, and thermal conductivity of wax.

ACKNOWLEDGEMENTS

The authors are extending their sincere thanks to Mr. A.P. Arun Pravin and Mrs. M. Jeyashree of St. Peters College of Engineering and Technology Chennai for their technical support.

REFERENCES

Demirbas, F., 2006, "Thermal Energy Storage and Phase Change Material: An Overview" Taylor & Francis, Part B, 1:85-95. http:dx.doi.org/10.1080/009083190881481

Duffie JA., Beckman WA., 1991, "Solar Engineering of Thermal Processes". New York: Wiley.

Dunkle, R.V., 1961, "International development in heat transfer". In: ASME Proceedings. Inter. Heat Transfer, Part V, University of Colorado; p. 895.

El-Sebaii, A.A., Al-Ghamdi, F.S., Al-Hazmi, Adel S. Faidah, 2009, "Thermal Performance of A Single Basin Solar Still with PCM as Storage Medium," *Journal of Applied Energy,* **86**, 1187 -1195. http:dx.doi.org/10.1016/j.apenergy.2008.10.014

Farell, A.J., Norton.B., Kennedy, D.M., 2006, "Corrosive Effects Of Salt Hydrated Phase Change Materials Used With Copper And Aluminum," *Journal of Material Processing and Technology*, 175 ,198-205. http:dx.doi.org/10.1016/j.jmatprotec.2005.04.058

Haji-Sheikh, A., Eftekhar, J., Lou, D.Y.S., 1982, "Some Thermo Physical Properties of Paraffin Wax As A Thermal Storage Medium," 3rd Joint Thermophysics, Fluids, Plasma and Heat Transfer Conference, St. Louis, MO.

Jinjia, W., Kawaguchi, Y., Hirano, S., Takeuchi, H., 2005, "Study on A PCM Heat Storage System for Rapid Heat Supply," Journal *of Applied Thermal Energy,* **25**, 2903 – 2920. http:dx.doi.org/10.1016/j.applthermaleng.2005.02.014

Malik, M.A.S., Tiwari G.N., Kumar, A., Sodha M.S., "Solar Distillation".UK: Pergamon Press; 1982.

Naim, M.M., 2002, "Non-Conventional Solar Stills with Energy Storage Element" Desalination, **153**, 71-80. http:dx.doi.org//10.1016/S0011-9164 (02)01095-0

Prakash. J, Kavathekar, A.K., 1986, "Performance Prediction of A Regenerative Solar Still," *Solar & Wind Technology*, **3**(2), 119–125

Radhwan AM., 2004, "Transient Performance of a Steeped Solar Still with Built-in Latent Heat Thermal Energy Storage," *Desalination,* **171**, 61–76. http:dx.doi.org/10.1016/j.desa1.2003.12.010

RaviShankar, S., Nagarajan, P.K., Vijayakumar, D., and Jawahar, M.K. 2013, "Phase Change Material on Augmentation of Fresh Water Production Using Pyramid Solar Still," *Int. Journal of Renewable Energy Development,* **2**(3), 115-120.

Sharma, V.B., Mullick, S.C., 1993, "Calculation of hourly output of a solar still," *J. Solar Energy Eng,* **115**, 231–236. http://dx.doi.org/10.1115/1.2930055

Tabrizi, F.F., Dashtban, M., Moghaddam, H., 2010, "Effect of Water Flow Rate on Internal Heat and Mass Transfer and Daily Productivity of a Weir- Type Cascade Solar Still," *Desalination*, **260**(1-3), 248-253. http://dx.doi.org/10.1016/j.desal.2010.03.033

Tabrizi, F.F., Dashtban, M., Moghaddam, H., 2010, "Experimental Investigation of a Weir – Type Cascade Solar Still with Built-in Latent Heat Thermal Energy Storage," *Desalination*, **260**(1-3), 239-247. http://dx.doi.org/10.1016/j.desal.2010.03.037

Wong, H.Y., 1977, *Handbook of Essential Formula and Data on Heat Transfer for Engineers*, Longman, London.

Zurigat, Y.H., Abu-Arabi M.K., 2004, "Modelling and Performance Analysis of A Regenerative Solar Desalination Unit," *Applied Thermal Engineering,* **24**, 1061-72. http:dx.doi.org/10.1016/j.applthermaleng.2003.11.010

5

EXPERIMENTS ON HEAT TRANSFER CHARACTERISTICS OF SHEAR-DRIVEN LIQUID FILM IN CO-CURRENT GAS FLOW

Tomoki Hirokawa[a,*] Masahiko Murozono[a], Oleg Kabov[b,c] and Haruhiko Ohta[a]

[a] *Department of Aeronautics and Astronautics, Kyushu University, Fukuoka, 813-0385, Japan*
[b] *Institute of Thermophysics, Russian Academy of Science, Siberian branch, Novosibirsk, 630090, Russia*
[c] *Tomsk Polytechnic University, Tomsk, 634050, Russia*

ABSTRACT

Experiments are performed to study the liquid film behavior and corresponding local heat transfer to shear-driven liquid film flow of water in the co-current nitrogen gas flow. The heated channel has a cross section of 30mm in width and 5mm in height, where the bottom is operated as a heating surface of 30mm in width and 100mm in length. The heated section is divided into segments to evaluate the local heat transfer coefficients. Under most gas Reynolds numbers, the local heat transfer coefficients are increased with increasing heat flux, where three mechanisms are important; (i) increase of areas along the three-phase interline around dried areas, (ii) rewetting of dried areas by the transverse liquid flow pushed by the generation of bubbles at the side edges of duct, (iii) microlayer evaporation during nucleate boiling in the film flow. The existence of duct corners makes the phenomena more unsteady and non-uniform in the transverse direction.

Keywords: *Evaporation, Shear-driven liquid film, Heat transfer coefficient*

1. INTRODUCTION

The density of dissipated heat from semiconductors tends to increase by the progress in electronic technology. The technical improvement is required for the existing thermal management systems using convective air or liquid. In recent years, the cooling systems utilizing phase change attract much attention for the cooling of power electronics in addition to the small semiconductor chips. In nucleate boiling and two-phase forced convection, the formation or the consumption of liquid film underneath bubbles and around the vapor core flow, respectively, becomes a key factor which determines the limitation of heat transfer. In such systems, however, the distribution and the behavior of liquid-vapor interface cannot be controlled directly to satisfy the various cooling requirements but indirectly through the dynamic behaviors of the vapor phase. For usual cooling systems by forced convection of liquid, the location of heat generation is separated from that of final heat dissipation. In the proposed system, behaviors of liquid film flow are regulated by the interfacial shear stress exerted by the vapor flow, where multiple cooling units are combined to obtain the desired vapor velocities. To study the interaction between the phases under variable combinations of flow rates independent of heat flux level, the shear-driven liquid film flow is realized by the co-current gas flow of the component different from that of liquid in the present study.

The behavior and heat transfer in the liquid film e.g. liquid film falling down by gravity, evaporating meniscus and shear-driven liquid film have been investigated by a number of researchers. Kabov *et al.* (2002) investigated experimentally the heat transfer from a local heat source to a liquid film falling down by gravity along a vertical surface. Aqueous solution of 25% ethyl alcohol was used as a working fluid. Beyond the instability threshold, the existence of a thermocapillary counter flow produced a stagnation line of horseshoe shape at the upper edge, and the heat transfer was decreased with increasing film Reynolds number. Kabov and Chinnov (1997) investigated the same configuration by using a dielectric liquid, Perfluorine-triethyl-amine. The emergence of a horizontal wave was confirmed during the evaporation of subcooled liquid film at a threshold value of heat flux. At the downstream part of the wave, the film was divided into rivulets flowing down keeping certain horizontal distances. The distance was a function of liquid capillary constant and was depended weakly on the film Reynolds number. Lei *et al.* (2007, 2008 and 2009) investigated the formation and development of quasi-stable film structures in laminar wavy falling films by using IR-thermography. The heating surface was made of copper, where the length and the width of heater were 70mm and 130mm, respectively. The test fluid was Polydimethylsiloxane. The results showed that waves along the transverse direction were appeared by Rayleigh-Taylor instability. They also investigated experimentally the development of thermal entry length and the heat transfer across laminar wavy falling films averaged over the entire heating length. They concluded that, at low Reynolds numbers, the thermal entry length agreed with the predicted ones by the correlation in the previous work and it deviated from the correlation with increasing Reynolds numbers and/or heat fluxes.

The heat transfer characteristics of a thin liquid film in a meniscus region are influenced by the capillary pressure and the disjoining pressure. Stephan *et al.* (1995) investigated a model describing the influence of capillary pressure on the evaporation of liquid film along a vertical plate. The numerical results were compared with the classical analytical solution by the Nusselt's film condensation theory. The deviation from the Nusselt theory became outstanding for thinner liquid film, which could be attributed to the ignored effect of capillary pressure on the distribution of the liquid film thickness. Wang *et al.* (2007) analytically investigated characteristics of heat and mass transfer from an evaporating meniscus in a two-dimensional micro-gap by using

* *Corresponding author. Email: hirokawa@aero.kyushu-u.ac.jp*

the extended Young-Laplace model and kinetic theory-based expression for mass transport across the liquid-vapor interface. The contribution of the thin-film to the heat transfer was decreased with increasing channel size and/or wall superheat, and the importance of meniscus for the heat transfer was confirmed. They concluded that the capillary pressure played an important role for the determination of liquid film distribution along the meniscus. Du and Zhao (2012) proposed a conjugated heat transfer model for micro channels taking account of the interaction between the evaporating thin film in the contact line region and the heat conduction in the solid wall. High heat flux was confirmed in a narrow region of the evaporating thin film and in the corresponding region of the wall resulting in the existence of local minimal value of wall temperature. The apparent value of contact angle in the conjugated heat transfer model was smaller than that predicted by the model with a constant substrate temperature. Both of the peak and the total heat flow rates obtained by the conjugated heat transfer model was larger than those by the model of constant substrate temperature. Park *et al.* (2003) proposed a mathematical model of the flow and heat transfer characteristics for a thin film region in a micro-channel taking account of the gradient of vapor pressure and capillary force. They discussed the effects of channel height, heat fluxes and slip boundary conditions at the solid-liquid interface. The results indicated that the shape of the thin film in a microchannel was influenced by the gradient of vapor pressure. They also concluded that with increasing heat fluxes the length and the maximum thickness of the thin film decreased exponentially and the local evaporating mass flux increased linearly. The decrease of the channel height had little effect on the shape and thickness of the film. In the case of the slip condition, the length and film thickness of the thin film were decreased compared with those of the no-slip condition, and the decreased film thickness caused the increase of capillary and disjoining pressures. Park and Lee (2003) also proposed a model of the heat and mass transfer for the evaporation across the meniscus region in a two-dimensional micro-capillary channel. In the thin film region, extremely larger values of local heat transfer coefficient were reported, while the heat transfer rate was smaller than that in the meniscus region. Ibrahem *et al.* (2010) experimentally investigated the local heat flow at the three-phase contact line using a micro-scale temperature measurement technique (Höhmann and Stephan, 2002). The test section was composed of two vertical parallel flat plates to form a liquid-vapor meniscus due to capillary forces. The surface of inner wall was made by thin metallic foils employed as resistance heaters. The test fluid was HEF7100 evaporated between the plates under steady-state conditions. They observed two-dimensional microscale temperature fields on the back side of the heating foil by using an infrared camera with a spatial resolution of 14.8μm × 14.8μm. The local heat fluxes from the heater to the evaporating meniscus are calculated from the measured wall temperatures using an energy balance for each pixel element. The local heat fluxes at the contact line area were found to be about 5.4-6.5 times higher than the mean input heat fluxes at the foil. Kunkelmann *et al.* (2012) investigated experimentally and analytically the influence of three-phase contact line speed on the local heat transfer during the evaporation of the meniscus in the contact line region, where a peak of local heat transfer rate was observed. In the case of a receding contact line, the local peak was almost independent of its moving speed, while the heat transfer rate was significantly increased with increasing speed of advanced contact line.

Instability and rupture of liquid film causes the extension of dry patches and heat transfer crisis. Williams and Miles (1982) analyzed nonlinear effects on the liquid film rupture on a flat surface by using Navier-Stokes equations including a body force term describing the London-van-der-Waals dispersion forces. Ajaev *et al.* (2011, 2012) investigated the stability and the breakup of a thin liquid film on a solid surface with grooves located at regular intervals which were filled by gas under the action of disjoining pressure. Their mathematical model took the effect of slip along the menisci separating the air entrapped in the grooves from the liquid film flow, the effect of the deformation of

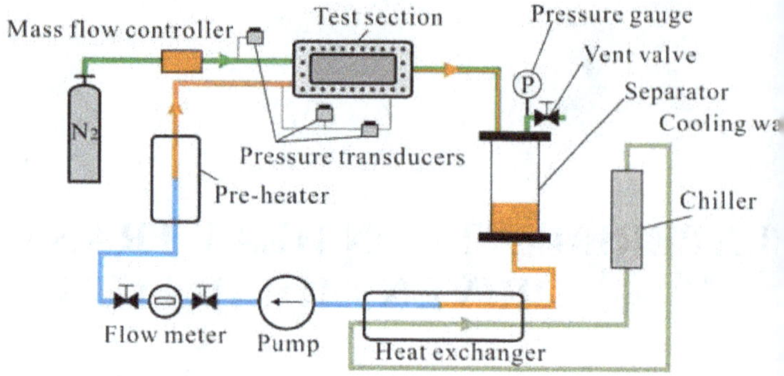

Fig. 1 Outline of test loop.

menisci due to local variations of pressure in the liquid film and the effect of the nonuniformities of the Hamaker constant into account. In the case of negligible variations of Hamaker constant, the effect of surface structure was destabilized compare to the results by the linear stability analysis for the flat plate. Numerical simulations in the strongly nonlinear regime indicated that the rupture time was decreased significantly due to the effect of grooves. They indicated the decrease of rupture time with increasing the ratio of the maximum and the minimum values of Hamaker constant.

Recently, the behavior and heat transfer characteristics of shear-driven liquid film in co-current gas flow have been investigated in detail. Liu and Kabov (2012) studied a two layer system consisting of a horizontal liquid layer contacting with its vapor. The Rayleigh instability and Marangoni instability were analyzed taking into account of the effects of buoyancy, thermocapillarity and interfacial shear forces and the evaporation of liquid. Kabov *et al.* (2007, 2011) investigated the flow of a locally heated liquid film moving under the shear force exerted by the gas flow. Water and FC72 were used as working liquids and air and nitrogen were used as gases. The critical heat flux by the formation of an initial stable dry patch in the liquid film was higher by several times than that for a vertical falling liquid film. The temperature distribution along the film surface was measured by the infrared scan. The thermocapillary tangential stress could exceed tangential shear stress caused by the friction of gas flow. Gatapova *et al.* (2003, 2004) and Gatapova and Kabov (2008) studied a two-dimensional model of steady laminar film flow and co-current gas flow along a horizontal plane wall of the channel with a heat source. An analytical solution for the temperature distribution in the locally heated liquid film was obtained for the linear velocity profile. A liquid bump caused by the thermocapillary force was obtained just above the heated zone. Kabova *et al.* (2008, 2009) investigated the thin non-isothermal liquid film flowing down by gravity and co-current gas flow in an inclined mini-channel. They developed 3D time-dependent mathematical model. The effect of gravity as well as the effect of gas velocity was studied to define the features of film dynamics. They found that the gravity force had a significant effect on the film deformation. At lower gravity, 3D liquid film pattern was changed in the transverse direction and the generation of a middle wave between two lateral waves was observed, where the speed of film deformation became higher and the time required for the stabilization of flow became longer compared to those at normal gravity. Independent of gravity level, the increase of heat flux resulted in the emergence of liquid film deformation. Cheverda *et al.* (2013) investigated gas shear-driven rivulet flows in a minichannel with a height of 1.4mm and a width of 30mm under the variation of gravity levels. They reported the reduction of rivulet width with decreasing gravity level and its increase with increasing gas and/or liquid flow rates. Lyulin and Kabov (2014) experimentally investigated the evaporation rates and gas-liquid interfacial behaviors for the horizontal evaporating liquid layer under the existence of the shear-stress exerted by the gas flow. The test fluids were HFE-7100 for the liquid layer and nitrogen gas. The area of a square liquid reservoir was 50mm×50mm,

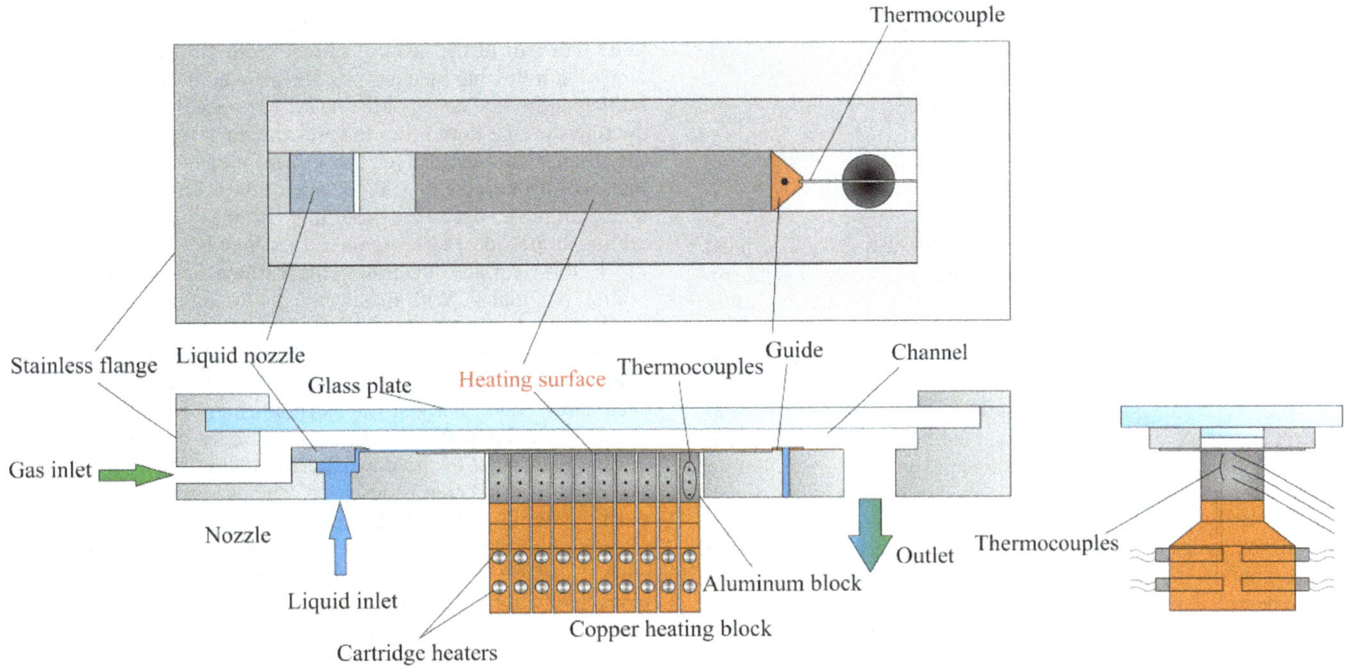

Fig. 2 Structure of test section.

and its depth was varied from 1.5 to 8mm. The results indicated that, with increasing gas and liquid temperatures from 20 to 40°C, the depth for the local maximum of evaporation rate was shifted from 3 to 5mm. At the maximum evaporation flow rate, a stable and uniform convective flow induced by the thermocapillary force in the direction opposite to the gas flow was observed in the liquid layer. Houshmand and Peles (2013) investigated the heat transfer characteristics of a liquid film flow in a horizontal rectangular microchannel with a 1mm × 1mm heater, where water is introduced from a 350μm circular hole to the gas stream of nitrogen. Considerable improvement of the heat transfer was observed in despite of no appreciable change in pressure drop compared to that for water flow without gas. At high heat fluxes, the heat transfer was enhanced at the location before the rupture of liquid film.

In the present study, to investigate the heat transfer during the evaporation and boiling of liquid film, the behaviors of film flow is varied by the interfacial shear stress exerted by the gas flow under different heat flux levels. To extend the ranges of flow rates of both phases, the two-component system is studied, where the gas flow rate is adjusted independently from the rate of film flow.

2. EXPERIMENTAL APPARATUS

Figure 1 shows the test loop composed of pump, flow meter, pre-heater, test section, separator and heat exchanger for the cooling of test liquid. Distilled water and nitrogen are selected as liquid and gas phases, respectively. The nitrogen gas is introduced directly to the test section by a mass flow controller. The test loop is always exposed to the atmosphere at the vent valve to exhaust nitrogen gas and evaporated water vapor in the downstream of the test section, and only liquid water is circulated again. To evaluate the performance of heat transfer, the gas temperature at the inlet and the liquid temperatures at the inlet and the outlet of test section are measured by K-type thermocouples inserted in the loop. The pressure drop across the test section is measured by pressure transducers.

Figure 2 shows the structure of test section assembled by an aluminum block with a heating surface, copper heating blocks, stainless steel flanges and a Pyrex glass plate for an observation window. The heated channel with a cross section of 30mm in width and 5mm in height is located horizontally. The top surface of the aluminum block is used as a heating surface with an area of 30mm in width and 100mm in length. Its bottom surface is contacted with copper heating blocks, where cartridge heaters are inserted at lower parts in order to supply heat flux by heat conduction. The heating assembly is divided into 10 segments along the flow direction as shown in the figure. They are thermally isolated each other by the gaps between the heated segments except the top part with a small thickness just beneath the heating surface. The segment structure makes possible the measurement of heat transfer data in the upstream location of the heating surface even under the conditions of dryout in the downstream by switching off a part of cartridge heaters. In addition, the evaluation of local heat transfer along the flow direction becomes possible. The liquid flow is introduced from the bottom of the test section by a nozzle of thin slit. The nozzle height h_N is 1.0mm, whereas the overall height of the test section H is 5.0mm. In each segment of aluminum block, 3 thermocouples are inserted at the depths of 1.5mm, 7.5mm and 13.5mm from the heating surface to evaluate local heat fluxes and local surface temperatures from the temperature gradients. At the outlet, a guide plate concentrating liquid film flow is located to measure the mean liquid film temperature by using a thermocouple. Experiments are conducted under the atmospheric pressure. Gas flow rate is varied as a parameter keeping the liquid flow rate constant. The gas Reynolds number defined by the superficial gas velocity and the hydraulic diameter is varied from 0 to 3175, and the film Reynolds number is fixed at 27.6. Both temperatures of subcooled liquid and gas flow at the inlet of the test section is kept at 25°C. For the detailed observation of liquid film flow, images of top view are recorded by using a high-speed video camera through the glass plate. Experimental conditions are shown in Table 1.

Table 1 Experimental conditions

Channel height	H	5mm
Nozzle height	h_N	1mm
Test section outlet pressure	P_{out}	1atm
Gas inlet temperature	$T_{g,in}$	25°C
Liquid inlet temperature	$T_{l,in}$	25°C
Liquid film Reynolds number	Re_l	27.6
Gas Reynolds number	Re_g	0 - 3175
Heat flux	q	0 - 280kW/m^2

3. DATA PROCESSING

A local heat transfer coefficient is defined by

$$\alpha_i = \frac{q_i}{T_{w,i} - T_{l,i}} \quad (1)$$

where α_i: local heat transfer coefficient, q_i: local heat flux, $T_{w,i}$: local surface temperature, $T_{l,i}$: local mean fluid temperature at each segment. The local heat fluxes and the local surface temperatures are evaluated from the linear approximation of temperatures indicated by three thermocouples inserted in the aluminum block at different depth. The local liquid temperature is evaluated by the following procedure. The heat loss from the liquid film to the gas flow at the unheated section located upstream of the heating surface cannot be negligible under the conditions of a large difference between the inlet temperatures of liquid film and gas flows. An unheated preliminary experiment is performed at the same inlet liquid and gas temperatures and gas and liquid film Reynolds numbers as those for the heated experiments. By the temperature gradient of liquid film along the flow direction obtained from the unheated experiment, the liquid film temperature at the 1st segment is evaluated during the heated experiments. On the other hand, the liquid film temperature at the 10th segment is equated to the temperature of liquid film at the outlet of the test section. The liquid film temperatures from the 2nd to 9th segments are evaluated by the linear interpolation of the temperatures at the 1st and 10th segments.

All of the measurement systems are carefully calibrated. The uncertainty in temperature is ±0.25K, in pressure measurement ±0.26kPa, in liquid flow rate ±1% and in gas flow rate ±1.2*l*/min even at the minimum experimental value of 20 *l*/min.

4. EXPERIMENTAL RESULTS AND DISCUSSION

Figures 3(a)-(c) show liquid film behaviors and its rupture for the different gas Reynolds numbers of 0 (no gas flow), 1270 and 3175 at the liquid inlet temperature 25°C and heat flux 50kW/m^2. In all experiments described here, the film Reynolds number is kept at 27.6. At gas Reynolds number 0, i.e. no shear stress on the surface of the liquid film flow, bubbles of dissolved air but not of evaporated vapor are observed along the heating surface. At gas Reynolds number 1270, the rupture of liquid film is initiated at the center of the transverse direction in the downstream of the heating surface. With increasing gas Reynolds number, the dried area is expanded toward the upstream (cf. Figs.3 (b), (c)). The rupture is promoted by two reasons. The increased interfacial shear stress exerted by the gas flow with higher velocity makes the liquid film thinner. In addition, toward the side edges of the duct by the formation of meniscus, the pressure difference between the gas and liquid is increased, i.e. the liquid pressure is decreased, and the liquid film is squeezed into the side edges. The dried area is initiated in the downstream because the liquid film thickness becomes thinner by the evaporation. Surface temperature distribution and local heat transfer coefficient at heat flux 50kW/m^2 are shown in Figs. 3(d) and (e), where the data for both of the 1st and the 10th segments are omitted because of larger heat losses to the flanges surrounding the heating surface. At gas Reynolds number 1270, the local heat transfer coefficient is larger than those for other gas Reynolds numbers in all segments because of the enhanced evaporation by the reduced liquid film thickness. At gas Reynolds numbers larger than 1270, heat transfer deterioration occurs owing to the excessive disappearance of liquid film promoting the large extension of dried area, where lower heat transfer coefficients are obtained for higher gas Reynolds number.

Figures 4(a)-(c) show liquid film behaviors at heat flux 150kW/m^2. At gas Reynolds number 0, nucleate boiling occurs in the downstream, but no dried area is observed on the entire heating surface. At gas Reynolds number 1270, the liquid film rupture occurs and the dried area is observed for all segments of the heating surface. The oscillatory behaviors of the rewetting and drying in the transverse direction on the latter half part of the heating surface from 5th to 10th segment occurs along the following mechanism. The growth of bubbles due to nucleate boiling at the side edges of the heating surface, i.e. the side edges of the duct, pushes the liquid film towards the transverse direction and rewets the dried area extended in the center of the heating surface. At the instance of the rewetting, the evaporation is enhanced and the surface temperature is decreased resulting in the suppression of nucleate boiling at the side edges. Again a dried area is extended in the center of the heating surface increasing the surface temperature and nucleate boiling is initiated. With increasing gas Reynolds number, the iteration of drying and rewetting is still observed in the midstream of the heating surface. However, the liquid film flowing along the side edges of the heating surface is accelerated towards the downstream by the enhanced shear stress, and finally at gas Reynolds number 3175, the coalescence of squeezed liquid films from the side edges occurs. At the unheated section connected to the downstream edge of the heating surface, the extension of dried area is observed. This is because there is no force to push the liquid film towards the center of the heating surface under no generation of bubbles at the side edges of the duct. Figures 4(d) and (e) show surface temperature and local heat transfer coefficient respectively at heat flux 150kW/m^2. The heat transfer coefficient for no gas flow is the highest because of no emergence of dried area on the entire heating surface. The heat transfer coefficient becomes larger along the flow direction with decreasing liquid film thickness. With gas flows, on the other hand, the heat transfer coefficients are smaller than the value without gas flow owing to the periodical emergence of dried area in the center of transverse direction. Also in this case, the decrease of film thickness due to the evaporation and increased gas flow rate containing the evaporated vapor increases the heat transfer coefficient along the flow direction. The heat transfer coefficient tends to decrease at higher gas Reynolds number by the excessive extension of the dried area.

Figures 5(a)-(c) show liquid film behaviors at heat flux 250kW/m^2. Under no gas flow, no rupture of liquid film is observed and nucleate boiling occurs as observed at heat flux 150kW/m^2. At gas Reynolds numbers larger than 1270, liquid films from the side edges of duct are coalesced, which is similar to the case of 150kW/m^2. And also two dried areas emerge on the upstream part of the heating surface and on the downstream unheated section. With increasing gas Reynolds number up to 3175, the dried area on the heating surface are extended widely because of further reduction of liquid film thickness and enhancement of evaporation. The surface temperature distribution and local heat transfer coefficient are shown in Figs. 5(d) and (e). Without gas flow, the surface temperature in the upstream part of the heating surface becomes higher because of the boiling with smaller nucleation site densities in subcooled liquid. In the flow direction, the surface temperature decreases with increasing liquid temperature on the upstream part of the heating surface. On the other hand, at gas Reynolds numbers larger than 1270, the extension of dried area is already observed on the upstream part of the heating surface and the surface temperatures become higher at larger gas Reynolds number. The surface temperature is decreased in the flow direction due to the iteration of rewetting by the liquid film pushed towards transvers direction by the generation of bubbles at the side edges of the duct. In the midstream of the heating surface, the coalescence of liquid films from the side edges occurs and three-phase interlines disappear resulting in the slight increase of surface temperature independent of gas Reynolds numbers including the case of no gas flow. In the downstream of the heating surface, the liquid film covering the entire width of the heating surface becomes thinner due to the evaporation resulting in the acceleration of gas flow rate, and the surface temperatures decrease again. In the case of no gas flow, however, the surface temperature increases in the downstream of the heating surface, which seems to be caused by the emergence of small dried areas, however, not clearly observed. In despite of complicated behaviors of surface temperatures along the flow direction, the heat transfer

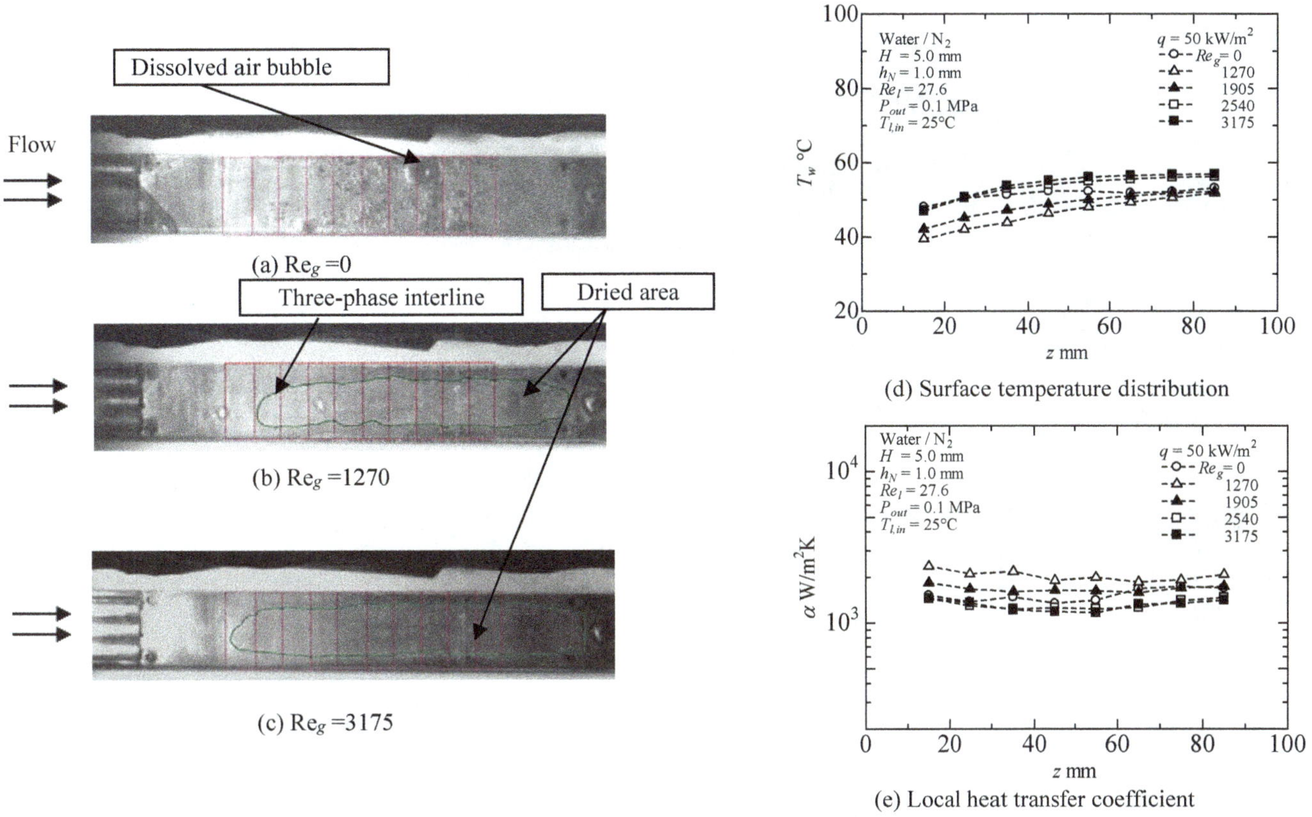

Fig. 3 Liquid film behavior, surface temperature distribution and local heat transfer at q=50kW/m^2.

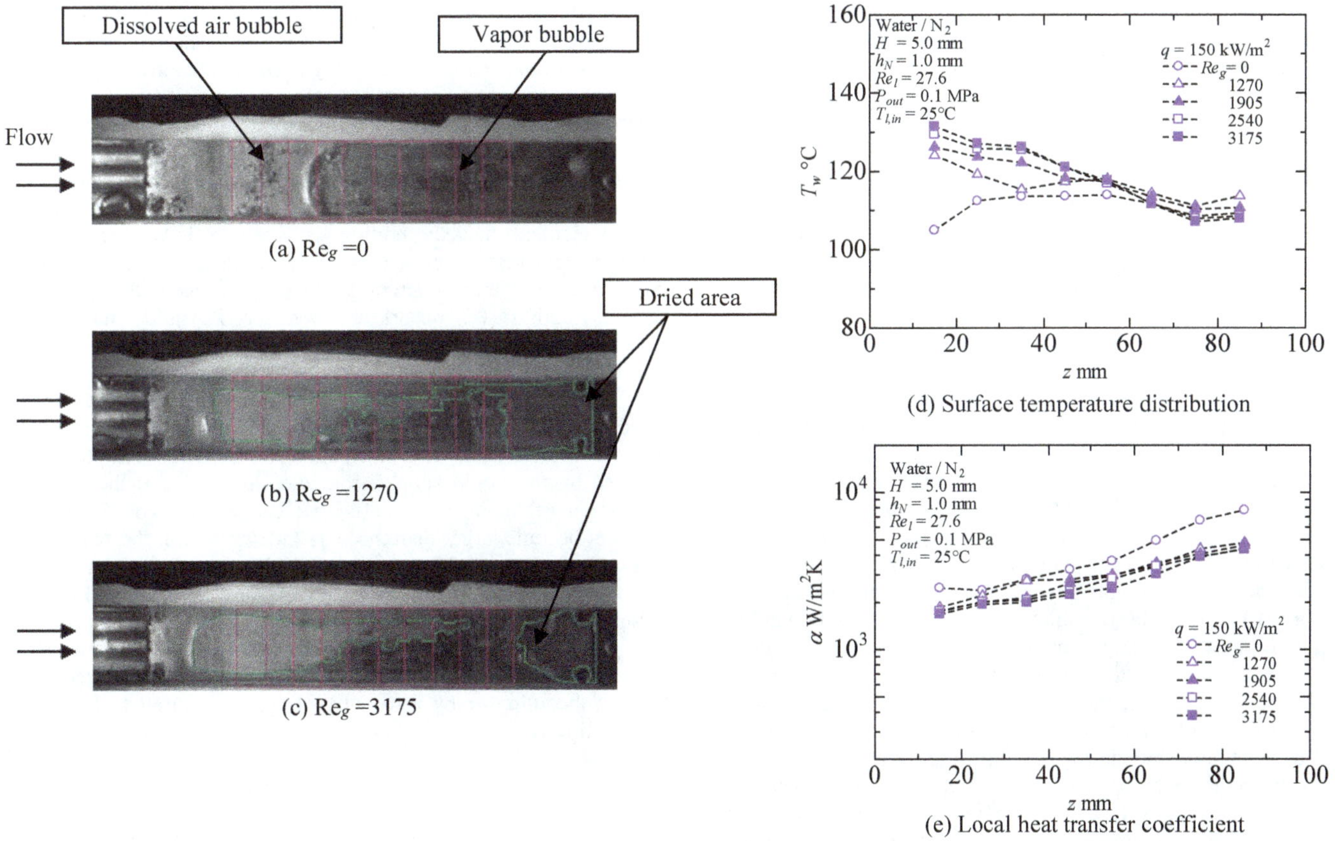

Fig. 4 Liquid film behavior, surface temperature distribution and local heat transfer at q=150kW/m^2.

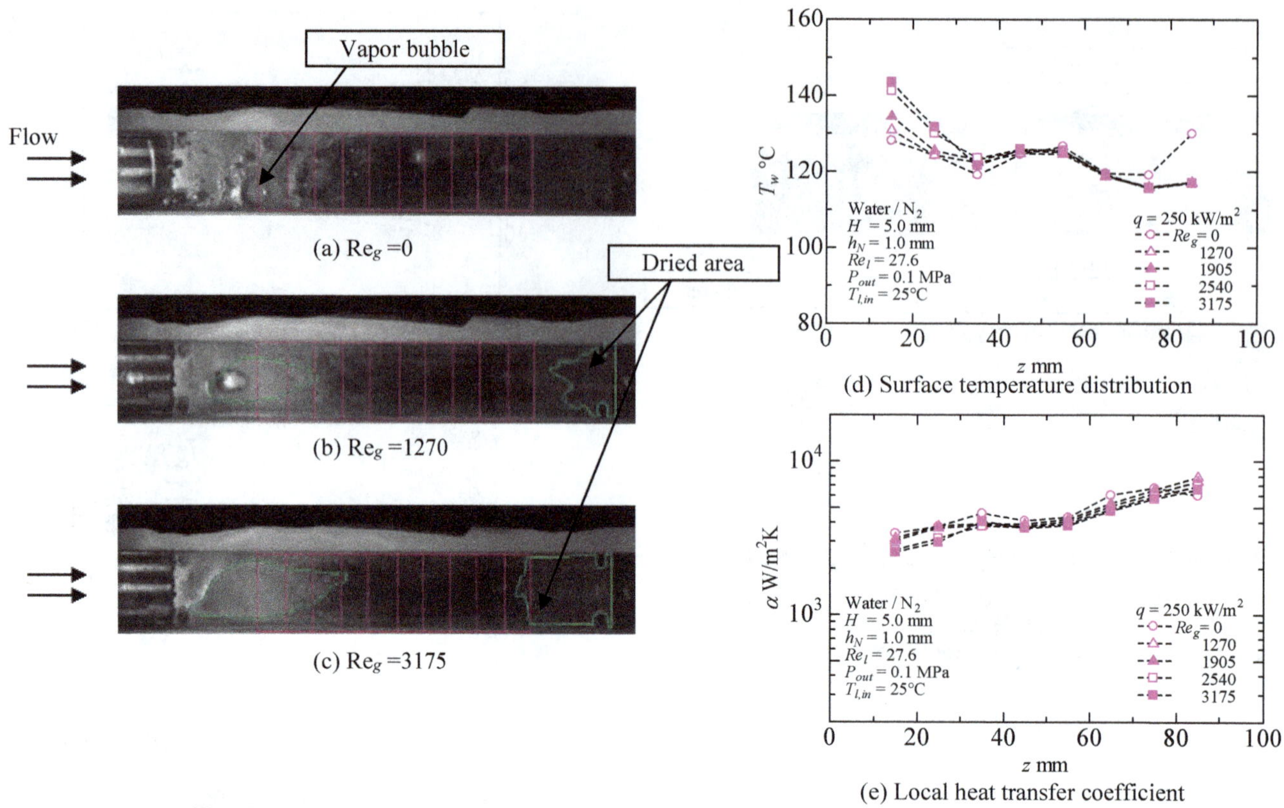

Fig. 5 Liquid film behavior, surface temperature distribution and local heat transfer at q=250kW/m^2.

coefficients increase almost monotonously towards the downstream for all gas Reynolds numbers including the case of no gas flow. The variation in the levels of heat transfer coefficient with Reynolds number is quite small compared with the case of 150kW/m^2, which implies that the liquid-vapor behaviors are dominated not by the gas flow rate but by the elevated heat flux via the phenomena such as the extension of dried area, the generation of bubbles and the generation of vapor due to enhanced evaporation. Figures 6 (a)-(c) show the liquid film behaviors and corresponding values of local heat transfer coefficients under the combinations of gas Reynolds number and heat flux for the 2nd, 5th and 9th segments, respectively. The liquid film behaviors along the center line of the heating surface are classified into the following patterns, i.e. film flow without nucleate boiling (●) or with nucleate boiling (▲), oscillation of drying and rewetting (△), rupture and extension of dried area without rewetting (○). For all of three locations, dryout accompanied by the temperature excursion (represented by white areas) was observed at heat flux larger than 300kW/m^2 and no heat transfer data was acquired. In Fig. 6(a), (a-i) the film flow without rupture (●▲) is observed for either of no gas flow or of the lowest heat flux, where no interfacial shear stress or almost no evaporation to reduce the film thickness. (a-ii)Subcooled nucleate boiling (▲) starts at high heat flux under no gas flow and the rapture of liquid film (○) occurs at high gas Reynolds number under the smallest heat flux. (a-iii) For the other combinations of gas Reynolds number and heat flux, liquid film behaviors are not influenced by gas Reynolds number but only by heat flux level, where the rupture of liquid film (○) turns to the iteration of extending dried area and its rewetting (△) with increasing heat flux. (a-iv) Further increase of heat flux results in CHF conditions accompanied by the rapid temperature rise (indicated by white area). It seems to be a contradictory trend that the rupture of liquid film occurred at low heat flux becomes periodically rewetted with increasing heat flux. Increase of heat flux promotes the nucleation of bubbles along the side edges of the duct, where a plenty of liquid is still flowing forming menisci. The generation of bubbles pushes the liquid accumulated along the side edges of the duct towards the center and the dried area is rewetted periodically synchronized by the bubble generation. (a-v) Local heat transfer coefficients become larger when the drying and rewetting are iterated at higher heat fluxes up to CHF values at around 300kW/m^2. (a-vi) Furthermore, lower gas Reynolds number or zero gas flow tends to increase the heat transfer coefficient, where the ratio of the area along three-phase interlines to the dried area becomes larger. In Fig.6(b) for 5th segment, (b-i) the film flow without rupture (●▲) is still observed for either of no gas flow or of the lowest heat flux. Because of the reduction of liquid subcooling from the value of 2nd segment, nucleate boiling (▲) starts at lower heat flux for no gas flow. (b-ii) Because the rupture of the liquid film extends from the downstream to the upstream on the heating surface with increasing heat flux, the rapture (○) occurs at lower gas Reynolds number at this location under the smallest heat flux. (b-iii) For the other combinations of gas Reynolds number and heat flux, with increasing heat flux up to CHF values, the rupture of liquid film (○) turns to the extension of dried area and its rewetting (△) and finally to the liquid film flow without rupture but with nucleation of bubbles. The transition is realized by the liquid supply from the side edges of the duct by the nucleation of bubbles there. (b-iv) Heat transfer coefficients at 5th segment become larger than those at 2nd segment at the same heat flux level. The increase of heat transfer at high heat flux is caused by the enhanced nucleate boiling at smaller liquid subcooling. Beneath the bubbles, there are microlayers with liquid film thickness far smaller than the thickness of the film flowing in the duct. The trends (b-i)-(b-iv) are more emphasized at 9th segment by the further reduction of liquid subcooling or by zero subcooling, i.e. saturated, depending on the conditions.

5. CONCLUSIONS

To investigate the performance of cooling system by the evaporation of thin liquid film, the behaviors of liquid film and heat transfer characteristics of co-current gas and liquid film flow were studied experimentally by the independent control of both flow rates.

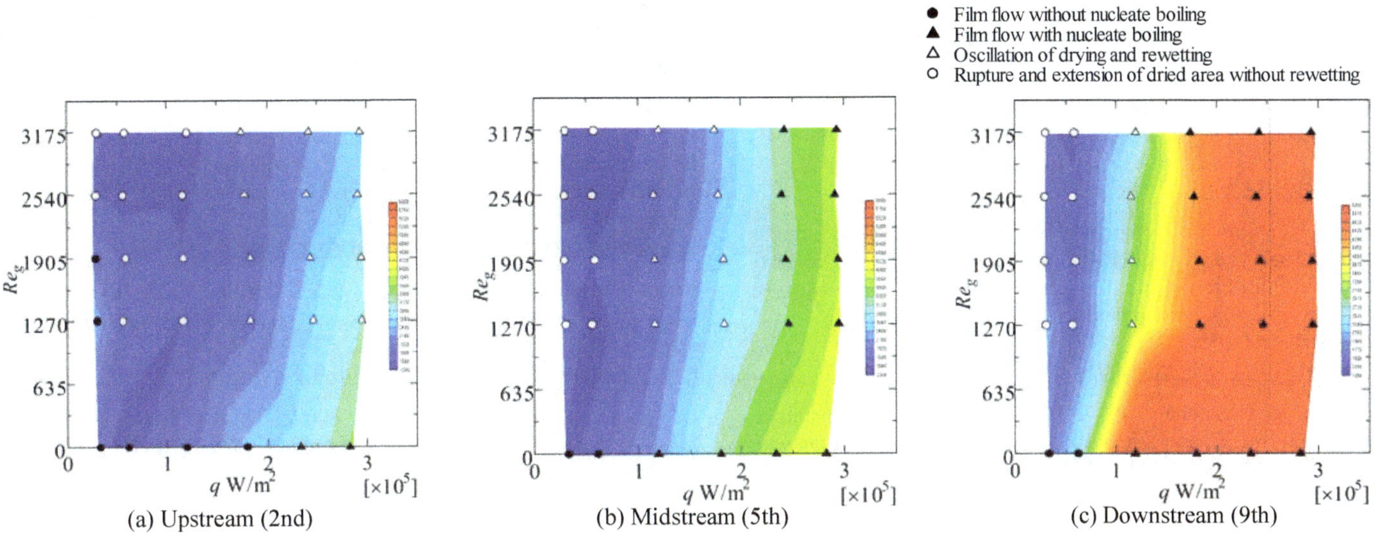

Fig. 6 Liquid film behavior and local heat transfer coefficient at $T_{l,in}$=25°C.

Experiments were performed by using water and nitrogen gas under the atmospheric pressure. Liquid film Reynolds number was fixed at 27.6, while gas Reynolds number was varied from 0 to 3175. The heat flux was increased up to 300kW/m^2. The followings were clarified.

1. Local heat transfer corresponding to the behaviors of liquid film especially of its rupture along the flow direction were clarified by the heating surface with the devised segment structure.

2. Liquid film behaviors along the center line of the heating surface were classified into four patterns; film flow without nucleate boiling or with nucleate boiling, iteration of drying and rewetting, rupture and extension of dried area without rewetting.

3. In low heat flux region, for all gas Reynolds numbers including the case of no gas flow, the local heat transfer coefficients were increased at higher heat flux because of larger extension of areas influenced by the three-phase interlines.

4. In high heat flux region, with increasing heat flux, local heat transfer coefficients were also increased not by the uniform reduction of liquid film thickness due to enhanced evaporation but by the iterated rewetting of dried area. The rewetting was promoted at higher heat flux where liquid supply from the side edges of duct was enhanced pushed by the bubble generation there. By further increase of heat flux results in nucleation of bubbles from the center of heating surface and the local heat transfer coefficients became higher by the evaporation of microlayer which was far thinner than the thickness of flowing liquid film.

5. For most combinations of gas Reynolds number and heat flux, local heat transfer coefficients were decreased with increasing gas Reynolds number because of the liquid film rupture and the extension of dried area by the excessive interfacial shear stress exerted by the gas flow.

It was clarified that the behaviors of liquid film and corresponding heat transfer were strongly influenced by the existence of duct corners and the phenomena became more unsteady or periodical and not uniform in the transverse direction perpendicular to the flow.

ACKNOWLEDGEMENTS

We gratefully acknowledge the support by the Ministry of Education and Science of Russia (Project identifier RFMEFI61314X0011).

NOMENCLATURE

h_N	Nozzle height (m)
H	Channel height (m)
P	Pressure (Pa)
q	Heat flux (W/m^2)
Re	Reynolds number
T	Temperature (K)

Greek Symbols

α	Heat transfer coefficient (W/m^2 K)

Subscripts

g	Gas
in	Inlet
l	Liquid
out	Outlet

REFERENCES

Ajaev, V.S., Gatapova, E.Y., and Kabov, O.A., 2011, "Rupture of Thin Liquid Films on Structured Surfaces," *Physical Review E.*, **84**, 041606. http://dx.doi.org/10.1103/PhysRevE.84.041606

Ajaev, V.S., Gatapova, E.Y., and Kabov, O.A., 2012, "Stability of a Liquid Film on a Surface with Periodic Array of Gas-filled Grooves," *Microgravity Science and Technology*, **24**, 33-37. http://dx.doi.org/10.1007/s12217-011-9288-z

Cheverda, V.V., Glushchuk, A., Queeckers, P., Chikov, S.B., and Kabov, O.A., 2013, "Liquid Rivulets Moved by Shear Stress of Gas Flow at Altered Levels of Gravity," *Microgravity Science and Technology*, **25**, 73-81. http://dx.doi.org/10.1007/s12217-012-9335-4

Du, S.Y., and Zhao, Y.H., 2012, "Numerical Study of Conjugated Heat Transfer in Evaporating Thin-films near the Contact Line," *International Heat and Mass Transfer*, **55**, 61-68. http://dx.doi.org/10.1016/j.ijheatmasstransfer.2011.08.039

Gatapova, E.Y., and Kabov, O.A., 2008, "Shear-driven Flows of Locally Heated Liquid Films," *International Journal of Heat and Mass*

Transfer, **51**, 4794-4810.
http://dx.doi.org/10.1016/j.ijheatmasstransfer.2008.02.038

Gatapova, E.Y., Kabov, O.A., and Marchuk, I.V., 2004, "Thermocapillary Deformation of a Locally Heated Liquid Film Moving under the Action of a Gas Flow," *Technical Physics Letters*, **30**, 418-421.
http://dx.doi.org/10.1134/1.1760873

Gatapova, E.Y., Lyulin, Y.V., Marchuk, I.V., Kabov, O.A., and Legros, J.C., 2003, "The Thermocapillary Convection in Locally Heated Laminar Liquid Film Flow Caused by Co-current Gas Flow in Narrow Channel," *International Conference on Microchannels and Minichannels*, **1**, 457-465.
http://dx.doi.org/10.1115/ICMM2003-1055

Höhmann, C., and Stephan, P., 2002, "Microscale Temperature Measurement at an Evaporating Liquid Meniscus," *Experimental Thermal and Fluid Science*, **26**, 157-162.
http://dx.doi.org/10.1016/S0894-1777(02)00122-X

Houshmand, F., and Peles, Y., "Convective Heat Transfer to Shear-driven Liquid Film Flow in a Microchannel," *International Journal of Heat and Mass Transfer*, **64**, 42-52.
http://dx.doi.org/10.1016/j.ijheatmasstransfer.2013.04.012

Ibrahem, K., Rabbo, M.F.A., Roisman, T.G., and Stephan, P., 2010, "Experimental Investigation of Evaporative Heat Transfer Characteristics at the 3-phase Contact Line," *Experimental Thermal and Fluid Science*, **34**, 1036-1041.
http://dx.doi.org/10.1016/j.expthermflusci.2010.02.014

Kabov, O.A., and Chinnov, E.A., 1997, "Heat Transfer from a Local Heat Source to a Subcooled Falling Liquid Film Evaporating in a Vapour-gas Medium," *Russian Journal of Engineering Thermophysics*, 7, 1-34.

Kabov, O.A., Lyulin, Y.V., Marchuk, I.V., and Zaitsev, D.V., 2007, "Locally Heated Shear-driven Liquid Films in Microchannels and Minichannels," *International Journal of Heat and Fluid Flow*, **28**, 103-112.
http://dx.doi.org/10.1016/j.ijheatfluidflow.2006.05.010

Kabov, O.A., Scheid, B., Sharina, I.A., and Legros, J.C., 2002, "Heat Transfer and Rivulet Structures Formation in a Falling Thin Liquid Film Locally Heated," *International Journal of Thermal Sciences*, **41**, 664-672.
http://dx.doi.org/10.1016/S1290-0729(02)01361-3

Kabov, O.A., Zaitsev, D.V., Cheverda, V.V., and Bar-Cohen, A., 2011, "Evaporation and Flow Dynamics of Thin, Shear-driven Liquid Films in Microgap Channels," *Experimental Thermal and Fluid Science*, **35**, 825-831.
http://dx.doi.org/10.1016/j.expthermflusci.2010.08.001

Kabova, Y.O., Kuznetsov, V.V., and Kabov, O.A., 2008, "Gravity Effect on the Locally Heated Liquid Film Driven by Gas Flow in an Inclined Minichannel," *Microgravity Science and Technology*, **20**, 187-192.
http://dx.doi.org/10.1007/s12217-008-9032-5

Kabova, O.Y., Kuznetsov, V.V., and Kabov, O.A., 2009, "The Effect of Gravity and Shear Stress on a Liquid Film Driven in a Horizontal Minichannel at Local Heating," *Microgravity Science and Technology*, **21**, 145-152.
http://dx.doi.org/10.1007/s12217-009-9154-4

Kunkelmann, C., Ibrahem, K., Schweizer, N., Herbert, S., and Stephan, P., 2012, "The Effect of Three-phase Contact Line Speed on Local Evaporative Heat Transfer: Experimental and Numerical Investigations," *International Journal of Heat and Mass Transfer*, **55**, 1896-1904.
http://dx.doi.org/10.1016/j.ijheatmasstransfer.2011.11.044

Lel, V., Stadler, H., Pavlenko, A., and Kneer, R., 2007, "Evolution of Metastable Quasi-regular Structures in Heated Wavy Liquid Films," *Heat Mass Transfer*, **43**, 1121-1132.
http://dx.doi.org/10.1007/s00231-006-0187-6

Lel, V.V., Al-Sibai, F., Knner, R., 2009, "Thermal Entry Length and Heat Transfer Phenomena in Laminar Wavy Falling Films," *Microgravity Science and Technology*, **21**, Suppl 1, S215-S220.
http://dx.doi.org/10.1007/s12217-009-9141-9

Lel, V.V., Kellermann, A., Dietze, G., Kneer, R., and Pavlenko, A.N., 2008, "Investigations of the Marangoni Effect on the Regular Structures in Heated Wavy Liquid Films," *Experimental Fluids*, **44**, 341-354.
http://dx.doi.org/10.1007/s00348-007-0408-x

Liu, R., and Kabov, O.A., 2012, "Instabilities in a Horizontal Liquid Layer in Cocurrent Gas Flow with an Evaporating Interface," *Physical Review E*, **85**, 066305.
http://dx.doi.org/10.1103/PhysRevE.85.066305

Park, K., Noh, K.J., and Lee, K.S., 2003a, "Transport Phenomena in the Thin-film Region of a Micro-channel," *International Journal of Heat and Mass Transfer*, **46**, 2381-2388.
http://dx.doi.org/10.1016/S0017-9310(02)00541-0

Park, K., and Lee, K.S., 2003b, "Flow and Heat Transfer Characteristics of the Evaporating Extended Meniscus in a Micro-capillary Channel," *International Journal of Heat and Mass Transfer*, **46**, 4587-4594.
http://dx.doi.org/10.1016/S0017-9310(03)00306-5

Stephan, K., Zhong, L.-C., and Stephan, P., 1995, "Influence of Capillary Pressure on the Evaporation of Thin Liquid Films," *Heat and Mass Transfer*, **30**, 467 - 472.
http://dx.doi.org/10.1007/BF01647453

Lyulin, Y., and Kabov, O., 2014, "Evaporative Convection in a Horizontal Liquid Layer under Shear-stress Gas Flow," *International Journal of Heat and Mass Transfer*, **70**, 599-609.
http://dx.doi.org/10.1016/j.ijheatmasstransfer.2013.11.039

Wang, H., Garimella, S.V., and Murthy, J.Y., 2007, "Characteristics of an Evaporating Thin Film in a Microchannel," *International Journal of Heat and Mass Transfer*, **50**, 3933-3942.
http://dx.doi.org/10.1016/j.ijheatmasstransfer.2007.01.052

Williams, M.B., and Davis, S.H., 1982, "Nonlinear Theory of Film Rupture," *Journal of Colloid and Interface Science*, **90**, 1, 220-228.
http://dx.doi.org/10.1016/0021-9797(82)90415-5

6

AN EXPERIMENTAL STUDY OF THE EFFECT OF PRESSURE INLET GAS ON A COUNTER-FLOW VORTEX TUBE

Mahyar Kargaran,* Mahmood Farzaneh-Gord

Faculty of Mechanical Engineering, Shahrood University of Technology, Shahrood, Iran

ABSTRACT

Vortex tube is a simple device which separate an inlet gas with a proper pressure into hot and cold flows .This device is well-suited for generating cooling load gas because it provides the cold gas without using any refrigerants . Many research works has been carried out in order to identify the factors which contribute to Vortex tube performance. Here, an experimental study has been made to determine the effect of geometrical (length of vortex tube) and thermo-physical (pressure) parameters on vortex tube performance and air also used as a working fluid.

Keywords: Vortex tube ,*geometrical parameters ,thermo-physical parameter* .

1. INTRODUCTION

Vortex tube is a simple device which separate an inlet gas with a proper pressure into hot and cold flows. This device is well-suited for generating cooling load gas because it provides the cold gas without using any refrigerants. Many research works has been carried out in order to identify the factors which contribute to Vortex tube performance. Here, an experimental study has been made to determine the effect of geometrical (length of vortex tube) and thermo-physical (pressure) parameters on vortex tube performance and air also used as a working fluid.

The vortex tube is a device without a moving part with the ability of separating hot and cold air from a higher pressure inlet gas which is tangentially blown into vortex chamber. Such a separation of the flow into regions of low and high total temperature is referred to as the temperature (or energy) separation effect. It contains the following parts: one or more inlet nozzles, a vortex chamber, a cold-end orifice, a hot-end control valve and a tube. When high-pressure gas is tangentially injected into the vortex chamber via the inlet nozzles, a swirling flow is created inside the vortex chamber. When the gas swirls to the center of the chamber, it is expanded and cooled. In the vortex chamber, part of the gas swirls to the hot end, and another part exists via the cold exhaust directly. Part of the gas in the vortex tube reverses for axial component of the velocity and move from the hot end to the cold end. At the hot exhaust, the gas escapes with a higher temperature, while at the cold exhaust, the gas has a lower temperature compared to the inlet temperature. The vortex tube was first discovered by Ranque, a metallurgist and physicist who was granted a French patent for the device in 1932, and a United States patent in 1934. The initial reaction of the scientific and engineering communities to his invention was disbelief and apathy. Since the vortex tube was thermodynamically highly inefficient, it was abandoned for several years. Interest in the device was revived by Hilsch, a German engineer, who reported an account of his own comprehensive experimental and theoretical studies aimed at improving the efficiency of the vortex tube.

Separating cold and hot flows by using the principles of the vortex tube can be applied to industrial applications such as cooling equipment in CNC machines, refrigerators, cooling suits, heating processes, etc. The vortex tube is well-suited for these applications because it is simple, compact, light, quiet, and does not use Freon or other refrigerants (CFCs/HCFCs). It has no moving parts and does not break or wear and therefore requires little maintenance. But, its low thermal efficiency is a main limiting factor for its application. Also the noise and availability of compressed gas may limit its application. In this research numerical and experimental method will be implemented to determine the significant factors on vortex tube behavior with an incompressible flow in order to boost vortex tube energy (thermal) efficiency.

Nimbalkar and Muller (2009) presented the results of a series of experiments focusing on various geometries of the ''cold end side" for different inlet pressures and cold fractions. The experimental results indicated that there is an optimum diameter of cold-end orifice for achieving maximum energy separation. It was observed that for cold fraction less or equal than 60%, the effect of cold end orifice diameter is negligible and above 60% cold fraction it becomes prominent. The results also show that the maximum value of performance factor was always reachable at a 60% cold fraction irrespective of the orifice diameter and the inlet pressure.

Dincer et al. (2009) have studied the effects of position, diameter and angle of a mobile plug, located at the hot outlet side experimentally to get the best performance. The most efficient (maximum temperature difference) combination of parameters is obtained for a plug diameter of 5 mm, and tip angle of 30^0 or 60^0, by keeping the plug in the same position, and letting the air enter into the vortex tube through 4 nozzles. Increasing the inlet pressure beyond 380 kPa did not cause any appreciable improvement in the performance. Stephan et al. (1983) measured the temperature profiles at different positions along a vortex tube axis and concluded that the length of the vortex tube would have an important influence on the transport mechanism inside.

Saidi and Valipour (2003) presented information data on the classification of the parameters affecting vortex tube operation. In their study, the thermophysical parameters such as inlet gas pressure, type of gas and cold gas mass ratio, moisture of inlet gas, and the geometrical parameters, i.e., diameter and length of main tube diameter of the outlet orifice and shape of the entrance nozzle, were designated and studied.

* Corresponding Author. E-mail: m.kargaran@gmail.com

Orhan and Muzaffer (2006) have carried out a series of experiments to investigate the effects of the length of the pipe, the diameter of the inlet nozzle, and the angle of the control valve on the performance of the counterflow vortex tubes for different inlet pressures. Experiments showed that the higher the inlet pressure, the greater the temperature difference of the outlet streams. It was also shown that the cold fraction is an important parameter influencing the performance of the energy separation in the vortex tube. Optimum values for the angle of the control valve, the length of the pipe and the diameter of the inlet nozzle were obtained.

There have been other media than air used as the working fluid. Balmer (1988) applied liquid water as the working medium. It was found that when the inlet pressure is high, for instance 20~50 bar, the energy separation effect still exists. So it proves that the energy separation process exists in an incompressible vortex flow as well. Eiamsa-ard and Promvonge (2008) presented a complete overview of the past investigations of the mean flow and temperature behaviours in a turbulent vortex tube in order to understand the nature of the temperature separation or Ranque–Hilsch effect. They have proposed optimum values for the cold orifice to the VT inlet diameter (d/D) of 0.5, the angle of the conical control valve of 50 degrees, the length of the vortex tube to the VT inlet diameter (L/D) of 20 and the diameter of the inlet nozzle to the VT inlet diameters (δ/D) of 0.33 for air as the working fluid.

Providing cooling load has always been one of the major challenges and one the other hand ,using refrigerants such as Freon (which causes global warming) has been increasing .This device is well-suited for generating cooling load gas because not only it is light ,simple and compact but also provides the cold gas without using any refrigerants In this research an experimental investigation has been carried out to investigate The effects of the VT tube length and the pressure of inlet air on the VT thermal separation. Further, the amount of cooling capacity created as air pass through a VT has been calculated and compared when air with 4.2 and 5.8 bar were injected into vortex tube. As discussed above, it will have potential applications in refrigerators.

2. THE VORTEX TUBE PARAMETERS

There are a few important parameters affecting the VT thermal behavior which should firstly be introduced

2.1 The geometrical parameters

Figure 1 shows a schematic diagram of a counter flow vortex tube which was constructed and used in this study. As shown in Fig.3, the geometrical parameters are inlet VT diameters (D), cold orifice diameter (d), inlet nozzle diameter (δ), conical controlling valve angle (Φ), cold tube length (L_c) and hot tube length (L_h) and N is the number of nozzle . Table 1 shows the detailed geometrical parameters dimensions used in this study. These values are selected based on proposed optimum values by Eiamsa-ard and Promvonge (2008). As can be seen, the cold orifice diameter was 17.1 mm and hot tube length was varied from 250 to 769 mm.

Table 1 The geometrical parameters and their values

Parameters	*value*		
D	25mm		
d	17.1mm		
Φ	50 degree		
δ	8mm		
Lc	50 mm		
N	1		
L_h	250 mm	519 mm	769mm

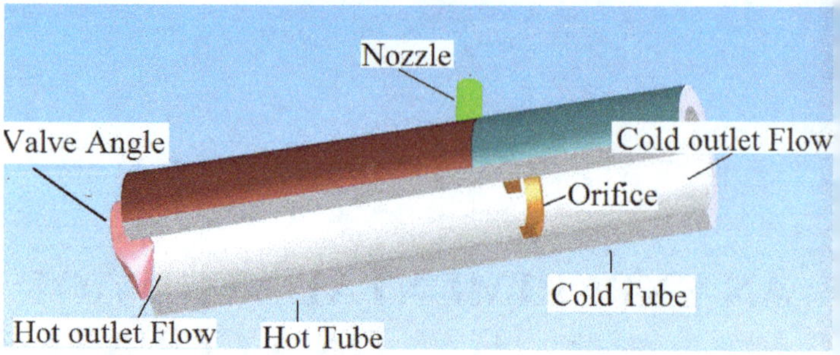

Fig. 1 The schematic diagram of a counter flow vortex tube

2.2 The flow parameters

As mentioned by Eiamsa-ard and Promvonge (2008), the most important flow parameter is believed to be cold mass fraction ,defined as:

$$\mu_c = \frac{\dot{m}_c}{\dot{m}_i} \quad (1)$$

where $\dot{m}_c$ and $\dot{m}_i$ are the mass flow rates at the inlet of the vortex tube and at the cold outlet, respectively. The other flow parameters are:

a) The cooling (ΔT_c) and the heating (ΔT_h) effects of the vortex tube are defined as follows , respectively:

$$\Delta T_c = T_i - T_c \quad (2)$$

$$\Delta T_h = T_h - T_i \quad (3)$$

where T_i is the inlet stream temperature, T_c is the outlet stream temperature of the cold end and T_h is the outlet stream temperature of the hot end.

b) The performance of the vortex tube was defined as the difference between the heating effect and the cooling effect. Subtracting Eq. 2 from Eq. 3 gives the vortex tube performance equation as follows (Eq. 4):

$$\Delta T = T_h - T_c \quad (4)$$

c) the cooling capacity which is defined as:

$$\dot{Q}_C = \dot{m}_c \Delta h_c = \dot{m}_c (h_i - h_c) \quad (5)$$

For the case of an ideal gas, the cooling capacity may be defined as:

$$\dot{Q}_C = \dot{m}_c \Delta h_c = \dot{m}_c c_p (T_i - T_c) = \dot{m}_c c_p \Delta T_c \quad (6)$$

Considering above definitions, the specific cooling capacity can be derived as follow:

$$q_c = \frac{\dot{Q}_c}{\dot{m}_i} = \mu_c \Delta h_c \quad (7)$$

3. EXPERIMENTAL APPARATUS

Figure 2 shows a schematic diagram of the experimental apparatus and measuring devices. High pressure air from compressor is directed tangentially into the vortex tube. The high pressure gas expands in the vortex tube and separates into cold and hot streams. The cold gas leaves the central orifice near the entrance nozzle, while the hot gas discharges the periphery at the far end of the tube. The control valve is being used to control the flow rate of the hot stream. This would help to regulate cold mass friction. Two orifice flow meters constructed according to ISO5167 are employed to measure the mass flow rate of the hot and cold streams. 3 PT100 temperature sensors are installed to measure inlet, hot and cold stream temperatures. 2 pressure transmitters are utilized to quantify inlet pressure and outlet pressure of hot streams.

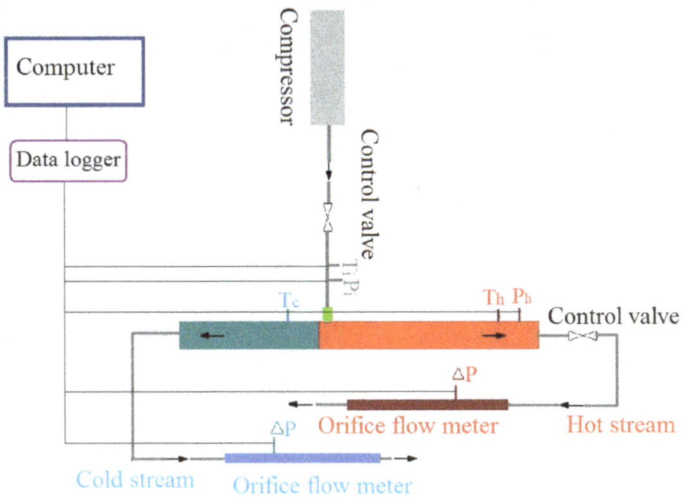

Fig. 2 A schematic diagram of experimental layout

Figure 3 shows the experimental test bed has been conducted at Koolab Toos Company to investigate thermal separation of air as working fluid. The inlet pressure was varied during experiments from 4.2 to 5.8 bar which are gauge pressure and the inlet temperature was 25.4°C.Noting from the figure, the hot length of the VT and hot stream flow meter were painted in red. In other hand, the cold length of the VT and cold stream flow meters were painted in blue. The VT was made from steel with inlet diameter of 25 mm. During the tests, the hot tube length of the VT was varied among 3 available as detailed in Table 1.

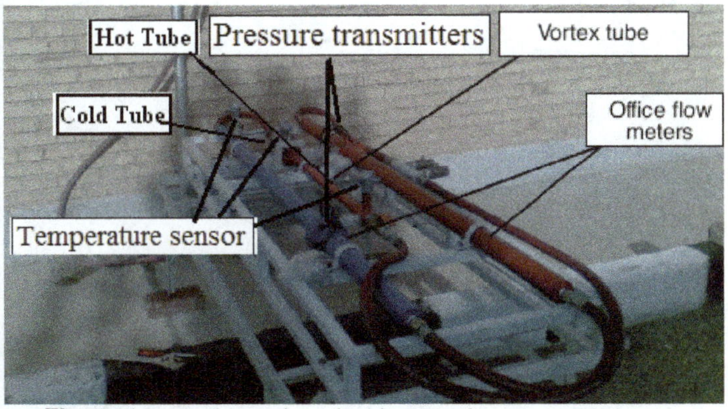

Fig. 3 The experimental test bed in operation

4. ERROR ANALYSIS

The errors associated with temperature measurements are computed in this section. The maximum possible errors in various measured parameters ;namely ,temperature and pressure ,were estimated by using the method proposed by Moffat (1985). Errors were estimated from the minimum values of output and the accuracy of the instrument. This method is based on careful specification of the uncertainties in the various experimental measurements. If an estimated quantity, Y depends ,depends on independent variables like x_i then the error in the value of "Y" is given by

$$\frac{\partial Y}{Y}=\sqrt{\sum_{1}^{n}\left(\frac{\partial x_i}{x_i}\right)^2} \qquad (8)$$

where $\frac{\partial x_i}{x_i}$ are the errors in the independent variables.

∂x_i = Accuracy of the measuring instrument

x_i =Minimum Value of the output measured

4.1 Error in temperature measurement

PT100 temperature sensors were used to measure the gas temperatures. Temperatures are logged in file with accuracy of 0.1^0c. The maximum possible error in the case of temperature measurement was calculated from the minimum values of the temperature measured and accuracy of the instrument. The error in the temperature measurement is:

$$\frac{\partial T}{T}=\sqrt{\left(\frac{\partial T_{PT100}}{T_{\min}}\right)^2+\left(\frac{\partial T_{\log}}{T_{\min}}\right)^2}=\sqrt{\left(\frac{.5}{12}\right)^2+\left(\frac{.1}{12}\right)^2}\approx 0.04=4\% \qquad (9)$$

4.2 Error in pressure measurement

Pressure transmitters were used to measure the gas pressure. Pressures directly are logged in file with accuracy 0f 0.01 bar. The error in the pressure measurement is:

$$\frac{\partial P}{P}=\sqrt{\left(\frac{\partial P_{\max}}{P_{\min}}\right)^2+\left(\frac{\partial P_{\log}}{P_{\min}}\right)^2}=\sqrt{\left(\frac{.01}{1.33}\right)^2+\left(\frac{.0.01}{1.33}\right)^2}\approx 0.01=1\% \qquad (10)$$

4.3 Error in flow rate measurement

Flow measurement has been made using orifice flow meters. Uncertainty analysis was conducted according to the standard procedures reported in ISO5167. The analysis shows that the error in the flow rate measurement is 4.5%.

5. RESULT AND DISCUSSION

Figure 4 shows the effects of tube length on non-dimensional hot temperature differences when p=5.8 bar. As the graph shows , the tube with L_h =769 mm creates the highest cold temperature differences. It should be also noticed that there is a μ_c in each case which causes the temperature differences to be maximized. For this case, at $\mu_c \approx 0.59$ the highest cold temperature differences are generated .

Figure 5 shows the effects of tube length on non-dimensional hot temperature differences when p=5.8 bar. As the graph demonstrates, the tube with L_h=250 mm produced maximum hot temperature differences and at $\mu_c \approx .8$ the highest hot temperature differences are produced. Figure 6 represents the effects of tube length on non-dimensional cold temperature differences when p=4.2 bar. As can be seen , the tube with L_h =769 mm creates the highest cold temperature differences. It should be also noticed that there is a μ_c in each case which causes the temperature differences to be maximized. For this case, at $\mu_c \approx 0.58$ the highest cold temperature differences are generated.

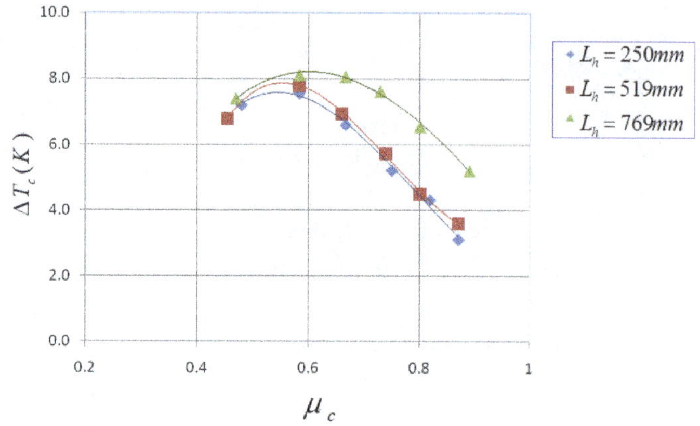

Fig. 4 Effect of L_h on cold temperature difference

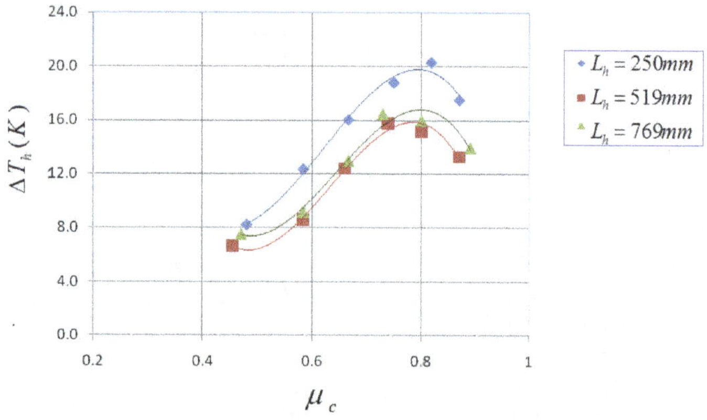

Fig. 5 Effect of L_h on hot temperature difference

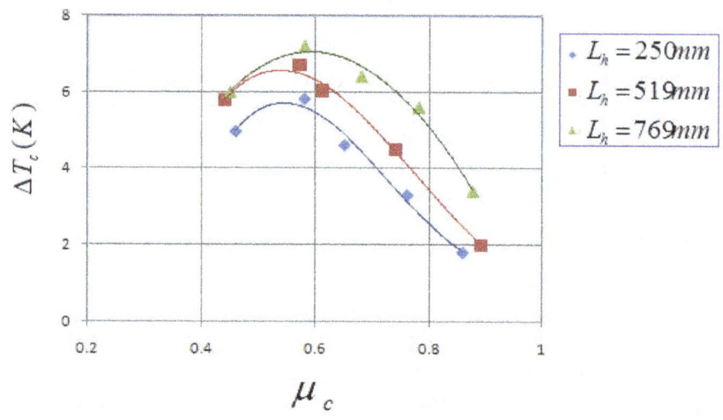

Fig. 6 Effect of L_h on cold temperature difference

Figure 7 indicates the effects of tube length on non-dimensional hot temperature differences when p=4.2 bar. As the graph demonstrates, the tube with L_h =250 mm produced maximum hot temperature differences and at $\mu_c \approx 0.75$ the highest hot temperature differences are produced. Figure 8 shows the effects of hot tube length on specific cooling capacity when p=5.8 bar. As can be seen, the tube with L_h =769 mm produces the highest specific cooling capacity at $\mu_c \approx 0.65$ while the tube with L_h =250 mm creates the lowest amount of cooling capacity.

Nikolaev et al. (1995) found that the maximum refrigeration capacity of the vortex tube falls within the range from 60% to 70% cold fraction. Poshernev and Khodorkov (2003) mentioned that within their range of input parameters the refrigerating capacity has a distinct maximum at a cold fraction of about 50%–60%. Nimbalkar and Muller (2009) mentioned that the maximum value of performance factor was always reachable at a 60% cold fraction irrespective of the orifice diameter. As can be seen from figure 8, our results also show that maximum cooling capacity is encountered at about 65% cold fraction regardless of the orifice diameter. This phenomenon can be explained on the basis of pressure balance inside the vortex tube, discussed by Love (1974), and Piralishvili and Fuzeeva (2005).

Figure 9 shows the effects of hot tube length on specific cooling capacity when p=4.2 bar. The graph indicates that the tube with L_h =769 mm produces maximum specific cooling capacity (like the previous diagram) at $\mu_c \approx 0.58$. On the other hand, tube with L_h =250 mm has the lowest q_c in this regard.

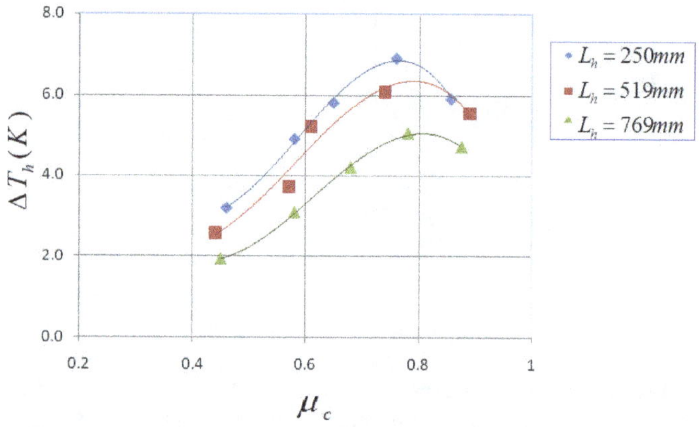

Fig. 7 Effect of L_h on hot temperature difference

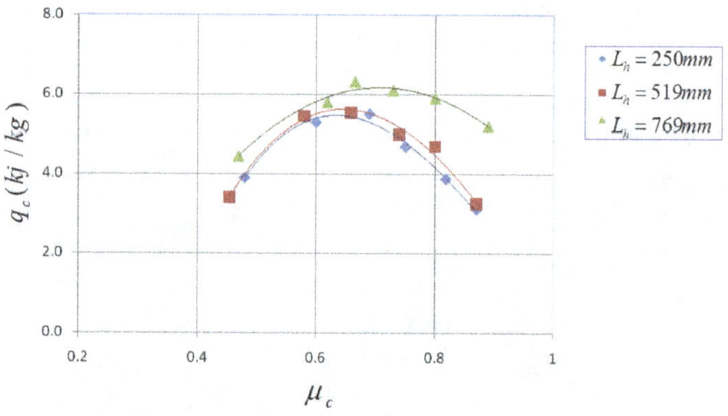

Fig. 8 Effect of L_h on specific cooling capacity

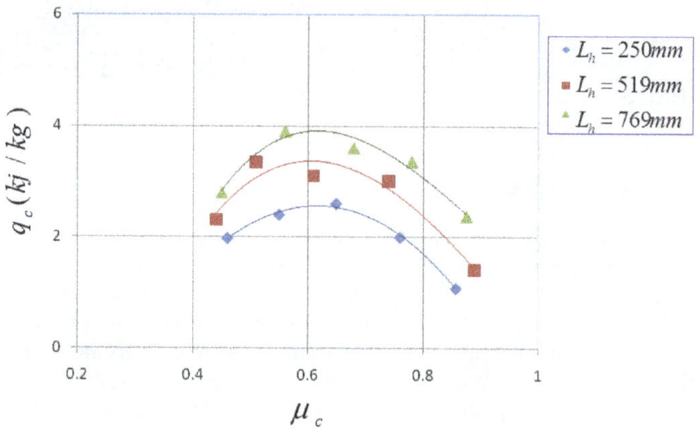

Fig. 9 Effect of L_h on specific cooling capacity

6. CONCLUSION

An experimental study has been carried out to offer optimum values for the length of VT to its inlet diameter (L_h / D) and assessed the pressure inlet gas factor on vortex tube performance .The comparison between Figures 4-7 demonstrated that cold temperature difference slightly and hot temperature difference dramatically increases as the inlet pressure increases .In this study , ΔT_c reached about 8 degree centigrade and ΔT_h went over 20 degree centigrade for p=5.8 bar. The results also show that hot temperature difference is maximized for a specific L_h. In this regard, $L_h / D = 250 / 25 = 10$. As far as cooling capacity and hot temperature difference are concerned ,the study indicates that for $L_h / D = 769 / 25 = 30.76$ we have maximum proficiency. As for μ_c value, at $\mu_c \approx 0.6$, we witness the maximum of cold temperature difference and cooling capacity , while for hot temperature difference μ_c is about .75.

ACKNOWLEDGEMENTS

This study has been supported by Semnan (Iran) Gas Company. Special thanks also go to Koolab Toos company. The authors are also grateful to the reviewers of this paper for their time and valuable comments .

NOMENCLATURE

D	vortex tube inlet diameter (mm)
d	cold orifice diameter (mm)
L	vortex tube length (mm)
L_h	hot tube length (mm)
L_c	cold tube length (mm)
P	pressure (bar)
$\dot{Q}_c$	cooling capacity (Kw)
$\dot{q}_c$	specific cooling capacity (kj/kg)
RHVT	Ranque-Hilsch vortex tube
T	temperature (K)
ΔT	temperature difference (K)
VT	Vortex tube
$\dot{m}$	mass flow
μ_c	cold mass fraction tube

Greek Symbols

δ	inlet nozzle diameter(mm)
Φ	conical controlling valve angle

Superscripts

1	inlet gas condition of the pressure drop
2	outlet gas condition of the pressure drop

Subscripts

c	cold stream
h	hot stream
i	inlet stream

REFERENCES

Balmer R.T.,1998, "Pressure-driven Ranque-Hilsch Temperature Separation in Liquids," *ASME, J. Fluids Engineering* 110, 161–164. http://dx.doi.org/10.1115%2F1.3243529

Cockerill T., 1995, "The Ranque–Hilsch Vortex Tube", *PhD thesis, Cambridge University, Engineering Department, Sunderland.*

Collins R.L., Lovelace R.B. 1997, "Experimental Study of Two-phase propane expanded through the Ranque-Hilsch tube," Trans. *ASME, J. Heat Transfer* 101, 300–305. http://dx.doi.org/10.1115%2F1.3450964

Dincera K., Baskayab S., Uysalc B.Z., Ucguld I., 2009, " Experimental Investigation of The Performance of a Ranque–Hilsch vortex tube with regard to a plug located at the hot outlet," *International journal of refrigeration 32,* 87–94. http://dx.doi.org/10.1016%2Fj.ijrefrig.2008.06.002

Eiamsa-ard S., Promvonge P. ,2008," Review of Ranque–Hilsch Effects in Vortex Tubes," *Renewable and Sustainable Energy Reviews* 12, 1822–1842. http://dx.doi.org/10.1016%2Fj.rser.2007.03.006

Hilsch R., 1947, " The Use of Expansion of Gases in a Centrifugal Field as a Cooling Process," *Rev Sci Instrum.* 18, 2, 108–13.

ISO-51671, *Measurement of fluid flow by means of pressure differential devices inserted in circular-cross section conduits running full* - Part1: *General principles and requirements.* & Part 2: *Orifice Plates.*

Khodorkov L., Poshernev N.V., Zhidkov. M.A. (203) "The vortex Tube—a Universal Device for Heating, Cooling, Cleaning, and Drying Gases and Separating Gas Mixtures," *Chemical and Petroleum Engineering* 39, 7-8, 409–415.

Kulkarni M.R., Sardesai C.R.,2002, "Enrichment of Methane Concentration Via Separation of Gases Using Vortex Tubes," J. *Energy Engrg* 128, 1, 1–12. http://dx.doi.org/10.1061%2F%28ASCE%290733-9402%282002%29128%3A1%281%29

Lin S., Chen J.R., Vatistas G.H. ,1990, "A Heat Transfer Relation for Swirl Flow in a Vortex Tube", *Can J Chem Eng* 68, 6, 944–7. http://dx.doi.org/10.1002%2Fcjce.5450680608

Lewins J., Bejan A., 1995, "Vortex Tube Optimization Theory", *Energy* 24, 931–943. http://dx.doi.org/10.1016%2FS0360-5442%2899%2900039-0

Love W.J. ,1974, "Prediction of Pressure Drops in Straight Vortex Tube," AIAA J. 12, 7. http://dx.doi.org/10.2514%2F3.49387

M.Farzaneh –Gord ,M.Kargaran, 2010,"Recovering Energy at Entry of Natural Gas into Customer Premises by Employing a Counter-Flow Vortex Tube ,"*Oil,Gas Science and Technology-Revue de l'IFP ,Vol.65,* No .6,pp.903-912 http://dx.doi.org/10.2516%2Fogst%2F2009074

Farzaneh–Gord, M., Kargaran, M., Bayat, Y., Hashemi, SH, 2012, "Investigation of Natural Gas Thermal Separation through a Vortex Tube," *Enhanced Heat Transfer,*Vol.19, No .187-94. http://dx.doi.org/10.1615%2FJEnhHeatTransf.2012001545

Moffat, R.J., 1985, "Using Uncertainty Analysis in the Planning of an Experiment", Trans. *ASME, J. Fluids Eng.* 107, 173-178. http://dx.doi.org/10.1115%2F1.3242452

Nikolaev V.V., Ovchinnikov V.P., Zhidkov M.A. ,1995, " Experience from the Operation of a Variable Vortex Tube in a Gas Separating Station," *Gaz. Prom* 10, 13.

Nimbalkar S.U., Muller M.R. ,2009, " An Experimental Investigation of the Optimum Geometry for the Cold end Orifice of a Vortex Tube," *Applied Thermal Engineering* 29, 509–514. http://dx.doi.org/10.1016%2Fj.applthermaleng.2008.03.032

Orhan A., Baki Muzaffer ,2006, " An Experimental Study on the Design Parameters of a Counterflow Vortex Tube," *Energy* 31, 2763–2772.

Piralishvili S.A., Polyaev V.M. ,1996, "Flow and Thermodynamic Characteristics of Energy Separation in a Double-Circuit Vortex Tube-an Experimental Investigation," *Exp Thermal Fluid Sci* 12, 4, 399–410. http://dx.doi.org/10.1016%2F0894-1777%2895%2900122-0

Piralishvili A., Fuzeeva A.A.,2005, "Hydraulic Characteristics of Ranque–Hilsch Energy Separators,"High Temp 43, 6, 900–907. http://dx.doi.org/10.1007%2Fs10740-005-0137-x

Ranque G.J.,1993, "Experiments on Expansion in a Vortex with Simultaneous Exhaust of Hot Air and Cold Air," *J Phys Radium (Paris),* 4:112–4 S-115, June

Ranque G.J.,1934, "Method and Apparatus for Obtaining from a Fluid under Pressure Two Outputs of Fluid at Different Temperatures," US patent 1:952,281.

Saidi M.H., Valipour M.S. ,2003, "Experimental Modeling of Vortex Tube Refrigerator,"*Applied Thermal Engineering* 23, 1971–1980. http://dx.doi:10.1016/S1359-4311(03)00146-7.

Stephan K., Lin S., Durst M., Huang F., Seher D.,1983, "An Investigation of Energy Separation in a Vortex Tube," *Heat Mass Transfer* 26, 341–8. http://dx.doi.org/10.1016%2F0017-9310%2883%2990038-8

Takahama H., Kawamura H., Kato S., Yokosawa H. ,1979, "Performance Characteristics of Energy Separation in a Steam-Operated Vortex Tube," *Int J Eng Sci* 17, 735–44. http://dx.doi.org/10.1016%2F0020-7225%2879%2990048-X

7

EFFECT OF WALL THERMAL CONDUCTIVITY ON MICRO-SCALE COMBUSTION CHARACTERISTICS OF HYDROGEN-AIR MIXTURES WITH DETAILED CHEMICAL KINETIC MECHANISMS IN Pt/γ-Al_2O_3 CATALYTIC MICRO-COMBUSTORS

Junjie Chen*, Longfei Yan, Wenya Song

School of Mechanical and Power Engineering, Henan Polytechnic University, Jiaozuo, Henan, 454000, China

ABSTRACT

To understand the effect of different thermal conductivities on catalytic combustion characteristics, effect of thermal conductivity on micro-combustion characteristics of hydrogen-air mixtures in Pt/γ-Al_2O_3 catalytic micro-combustors were investigated numerically with detailed chemical kinetics mechanisms. Three kinds of wall materials (100, 7.5, and 0.5 W/m·K) were selected to investigate the effect of heat conduction on the catalytic combustion. The simulation results indicate that the catalytic reaction restrains the gas phase reaction in Pt/γ-Al_2O_3 catalytic micro-combustors. The gas phase reaction restrained by Pt/γ-Al_2O_3 catalysts is sensitive to thermal boundary condition at the wall. For most conditions, the gas phase reaction cannot be ignored in Pt/γ-Al_2O_3 catalytic micro-combustors. For low thermal conductivity, the higher temperature gradient on the wall will promote the gas phase reaction shift upstream; high temperature gradient exists on the wall, and the hot spot can cause the material to melt or degrade the catalyst. Due to the gas phase reaction is ignited and sustained in micro-combustors by the heat from the catalytic reaction, the effect of thermal conductivity on micro-scale combustion characteristics is not as obvious as it is in micro-combustors without Pt/γ-Al_2O_3 catalysts.

Keywords: *thermal conductivity; combustion characteristics; catalytic combustion; wall material; micro-scale; Pt/γ-Al_2O_3.*

1. INTRODUCTION

With the increasing demands on micro devices such as micro-satellite thrusters, micro unmanned aerial vehicles, and chemical micro-sensors and reactors, the needs for the micro-power supply are also increasing, especially on the micro-power source with high-energy density. The alkaline and lithium batteries have quite low specific energy compared with hydrocarbon fuels (Shirsat and Gupta, 2011). Development of micro-combustion based power generating devices, even with relatively inefficient conversions of hydrocarbon fuels to power, would result in reduced weight and increased lifetime of a mechanical or electronic system that currently uses lithium batteries for power. Recent advances in the fields of silicon-based MEMS and silicon micro-fabrication techniques have caused the possibility of the new generation of micro-engines for power generation. Micro-combustor is the critical component for micro-power generation systems using hydrocarbon fuels as the energy source. Several types of chemical micro-reactor and micro-combustor are currently under development (Hua et al., 2005).

In fact, there are some challenges to maintain the stable combustion in micro-combustors. Extinction occurs easily at the micro-scale, because the large area-to-volume ratio increases the heat loss. Another important factor affecting the micro-combustion stability is the quenching distance, which equals to the dimension of micro-combustor (Zhang et al., 2007; Wang et al., 2011).

To improve the micro-combustion stability and the thermal efficiency, various efforts have been paid on the use of catalyst and the optimization of the thermal management, such as the excess enthalpy combustors (Chen et al., 2013b; Chein et al., 2013; Kocich et al., 2012; Zhong et al., 2012), the heat-recirculating combustors (Belmont and Ellzey, 2014; Chen et al., 2013a; Deshpande and Kumar, 2013; Kurdyumov and Matalon, 2011; Rana et al., 2014; Shirsat and Gupta, 2013), and the catalytic micro-combustors to suppress radical depletion and to enhance the reaction (Federici et al., 2009; Federici and Vlachos, 2011; Hsueh et al., 2010; Hsueh et al., 2011; Wang et al., 2010). Many micro-combustor studies have investigated the variation of dimension, thermal conductivity, inlet velocity, equivalence ratio, etc. (Belmont and Ellzey, 2014; Hua et al., 2005; Ju and Xu, 2005; Raimondeau et al., 2011; Wang et al., 2011; Zhong et al., 2011). Ju and Xu (2005) theoretically and experimentally studied the flame propagation and extinction in micro-channels. Their results showed that the wall thermal properties, flow velocity, and channel width have significant effects on the flame propagation, and cause extinction limits and multiple flame regimes. Leach et al. (2006) and Seyed-Reihani and Jackson (2004) performed a one-dimensional numerically investigated the effects of heat exchange on the reaction zone thickness of stoichiometric premixed hydrogen-air mixtures in micro-channel combustors. They presented an analytical model, which is used to predict the reaction zone thickness of hydrogen-air mixtures based on the thermal properties and the channel size. The heat exchange through the micro-combustors structure can lead to a broadening reaction zone. Vijayan and Gupta (2010 and 2011) performed simulations to highlight the importance of material thermal conductivity and determined the range of thermal conductivities over which combustion could be supported. The results showed that heat loss from the combustion chamber was lower for lower thermal conductivity materials with the overall heat loss being roughly independent of thermal conductivity. However, it is heat lost from the combustion chamber that promotes thermal quenching and not the overall heat loss. Lower thermal

* *Corresponding author. Email: comcjj@163.com*

conductivity materials also have higher rates of heat recirculation to the reactants.

While the micro-combustion has been studied, the effect of the catalytic wall on micro-combustion characteristics are still not fully understood. Most previous computational studies dealt with surface catalytic or gas phase reaction separately and concentrated on the flame stability or extinction limit (Chen et al., 2007; Deutschmann et al., 2012; Ju and Maruta, 2011; Kamijo et al., 2009; Karagiannidis and Mantzaras, 2010; Maruta, 2011). Although the catalyst is used to maintain the reaction and to decrease the heat loss, the effect of thermal conductivity on micro-combustion characteristics and the interaction between surface catalytic and gas phase reaction in micro-combustors are still not fully understood. In this work, micro-combustion characteristics of premixed hydrogen-air mixture in Pt/γ-Al_2O_3 catalytic micro-combustors are investigated numerically. Numerical simulations with detailed chemical kinetics mechanisms of hydrogen-air mixture combustions in Pt/γ-Al_2O_3 catalytic micro-combustors are investigated by using the FLUENT coupled with the DETCHEM (Deutschmann et al., 2013). The effect of different reaction models and thermal conductivities on micro-combustion characteristics are discussed.

2. NUMERICAL MODELS AND SIMULATION APPROACH

2.1 Model geometry and mesh

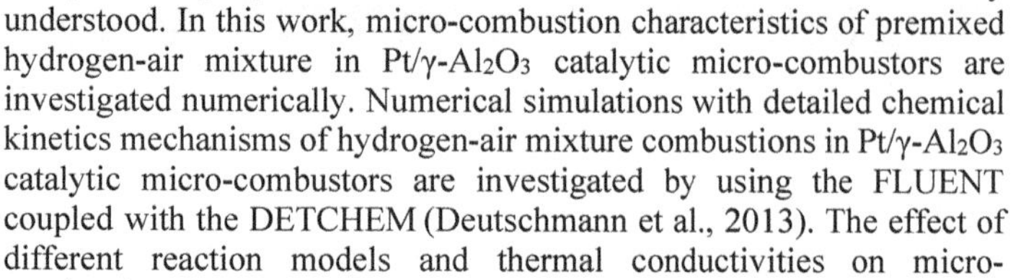

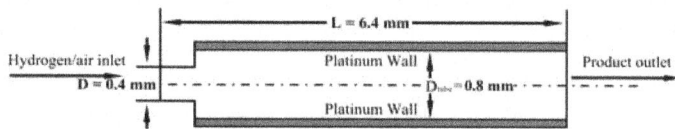

Fig. 1 Schematic diagram of micro-combustor

Table 1 The properties of Pt/γ-Al_2O_3 catalyst washcoat

Property	Value
catalyst surface site density Γ (mol/cm^2)	2.7×10^{-9}
average pore diameter d_{pore} (m)	2.08×10^{-8}
catalyst porosity ε_{cat}	0.4
catalyst tortuosity τ_{cat}	8.0

The geometry of the Pt/γ-Al_2O_3 catalytic micro-combustor is shown in Fig. 1. The micro-tube is modeled with the inner wall coated with Pt/γ-Al_2O_3 catalyst. The properties of Pt/γ-Al_2O_3 catalyst washcoat are shown in Table 1. The solid wall material is refractory ceramics (SiC). The aspect ratio of the micro-combustion chamber is kept the same to investigate the effect of dimensional scaling on micro-combustion characteristics. The inlet diameter (D), tube diameter (D_{tube}) and tube length (L) of the micro-tube are 0.4 mm, 0.8 mm and 6.4 mm, respectively. The wall thickness is 0.1 mm. Premixed; stoichiometric hydrogen-air mixtures is injected into the micro-combustor from the inlet located at one axial end with a step expansion as shown in Fig. 1. The geometry is modeled as a two-dimensional axisymmetric model because of the axial symmetry of the micro-combustor. For all scenarios analyzed, the same grid size of 2 μm is used to mesh the micro-combustion models for the numerical simulations. This fine mesh size will be able to provide good spatial resolution for the distribution of most variables in micro-combustors.

2.2 Fluid flow modeling

The reacting gas flow path in micro-combustors and the characteristic length of the combustion chamber, even for power MEMS systems, are still sufficiently larger than the molecular mean-free path of the gases flowing through the systems. Therefore, the fluid media can be reasonably considered as continuum in micro-combustors. The Navier-Stokes equation is solved for the fluid domain and no-slip condition on the wall is applied. FLUENT 6.3 is used to perform numerical simulations of the fluid flow by solving the conservation equations of energy, momentum, mass, species as well as the conjugated heat transfer condition in the walls. The laminar viscous flow in micro-combustors is considered, and the segregated solution solver of double-precision is applied to solve the above-mentioned set of governing equations. The fluid mixtures thermal conductivity, viscosity, and specific heat are calculated from a mass fraction weighted average of species properties. The fluid density in micro-combustors is calculated using the ideal gas law. In order to couple the heat transfer and the fluid dynamics of the gas-mixtures flow with the detailed chemical kinetics mechanisms (gas phase chemical kinetics and surface catalytic chemical kinetics), an external program DETCHEM 2.5 is applied as user-defined function to FLUENT to extend the modeling capabilities in simulating the detailed chemical kinetics. The DETCHEM software package is designed for modeling and simulation of reactive fluid flows including surface catalytic reaction on catalyst and can apply elementary and multi-step reaction mechanisms in the gas phase and on surfaces (Deutschmann et al., 2013). The CFD simulation convergence is judged upon the residuals of above-mentioned governing equations.

2.3 Chemical kinetics mechanisms

Chemical kinetics mechanisms are applied on surfaces and in the gas phase. The surface catalytic reaction mechanism presented by Deutschmann et al. (1996) were employed. The gas phase reaction mechanism consists of 19 reactions and 9 species, which are adopted from the mechanism presented by Miller and Bowman (1989). The above-mentioned chemical kinetics mechanisms have been applied in the previous study (Deutschmann et al., 2000), and the comparisons with experimental results are satisfactory. For hydrogen fuel, five surface species (H(s), O(s), OH(s), H_2O(s) and Pt(s)) describe the coverage of the surface with adsorbed species. Pt(s) denotes free surface sites which are available for adsorption. The chemical kinetics mechanisms with CHEMKIN format in the gas phase and chemical kinetics mechanisms with surface CHEMKIN on surfaces are imported into the code.

2.4 Boundary conditions

The concentration of the hydrogen-air mixtures is specified at the inlet section of micro-combustor. The mass fraction of hydrogen is 0.0283. The inlet temperature and the ambient temperature for the hydrogen-air mixture are 300 K. For all cases studied, a uniform inlet velocity is specified. At the solid wall, the thermal boundary condition is the heat loss to the ambient air. At the outer surface of the solid walls, heat losses to the surroundings are calculated through Eq. (1), in which both natural convection and thermal radiation are considered.

$$q = h(T_{w,o} - T_\infty) + \varepsilon\delta(T_{w,o}^4 - T_\infty^4) \qquad (1)$$

Where the heat transfer coefficient h is assumed be 20 W/m^2·K. $T_{w,o}$ and T_∞ are the outer wall and ambient temperatures, respectively. The emissivity ε of solid wall is 0.5 and δ is the Stefan-Boltzmann constant. At the exit, the far-field boundary condition is not adopted (Norton and Vlachos, 2003), and the pressure is specified at a constant pressure of 0.1 MPa and an extrapolation scheme is used for temperature and species.

3. RESULTS AND DISCUSSION

3.1 Micro-combustion characteristics for different reaction models

In the present work, the interaction between gas phase reaction and surface catalytic reaction of hydrogen-air in Pt/γ-Al_2O_3 catalytic micro-combustors is the main issue. In order to clearly identify and to highlight the effect of surface catalytic reaction on micro-combustion characteristics, three different reaction modes by different combinations of the gas phase and surface catalytic reaction mechanism are applied. They are gas phase reaction alone, surface catalytic reaction alone, both gas phase reaction and surface catalytic reaction mechanisms, respectively.

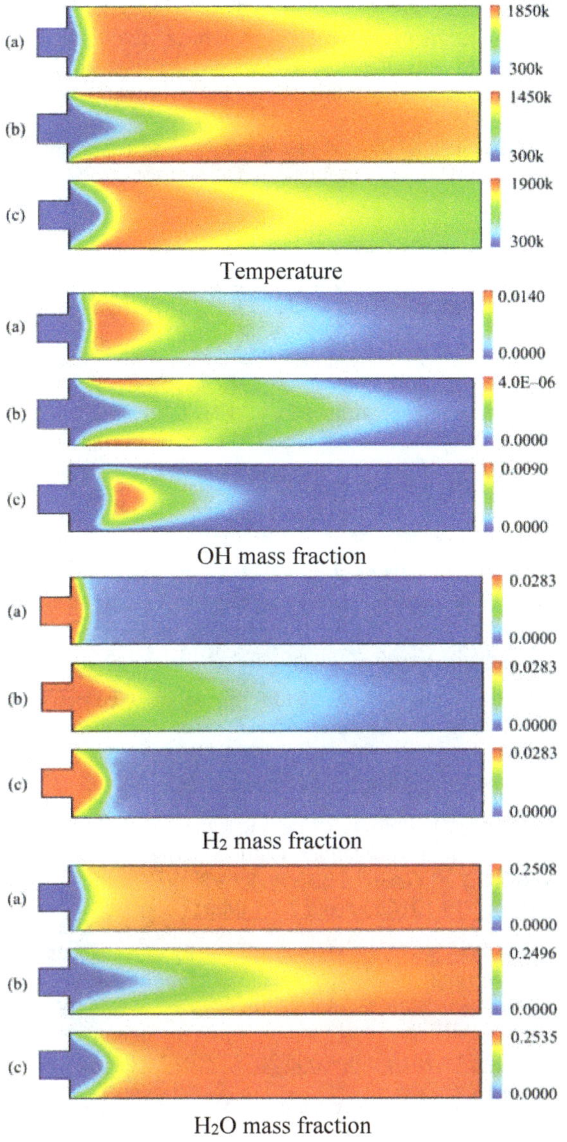

Fig. 2 The computed temperature, OH, H_2 and H_2O mass fraction contours for different reaction models: (a) gas phase reaction alone; (b) surface catalytic reaction alone; (c) both gas phase reaction and surface catalytic reaction

For this case, the wall thickness is ignored. The inlet velocity and diameter of the tube are respectively set to 2 m/s and 0.4mm. The results shown in Fig.2 indicate that the micro-combustion characteristic for different reaction models. In all cases, the comparatively rapid temperature rises close to the inlet was attained, due to highly reactive nature of hydrogen-air mixtures. However, as observed, some differences were seen in these three cases. In the gas phase reaction alone case, the flame structure of micro-combustion displays a cone shape, and the highest temperature and highest OH mass fraction are in the fluid region. Along the radial direction, the temperature is decreased because of the heat is dissipated by the wall. Along the axial direction, the H_2 mass fraction displays a sharp decrease. On the assumption of an inert wall, the OH concentration near the wall reaches a certain value. For the hydrogen and hydrocarbon oxidation, the OH radical is one of significant radicals, and the existence of OH radical generally indicated the high temperature regions and reaction zone. Consequently, OH concentration is usually used to delineate the gas reaction in Pt/γ-Al_2O_3 catalytic micro-combustors (Deutschmann et al., 2000). For the surface catalytic reaction alone case, chemical reactions can only occur on the wall. Therefore, the highest temperature is found on the wall near the inlet, and heat transports downstream by convection. The highest temperature is also lower than these for the gas phase reaction alone case or both gas phase reaction and surface catalytic reaction cases. The highest OH concentration also appears on the wall, and its strength is apparently weak since OH radical has high absorption ability. The difference of OH concentration also reveals the variation of surface catalytic reaction and gas phase reaction. For the both gas phase reaction and surface catalytic reaction cases, the high temperature regions exist both in the fluid region and on the wall near the entrance. The high temperature regions are not simply heat convection from the wall, and both gas phase reaction and surface catalytic reaction exist in Pt/γ-Al_2O_3 catalytic micro-combustors. When compared with the above-mentioned gas phase reaction alone case, the gas phase reaction occurs in the fluid region of micro-combustor, but it shifts slightly downstream. As observed in Fig.2, the gas phase reaction is obviously restrained by the presence of surface catalytic reaction, because most prompt temperature increases are seen in gas phase reaction alone case, not in both gas phase reaction and surface catalytic reaction cases. The above-mentioned tendency can be more evidently seen from the absence of OH radical in the vicinity of inlet, especially in a region close to the Pt/γ-Al_2O_3 wall for both gas phase reaction and surface catalytic reaction cases. For surface catalytic reaction alone case, OH concentration is higher at wall vicinity (maximum value is 3.96×10^{-6}) than that for both gas phase reaction and surface catalytic reaction cases (maximum value is 4.77×10^{-8}). More significantly, for surface catalytic reaction alone case, OH radical is only seen in the vicinity of the Pt/γ-Al_2O_3 wall because chemical reaction only occurs on the wall. Product (H_2O) formation and fuel (H_2) consumption also indicated that the inhibition of gas phase reaction by Pt/γ-Al_2O_3. Note that reaction inhibition by Pt/γ-Al_2O_3 is sensitive to thermal boundary condition at the wall. For most conditions, the simulation results indicate that the gas phase reaction cannot be ignored in Pt/γ-Al_2O_3 catalytic micro-combustors.

3.2 Effect of different wall thermal conductivities

For the materials integrity and flame stability of micro-combustor, the wall thermal conductivity plays a vital role. In micro-combustor, the wall plays a dual, competing role in the overall heat transfer. On one hand, the wall allows exterior heat losses, which can delay ignition and cause extinction. On the other hand, it provides a route for heat transfer from the post-combustion region to upstream for preheating that can improve flame stability and thermal efficiency. The reported simulation results indicated that moderate wall thermal conductivity is essential for flame ignition and stabilization near the entrance of micro-combustor (Norton and Vlachos, 2004).

In the present work, the effect of heat conduction within the wall of combustion chamber on micro-combustion characteristics is investigated. Three kinds of wall materials are selected to study the effect of heat conduction on micro-combustion characteristics. First, the low thermal conductivity material such as insulation material is selected. It has relatively low thermal conductivity around 0.5 W/m·K. Second, the medium thermal conductivity material such as ceramic is chosen. Its thermal conductivity is about 7.5 W/m·K. Moreover, the third kind of the selected material has higher thermal conductivity at the level of 100 W/m·K such as metal and silicon. For this case, the wall thickness is 0.1 mm, and the inlet velocity and diameter of the tube are set to 2 m/s and 0.4mm, respectively.

The material thermal conductivity affects the possibility of hot spots and the temperature profile within the wall of combustion chamber. For different material thermal conductivities, the temperature profiles along the interior wall are showed in Fig.3 (a). For lower thermal conductivity materials, hotspot temperatures in excess of 1700 K can occur, an undesirable situation, as it exceeds the maximum operating temperatures of most construction materials. In addition, significant axial temperature gradients are observed in Fig.3 (a). For low wall thermal conductivity, high temperature gradient exists on the wall, and the hot spot can cause degrade the Pt/γ-Al_2O_3 catalysts or the material to melt. As the wall thermal conductivity is increased, the wall hot spot is eliminated, and the wall temperature profiles become more uniform. Despite the apparent advantages of higher wall thermal conductivity for material stability, most materials that offer high thermal conductivity are metals, and therefore would not be inert to radical quenching. A more reasonable solution would be thicker walls of a more inert material that may have a lower thermal conductivity.

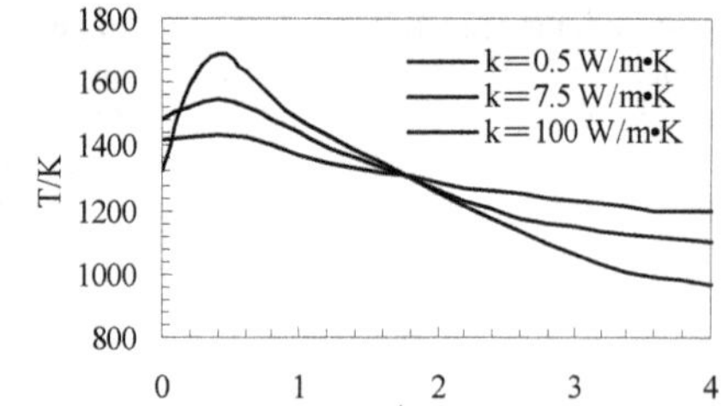

(a) Temperature profile along the interior wall

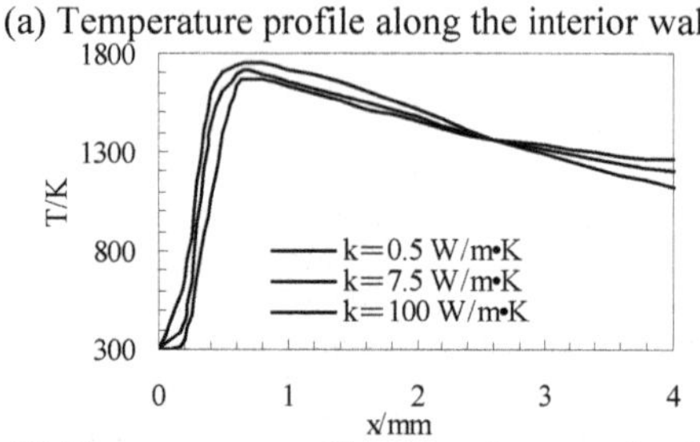

(b) Temperature profile along the central axis

Fig. 3 Temperature profile along the interior wall and central axis for different wall thermal conductivities.

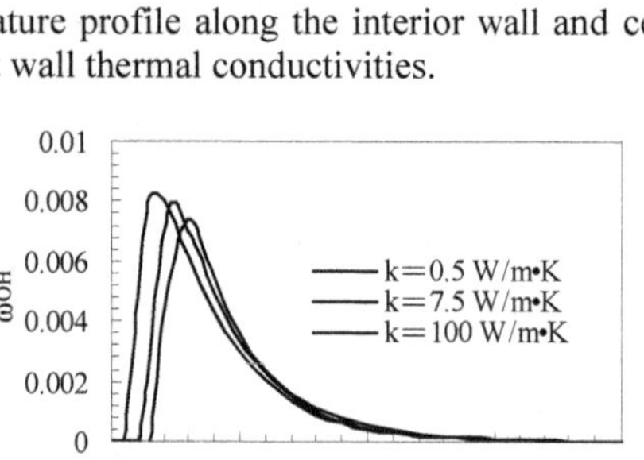

Fig.4 OH mass fraction profile along the central axis for different wall thermal conductivities

The temperature distribution along the central axis for different wall thermal conductivities is shown in Fig.3 (b). As observed, it shows a higher temperature for low wall thermal conductivity. The OH distribution along the central axis for different wall thermal conductivities is shown in Fig.4. For high thermal wall conductivity, the flame location shifts slightly downstream, and a lower OH concentration is found in combustion chamber. A high temperature gradient on the wall will make the gas phase combustion shift upstream, and the combustion chamber will have a higher peak temperature. However, the above-mentioned behavior is different from that in micro-combustors without Pt/γ-Al_2O_3 catalyst walls. In that case (Hua et al., 2005), as the wall thermal conductivity is decreased, the flame core in combustion chamber will shift downstream. The simulation results indicate that the effect of wall thermal conductivity on micro-combustion characteristics is not as evident as that in micro-combustors without Pt/γ-Al_2O_3 catalyst walls. The effect of different wall thermal conductivities on micro-combustion characteristics is not so obvious, because the heat to ignite the gas phase reaction is primarily from the surface catalytic reaction on the wall, not from the upstream heat conduction.

4. CONCLUSIONS

In this study, micro-combustion characteristics for different combustion models and the effects of different wall thermal conductivities on surface catalytic combustion of hydrogen-air mixtures in Pt/γ-Al_2O_3 catalytic micro-combustors were investigated numerically with detailed chemical kinetics mechanisms. From this study, the following conclusions are obtained.

For most conditions, the gas phase combustion cannot be ignored in Pt/γ-Al_2O_3 catalytic micro-combustors. The surface catalytic combustion restrains the gas phase combustion in Pt/γ-Al_2O_3 catalytic micro-combustors. The gas phase combustion inhibition by Pt/γ-Al_2O_3 catalysts is sensitive to thermal boundary condition at the wall.

For low wall thermal conductivity, the higher temperature gradient within the wall of combustion chamber will promote the gas phase combustion shift upstream, and will result in higher temperature distribution in combustion chamber. The effect of different wall thermal conductivities on micro-combustion characteristics is not as obvious as it is in micro-combustors without Pt/γ-Al_2O_3 catalyst walls, because the gas phase combustion is ignited and sustained by the heat from the surface catalytic combustion.

REFERENCES

Belmont, E.L., and Ellzey, J.L., 2014, "Lean Heptane and Propane Combustion in a Non-catalytic Parallel-plate Counter-flow Reactor," *Combustion and Flame*, **161**(4), 1055-1062.
http://dx.doi.org/10.1016/j.combustflame.2013.10.026

Chen, C.H., and Ronney, P.D., 2013a, "Scale and Geometry Effects on Heat-Recirculating Combustors," *Combustion Theory and Modelling*, **17**(5), 888-905.
http://dx.doi.org/10.1080/13647830.2013.812807

Chen, G.B., Chen, C.P., Wu, C.Y., and Chao, Y.C., 2007, "Effects of Catalytic Walls on Hydrogen/air Combustion inside a Micro-Tube," *Applied Catalysis A: General*, **332**(1), 89-97.
http://dx.doi.org/10.1016/j.apcata.2007.08.011

Chen, W.H., Cheng, Y.C., and Hung, C.I., 2013b, "Enhancement of Heat Recirculation on the Hysteresis Effect of Catalytic Partial Oxidation of Methane," *International Journal of Hydrogen Energy*, **38**(25), 10394-10406.
http://dx.doi.org/10.1016/j.ijhydene.2013.05.057

Chein, R.Y., Chen, Y.C., Chang, C.M., and Chung, J.N., 2013, "Experimental Study on the Performance of Hydrogen Production from Miniature Methanol-steam Reformer Integrated with Swiss-roll Type Combustor for PEMFC," *Applied Energy*, **105**, 86-98.
http://dx.doi.org/10.1016/j.apenergy.2012.12.040

Deshpande, A.A., and Kumar, S., 2013, "On the Formation of Spinning Flames and Combustion Completeness for Premixed Fuel-air Mixtures In Stepped Tube Microcombustors," *Applied Thermal Engineering*, **51**(1-2), 91-101.
http://dx.doi.org/10.1016/j.applthermaleng.2012.09.013

Deutschmann, O., Maier, L.I., Riedel, U., Stroemman, A.H., and Dibble, R.W., 2000, "Hydrogen Assisted Catalytic Combustion of Methane on Platinum," *Catalysis Today*, **59**(1-2), 141-150.
http://dx.doi.org/10.1016/S0920-5861(00)00279-0

Deutschmann, O., Schmidt, R., Behrendt, F., and Warnatz, J., 1996, "Numerical Modeling of Catalytic Ignition," *Symposium (International) on Combustion*, **26**(1), 1747-1754.
http://dx.doi.org/10.1016/S0082-0784(96)80400-0

Deutschmann, O., Tischer, S., Kleditzsch, S., Janardhanan, V.M., Correa, C., Chatterjee, D., Mladenov, N., Mihn, H.D., and Karadeniz, H., 2013,

"DETCHEM Software Package, 2.4 Ed.," http://www.detchem.com, Karlsruhe (accessed November 22, 2014).

Federici, J.A., and Vlachos, D.G., 2011, "Experimental Studies on Syngas Catalytic Combustion on Pt/Al_2O_3 in a Microreactor," *Combustion and Flame*, **158**(12), 2540-2543. http://dx.doi.org/10.1016/j.combustflame.2011.05.003

Federici, J.A., Wetzel, E.D., Geil, B.R., and Vlachos, D.G., 2009, "Single Channel and Heat Recirculation Catalytic Microburners: An Experimental and Computational Fluid Dynamics Study," *Proceedings of the Combustion Institute*, **32**(2), 3011-3018. http://dx.doi.org/10.1016/j.proci.2008.07.005

Hsueh, C.Y., Chu, H.S., Yan, W.M., and Chen, C.H., 2010, "Numerical Study of Heat and Mass Transfer in a Plate Methanol Steam Micro Reformer with Methanol Catalytic Combustor," *International Journal of Hydrogen Energy*, **35**(12), 6227-6238. http://dx.doi.org/10.1016/j.ijhydene.2010.03.036

Hsueh, C.Y., Chu, H.S., Yan, W.M., Leu, G.C., and Tsai, J.I., 2011, "Three-Dimensional Analysis of a Plate Methanol Steam Micro-Reformer and a Methanol Catalytic Combustor with Different Flow Channel Designs," *International Journal of Hydrogen Energy*, **36**(21), 13575-13586. http://dx.doi.org/10.1016/j.ijhydene.2011.07.099

Hua, J., Wu, M., and Kumar, K., 2005, "Numerical Simulation of the Combustion of Hydrogen-air Mixture in Micro-Scaled Chambers. Part I: Fundamental Study," *Chemical Engineering Science*, **60**(13), 3497-3506. http://dx.doi.org/10.1016/j.ces.2005.01.041

Ju, Y., and Maruta, K., 2011, "Microscale Combustion: Technology Development and Fundamental Research," *Progress in Energy and Combustion Science*, **37**(6), 669-715. http://dx.doi.org/10.1016/j.pecs.2011.03.001

Ju, Y., and Xu, B., 2005, "Theoretical and Experimental Studies on Mesoscale Flame Propagation and Extinction," *Proceedings of the Combustion Institute*, **30**(2), 2445-2453. http://dx.doi.org/10.1016/j.proci.2004.08.234

Kamijo, T., Suzuki, Y., Kasagi, N., and Okamasa, T., 2009, "High-Temperature Micro Catalytic Combustor with Pd/Nano-Porous Alumina," *Proceedings of the Combustion Institute*, **32**(2), 3019-3026. http://dx.doi.org/10.1016/j.proci.2008.06.118

Karagiannidis, S., and Mantzaras, J., 2010, "Numerical Investigation on the Start-up of Methane-Fueled Catalytic Microreactors," *Combustion and Flame*, **157**(7), 1400-1413. http://dx.doi.org/10.1016/j.combustflame.2010.01.008

Kocich, R., Bojko, M., Machackova, A., and Kleckova, Z., 2012, "Numerical Analysis of the Tubular Heat Exchanger Designed for Co-Generating Units on the Basis of Microturbines," *International Journal of Heat and Mass Transfer*, **55**(19-20), 5336-5342. http://dx.doi.org/10.1016/j.ijheatmasstransfer.2012.05.050

Kurdyumov, V.N., and Matalon, M., 2011, "Analysis of an Idealized Heat-Recirculating Microcombustor," *Proceedings of the Combustion Institute*, **33**(2), 3275-3284. http://dx.doi.org/10.1016/j.proci.2010.07.041

Leach, T.T., Cadou, C.P., and Jackson, G.S., 2006, "Effect of Structural Conduction and Heat Loss on Combustion in Micro-Channels," *Combustion Theory and Modelling*, **10**(1), 85-103. http://dx.doi.org/10.1080/13647830500277332

Maruta, K., 2011, "Micro and Mesoscale Combustion," *Proceedings of the Combustion Institute*, **33**(1), 125-150. http://dx.doi.org/10.1016/j.proci.2010.09.005

Miller, J.A., and Bowman, C.T., 1989, "Mechanism and Modeling of Nitrogen Chemistry in Combustion," *Progress in Energy and Combustion Science*, **15**(4), 287-338. http://dx.doi.org/10.1016/0360-1285(89)90017-8

Norton, D.G., and Vlachos, D.G., 2003, "Combustion Characteristics and Flame Stability at the Microscale: a CFD Study of Premixed Methane/Air Mixtures," *Chemical Engineering Science*, **58**(21), 4871-4882. http://dx.doi.org/10.1016/j.ces.2002.12.005

Norton, D.G., and Vlachos, D.G., 2004, "A CFD Study of Propane/Air Microflame Stability," *Combustion and Flame*, **138**(1-2), 97-107. http://dx.doi.org/10.1016/j.combustflame.2004.04.004

Raimondeau, S., Norton, D., Vlachos, D.G., and Masel, R.I., 2002, "Modeling of High-Temperature Microburners," *Proceedings of the Combustion Institute*, **29**(1), 901-907. http://dx.doi.org/10.1016/S1540-7489(02)80114-6

Rana, U., Chakraborty, S., and Som, S.K., 2014, "Thermodynamics of Premixed Combustion in a Heat Recirculating Micro Combustor," *Energy*, **68**, 510-518. http://dx.doi.org/10.1016/j.energy.2014.02.070

Seyed-Reihani, S.A., and Jackson, G.S., 2004, "Effectiveness in Catalytic Washcoats with Multi-step Mechanisms for Catalytic Combustion of Hydrogen," *Chemical Engineering Science*, **59**(24), 5937-5948. http://dx.doi.org/10.1016/j.ces.2004.07.028

Shirsat, V., and Gupta, A.K., 2011, "A Review of Progress in Heat Recirculating Meso-Scale Combustors," *Applied Energy*, **88**(12), 4294-4309. http://dx.doi.org/10.1016/j.apenergy.2011.07.021

Shirsat, V., and Gupta, A.K., 2013, "Extinction, Discharge, and Thrust Characteristics of Methanol Fueled Meso-Scale Thrust Chamber," *Applied Energy*, **103**, 375-392. http://dx.doi.org/10.1016/j.apenergy.2012.09.058

Vijayan, V., and Gupta, A.K., 2011, "Thermal Performance of a Meso-Scale Liquid Fueled Combustor," *Applied Energy*, **88**(7), 2335-2343. http://dx.doi.org/10.1016/j.apenergy.2011.01.012

Vijayan, V., and Gupta, A.K., 2010, "Flame Dynamics of a Meso-Scale Heat Recirculating Combustor," *Applied Energy*, **87**(12), 3718-3728. http://dx.doi.org/10.1016/j.apenergy.2010.06.003

Wang, Y., Zhou, Z., Yang, W., Zhou, J., Liu, J., Wang, Z., and Cen, K., 2010, "Combustion of Hydrogen-Air in Micro Combustors with Catalytic Pt Layer," *Energy Conversion and Management*, **51**(6), 1127-1133. http://dx.doi.org/10.1016/j.enconman.2009.12.021

Wang, Y., Zhou, Z., Yang, W., Zhou, J., Liu, J., Wang, Z., and Cen, K., 2011, "Instability of Flame in Micro-Combustor under Different External Thermal Environment," *Experimental Thermal and Fluid Science*, **35**(7), 1451-1457. http://dx.doi.org/10.1016/j.expthermflusci.2011.06.003

Zhang, Y., Zhou, J., Yang, W., Liu, M., and Cen, K., 2007, "Effects of Hydrogen Addition on Methane Catalytic Combustion in a Microtube," *International Journal of Hydrogen Energy*, **32**(9), 1286-1293. http://dx.doi.org/10.1016/j.ijhydene.2006.07.023

Zhong, B.J., Yang, F., and Yang, Q.T., 2012, "Catalytic Combustion of n-C_4H_{10} and DME in Swiss-roll Combustor with Porous Ceramics," *Combustion Science and Technology*, **184**(5), 573-584. http://dx.doi.org/10.1080/00102202.2011.651231

8

BOUNDARY LAYER STAGNATION-POINT FLOW OF CASSON FLUID AND HEAT TRANSFER TOWARDS A SHRINKING/STRETCHING SHEET

Krishnendu Bhattacharyya*

Department of Mathematics, The University of Burdwan, Burdwan-713104, West Bengal, India

ABSTRACT

The steady boundary layer stagnation-point flow of Casson fluid and heat transfer towards a shrinking/stretching sheet is studied. Appropriate similarity transformations are employed to transform the governing partial differential equations into the self-similar ordinary differential equations and those are then solved numerically using very efficient shooting method. The numerical computations are carried out for several values of parameters involved (especially, velocity ratio parameter and Casson parameter) to know the possibility of similarity solution for the boundary layer stagnation-point flow. It is found that the range of velocity ratio parameter for which similarity solution exists is unaltered for any change in Casson parameter, though the skin friction changes with Casson parameter. Thus, the possibility of similarity solution for Casson fluid flow is same as that of Newtonian fluid flow.

Keywords: *Boundary layer stagnation-point flow, Casson fluid, heat transfer, shrinking/stretching sheet.*

1. INTRODUCTION

Derivation of boundary layer equations for the flow and their solutions using similarity transformations is among the most successful idealization in the history of fluid mechanics [Schlichting and Gersten (2000)]. With the help of this boundary layer theory, the flows of various types of fluids (Newtonian and different non-Newtonian fluids) have been successfully mathematically modeled and the derived equations are solved. The obtained results are in excellent agreement with experimental observations. However, many fluids of industrial importance are of non-Newtonian type. It is now generally recognized that, in real industrial applications, non-Newtonian fluids are more appropriate than Newtonian fluids, due to their applications in petroleum drilling, polymer engineering, certain separation processes, manufacturing of foods and paper and some other industrial processes [Mustafa *et al.* (2011), Cortell (2008)]. Therefore, the analysis of flow dynamics of non-Newtonian fluids is extremely important.

For non-Newtonian fluids, various types of nonlinear relationship between stress and the rate of strain are observed and it is difficult to express all those properties of several non-Newtonian fluids in a single constitutive equation. Consequently, several non-Newtonian fluid models [Fox et al. (1969), Wilkinson (1970), Djukic (1974), Rajagopal (1980), Rajagopal and Gupta (1981), Dorier and Tichy (1992), Zhou and Gao (2007), Cui *et al.* (2010) and Bhattacharyya and Layek (2011a)] have been proposed depending on various physical characters. Casson fluid is one of such non-Newtonian fluids, which behaves like an elastic solid and for this fluid, a yield shear stress exists in the constitutive equation. Fredrickson (1964) investigated the steady flow of a Casson fluid in a tube. Mustafa *et al.* (2011) studied the unsteady boundary layer flow and heat transfer of a Casson fluid over a moving flat plate with a parallel free stream using homotopy analysis method (HAM).

On the other hand, boundary layer flows of non-Newtonian fluids caused by a stretching sheet have vast applications in several manufacturing processes such as extrusion of molten polymers through a slit die for the production of plastic sheets, hot rolling, wire and fiber coating, processing of food stuffs, metal spinning, glass-fiber production and paper production [Hayat *et al.* (2008a)]. During the processes, the rate of cooling has an important bearing on the properties of the final product. Hence, the quality of the final product depends on the rate of heat transfer from the stretching surface. The viscous fluid flow due to a stretching flat sheet was first investigated by Crane (1970). The pioneering work of Crane was extended by Rajagopal *et al.* (1984) by taking viscoelastic fluid and also Siddappa and Abel (1985) discussed some other important aspects of flow of non-Newtonian fluid over stretching sheet. Sankara and Watson (1985) studied micropolar fluid flow over a stretching sheet. Troy *et al.* (1987) established the uniqueness of solution of the flow of second order fluid over a stretching sheet. Andersson and Dandapat (1991) reported the flow behaviour of a non-Newtonian power-law fluid over a stretching sheet.

Hiemenz (1911) first reported the stagnation point flow towards a flat plate. It is worthwhile to note that the stagnation flow appears whenever the flow impinges to any solid object and the local fluid velocity at a point (called the stagnation-point) is zero. Chiam (1994) extended the works of Hiemenz (1911) replaced the solid body a stretching sheet with equal stretching and straining velocities and he was unable to obtain any boundary layer near the sheet. Whereas, Mahapatra and Gupta (2001) re-investigated the stagnation-point flow towards a stretching sheet considering different stretching and straining velocities and they found two different kinds of boundary layers near the sheet depending on the ratio of the stretching and straining constants. Some other important aspects of stagnation-point flow of Newtonian fluid are discussed by Nazar *et al.* (2004), Layek *et al.* (2007), Nadeem *et al.* (2010), Bhattacharyya *et al.* (2011a,2012a,b) and

* *Email: krish.math@yahoo.com; Fax: (+91)342 2530452, Tel: (+91)9474634200*

Salem and Fathy (2012). Mahapatra *et al.* (2009) studied the boundary layer magnetohydrodynamic (MHD) stagnation-point flow of an electrically conducting non-Newtonian power-law fluid towards a stretching surface. Van Gorder and Vajravelu (2010) explained the hydromagnetic stagnation point flow of a second grade fluid over a stretching sheet. Recently, Hayat *et al.* (2012) analyzed the mixed convection stagnation-point flow of a non-Newtonian Casson fluid over a stretching sheet considering convective boundary conditions.

In contrast, the flow due to a shrinking sheet [Wang (1990)] exhibits quite different behaviour from the forward stretching sheet flow. This shrinking sheet flow is essentially one of backward flows described by Goldstein (1965). The generated vorticity due to shrinking makes the nature of the flow interesting. In their study, Miklavčič and Wang (2006) established the requirement of adequate amount of fluid mass suction through the porous sheet to maintain the steady boundary layer flow of Newtonian fluid due to shrinking of porous flat sheet. Actually, fluid mass suction suppresses the vorticity generated due to shrinking of the sheet, inside the boundary layer. Later, numerous important properties of shrinking sheet flows of Newtonian fluid are discussed by Fang and Zhang (2009), Fang *et al.* (2009,2010), Bhattacharyya and Pop (2011) and Bhattacharyya (2011a,b,c). Hayat *et al.* (2008b,2010) reported the MHD flow and mass transfer of a upper-convected Maxwell fluid over a porous shrinking sheet in presence of chemical reaction and they also obtain an analytic solution of flow of non-Newtonian second grade fluid due to shrinking sheet in a rotating frame. Bhattacharyya *et al.* (2012c) showed the effects of thermal radiation on micropolar fluid flow and heat transfer on a porous shrinking sheet. The non-Newtonian power-law fluid flow past a permeable shrinking sheet with fluid mass transfer was studied by Fang *et al.* (2012). Most importantly, Bhattacharyya *et al.* (2013a,b) recently investigated the boundary layer flow of Casson fluid over a permeable stretching/shrinking sheet without and with magnetic field effect.

The boundary layer stagnation-point flow on a shrinking sheet is interesting for its unusual nature and Wang (2008) illustrated those characters by his study of two-dimensional stagnation point flow of Newtonian fluid towards a shrinking sheet. Later, Bhattacharyya and Layek (2011b) explained the effects of suction/blowing and thermal radiation on boundary layer stagnation-point flow and heat transfer past a shrinking sheet and Bhattacharyya *et al.* (2011b) reported the slip effects on steady stagnation-point flow and heat transfer over a shrinking sheet. The influence of external magnetic field on stagnation-point flow over a shrinking sheet was described by Mahapatra *et al.* (2011) and Lok *et al.* (2011). Fan *et al.* (2010) obtained analytic homotopy solutions of unsteady stagnation-point flow and heat transfer over a shrinking sheet, where as, Bhattacharyya (2011d,2013) found the numerical solutions of that flow problem. In addition, Bhattacharyya (2011e), Bachok *et al.* (2011), Rosali *et al.* (2011), Bhattacharyya and Vajravelu (2012), Bhattacharyya *et al.* (2012d), Van Gorder *et al.* (2012) and Mahapatra *et al.* (2012) and Mahapatra and Nandi (2013) explored various important aspects of stagnation-point flow due to shrinking sheet for Newtonian fluid. From literature, it can be found that not much attention is given to the stagnation-point flow of non-Newtonian fluid on shrinking sheet. Ishak *et al.* (2010) and Yacob *et al.* (2011) discussed the steady boundary layer stagnation-point flow of micropolar fluid past a stretching/shrinking sheet. Nazar *et al.* (2011) presented the stagnation-point flow and heat transfer towards a shrinking sheet in a nanofluid. Khan *et al.* (2012) proposed a mathematical model for the unsteady stagnation point flow of a linear viscoelastic fluid bounded by a stretching/shrinking sheet.

The increasing use of several non-Newtonian fluids in processing industries has given a strong motivation to understand their behavior in several transport processes. Therefore, in this investigation, the steady boundary layer stagnation-point flow of an incompressible Casson fluid and heat transfer towards a shrinking/stretching sheet are studied. The governing partial differential equations are converted into the nonlinear ordinary differential equations using the suitable similarity transformations. The transformed self-similar ODEs are solved by shooting method, an efficient numerical method [Mahapatra and Nandi (2013), Ishak *et al.* (2010), Yacob *et al.* (2011)] for solving boundary value problem. Then a graphical analysis is presented to show the existence and uniqueness of solution and to elaborately discuss the characters of the flow and heat transfer for the variation of physical parameters.

2. FLOW ANALYSIS

Consider the steady two-dimensional stagnation-point flow of incompressible Casson fluid induced by a shrinking/stretching sheet located at y=0, the flow being confined in y>0 (Fig. 1). It is assumed also that the rheological equation of state for an isotropic and incompressible flow of a Casson fluid can be written as [Nakamura and Sawada (1988), Mustafa *et al.* (2011), Bhattacharyya *et al.* (2013a)]:

$$\tau_{ij} = \begin{cases} \left(\mu_B + p_y/\sqrt{2\pi}\right)2e_{ij}, & \pi > \pi_c \\ \left(\mu_B + p_y/\sqrt{2\pi_c}\right)2e_{ij}, & \pi < \pi_c, \end{cases} \tag{1}$$

where μ_B is plastic dynamic viscosity of the non-Newtonian fluid, p_y is the yield stress of fluid, π is the product of the component of deformation rate with itself, namely, $\pi=e_{ij}e_{ij}$, e_{ij} is the (i,j)-th component of the deformation rate and π_c is critical value of π based on non-Newtonian model.

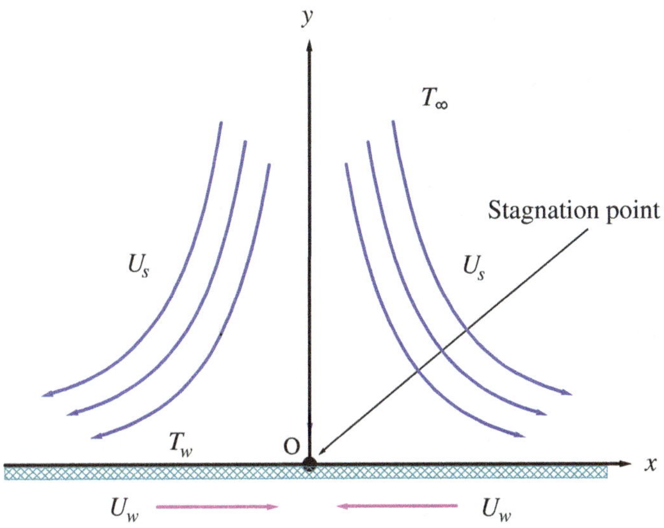

Fig. 1 Physical sketch of the problem.

Under above conditions the boundary layer equations for steady stagnation-point flow towards a shrinking/stretching sheet can be written as:

$$\frac{\partial u}{\partial x} + \frac{\partial v}{\partial y} = 0, \tag{2}$$

$$u\frac{\partial u}{\partial x} + v\frac{\partial u}{\partial y} = U_s\frac{dU_s}{dx} + \upsilon\left(1+\frac{1}{\beta}\right)\frac{\partial^2 u}{\partial y^2}, \tag{3}$$

where u and v are the velocity components in x and y directions respectively, x is distance along the sheet, y is distance perpendicular to the sheet, U_s=ax is the straining velocity of the stagnation-point flow with a(>0) being the straining constant, υ is the kinematic fluid viscosity and $\beta = \mu_B\sqrt{2\pi_c}/p_y$ is the non-Newtonian or Casson parameter.

The boundary conditions for the velocity components are

$$u = U_w,\ v = 0 \text{ at } y = 0;\ u \to U_s \text{ as } y \to \infty, \tag{4}$$

where U_w=cx is shrinking/stretching velocity of the sheet with c being the shrinking/stretching constant and c<0 corresponds to shrinking and c>0 corresponds to stretching.

Table 1 Values of $f''(0)$ for several values of *c/a* for β=∞ (Newtonian fluid case) for stretching sheet.

c/a	Present study	Wang (2008)	Ishak *et al.* (2010)
0	1.2325878	1.232588	1.232588
0.1	1.1465608	1.14656	1.146561
0.2	1.0511299	1.05113	1.051130
0.5	0.7132951	0.71330	0.713295
1	0	0	0

Table 2 Values of $f''(0)$ for several values of *c/a* for β=∞ (Newtonian fluid case) for shrinking sheet.

c/a	Present study		Wang (2008)		Ishak *et al.* (2010)	
	First solution	Second solution	First solution	Second solution	First solution	Second solution
–0.25						
–0.5	1.4022405		1.40224		1.402241	
–0.75	1.4956697		1.49567		1.495670	
–1	1.4892981		1.48930		1.489298	
–1.1	1.3288169	0	1.32882	0	1.328817	0
–1.15	1.1866806	0.0492286			1.186681	0.049229
–1.2	1.0822316	0.1167023	1.08223	0.116702	1.082231	0.116702
–1.24	0.9324728	0.2336491			0.932474	0.233650
–	0.7066020	0.4356712				
1.2465	0.5842915	0.5542856	0.55430		0.584295	0.554283
–	0.5745268	0.5639987				
1.24657						

The following relations for u and v are introduced:

$$u=\frac{\partial\psi}{\partial y} \text{ and } v=-\frac{\partial\psi}{\partial x}, \tag{5}$$

where ψ is the stream function.

For relations in (5), the equation (2) is satisfied automatically and the equation (3) takes the following form:

$$\frac{\partial\psi}{\partial y}\frac{\partial^2\psi}{\partial x\partial y}-\frac{\partial\psi}{\partial x}\frac{\partial^2\psi}{\partial y^2}=U_s\frac{dU_s}{dx}+\upsilon(1+1/\beta)\frac{\partial^3\psi}{\partial y^3}. \tag{6}$$

Also, the boundary conditions in (4) reduce to

$$\frac{\partial\psi}{\partial y}=U_w,\ \frac{\partial\psi}{\partial x}=0 \text{ at } y=0;\ \frac{\partial\psi}{\partial y}\to U_s \text{ as } y\to\infty. \tag{7}$$

The dimensionless variable for the stream function is introduced as:

$$\psi=\sqrt{a\upsilon}\,xf(\eta), \tag{8}$$

where the similarity variable η is given by $\eta=y\sqrt{a/\upsilon}$.

Using the relation (8) and the similarity variable, the equation (6) finally takes following self-similar form:

$$(1+1/\beta)f'''+ff''-f'^2+1=0, \tag{9}$$

where primes denote differentiation with respect to η.

The boundary conditions reduce to

$$f(\eta)=0, f'(\eta)=c/a \text{ at } \eta=0;\ f'(\eta)\to 1 \text{ as } \eta\to\infty, \tag{10}$$

where *c/a* is the velocity ratio parameter.

3. HEAT TRANSFER

For the temperature distribution in the flow field, the governing energy equation can be written as:

$$u\frac{\partial T}{\partial x}+v\frac{\partial T}{\partial y}=\frac{\kappa}{\rho c_p}\frac{\partial^2 T}{\partial y^2}, \tag{11}$$

where T is the temperature, κ is the thermal conductivity, ρ is the fluid density and c_p is the specific heat.

The appropriate boundary conditions are

$$T=T_w \text{ at } y=0;\ T\to T_\infty \text{ as } y\to\infty, \tag{12}$$

where T_w is the constant temperature at the sheet and T_∞ is the free stream temperature assumed to be constant.

Next, the dimensionless temperature θ is introduced as:

$$\theta(\eta)=\frac{T-T_\infty}{T_w-T_\infty}. \tag{13}$$

Using (8), (13) and the similarity variable, the equation (11) reduces to

$$\theta''+Pr\,f\theta'=0, \tag{14}$$

where primes denote differentiation with respect to η and $Pr=c_p\mu/\kappa$ is the Prandtl number.

The boundary conditions for θ are obtained from (12) as:

$$\theta(\eta)=1 \text{ at } \eta=0;\ \theta(\eta)\to 0 \text{ as } \eta\to\infty. \tag{15}$$

4. NUMERICAL METHOD FOR SOLUTION

The self-similar equations (9) and (14) along with boundary conditions (10) and (15) are solved using shooting method [Bhattacharyya *et al.* (2011c,d) and Bhattacharyya (2012)] by converting them to an initial value problem (IVP). In this method, it is necessary to choose a suitable finite value of $\eta\to\infty$, say η_∞. The following system is set:

$$\left.\begin{aligned} f'&=p,\\ p'&=q,\\ q'&=(p^2-fq-1)/(1+1/\beta),\end{aligned}\right\} \tag{16}$$

$$\left.\begin{aligned} \theta'&=z,\\ z'&=-Prfz\end{aligned}\right\} \tag{17}$$

with the boundary conditions

$$f(0)=0,\ p(0)=c/a,\ \theta(0)=1. \tag{18}$$

In order to integrate the IVP (16) and (17) with (18), the values for $q(0)$ i.e. $f''(0)$ and $z(0)$ i.e. $\theta'(0)$ are required, but no such values are given at the boundary. The suitable guess values for $f''(0)$ and $\theta'(0)$ are chosen and then integration is carried out. Then the calculated values for f' and θ at η_∞=15 (say) are compared with the given boundary conditions $f'(15)=1$ and $\theta(15)=0$ and the estimated values, $f''(0)$ and $\theta'(0)$ are adjusted to give a better approximation for the solution. A series of values for $f''(0)$ and $\theta'(0)$ are taken and the fourth order classical Runge-Kutta method with step-size $\Delta\eta$=0.01 is applied. The above procedure is repeated until the asymptotically converged results within a tolerance level of 10^{-5} are obtained.

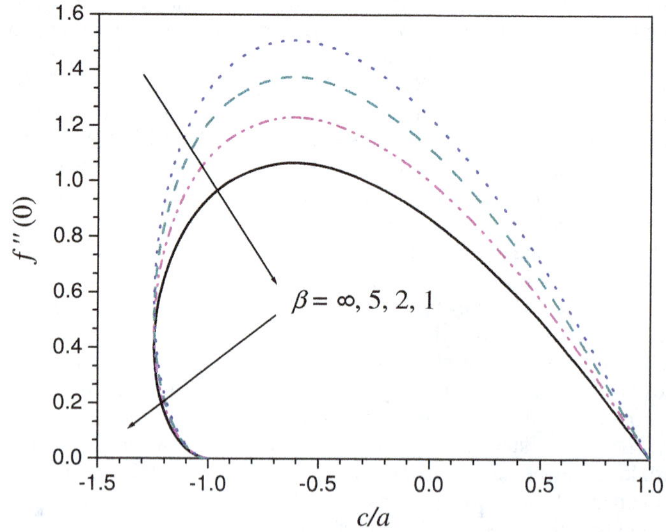

Fig. 2 The values of $f''(0)$ vs. *c/a* for different values of β.

5. RESULTS AND DISCUSSIONS

The numerical computations have been carried out using above-described shooting method for several values of the physical parameters arised in the study: such as, the velocity ratio parameter (*c/a*), the Casson parameter (β) and the Prandtl number (*Pr*). Then acquired results are presented in graphs (Fig. 2-Fig. 14) to explain the existence and uniqueness of solution for the flow, as well as, the variations in velocity and temperature fields. Also, to validate the numerical scheme, a comparison of results is made. Table 1 and Table 2 show that the values of $f''(0)$ for $\beta=\infty$ (i.e., Newtonian fluid case) are in a favorable agreement with previously published data in the literature by Wang (2008) and Ishak *et al.* (2010).

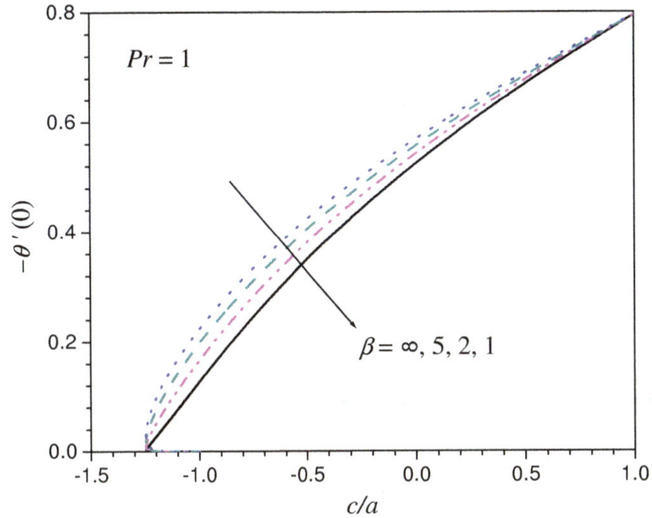

Fig. 3 The values of $-\theta'(0)$ vs. c/a for different values of β.

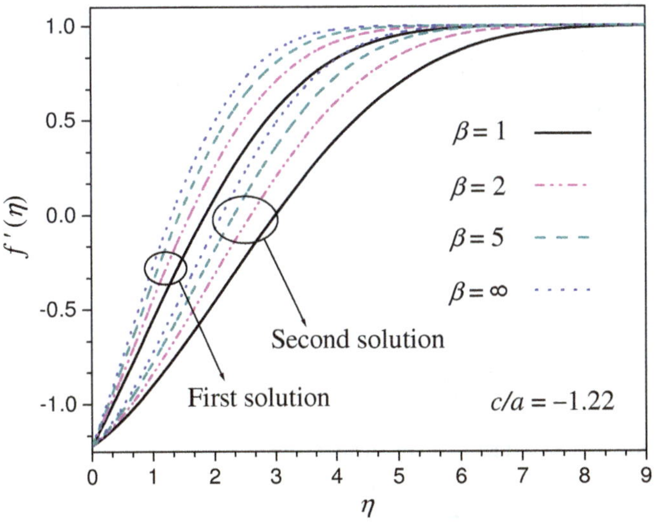

Fig. 4 The effects of β on dual velocity profiles $f'(\eta)$.

The stagnation-point flow over a shrinking sheet was investigated by Wang (2008) for Newtonian fluid and he found that the self-similar solution of boundary layer flow is possible only if the velocity ratio parameter c/a satisfies the inequality $c/a\geq-1.2465$. In this study, it is obtained that for $\beta=\infty$ i.e., for Newtonian fluid the boundary layer exists if $c/a\geq-1.24657$, which is similar to that of Wang (2008). It has also noted that for $\beta=\infty$ the solution is of dual nature for $-1.24657\leq c/a\leq 0$, the solution is unique for $c/a>0$ and for $c/a<-1.24657$ no solution is found. Due to decrease in β i.e., for Casson fluid, the existence range does not alter and the similarity solution of boundary layer is obtained when $c/a\geq-1.24657$. Therefore, it is worth noting that there is no change occurred in the solution range of c/a due to variation in the Casson parameter β. For all values of β, dual solutions exist for $-1.24657\leq c/a\leq 0$, the solution is unique for $c/a>0$ and no similarity solution is found for $c/a<-1.24657$. Thus, there exists dual self-similar solutions in some situations of shrinking sheet case and for stretching sheet case the solution is always unique. These all phenomena can be observed in Fig. 2 and Fig. 3 of $f''(0)$ and $-\theta'(0)$ vs. c/a, those are related to wall skin friction coefficient and the heat transfer coefficient respectively. Though there is no change in the solution range of c/a, but the value of $f''(0)$ decreases with decrease of β for first and second solutions in dual solutions case and unique solution case. Similar effects is observed for the values of $-\theta'(0)$. Hence, for Casson fluid flow over a shrinking sheet near a stagnation-point dual solutions and unique solution are found in the same ranges as that of Newtonian fluid case, but Casson parameter affects the wall skin-friction coefficient as well as the heat transfer coefficient. To know the detailed effects of Casson parameter on the flow, the dimensionless curves related to velocity, temperature and their gradients are plotted.

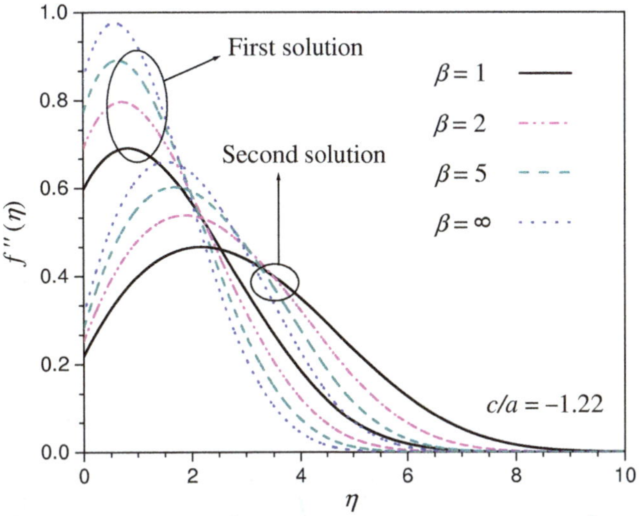

Fig. 5 The effects of β on dual velocity gradient profiles $f''(\eta)$.

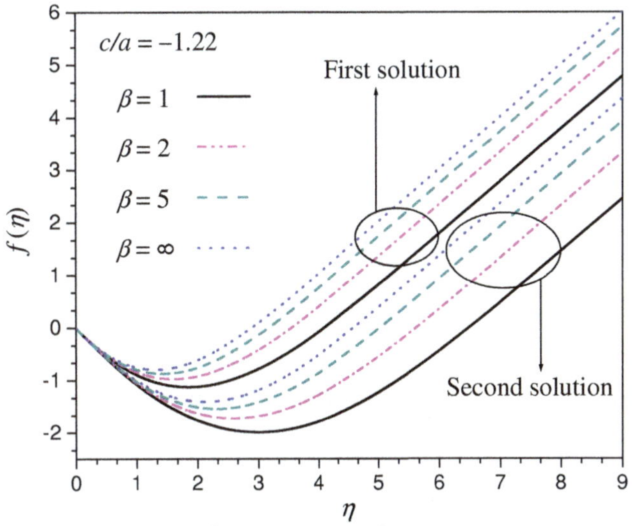

Fig. 6 The effects of β on $f(\eta)$.

The dimensionless velocity, velocity gradient, stream function, temperature and temperature gradient are depicted in Fig. 4-Fig. 10 for different values of Casson parameter β. The velocity at a point decreases with decreasing values of β for both solutions in dual solutions (Fig. 4) and also for unique solution (Fig. 9 and Fig. 11). The velocity gradient decreases near the sheet with decreasing β, but faraway from the sheet it increases (Fig. 5). Also, it is important to note that the decrease in Casson parameter makes the velocity boundary layer thickness larger. Thus, the velocity boundary layer thickness for Casson fluid is larger than that of Newtonian fluid. It happens because of plasticity of Casson fluid. When Casson parameter decreases the plasticity of the fluid increases, which causes the increment in velocity boundary layer thickness. The dimensionless stream function profiles (Fig. 6) show the back flow character of stagnation point flow over a shrinking sheet. The dimensionless temperature increases with the decrease in β for all cases, both solutions in dual solutions (Fig. 7) and unique solution (Fig. 10). Similar to the velocity boundary layer, due to increase in plasticity of fluid the thermal boundary layer thickness increases with decreasing β, which can be confirmed from the temperature gradient profiles in Fig. 8. Here also, in dual solutions the boundary layer thickness for second solution is thicker.

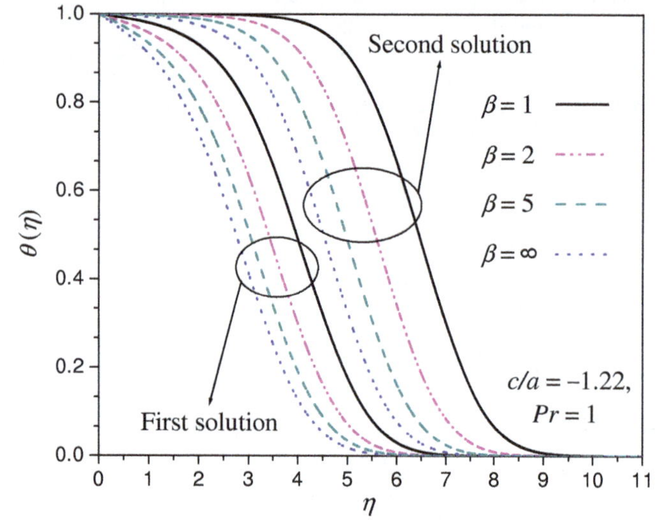

Fig. 7 The effects of β on dual temperature profiles $\theta(\eta)$.

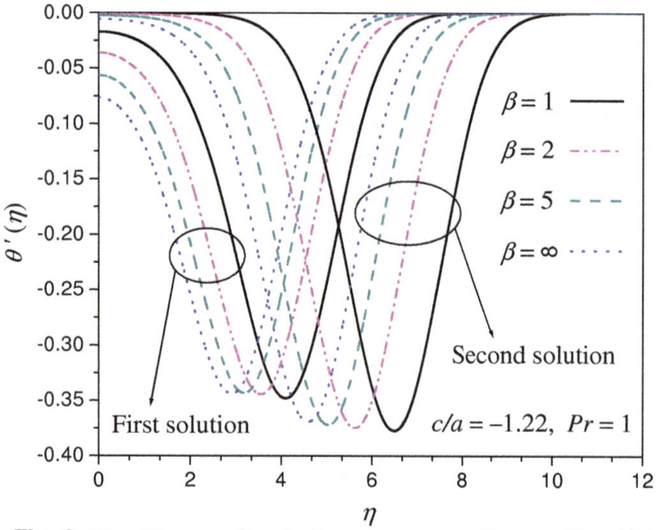

Fig. 8 The effects of β on dual temperature gradient profiles $\theta'(\eta)$.

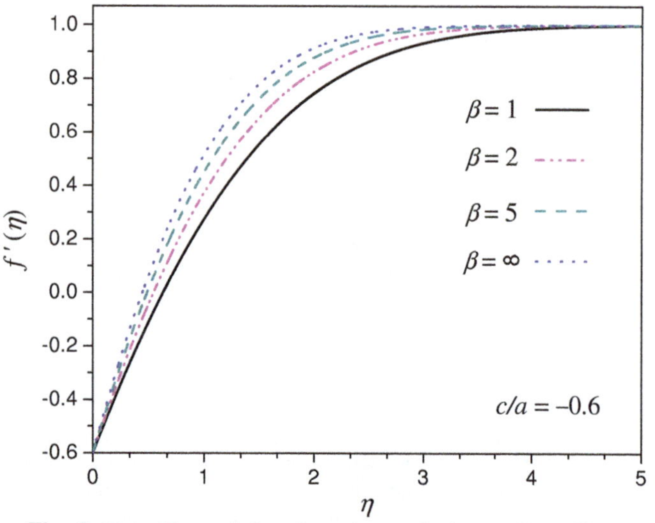

Fig. 9 The effects of β on the unique velocity profiles $f'(\eta)$ for shrinking sheet.

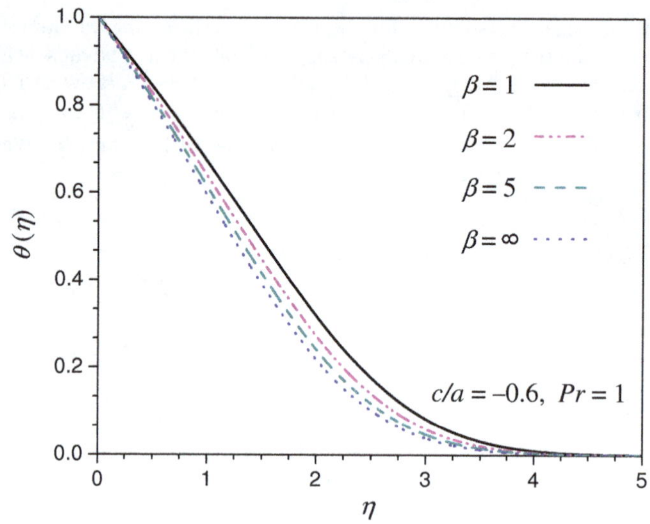

Fig. 10 The effects of β on the unique temperature profiles $\theta(\eta)$ for shrinking sheet.

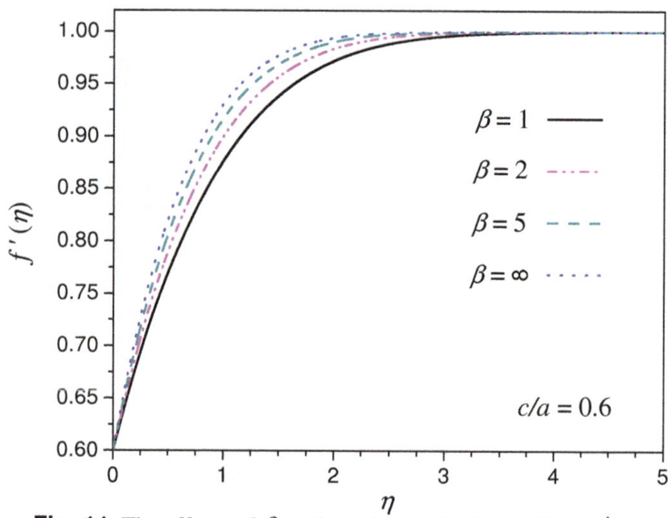

Fig. 11 The effects of β on the unique velocity profiles $f'(\eta)$ for stretching sheet.

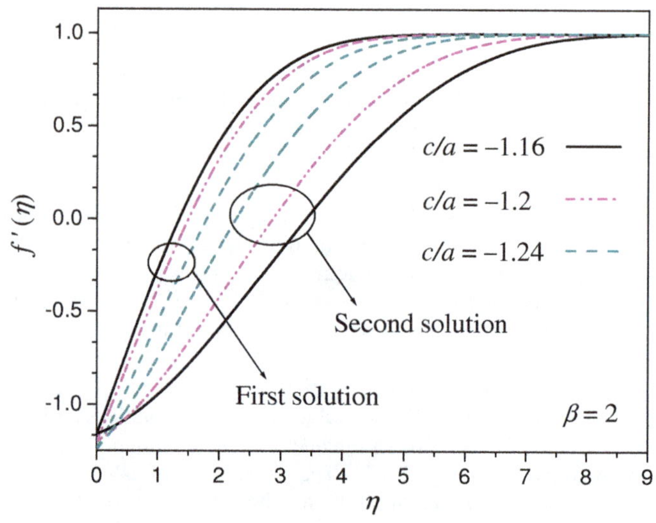

Fig. 12 The effects of c/a on dual velocity profiles $f'(\eta)$.

Next, the focus is concentrated to the effects of the velocity ratio parameter c/a and the Prandtl number Pr on the velocity and temperature distributions in Casson fluid flow. In Fig. 12 and Fig. 13 the influence of c/a on dual velocity and temperature profiles are presented respectively. Similar to Newtonian fluid case, here for Casson fluid two opposite effects are observed in two solutions. For first solution, boundary layer thicknesses (velocity and thermal) increase with increasing magnitude of c/a and in second solution those decrease. The dual temperature profiles for several values of Prandtl number are demonstrated in Fig. 14. The thermal boundary layer thickness decreases with increasing Prandtl number with a temperature crossing over.

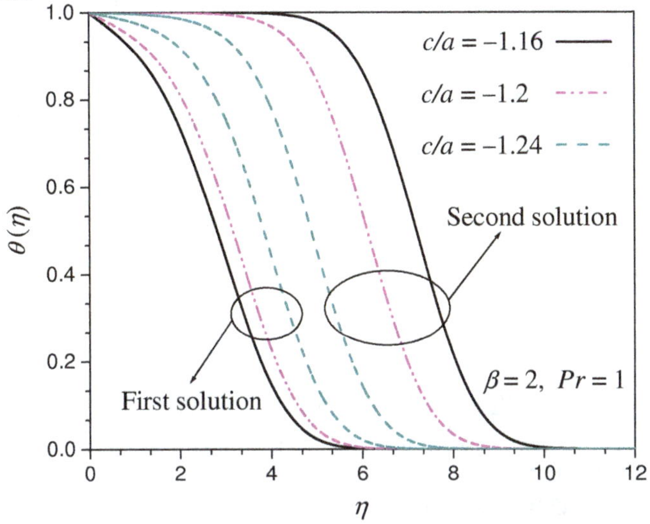

Fig. 13 The effects of a/c on dual temperature profiles $\theta(\eta)$.

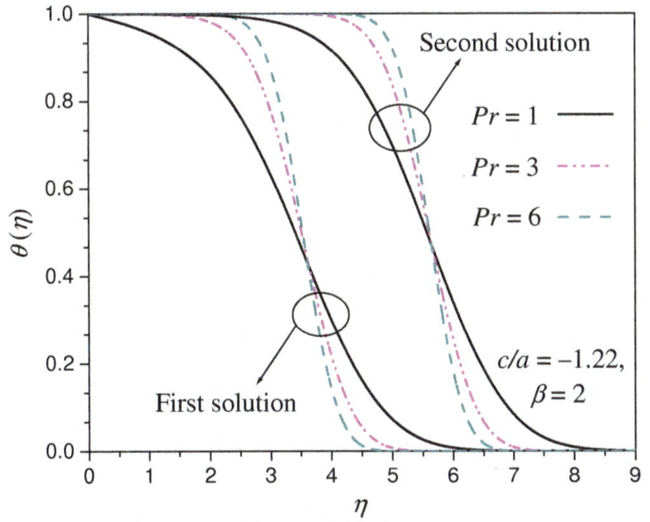

Fig. 14 The effects of Pr on dual temperature profiles $\theta(\eta)$.

6. CONCLUSIONS

The objective of this investigation is to emphasize on similarity solution properties of the boundary layer stagnation-point flow of Casson fluid and heat transfer towards a shrinking/stretching sheet. An analysis of existence and uniqueness of boundary layer self-similar solution of transformed equations is made based on numerical computations using shooting method. The study explores that similar to Newtonian case, the self-similar solution is of dual nature in some situations of shrinking sheet case, for stretching sheet case the solution is always unique. Also, it is found that the velocity and thermal boundary layer thicknesses are larger for Casson fluid than that of Newtonian fluid.

ACKNOWLEDGEMENTS

The author gratefully acknowledges the financial support of *National Board for Higher Mathematics* (NBHM), Department of Atomic Energy, Government of india for pursuing this work. The author is also thankful to the referee for his valuable comments and suggestions.

NOMENCLATURE

a	straining constant
c/a	velocity ratio parameter
c	shrinking/stretching constant
c_p	specific heat
f	dimensionless stream function
f'	dimensionless velocity
Pr	Prandtl number
p	a variable
p_y	yield stress of fluid
q	a variable
T	temperature
T_w	constant temperature at the sheet
T_∞	constant free stream temperature
U_s	straining velocity of the stagnation-point flow
U_w	shrinking/stretching velocity of the sheet
u	velocity component in x direction
v	velocity component in y direction
x	distance along the sheet
y	distance perpendicular to the sheet
z	a variable

Greek symbols

β	non-Newtonian/Casson parameter
η	similarity variable
η_∞	finite value of η
κ	thermal conductivity
μ_B	plastic dynamic viscosity of the non-Newtonian fluid
π	product of the component of deformation rate with itself
π_c	critical value of π
υ	kinematic fluid viscosity
ρ	fluid density
ψ	stream function
θ	dimensionless temperature

REFERENCES

Andersson, H.I., and Dandapat, B.S., 1991, "Flow of a Power-Law Fluid over a Stretching Sheet," *Stability Appl. Anal. Continuous Media*, **1**, 339-347.

Bachok, N., Ishak, A., and Pop, I., 2011, "On The Stagnation-Point Flow Towards a Stretching Sheet with Homogeneous–Heterogeneous Reactions Effects," *Commun. Nonlinear Sci. Numer. Simulat.*, **16**, 4296-4302.
http://dx.doi.org/10.1016/J.cnsns.2011.01.008

Bhattacharyya, K., 2011a, "Boundary Layer Flow and Heat Transfer over an Exponentially Shrinking Sheet," *Chin. Phys. Lett.*, **28**, 074701.
http://dx.doi.org/10.1088/0256-307X/28/7/074701

Bhattacharyya, K., 2011b, "Effects of Radiation and Heat Source/Sink On Unsteady MHD Boundary Layer Flow and Heat Transfer over a Shrinking Sheet with Suction/Injection," *Front. Chem. Sci. Eng.*, **5**, 376-384.
http://dx.doi.org/10.1007/S11705-011-1121-0

Bhattacharyya, K., 2011c, "Effects of Heat Source/Sink on MHD Flow and Heat Transfer over a Shrinking Sheet with Mass Suction," *Chem. Eng. Res. Bull.*, **15**, 12-17.

http://dx.doi.org/10.3329/Cerb.V15i1.6524

Bhattacharyya, K., 2011d, "Dual Solutions in Unsteady Stagnation-Point Flow over a Shrinking Sheet," *Chin. Phys. Lett.*, **28**, 084702. http://dx.doi.org/10.1088/0256-307X/28/8/084702

Bhattacharyya, K., 2011e, "Dual Solutions in Boundary Layer Stagnation-Point Flow and Mass Transfer with Chemical Reaction past a Stretching/Shrinking Sheet," in*t. Commun. Heat Mass Transfer*, **38**, 917-922.
http://dx.doi.org/10.1016/J.Icheatmasstransfer.2011.04.020

Bhattacharyya, K., and Layek, G.C., 2011a, "MHD Boundary Layer Flow of Dilatant Fluid in a Divergent Channel with Suction or Blowing," *Chin. Phys. Lett.*, **28**, 084705.
http://dx.doi.org/10.1088/0256-307X/28/8/084705

Bhattacharyya, K., and Layek, G.C., 2011b, "Effects of Suction/Blowing On Steady Boundary Layer Stagnation-Point Flow and Heat Transfer towards a Shrinking Sheet with Thermal Radiation," in*t. J. Heat Mass Transfer*, **54**, 302-307.
http://dx.doi.org/10.1016/J.Ijheatmasstransfer.2010.09.043

Bhattacharyya, K., and Pop, I., 2011, "MHD Boundary Layer Flow Due To an Exponentially Shrinking Sheet," *Magnetohydrodynamics*, **47**, 337-344.

Bhattacharyya, K., Mukhopadhyay, S., and Layek, G.C., 2011a, "Slip Effects On Unsteady Boundary Layer Stagnation-Point Flow and Heat Transfer Towards a Stretching Sheet," *Chin. Phys. Lett.*, **28**, 094702. http://dx.doi.org/10.1088/0256-307X/28/9/094702

Bhattacharyya, K., Mukhopadhyay, S., and Layek, G.C., 2011b, "Slip Effects On Boundary Layer Stagnation-Point Flow and Heat Transfer Towards a Shrinking Sheet," in*t. J. Heat Mass Transfer*, **54**, 308-313. http://dx.doi.org/10.1016/J.Ijheatmasstransfer.2010.09.041

Bhattacharyya, K., Mukhopadhyay, S., and Layek, G.C., 2011c, "MHD Boundary Layer Slip Flow and Heat Transfer over a Flat Plate," *Chin. Phys. Lett.*, **28**, 024701.
http://dx.doi.org/10.1088/0256-307X/28/2/024701

Bhattacharyya, K., Mukhopadhyay, S., and Layek, G.C., 2011d, "Steady Boundary Layer Slip Flow and Heat Transfer over a Flat Porous Plate Embedded in a Porous Media," *J. Petrol. Sci. Eng.*, **78**, 304-309. http://dx.doi.org/10.1016/J.Petrol.2011.06.009

Bhattacharyya, K., 2012, "Mass Transfer On a Continuous Flat Plate Moving in Parallel Or Reversely To a Free Stream in The Presence of a Chemical Reaction," in*t. J. Heat Mass Transfer*, **55**, 3482-3487. http://dx.doi.org/10.1016/J.Ijheatmasstransfer.2012.03.005

Bhattacharyya, K., and Vajravelu, K., 2012, "Stagnation-Point Flow and Heat Transfer over an Exponentially Shrinking Sheet," *Commun. Nonlinear Sci. Numer. Simulat.*, **17**, 2728-2734.
http://dx.doi.org/10.1016/J.Cnsns.2011.11.011

Bhattacharyya, K., Mukhopadhyay, S., and Layek, G.C., 2012a, "Reactive Solute Transfer in Magnetohydrodynamic Boundary Layer Stagnation-Point Flow over a Stretching Sheet with Suction/Blowing," *Chem. Eng. Commun.*, **199**, 368-383.
http://dx.doi.org/10.1080/00986445.2011.592444

Bhattacharyya, K., Mukhopadhyay, S., and Layek, G.C., 2012b, "Effects of Partial Slip On Boundary Layer Stagnation Slip Flow and Heat Transfer Towards a Stretching Sheet with Temperature Dependent Fluid Viscosity," *Acta Tech.*, **57**, 183-195.

Bhattacharyya, K., Mukhopadhyay, S., Layek, G.C., and Pop, I., 2012c, "Effects of Thermal Radiation On Micropolar Fluid Flow and Heat Transfer over a Porous Shrinking Sheet," *Int. J. Heat Mass Transfer*, **55**, 2945-2952.
http://dx.doi.org/10.1016/J.Ijheatmasstransfer.2012.01.051

Bhattacharyya, K., arif, M.G., ali, Pk,W., 2012d, "MHD Boundary Layer Stagnation-Point Flow and Mass Transfer over a Permeable Shrinking Sheet with Suction/Blowing and Chemical Reaction," *Acta Tech.*, **57**, 1-15.

Bhattacharyya, K., 2013, "Heat Transfer in Unsteady Boundary Layer Stagnation-Point Flow towards a Shrinking Sheet," *Ain Shams Eng. J.*, **4**, 259-264.
http://dx.doi.org/10.1016/J.Asej.2012.07.002

Bhattacharyya, K., Hayat, T., and alsaedi, A., 2013a, "Exact Solution For Boundary Layer Flow of Casson Fluid over a Permeable Stretching/Shrinking Sheet," *Z. Angew. Math. Mech.*, http://dx.doi.org/10.1002/Zamm.201200031

Bhattacharyya, K., Hayat, T., and alsaedi, A., 2013b, "Analytic Solution For Magnetohydrodynamic Boundary Layer Flow of Casson Fluid over a Stretching/Shrinking Sheet with Wall Mass Transfer," *Chin. Phys. B*, **22**, 024702.
http://dx.doi.org/10.1088/1674-1056/22/2/024702

Chiam, T.C., 1994, "Stagnation-Point Flow towards a Stretching Plate," *J. Phys. Soc. Jpn.*, **63**, 2443-2444.
http://dx.doi.org/10.1143/JPSJ.63.2443

Cortell, R., 2008, "Analysing Flow and Heat Transfer of a Viscoelastic Fluid over a Semi-Infinite Horizontal Moving Flat Plate," *Int. J. Non-Linear Mech.*, **43**, 772-778.
http://dx.doi.org/10.1016/J.Ijnonlinmec.2008.04.006

Crane, L.J., 1970, "Flow Past a Stretching Plate," *Z. Angew. Math. Phys.*, **21**, 645-647.
http://dx.doi.org/10.1007/BF01587695

Cui, Z.-W., Liu, J.-X., Yao, G.-J., and Wang, K.-X., 2010, "Borehole Guided Waves in a Non-Newtonian (Maxwell) Fluid-Saturated Porous Medium," *Chin. Phys. B*, **19**, 084301.
http://dx.doi.org/10.1088/1674-1056/19/8/084301

Djukic, D.S., 1974, "Hiemenz Magnetic Flow of Power-Law Fluids," *ASME J. appl. Mech.*, **41**, 822-823.
http://dx.doi.org/10.1115/1.3423405

Dorier, C., and Tichy, J., 1992, "Behavior of a Bingham-Like Viscous Fluid in Lubrication Flows," *J. Non-Newtonian Fluid Mech.*, **45**, 291-310. http://dx.doi.org/10.1016/0377-0257(92)80065-6

Fan, T., Xu, H., and Pop, I., 2010, "Unsteady Stagnation Flow and Heat Transfer Towards a Shrinking Sheet," *Int. Commun. Heat Mass Transfer*, **37**, 1440-1446.
http://dx.doi.org/10.1016/J.Icheatmasstransfer.2010.08.002

Fang, T., and Zhang, J., 2009, "Closed-Form Exact Solution of MHD Viscous Flow over a Shrinking Sheet," *Commun. Nonlinear Sci. Numer. Simulat.*, **14**, 2853-2857.
http://dx.doi.org/10.1016/J.Cnsns.2008.10.005

Fang, T., Zhang, J., and Yao, S., 2009, "Viscous Flow over an Unsteady Shrinking Sheet with Mass Transfer," *Chin. Phys. Lett.*, **26**, 014703.
http://dx.doi.org/10.1088/0256-307X/26/1/014703

Fang, T., Zhang, J., and Yao, S., 2010, "Slip Magnetohydrodynamic Viscous Flow over a Permeable Shrinking Sheet," *Chin. Phys. Lett.*, **27**, 124702.
http://dx.doi.org/10.1088/0256-307X/27/12/124702

Fang, T., Tao, H., and Zhong, Y.F., 2012, "Non-Newtonian Power-Law Fluid Flow over a Shrinking Sheet," *Chin. Phys. Lett.*, **29**, 114703. http://dx.doi.org/10.1088/0256-307X/29/11/114703

Fox, V.G, Erickson, L.E., and Fan, L.T., 1969, "The Laminar Boundary Layer on a Moving Continuous Flat Sheet Immersed in a Non-Newtonian Fluid," *AIChE J.*, **15**, 327-333.

http://dx.doi.org/10.1002/Aic.690150307

Fredrickson, A.G., 1964, *Principles and applications of Rheology*, Prentice-Hall, Englewood Cliffs, N.J.

Goldstein, S., 1965, "On Backward Boundary Layers and Flow in Converging Passages," *J. Fluid Mech.*, **21**, 33-45. http://dx.doi.org/10.1017/S0022112065000034

Hayat, T., Sajid, M., and Pop, I., 2008a, "Three-Dimensional Flow over a Stretching Surface in a Viscoelastic Fluid," *Nonlinear Anal. Real World appl.*, 9, 1811-1822. http://dx.doi.org/10.1016/J.NonrwA.2007.05.010

Hayat, T., abbas, Z., and ali, N., 2008b, "MHD Flow and Mass Transfer of a Upper-Convected Maxwell Fluid Past a Porous Shrinking Sheet with Chemical Reaction Species," *Phys. Lett. a*, **372**, 4698-4704. http://dx.doi.org/10.1016/J.PhysletA.2008.05.006

Hayat, T., Iram, S., Javed, T., and asghar, S., 2010, "Shrinking Flow of Second Grade Fluid in a Rotating Frame: an analytic Solution," *Commun. Nonlinear Sci. Numer. Simulat.*, **15**, 2932-2941. http://dx.doi.org/10.1016/J.Cnsns.2009.11.030

Hayat, T., Shehzad, S.A., alsaedi, A., alhothuali, M.S., 2012, "Mixed Convection Stagnation Point Flow of Casson Fluid with Convective Boundary Conditions," *Chin. Phys. Lett.*, **29**, 114704. http://dx.doi.org/10.1088/0256-307X/29/11/114704

Hiemenz, K., 1911, "Die Grenzschicht an Einem in Den Gleichformingen Flussigkeits-Strom Einge-Tauchten Graden Kreiszylinder," *Dingler's Poly. J.*, **326**, 321-324.

Ishak, A., Lok, Y.Y., and Pop, I., 2010, "Stagnation-Point Flow over a Shrinking Sheet in a Micropolar Fluid," *Chem. Eng. Commun.*, **197**, 1417-1427. http://dx.doi.org/10.1080/00986441003626169

Layek, G.C., Mukhopadhyay, S., and Samad Sk.A., 2007, "Heat and Mass Transfer analysis For Boundary Layer Stagnation-Point Flow towards a Heated Porous Stretching Sheet with Heat absorption/Generation and Suction/Blowing," Int*Commun. Heat Mass Transfer*, **34**, 347-356. http://dx.doi.org/10.1016/J.Icheatmasstransfer.2006.11.011

Khan, Y., Hussain, A., and Faraz, N., 2012, "Unsteady Linear Viscoelastic Fluid Model over a Stretching/Shrinking Sheet in The Region of Stagnation Point Flows," *Sci. Iran.*, **19**, 1541-1549. http://dx.doi.org/10.1016/J.Scient.2012.10.019

Lok, Y.Y., Ishak, A., and Pop, I., 2011, "MHD Stagnation-Point Flow Towards a Shrinking Sheet," *Int. J. Numer. Meth. Heat Fluid Flow*, **21**, 61-72. http://dx.doi.org/10.1108/09615531111095076

Mahapatra, T.R., and Gupta, A.S., 2001, "Magnetohydrodynamic Stagnation-Point Flow Towards a Stretching Sheet," *Acta Mech.*, **152**, 191-196. http://dx.doi.org/10.1007/BF01176953

Mahapatra, T.R., Nandy, S,K., and Gupta, a,S., 2009, "Magnetohydrodynamic Stagnation-Point Flow of a Power-Law Fluid Towards a Stretching Surface," *Int. J. Non-Linear Mech.*, **44**, 124-129. http://dx.doi.org/ 10.1016/J.Ijnonlinmec.2008.09.005

Mahapatra, T.R., Nandy, S.K., and Gupta, A.S., 2011, "Momentum and Heat Transfer in MHD Stagnation-Point Flow over a Shrinking Sheet," a*SME J appl. Mech.*, **78**, 021015. http://dx.doi.org/10.1115/1.4002577

Mahapatra, T.R., Nandy, S.K., and Gupta, A.S., 2012, "Oblique Stagnation-Point Flow and Heat Transfer Towards a Shrinking Sheet with Thermal Radiation," *Meccanica*, **47**, 1325-1335. http://dx.doi.org/10.1007/S11012-011-9516-Z

Mahapatra, T.R., and Nandy, S.K., 2013, "Stability of Dual Solutions in Stagnation-Point Flow and Heat Transfer over a Porous Shrinking Sheet with Thermal Radiation," *Meccanica*, **48**, 23-32. http://dx.doi.org/10.1007/S11012-012-9579-5

Miklavčič, M., and Wang, C.Y., 2006, "Viscous Flow Due a Shrinking Sheet," *Q. appl. Math.*, **64**, 283-290. http://dx.doi.org/10.1090/S0033-569X-06-01002-5

Mustafa, M., Hayat, T., Pop, I. and aziz, A., 2011, "Unsteady Boundary Layer Flow of a Casson Fluid Due To an Impulsively Started Moving Flat Plate," *Heat Transfer asian Res.*, **40**, 563-576.

http://dx.doi.org/10.1002/Htj.20358

Nadeem, S., Hussain, A., and Khan, M., 2010, "HAM Solutions For Boundary Layer Flow in The Region of The Stagnation Point Towards a Stretching Sheet," *Commun. Nonlinear Sci. Numer. Simulat.*, **15**, 475-481. http://dx.doi.org/10.1016/J.Cnsns.2009.04.037

Nakamura, M., and Sawada, T., 1988, "Numerical Study On The Flow of a Non-Newtonian Fluid Through an axisymmetric Stenosis," a*SME J. Biomechanical Eng.*, **110**, 137-143. http://dx.doi.org/10.1115/1.3108418

Nazar, R., amin, N., Filip, D., and Pop, I., 2004, "Unsteady Boundary Layer Flow in The Region of The Stagnation Point On a Stretching Sheet," *Int. J. Eng. Sci.*, **42**, 1241-1253. http://dx.doi.org/10.1016/J.Ijengsci.2003.12.002

Nazar, R., Jaradat, M., arifin, N.M., and Pop, I., 2011, "Stagnation-Point Flow Past a Shrinking Sheet in a Nanofluid," *Cent. Eur. J. Phys.*, **9**, 1195-1202. http://dx.doi.org/10.2478/S11534-011-0024-5

Rajagopal, K.R., 1980, "Viscometric Flows of Third Grade Fluids," *Mech. Res. Commun.*, **7**, 21-25. http://dx.doi.org/ 10.1016/0093-6413(80)90020-8

Rajagopal, K.R., and Gupta A.S., 1981, "On a Class of Exact Solutions To The Equations of Motion of a Second Grade Fluid," *Int. J. Eng. Sci.*, **19**, 1009-1014. http://dx.doi.org/10.1016/0020-7225(81)90135-X

Rajagopal, K.R., Na, T.Y., and Gupta, A.S., 1984, "Flow of Viscoelastic Fluid over a Stretching Sheet," *Rheol. acta*, **23**, 213-215. http://dx.doi.org/10.1007/BF01332078

Rosali, H., Ishak, A., Pop, I., 2011, "Stagnation Point Flow and Heat Transfer over a Stretching/Shrinking Sheet in a Porous Medium," Int*Commun. Heat Mass Transfer*, **38**, 1029-1032. http://dx.doi.org/10.1016/J.Icheatmasstransfer.2011.04.031

Salem, A,M., and Fathy, R., 2012, "Effects of Variable Properties On MHD Heat and Mass Transfer Flow Near a Stagnation Point Towards a Stretching Sheet in a Porous Medium with Thermal Radiation," *Chin. Phys. B*, **21**, 054701. http://dx.doi.org/10.1088/1674-1056/21/5/054701

Sankara, K.K., and Watson, L.T., 1985, "Micropolar Flow Past a Stretching Sheet," *Z. angew. Math. Phys.*, **36**, 845-853. http://dx.doi.org/10.1007/BF00944898

Schlichting, H., and Gersten, K., 2000, *Boundary Layer Theory*, Springer, Berlin.

Siddappa, B., and Abel, M.S., 1985, "Non-Newtonian Flow past a Stretching Plate," *Z. Angew. Math. Phys.*, **36**, 890-892. http://dx.doi.org/10.1007/BF00944900

Troy, W.C., Overman, E.A., Ermentrout, H.G.B., and Keerner, J.P., 1987, "Uniqueness of the Flow of a Second Order Fluid past a Stretching Sheet," *Quart. appl. Math.*, **44**, 753-755. MR 872826 (87m:76009)

Van Gorder, R.A., and Vajravelu, K., 2010, "Hydromagnetic Stagnation Point Flow of a Second Grade Fluid over a Stretching Sheet," *Mech. Res. Commun.*, **37**, 113-118.
http://dx.doi.org/10.1016/J.Mechrescom.2009.09.009

Van Gorder, R.A., Vajravelu, K., and Pop, I., 2012, "Hydromagnetic Stagnation Point Flow of a Viscous Fluid over a Stretching Or Shrinking Sheet," *Meccanica*, **47**, 31-50.
http://dx.doi.org/10.1007/S11012-010-9402-0

Wang, C.Y., 1990, "Liquid Film on an Unsteady Stretching Sheet," *Q. Appl. Math.*, **48**, 601-610.

Wang, C.Y., 2008, "Stagnation Flow towards a Shrinking Sheet," *Int. J. Non-Linear Mech.*, **43**, 377-382.
http://dx.doi.org/10.1016/J.Ijnonlinmec.2007.12.021

Wilkinson, W., 1970, "The Drainage of a Maxwell Liquid Down a Vertical Plate," *Chem. Eng. J.*, **1**, 255-257.
http://dx.doi.org/10.1016/0300-9467(70)80008-9

Yacob, N.A., Ishak, A., and Pop, I., 2011, "Melting Heat Transfer in Boundary Layer Stagnation-Point Flow Towards a Stretching/Shrinking Sheet in a Micropolar Fluid," *Comput. Fluids*, **47**, 16-21.
http://dx.doi.org/10.1016/J.Compfluid.2011.01.040

Zhou, X.-F., and Gao L., 2007, "Effect of Multipolar interaction on The Effective Thermal Conductivity of Nanofluids," *Chin. Phys. B*, **16**, 2028-2032.
http://dx.doi.org/10.1088/1009-1963/16/7/037

9

MODELLING AND SIMULATION OF AU-WATER NANOFLUID FLOW IN WAVY CHANNELS

Suripeddi Srinivas[a], Akshay Gupta[b*], Ashish Kumar Kandoi[b]

[a] *Fluid Dynamics Division, School of Advance Sciences, VIT University, Vellore, Tamil Nadu, 632014, India*
[b] *School of Mechanical and Building Sciences, VIT University, Vellore, Tamil Nadu,632014, India*

ABSTRACT

The present work deals with the flow and thermal analysis of nanofluid in the wavy channels. The governing flow equations are solved numerically using CFD package assuming single phase approach. To study the effect of the concentration and size variation of the nanoparticle, the concentration and size are varied from 0% - 5% and 25 nm - 100 nm respectively over the Reynolds number range of 250-1500 for Au-water nanofluid. The effect on heat transfer enhancement because of corrugation of wavy channel is analyzed on four different shapes (sinusoidal, triangular, trapezoidal and square) channels. The effect of amplitude of the wavy channel is analyzed by varying it from 1 mm-3 mm over the considered Reynolds number range. Also the effect of phase difference (ϕ *) between the top and bottom wavy walls is investigated and its effect on the thermal and flow parameters are reported.

Keywords: Sinusoidal, triangular, trapezoidal, square channels, nano fluid, single phase approach, amplitude and Phase difference.

1. INTRODUCTION

For a heat dissipating device used in a heat engine operating between two temperature limits, its efficiency enhances if more heat is extracted from the medium at the sink. For this task passive method of introducing wavy corrugation on the walls of heat exchanger has already been found to be an effective method. The wavy channel helps in creating turbulence in the channel hence leading to better mixing of the fluid and enhancing the heat transfer. Nanotechnology has found its application in multitude streams of engineering ranging from electronics, information technology, energy, medical science, tribology and many more fields of science and technology. Use of nanofluid to enhance the heat transfer characteristics for heat exchanging devices can lead to further cost and size reduction of the heat exchanging devices.

Nanofluid contains extremely minute particles in the size range of nanometers, thus the initial researches suggested them to be homogenous in nature and hence single phase model came into existence. For single phase model approximation, works by Xuan *et al.* (2000), Brinkman (1952) and Maxwell (1904) acted as the cornerstone in the field which gave formulations for fundamental properties like specific heat, viscosity and conductivity respectively. The Maxwell equation used only the conductivities of the base fluid, nanoparticle and concentration of nanoparticle for the calculation of effective conductivity. There after attempts were made to refine the model by considering particle shape factor in Hamilton-Crosser (1962) model and in Yu and Choi (2003) model they considered a nanolayer surrounding the particle having conductivity greater than base fluid but in all these models the random Brownian motion of the nanoparticle was not considered hence all these models use to underestimate the conductivity. Thus in the present work the thermal conductivity is determined from Patel *et al.* (2005) model correlation which considers the Brownian motion and diameter of the particle. In recent times the work on multiphase model has narrowed the error between the experimental and numerical results. Lotfi *et al.* (2010) compared the experimental results of forced convective heat transfer for Al_2O_3 in a straight horizontal channel with single phase, two-phase Eulerian model and mixture model and concluded that two-phase model being a more precise model. Further Kalteh *et al.* (2011) numerically analyzed the forced laminar flow of Cu-water nanofluid through an isothermally heated micro channel and compared the results of single phase and two-phase Eulerian approach and found that the relative velocity and temperature between the phases was very negligible and the nanoparticle concentration distribution was nearly uniform in Eulerian model. A recent extensive study on multiphase approach by Moraveji *et al.* (2013) in which the author have compared single phase, VOF, mixture and Eulerian models for the flow of Al_2O_3 based nanofluid numerically and found that all multiphase approaches deviates slightly from the experimental results. Yang *et al.* (2014) have worked on sinusoidal corrugated channel and numerically compared the results of flow of Cu-water, Al_2O_3-water and CuO-water by multiphase and single phase approaches and reported the discrepancy with single phase results were within 8%.

The effective thermo physical properties of nanofluid are dependent on the flow regime of nanofluid, concentration and size of the nanoparticles. Some authors (Santra *et al.*, 2009; Davarnejad *et al.*, 2013; Moraveji *et al.*, 2013; Hussein *et al*, 2013) have showed the effect of concentration and size of nanoparticle in nanofluid and

* Corresponding Author : akshay.gupta2010@vit.ac.in

found that enhancement due to addition of nanoparticles is significant without significant increase in pressure and also the effect of enhancement is more visible at higher Reynolds number which is associated with higher turbulence associated hence increasing the overall conductivity.

The corrugations on the walls of the channel carrying coolant have found to enhance the heat transfer significantly, Hossain *et al.* (2007) have compared sinusoidal, triangular and arc shaped channels for heat transfer characteristics with same geometric dimensions, and reported that after the critical Reynolds number which depends on geometric configuration the flow starts mixing up and enhances heat transfer coefficient i.e. after a laminar flow regime when the flow starts to get turbulent. Using nanofluids in corrugated channels can enhance it even further, Pehlivan *et al.* (2013) studied the effect of corrugation angles and fin height on the heat transfer coefficient for constant heat flux at the boundary walls for a convergent-divergent section and found that by increasing the angle the nusselt number increases. Heidary *et al.* (2010) discussed the influence of nanoparticles on forced convection in a sinusoidal wavy channel for varying amplitude over a range of particle concentration and Reynolds number for Cu-water nanofluid using CFD package and reported the increase in enhancement with increasing amplitude, particle concentration and particle size. Ahmed *et al.* (2011, 2012, and 2013) in series of works have found the similar findings for triangular, trapezoidal and sinusoidal corrugations.

The Nusselt number evaluation can also vary by changing the relative positions of the top and bottom wavy walls. Naphon (2009) researched the effects of wavy plate configuration on temperature and flow distribution on channels of different shaped corrugations like triangular and trapezoidal for different phase orientation of the top and the bottom wavy plate using CFD. Vanaki *et al.* (2014) reported the heat transfer enhancement and pressure drop in a wavy channel with different phase shifts over turbulent flow regime for SiO_2 nanoparticle with different shaped particles-blades, platelets, cylindrical, bricks, and spherical in a sinusoidal channel and concluded the best results for SiO_2-ethelyne glycol nanofluid with platelets shaped particles. Jixiang *et al.* (2012) analyzed the effect of phase shift for sinusoidal channel on flow and heat transfer characteristics over turbulent flow regime. The author also has seen the effect of shear stress and friction factor with different phase shift angles.

Conventionally for increasing the heat transfer in a heat exchanger, one has to use high viscous fluid but that leads to increase in pressure drop across the channel. The effect of corrugated geometry and nanoparticle is such that by adding nanoparticles the enhancement is a lot more significant without significant increase in the pressure drops. Few authors (Arani and Amani ,2012; Kheram ,2011; Fakoor *et al.*, 2013; Aliabadi *et al.*, 2014; Youssef *et al., 2012*) have suggested the effect of geometry of the channel and nanoparticle size and concentration on the pressure drop across the channel. Vatani *et al.* (2013) discussed the effect on heat transfer enhancement and friction factor for varying nanoparticle type, size and concentration for turbulent flow in triangular, square and arc shaped ribbed channel.

In the present work, an exhaustive study is performed for four different corrugated wavy-shaped channels-sinusoidal, square, triangular and trapezoidal with nanofluid flowing through it. The analyses are done in Ansys Fluent CFD package using single phase approach. The effect of nanoparticle size and particle concentration is observed over a Reynolds number range of 250-1500 with Au nanoparticles. Further, the effect of geometrical parameters on heat transfer and pressure drop is carried out for different wavy amplitude and phase difference (ϕ^*) for different shaped corrugations.

2. MATHEMATICAL MODELLING

2.1 Problem Statement

In the present problem a single phase model assumption is used to model nanofluid flow in different wavy-shaped channels. The physical model and geometric configuration for the problem is shown in Fig 6. It consists of a straight channel with 20 mm hydraulic diameter and two parallel isothermally heated walls of 80 mm length at the entrance for the flow to be fully developed in the wavy part. Six wave numbers of wavelength 22 mm are taken in the wavy channel which is also the heated section. Total length of the channel is 400 mm. In the analyses the amplitude of the wavy part is varied as 1 mm, 2 mm and 3 mm and the ϕ^* of 0°, 45°, 90°, 135° and 180° is applied between the top and bottom wavy parts.

2.2 Governing Equation

The governing equation in non-dimensional for while considering single phase approach for the mixture of base fluid and nanoparticles are as follows (Santra *et al.*, 2009; 33):

Continuity Equation:

$$\frac{\partial U}{\partial X}+\frac{\partial V}{\partial Y}=0 \tag{1}$$

X-Momentum:

$$\frac{\partial U}{\partial \tau}+U\frac{\partial U}{\partial X}+V\frac{\partial V}{\partial Y}=-\frac{\partial P}{\partial X}+\frac{1}{\text{Re}}\frac{\rho_f}{\rho_{nf}}\frac{1}{(1-\phi)^{2.5}}\left[\frac{\partial^2 U}{\partial X^2}+\frac{\partial^2 U}{\partial Y^2}\right] \tag{2}$$

Y-Momentum:

$$\frac{\partial V}{\partial \tau}+U\frac{\partial V}{\partial X}+V\frac{\partial V}{\partial Y}=-\frac{\partial P}{\partial Y}+\frac{1}{\text{Re}}\frac{\rho_f}{\rho_{nf}}\frac{1}{(1-\phi)^{2.5}}\left[\frac{\partial^2 V}{\partial X^2}+\frac{\partial^2 V}{\partial Y^2}\right] \tag{3}$$

Energy:

$$\frac{\partial \theta}{\partial \tau}+U\frac{\partial \theta}{\partial X}+V\frac{\partial \theta}{\partial Y}=\frac{1}{\text{RePr}}\frac{K_{nf}}{K_f}\frac{(\rho C_P)_f}{(\rho C_P)_{nf}}\left[\frac{\partial^2 \theta}{\partial X^2}+\frac{\partial^2 \theta}{\partial Y^2}\right] \tag{4}$$

The non-dimensional parameters are:

$$X=\frac{x}{D_h},\quad Y=\frac{y}{D_h},\quad U=\frac{u}{u_{in}},$$
$$V=\frac{v}{u_{in}},\quad \theta=\frac{T-T_{in}}{T_w-T_{in}},\quad P=\frac{p}{\rho_f \mu_f^{\,2}}$$
$$\tau=\frac{tu_{in}}{D_h},\quad \text{Re}=\frac{\rho_f u_{in} D_h}{\mu_f},\quad \text{Pr}=\frac{u_f C_{pf}}{k_f} \tag{5}$$

The solution of the governing equations gives the flow and temperature field which are used to calculate the Nusselt number for the flow as follows. The local and the average Nusselt number for the corrugated walls are defined as follows (Santra *et al.*, 2009).

$$Nu_l=\frac{1}{x_e-x_s}\int_{x_s}^{x_e} Nu_x dx \tag{6}$$

$$Nu_x=\frac{h_x D_h}{k_{nf}}=\left|-\frac{k_f}{k_{nf}}\frac{\partial \theta}{\partial Y}\right|_{x,0} \tag{7}$$

The pressure drop for the flow in the corrugated channel is calculated as:

$$\Delta P = f\frac{(X_e - X_s)}{D_h}\frac{\rho_{nf} u_{in}^2}{2} \quad (8)$$

2.3 Thermo physical properties of nanofluid

Thermo physical properties of nanofluid are calculated from the Single phase model equations and are used in the governing equations:

Viscosity (Brinkman, 1952), Brinkman's model has been used:

$$\mu_{nf} = \frac{\mu_f}{(1-\phi)^{2.5}} \quad (9)$$

Density:

$$\rho = (1-\phi)\rho_f + \phi\rho_p \quad (10)$$

Specific Heat (Xuan *et al.*, 2000):

$$(\rho C_p)_{nf} = (1-\phi)(\rho C_p)_f + \phi(\rho C_p)_p \quad (11)$$

Conductivity (Patel *et al.*, 2005):

$$\frac{K_{eff}}{K_f} = 1 + \frac{K_p A_p}{K_f A_f} + cK_p Pe\frac{A_p}{K_f A_f} \quad (12)$$

Where:

$$\frac{A_p}{A_f} = \frac{d_f}{dp}\frac{\phi}{(1-\phi)} \qquad Pe = \frac{u_p d_p}{a_f} \quad (13)$$

Where u_p is the Brownian motion velocity of the particles which is defined as follows:

$$u_p = \frac{2K_p T}{\pi\mu_f d_p^2} \quad (14)$$

2.4 Boundary Conditions

Boundary Conditions are as follows:

1) At the inlet:

$$u = u_{in}, v = 0, T = T_{in} = 293 \quad (15)$$

2) Along the heated Section:

$$u = v = 0, T_w = 303K \quad (16)$$

3) At the Straight section:

$$u = v = 0 \text{ (no slip)}, \; T = T_{in} \quad (17)$$

4) Outlet boundary:

$$\frac{\partial u}{\partial x} = \frac{\partial v}{\partial x} = \frac{\partial T}{\partial x} = 0 \quad (18)$$

3. NUMERICAL TECHNIQUE

After the conversion of non-dimensional governing equations from Cartesian to computational domain, the discretization of time-independent incompressible Navier–Stokes governing equations was done using the finite-volume method (FVM). The diffusion term in the momentum and energy equations is approximated by second-order central difference and QUICK Scheme is adopted for the convective terms. The flow field was solved using the SIMPLEC algorithm this is an iterative solution procedure where the computation is initialized by guessing the pressure field. Then, the momentum equation is solved to determine the velocity components. The pressure is updated using the continuity equation. The convergence criteria for all the variables are set as 10^{-5}.The mesh independency test was carried for four different grid sizes 560 X 30, 653 X 40, 738 X 50 and 836 X 60 for water. Nusselt number and heat transfer coefficient (HTC) were found to vary by 0.3672% and 0.3653% respectively, thus indicating that results are independent of mesh size. For the optimal CPU time for computation in simulations, grid size of 738 X 50 was used for the CFD simulations.

4. VALIDATION

For validating the numerical method and the mathematical model used for the analysis, the results of the flow of Cu-water nanofluid in straight channel was compared with the work of Santra *et al.* (2009), as can be observed from Fig 1 and the results shows a slight deviation at Re=500 and Re=1500. Figure 2 shows the comparison of results obtained for Nusselt number in the present study from numerical simulation with that of Yang *et al.* (2014) for the flow of water through a sinusoidal channel with 3 mm amplitude and 180° phase difference for water.

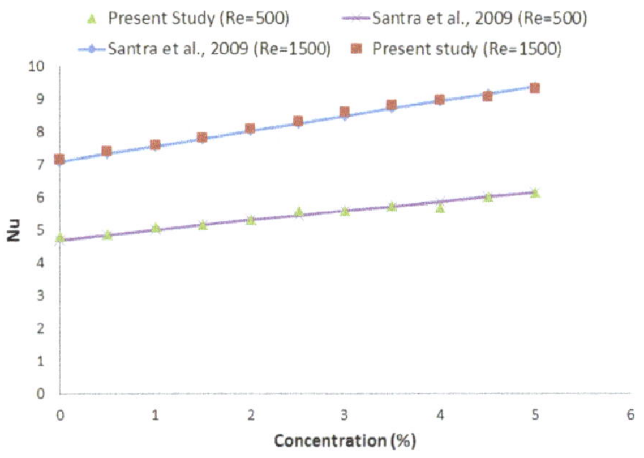

Fig. 1 Nusselt number at different concentrations

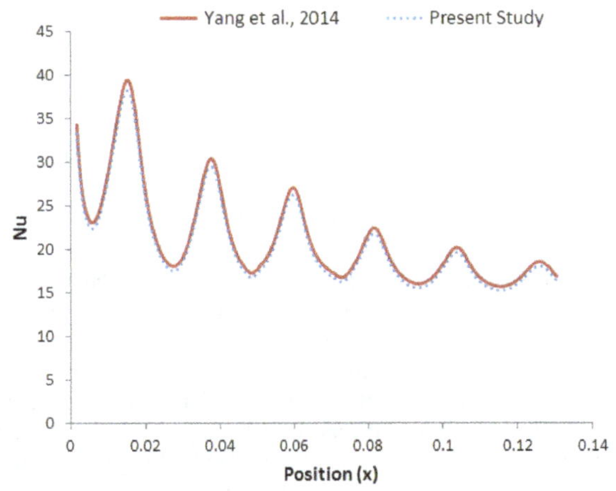

Fig. 2 Nusselt number at different sections of a sinusoidal channel

5. RESULT AND DISCUSSIONS

The analysis of the effect of nanoparticle concentration, size and geometrical variations were carried out for four different shaped channels-sinusoidal, triangular, trapezoidal and square for Au-water nanofluid. From the work of Kalteh (2013) , who compared the performance of nine different water based nanofluids (SiO_2, TiO_2, CuO, Al_2O_3, Fe, Cu, Ag, Au and diamond) in a straight channel at Re=100 and 1% concentration a comparative performance of different nanofluids can be understood,. Au-water nanofluid is selected for the present study as Au is widely known for its biocompatible properties (Bogliotti *et al.,* 2011; Arnida *et al., 2011.)*

5.1. Effect of nanoparticle size and concentration

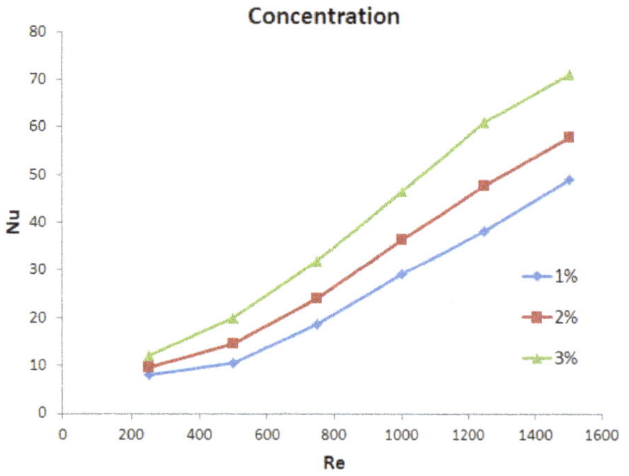

Fig. 3 Effect of nanoparticle concentration on Nusselt number

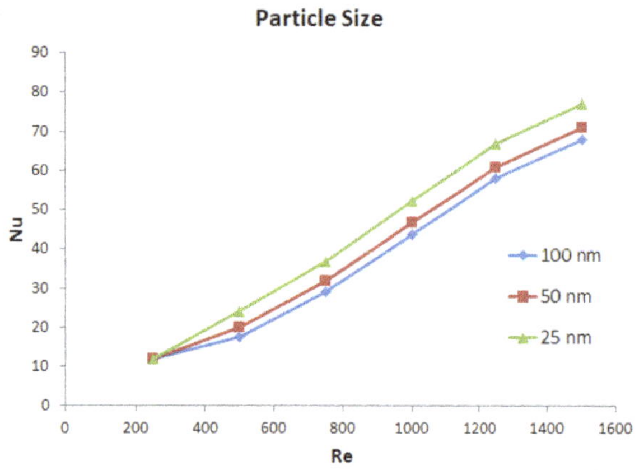

Fig. 4 Effect of nanoparticle size on Nusselt number.

Figure 3 is depicting the effect of varying Au particle concentration as 1%, 3% and 5% and Fig 4 is depicting the effect of Au nanoparticle size as 25 nm, 50 nm and 100 nm on Nusselt number in a sinusoidal channel with 180° phase difference and 3 mm amplitude for laminar flow regime with Reynolds number variation of 250-1500. The effect of increasing the concentration resulted in increase of Nusselt number due to increase in thermal conductivity of the nanofluid. It is observed that the difference between the 1% and 5% concentration for Nusselt number at Re=250 is just 3.85 and this difference grows to 22 at Re=1500, this shows that the enhancement at low Reynolds number is not reflected as clearly as at higher Reynolds number. This is because of randomness of the molecular motion causing the conductivity to increase and hence the enhancement at higher Reynolds number.

The effect of nanoparticle size variation can be seen in Fig 4 which suggests that the Nusselt number increases by decreasing the particle size as obvious because of increase in conductivity of nanofluid as a result of increased surface area and Brownian motion. It is observed that the difference of Nusselt number between the 25 nm and 100 nm Au nanoparticle size at Re=250 is 0.21 and this difference grows to 8.94 at Re=1500, this confirms that the enhancement at low Reynolds number is not reflected as clearly as at higher Reynolds number due to the effect of turbulence which increases the conductivity and hence the enhancement at higher Reynolds number.

5.2. Effect of amplitude on different shaped channels

The effects of amplitude of corrugations on different shaped wavy-channels are shown in Fig 5, it can be seen that as we increase the amplitude of the wavy channel, the turbulence and flow reversal increases. For the comparison of various geometries at 1 mm amplitude, from Fig 7 (a), it can be observed that the Nusselt number and pressure drop for a square channel is throughout greater than other shaped channels irrespective of the Reynolds number. The difference of Nusselt number between square and other shaped channels at lower Reynolds number is more than at higher Reynolds number, because in square channel, higher turbulence can be generated at lower Reynolds number itself whereas for other geometries the enhancement due to turbulence is generated at relatively higher Reynolds number. As we increase the amplitude it can be observed that at certain Reynolds number the enhancement in the heat transfer for other geometries goes beyond that of square but the pressure drop across the channel is always the highest for square and least for triangular and the plots of pressure drop are seen to diverge out more at higher Reynolds number with the increase of amplitude i.e. increase in pressure drop for square shaped corrugation at higher Reynolds number is more than the other channels. At higher Reynolds number thus it is favorable to operate at 3 mm amplitude in triangular corrugated channel so as to get the maximum heat transfer enhancement and least pressure drop.

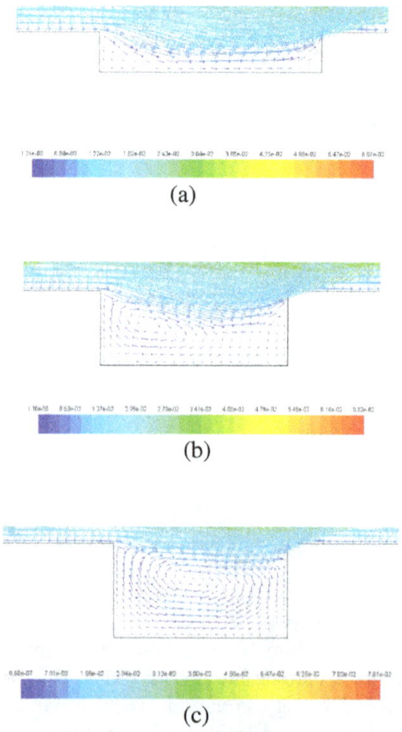

Fig. 5 Velocity vector for square channel at different amplitude at Re=1500: (a) 1 mm (b) 2 mm (c) 3 mm

5.3. Effect of phase difference on different shaped channels

The effect of phase difference (ϕ^*) between the top and the bottom waves of the corrugated channel is reported in Fig 8, Fig 9, Fig 10 and Fig 11. Five different phase shifts of 0°, 45°, 90°, 135° and 180° are considered for the analysis. As can be observed in Fig 8, the Nusselt number is minimum and pressure drop is maximum at 90° phase shift angle suggesting that it is not favorable for the heat dissipating devices such as radiators to operate at this particular phase shift angle.

Also as the phase angle is changing the maximum and minimum velocities across the section is changing because of the change in the effective cross section for the flow as can be seen in Fig 11, thus affecting the heat transfer characteristics of the channel.

In Fig 9 and Fig 10, the comparison for all the geometries for various Reynolds number with all the phase difference angles are presented. The observations are:

I. At 0° ϕ^* the Nusselt number is maximum for trapezoidal and least for square throughout the flow regime whereas the pressure drop is maximum for square and minimum for triangular thus suggesting that Square being the worse possible option for this orientation.

II. At 45°, 90° and 135° ϕ^* it is observed that throughout the Reynolds number range, triangular channel gives least pressure drop and maximum Nusselt number whereas square gives maximum pressure drop and minimum Nusselt number.

III. At 180°, the Nusselt number as well as the pressure drop is maximum for square geometry. Pressure drop is least for triangular and the Nusselt number is least for trapezoidal.

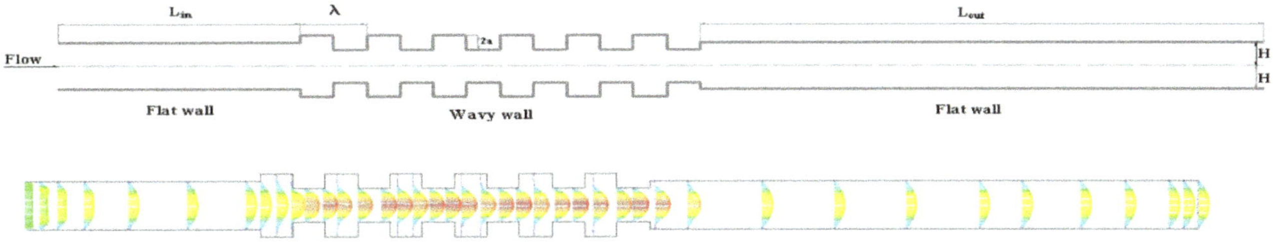

Fig. 6 Schematic diagram of the present problem

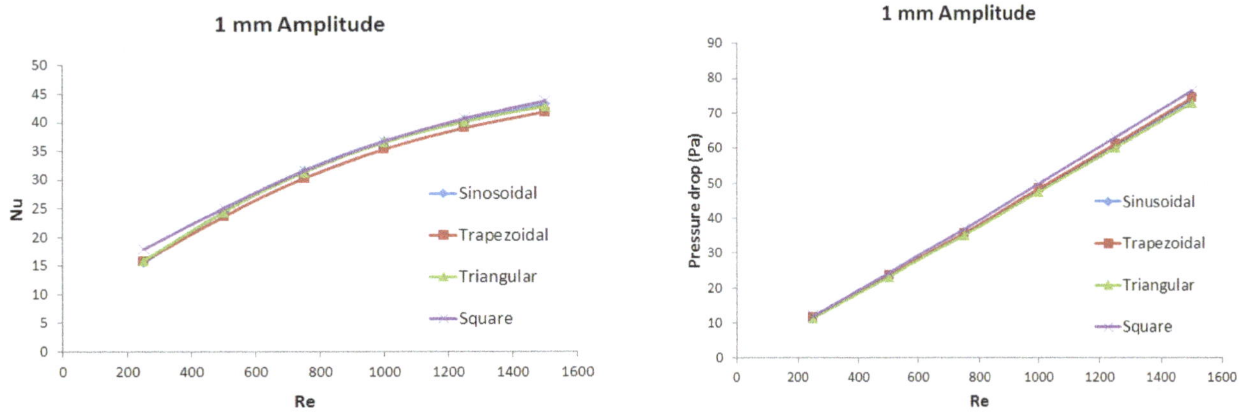

(a)

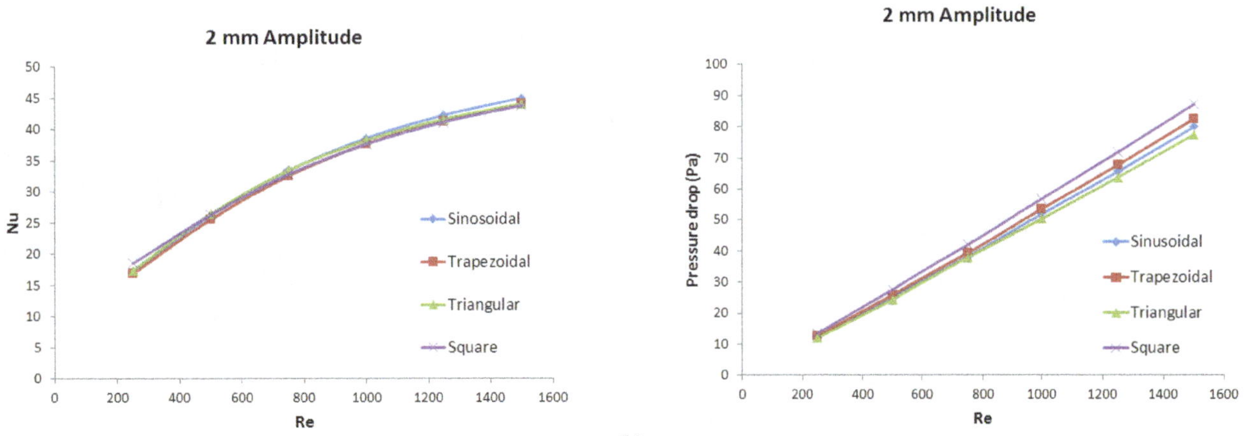

(b)

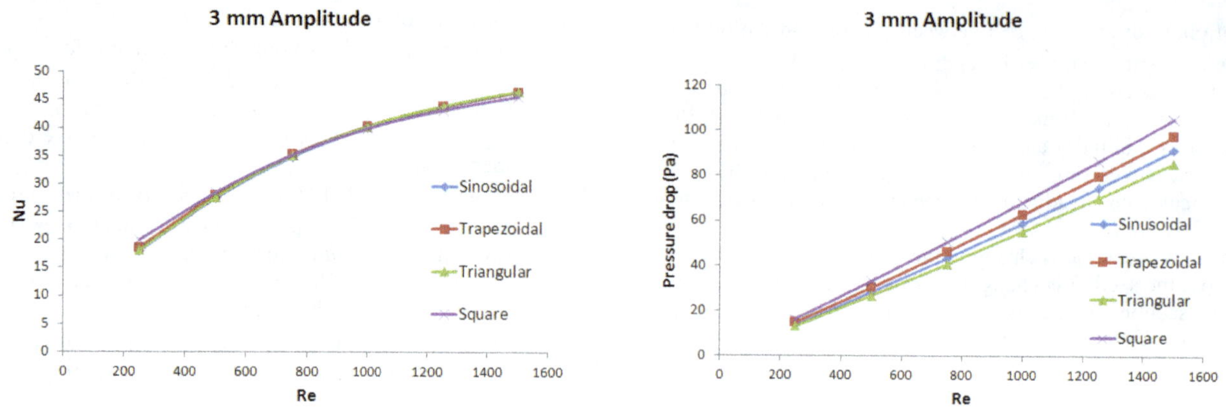

(c)

Fig. 7 Nusselt number and pressure drop (Pa) variation for (a) 1 mm (b) 2 mm (c) 3 mm amplitude.

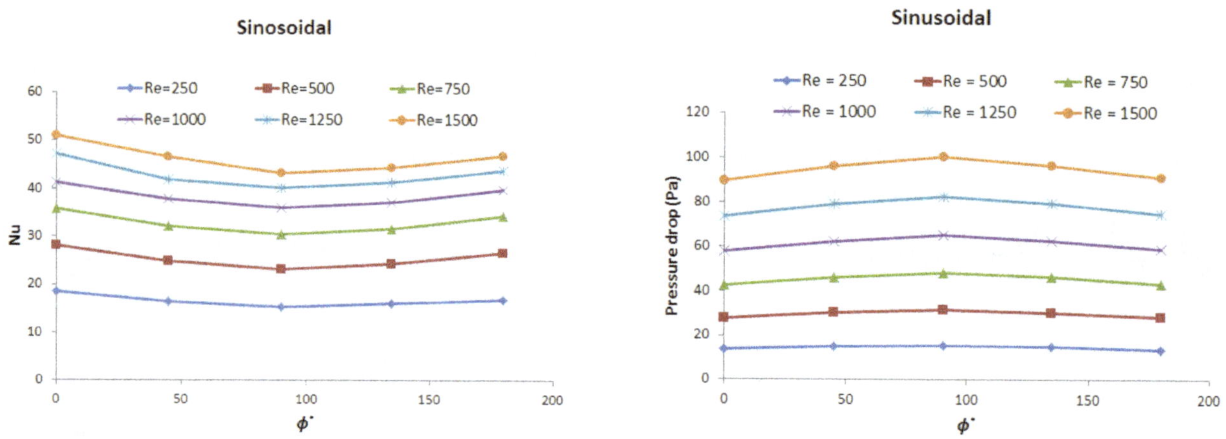

(a)

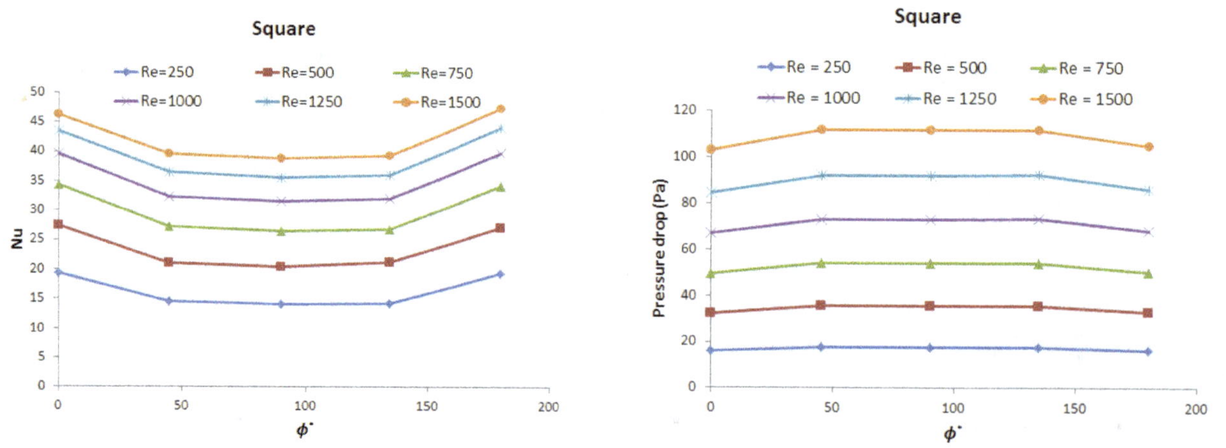

(b)

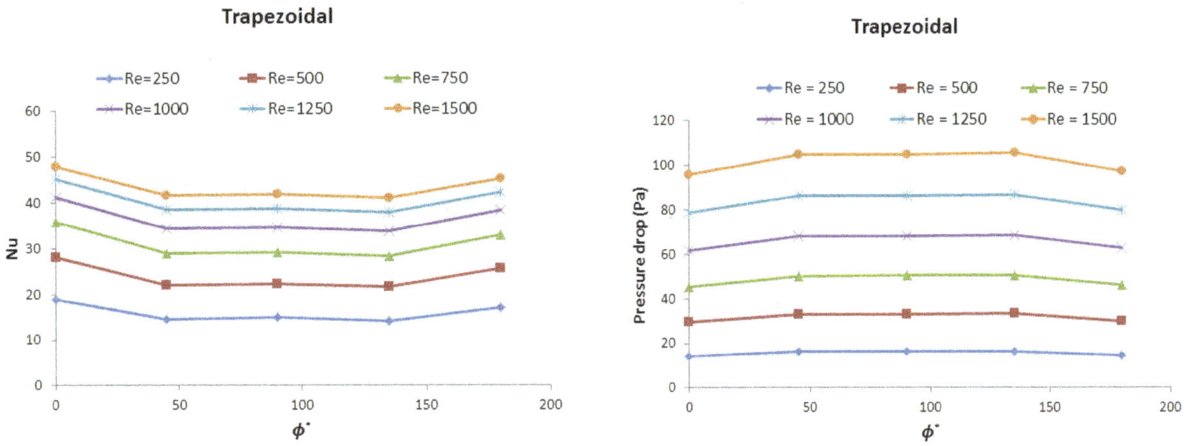

(c)

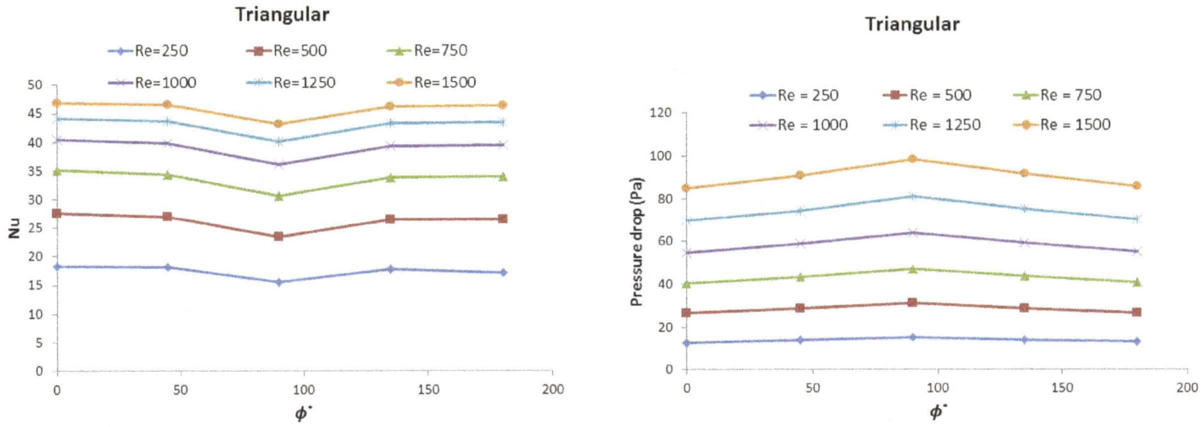

(d)

Fig. 8 Nusselt number and pressure drop (Pa) variation with ϕ^* (a) Sinusoidal (b) Square (c) Trapezoidal (d) Triangular

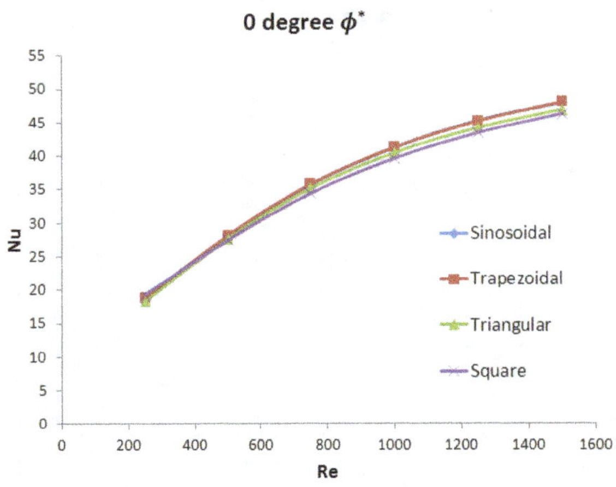

(a)

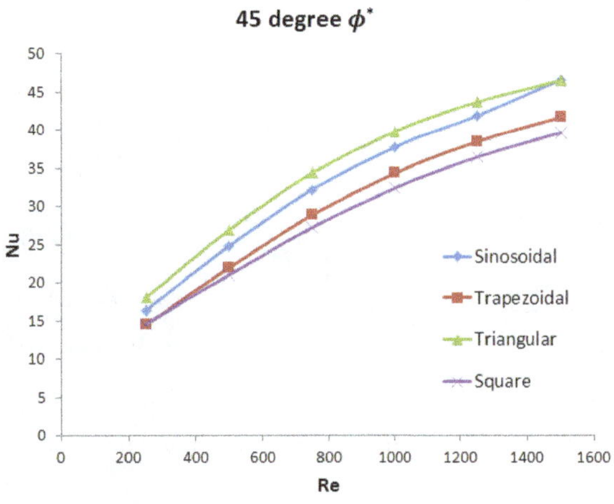

(b)

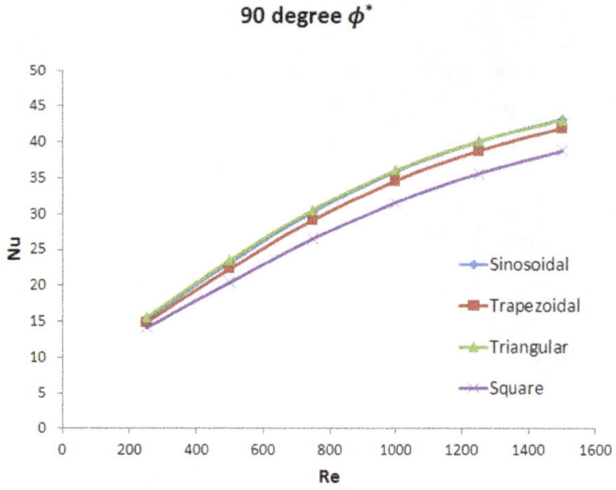

(c)

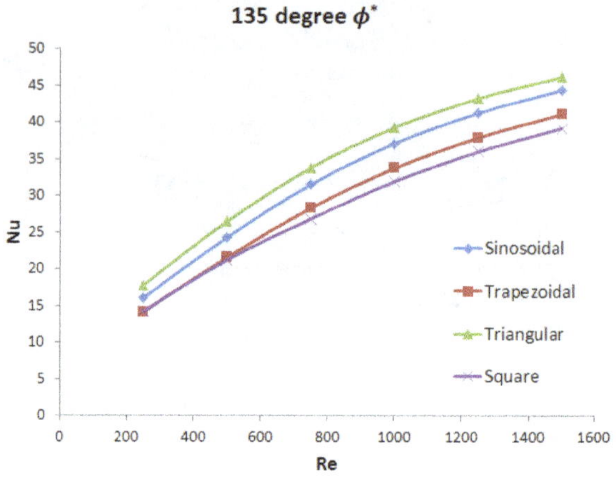

(d)

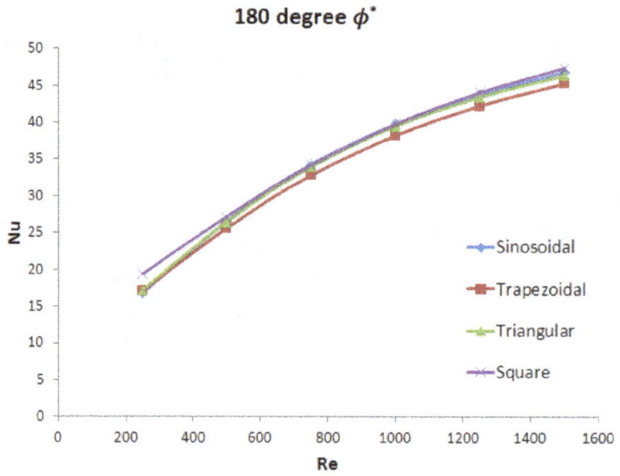

(e)

Fig. 9 Nusselt number variation for different geometries at different ϕ^*: (a) 0° (b) 45° (c) 90° (d) 135° (e) 180°

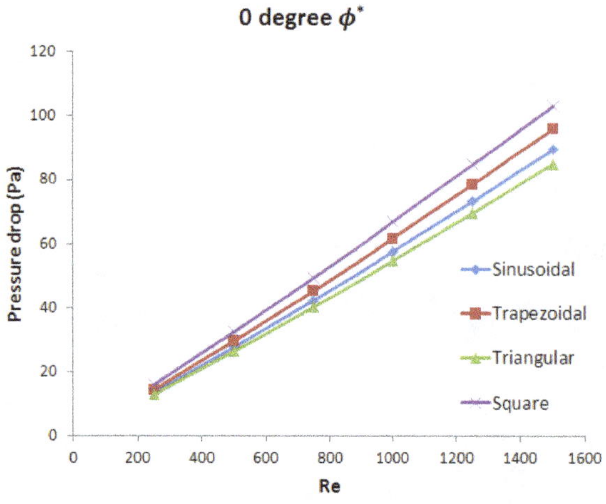

(a)

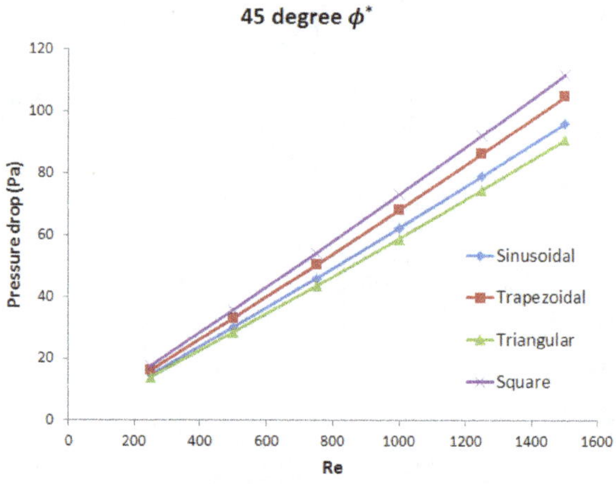

(b)

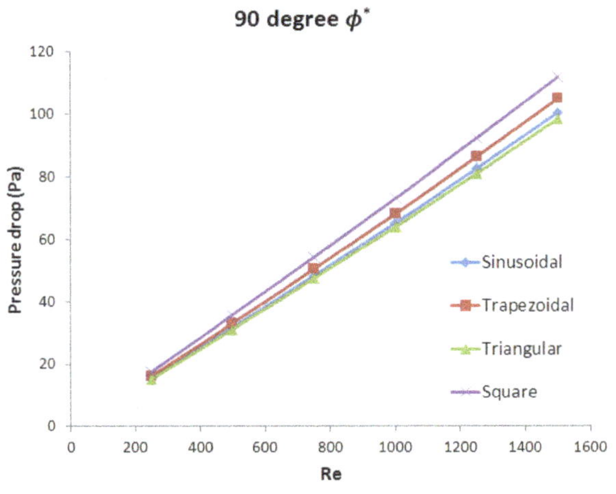

(c)

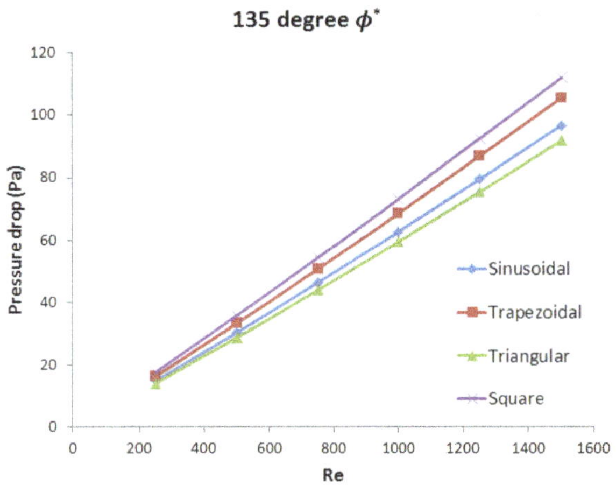

(d)

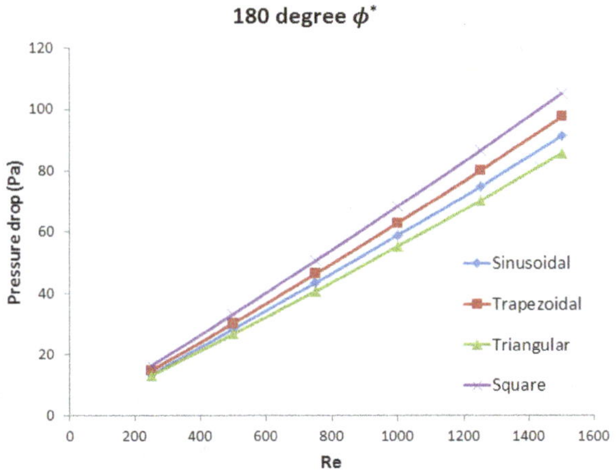

(e)

Fig. 10 Pressure drop (Pa) variation for different geometries at different ϕ^*: (a) 0° (b) 45° (c) 90° (d) 135° (e) 180°

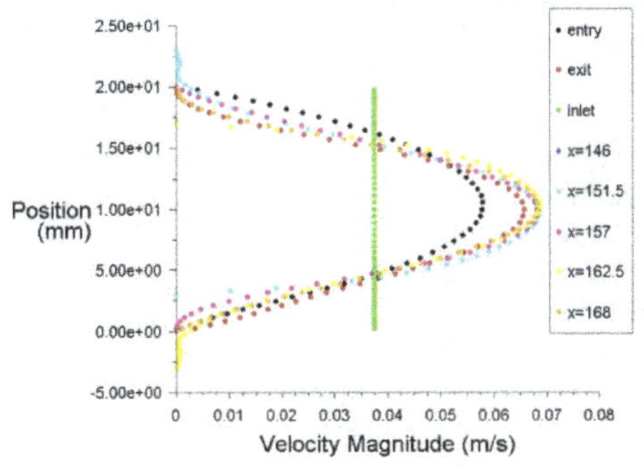

(a)

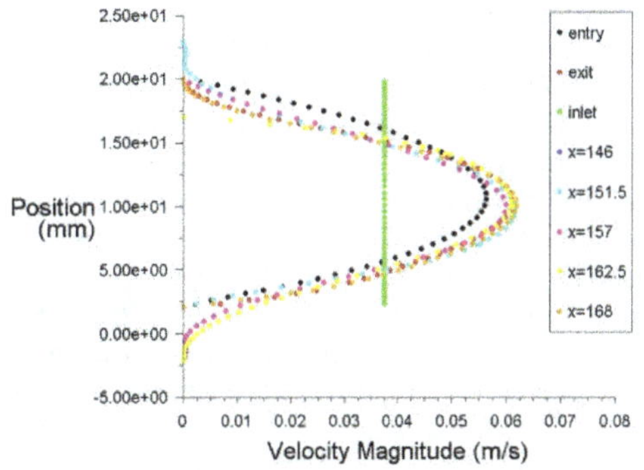

(b)

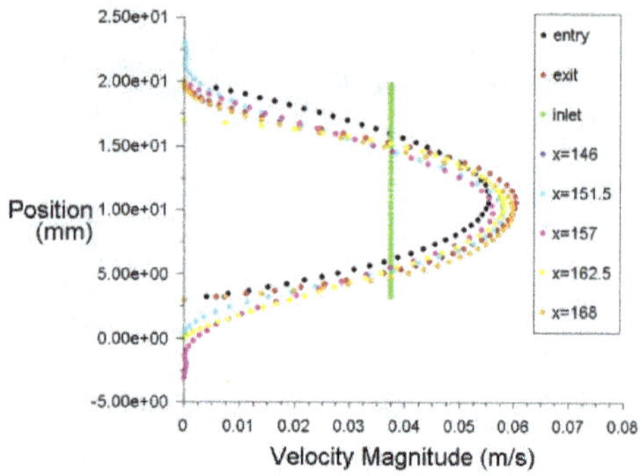

(c)

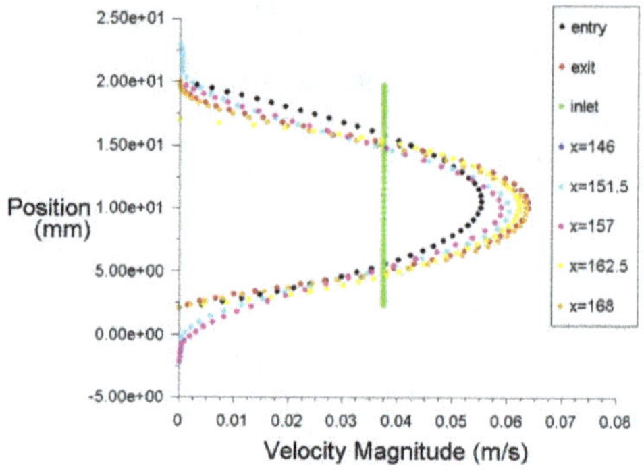

(d)

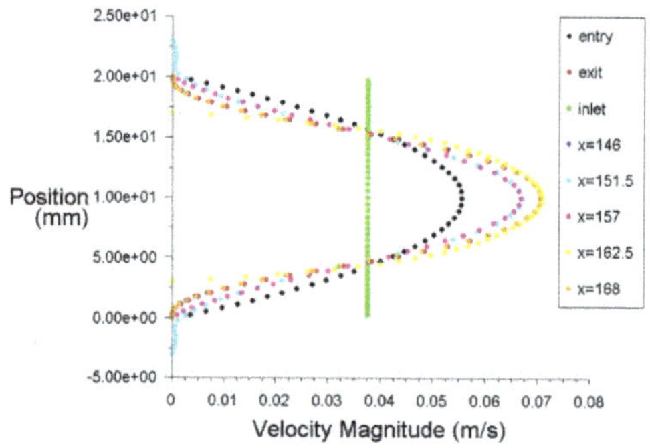

(e)

Fig. 11 Velocity magnitude plots for sinusoidal channel at different ϕ^*: (a) 0° (b) 45° (c) 90° (d) 135° (e) 180°

6. CONCLUSIONS

A detailed analysis pertaining to the flow of a nanofluid through wavy channel(s) is reported in the present work. The analysis of the effect of nanoparticle size and concentration on heat transfer enhancement for Au based nanofluid shows the enhancement to be significant for higher Reynolds number and also the effect of particle size and concentration is more predominating at higher Reynolds number. The effect of geometrical parameters (amplitude and ϕ^*) on heat transfer enhancement and pressure drop characteristics have been studied for four different wavy shapes (sinusoidal, triangular, trapezoidal and square) of the channel with Au-water nanofluid with 5% concentration and 50 nm nanoparticle size. By increasing the amplitude, the results show a higher heat transfer enhancement but at the cost of increase in pressure drop. The selection of the required geometry is based on the particular needs and conditions and there is a tradeoff between heat transfer enhancement and pressure drop. The effect of phase difference between the top and the bottom wavy walls for all the four shapes (sinusoidal, trapezoidal, triangular and square) are compared. The analysis suggests that 90° phase shift of the wavy channel is not favorable for heat dissipating devices. To conclude, the favorable results for heat transfer enhancement are obtained for triangular channel followed by sinusoidal at 45°, 90° and 135° phase shift. But if pressure drop is not of much consideration as compared to heat transfer enhancement then square channel at 180° and trapezoidal channel at 0° can be considered for getting maximum heat transfer rate.

ACKNOWLEDGEMENTS

The authors gratefully acknowledge NBHM, Govt. of India for sanctioning a major research project under the grant number 2/48(19)/2012/NBHM(R.P.)/R&D II/9137.

NOMENCLATURE

a	wavy amplitude (mm)
C_p	specific heat (J/kg K)
d	molecular diameter (nm)
D_h	hydraulic diameter (mm)
H	half of channel height (mm)
K	thermal conductivity (W/m °C)
Nu	Nusselt number
P	static pressure (Pa)
Pe	Peclet number
Pr	Prandtl number
ΔP	pressure drop (Pa)
Re	Reynolds number
t	time (s)
T	temperature (°C)
x, y	2D Cartesian coordinates (m)
X, Y	dimensionless cartesian coordinate
u, v	velocity components (m/s)
U, V	dimensionless velocity component

Greek Symbols

α	thermal diffusivity (m^2/s)
λ	wavelength (mm)

ϕ	volume fraction of nanoparticles (%)
ϕ^*	phase angle between top and bottom wavy walls
μ	dynamic viscosity ($N.S/m^2$)
ρ	density of the fluid (kg/m^3)
τ	dimensionless time
θ	dimensionless temperature

Subscripts

e	end point of wavy wall
eff	effective
f	base fluid
in	inlet
l	average value
nf	nanofluid
p	particles
s	starting point of wavy wall
w	wall
x	local value

REFERENCES

Arani A.A.A., Amani J., 2012, "Experimental Study on the Effect of TiO_2–Water Nanofluid on Heat Transfer and Pressure Drop," *Experimental Thermal and Fluid Science*, **42**, pp. 107-115. http://dx.doi.org/10.1016/j.expthermflusci.2012.04.017

Ahmed M.A., Shuaib N.H., Yusoff M.Z., Al-Falahi A.H., 2011, "Numerical Investigations of Flow and Heat Transfer Enhancement in a Corrugated Channel Using Nanofluid," *International Communications in Heat and Mass Transfer*, **38** (10), pp. 1368–1375. http://dx.doi.org/10.1016/j.icheatmasstransfer.2011.08.013

Ahmed M.A., Shuaib N.H., Yusoff M.Z., 2012, "Numerical Investigations on the Heat Transfer Enhancement in a Wavy Channel Using Nanofluid," *International Journal of Heat and Mass Transfer*, **55** (21-22), pp. 5891–5898. http://dx.doi.org/10.1016/j.ijheatmasstransfer.2012.05.086

Ahmed M.A., Yusoff M.Z., Shuaib N.H., 2013, "Effects of Geometrical Parameters on the Flow and Heat Transfer Characteristics in Trapezoidal-Corrugated Channel Using Nanofluid," *International Communications in Heat and Mass Transfer,* **42**, pp. 69–74. http://dx.doi.org/10.1016/j.icheatmasstransfer.2012.12.012

Arnida, Janat-Amsbury M.M., Ray A., Peterson C.M., Ghandehari H., 2011, "Geometry and Surface Characteristics of Gold Nanoparticles Influence Their Biodistribution and Uptake by Macrophages," *European Journal of Pharmaceutics and Biopharmaceutics*, **77**(3), pp. 417-423. http://dx.doi.org/10.1016/j.ejpb.2010.11.010

Vatani A., Mohammed H.A, 2013, "Turbuent Nanofluid Nanofluid over Periodic Rib-Grooved Channel," *Engineering Application of Computational Fluid Mechanics*, **7**(3), pp. 369-381. http://scival-expert.utm.my/pubDetail.asp?t=pm&id=84882303327&

Bogliotti N., Oberleitner B., Di-Cicco A., Schmidt F., Florent J.C., Semetey V, 2011, "Optimizing the Formation of Biocompatible Gold Nanorods for Cancer Research: Functionalization, Stabilization and Purification," *Journal of Colloid and Interface Science*, **357**(1), pp. 75-81. http://dx.doi.org/10.1016/j.jcis.2011.01.0 53

Brinkman H.C., 1952, "The Viscosity of Concentrated Suspensions and Solution," *Journal of Chemical Physics*, **20**, 571–581. http://dx.doi.org10.1063/1.1700493

Davarnejad R., Barati S., Kooshki M., 2013, "CFD Simulation of the Effect of Particle Size on the Nanofluids Convective Heat Transfer in the Developed Region in a Circular Tube," *Springer Plus*, **2**, pp. 192-198. http://dx.doi.org/10.1186/2193-1801-2-192

Fakoor-Pakdaman M., Akhavan-Behabadi M.A., Razi P., 2013, "An Empirical Study on the Pressure Drop Characteristics of Nanofluid Flow Inside Helically Coiled Tubes," *International Journal of Thermal Sciences*, **65**, pp. 206-213. http://dx.doi.org/10.1016/j.ijthermalsci.2012.10.014

Fluent 6.3 User's Guide, Fluent Inc., 2006

Hamilton R.L., Crosser O.K., 1962, "Thermal Conductivity of Heterogeneous Two-Component Systems," *Industrial & Engineering Chemistry Fundamentals*, **1**(3), pp. 187–191. http://dx.doi.org/10.1021/i160003a005

Heidary H., Kermani M.J., 2010, "Effect of Nano-Particles on Forced Convection in Sinusoidal-Wall Channel," *International Communications in Heat and Mass Transfer*, **37**(10), pp. 1520–1527. http://dx.doi.org/10.1016/j.icheatmasstransfer.2010.08.018

Hossain M.Z. and Sadrul Islam A.K.M., 2007, "Numerical Investigation of Fluid Flow and Heat Transfer Characteristics in Sine, Triangular, and Arc-Shaped Channels," *Thermal science*, **11**(1), pp. 17-26. http://dx.doi.org/10.2298/TSCI0701017H

Hussein A.M., Sharma K.V., Bakar R.A., Kadirgama K., 2013, "A Review of Forced Convection Heat Transfer Enhancement and Hydrodynamic Characteristics of a Nanofluid," Renewable and Sustainable Energy Reviews, **29**, pp. 734-743. http://dx.doi.org/10.1016/j.rser.2013.08.014

Kalteh M., Abbassi A., Saffar-Avval M., Harting J., 2011, "Eulerian–Eulerian Two-Phase Numerical Simulation of Nanofluid Laminar Forced Convection in a Microchannel," *International Journal of Heat and Fluid Flow*, **32**(1), pp. 107–116. http://dx.doi.org/10.1016/j.ijheatfluidflow.2010.08.001

Kalteh M., 2013, "Investigating the Effect of Various Nanoparticle and Base Liquidtypes on the Nanofluids Heat and Fluid Flow in a Microchannel," *Applied Mathematical Modelling*, **37**(18-19), pp. 8600–8609. http://dx.doi.org/10.1016/j.apm.2013.03.067

Kheram M.A., 2011, "Numerical Study on Convective Heat Transfer for Water-Based Alumina Nanofluids," *International Journal of Nano Dimension*, **1**(4), pp. 297-304. http://www.hindawi.com/journals/stni/2012/928406/

Khoshvaght-Aliabadi M., Hormozi F., Zamzamian A., 2014, "Experimental Analysis of Thermal–Hydraulic Performance of Copper–Water Nanofluid Flow in Different Plate-Fin Channels," *Experimental Thermal and Fluid Science*, **52**, pp. 248–258. http://dx.doi.org/10.1016/j.expthermflusci.2013.09.018

Lotfi R., Saboohi Y., Rashidi A.M., 2010, "Numerical Study of Forced Convective Heat Transfer of Nanofluids: Comparison of Different Approaches," *International Communications in Heat and Mass Transfer*, **37**(1), pp. 74–78. http://dx.doi.org/10.1016/j.icheatmasstransfer.2009.07.013

Maxwell J.C., 1904, *Treatise on Electricity and Magnetism*, Oxford University Press London.

Moraveji M.K., Ardehali R.M., Ijam A., 2013, "CFD Investigation of Nanofluid Effects (Cooling Performance and Pressure Drop) in Mini-Channel Heat Sink," *International Communications in Heat and Mass Transfer*, **40**, pp. 58–66. http://dx.doi.org/10.1016/j.icheatmasstransfer.2012.10.021

Moraveji M.K., Ardehali R.M., 2013, "CFD Modeling (Comparing Single and Two-Phase Approaches) on Thermal Performance of Al_2O_3-Water Nanofluid in Mini-Channel Heat Sink," *International Communications in Heat and Mass Transfer*, **44**, pp. 157–164. http://dx.doi.org/10.1016/j.icheatmasstransfer.2013.02.012

Naphon P., 2009, "Effect of Wavy Plate Geometry Configurations on the Temperature and Flow Distributions," *International Communications in Heat and Mass Transfer*, **36**(9), pp. 942–946. http://dx.doi.org/10.1016/j.icheatmasstransfer.2009.05.007

Patel H.E., Sundarrajan T., Pradeep T., Dasgupta A., Das S.K., 2005, "A Micro-Convection Model for Thermal Conductivity of Nanofluid," *Pramana Journal of Phyics*, **65**(5), pp. 863–869. http://link.springer.com/article/10.1007%2FBF02704086

Pehlivan H., Taymaz I., İslamoğlu Y., 2013, "Experimental Study of Forced Convective Heat Transfer in a Different Arranged Corrugated Channel," *International Communications in Heat and Mass Transfer*, **46**, pp. 106–111. http://dx.doi.org/10.1016/j.icheatmasstransfer.2013.05.016

Rohsenow W.M., Hartnett J.P, Cho Y.I., 1998, *Handbook of Heat Transfer, 3rd edition*, McGraw-Hill, New York.

Santra A.K., Sen S., Chakraborty N., 2009, "Study of Heat Transfer Due to Laminar Flow of Copper–Water Nanofluid through Two Isothermally Heated Parallel Plates," *International Journal of Thermal Sciences*, **48**(2), pp. 391–400. http://dx.doi.org/10.1016/j.ijthermalsci.2008.10.004

Vanaki Sh.M., Mohammed H.A., Abdollahi A., Wahid M.A., 2014, "Effect of Nanoparticle Shapes on the Heat Transfer Enhancement in a Wavy Channel with Different Phase Shifts," *Journal of Molecular Liquids*, **196**, pp. 32–42. http://dx.doi.org/10.1016/j.molliq.2014.03.001

Wang C.C., Chen C.K., 2002, "Forced Convection in a Wavy-Wall Channel," *International Journal of Heat and Mass Transfer*, **45**(12), pp. 2587–2595. http://dx.doi.org/10.1016/S0017-9310(01)00335-0

Xuan Y., Roetzel W., 2000, "Conceptions for Heat Transfer Correlation of Nanofluid," *International Journal of Heat Mass Transfer, **43**(19)*, pp. 3701–3707. http://dx.doi.org/10.1016/S0017-9310(99)00369-5

Yang Y.T., Wang Y.H., Tseng P.K., 2014, "Numerical Optimization of Heat Transfer Enhancement in a Wavy Channel Using Nanofluids," *International Communications in Heat and Mass Transfer*, **51**, pp. 9–17. http://www.sciencedirect.com/science/article/pii/S073519331300239X

Yin J., Yang G.， Li Y., 2012, "The Effects of Wavy Plate Phase Shift on Flow and Heat Transfer Characteristics in Corrugated Channel," Energy Procedia, **14**, pp. 1566 – 1573. http://dx.doi.org/10.1016/j.egypro.2011.12.1134

Youssef M.S., Aly A.A., Zeidan E.S.B., 2012, "Computing the Pressure Drop of Nanofluid Turbulent Flows in a Pipe Using an Artificial Neural Network Model," *Open Journal of Fluid Dynamics*, **2**, pp. 130-136 http://dx.doi.org/10.4236/ojfd.2012.24013

Yu W., Choi S.U.S., 2003, "The Role of Interfacial Layers in the Enhanced Thermal Conductivity of Nanofluids: A Renovated Maxwell Model," *Journal of Nanoparticle Research*, **5**(1-2), pp. 167–171. http://link.springer.com/article/10.1023%2FA%3A1024438603801

10

ASSESSMENT OF TURBULENCE MODELS IN THE PREDICTION OF FLOW FIELD AND THERMAL CHARACTERISTICS OF WALL JET

Arvind Pattamatta[a, *], Ghanshyam Singh[b]

[a] *Assistant Professor, Indian Institute of Technology Madras, Chennai 600036, India*
[b] *Manager, Agni Biopower Energy Pvt. Ltd., Mohali 160062, India*

ABSTRACT

The present study deals with the assessment of different turbulence models for heated wall jet flow. The velocity field and thermal characteristics for isothermal and uniform heat flux surfaces in the presence of wall jet flow have been predicted using different turbulence models and the results are compared against the experimental data of Wygnanski *et al.* (1992), Schneider and Goldstein (1994), and AbdulNour et al. (2000). Thirteen different turbulence models are considered for validation, which include the Standard k-ε (SKE), Realizable k-ε (RKE), shear stress transport (SST), Sarkar & So (SSA), v^2-f, Reynolds stress Model (RSM), and Spalart Allmaras (SA) models. Both standard wall function (swf) and enhanced wall treatment (ewt) options available in a commercial CFD solver have been used for near wall treatment for the high Reynolds number models. From the study, it is observed that only a few models could accurately predict the complex flow and thermal features of the heated wall jet. The near wall velocity profile captured using Realizable k-ε (RKE) with enhanced wall treatment (ewt) shows the best agreement with the experimental data as compared to the other models. Considerable deviation has been observed using SKE with standard wall function (*swf)* whereas the models of v^2-f show good prediction of velocity and temperature profiles in the near field region. However, the v^2-f model is found to deviate from the data in the downstream region where the velocity profiles exhibit similarity. In the prediction of heat transfer coefficient, RSM followed by SA and RKE with ewt, is found to be in closer agreement with the experimental data compared to the rest of the models. The computational time required for RSM is substantially higher than that of the other RANS models. Therefore, in the case of gas turbine combustor, since flow field is much more complex, the RKE with ewt would be the preferred choice over the SA model.

Keywords: *wall jet, heated wall jet, turbulence models, enhanced wall treatment, Realizable* k-ε (RKE).

1. INTRODUCTION

In recent years aircraft engine gas turbine design is driven by the requirement of low emission and high-efficiency. In order to attain higher efficiency, the inlet temperatures and pressures are to be maintained high. On the contrary, the combustion zone temperature is to be as low as possible to minimize the emission level. Therefore, efficient cooling system is a very important aspect of the combustor design. A good cooling system must be able to maintain metal temperatures well within the acceptable limits for the most severe engine operating conditions. To achieve this, wall-jet has been found to be one of the popular methods employed in the cooling of gas turbine combustion chamber liner walls. Wall-jets are also the integral part of many engineering devices. Automobile and aircraft windshield, flow separation control in V/STOL aircraft, and turbine blade cooling are some of the other typical applications wherein wall-jets are used. Launder and Rodi (1981) have given a comprehensive overview of wall-jet flows. The most important feature of this type of flow is that the point at which the shear stress changes sign does not coincide with the position of the zero velocity gradient but lies closer to the wall. The performance of most of the two-equation turbulence models in predicting the spread of plane free jets and mixing layers is found to be satisfactory. However, these models fail to accurately predict near wall flows. The reason for this failure could be the use of wall function and equilibrium and isotropic assumptions behind the development of these models. Predictions for wall jet flows are even more difficult because of the interaction of shear and boundary layers. In the past, Gerodimos and So (1997) critically assessed some of the existing two-equation models for their ability to predict near-wall behavior and mixing characteristics of the outer and inner layers for the wall jets. They have also evaluated the merits of accounting for anisotropy effects as against to those due to non-equilibrium turbulence. Using the experimental data of Karlsson et al. (1992) and Wygnanski et al. (1992), it was demonstrated by Gerodimos and So (1997) that only the k-ε model of Sarkar and So (SSA) (1997) could predict the wall jet spread rate and axial decay of maximum velocity within acceptable accuracy. Kumar and Mongia (1999, 2000) have shown that for plane wall jets in external flows, predictions of heat transfer coefficient and film cooling effectiveness using the SSA turbulence model matches well with the test data as compared to the predictions using models of Yang and Shih (1993) and the standard *k*-ε model. The applicability of different scaling laws has been critically examined by them and it was shown that the 'law-of-wall' applies only to the viscous sub-layer. The extent of this region is much smaller in the case of wall jet flows than that for the boundary layers. Such flow feature limits the applicability of wall-function for wall-jet flows, which are based on the assumption of local equilibrium and isotropy of turbulence.

The previous studies in the literature, mainly considered the validity of standard *k*-ε model and its low Reynolds number variants for wall jet predictions. Also, the performance of these models was assessed for the flow field and the thermal field predictions, separately. Hence, a comprehensive study is required to understand the behavior of turbulence models for heated wall jet flow which takes in to account accuracy in the prediction of both flow field and the thermal

* *Corresponding author. Email: arvindp@iitm.ac.in*

characteristics. The main objective of the present work is to compare the performance of different turbulence models for heated wall jet flow. This paper is a first step toward comprehensive assessment of various state-of-the-art turbulence models including Large-Eddy-Simulation, LES, the results with the latter are planned in the future publications. This is done by carrying out comparison with the experimental measurements of Wygnanski *et al.* (1992), and Schneider and Goldstein (1994) for the velocity field and with the experimental measurements of AbdulNour et al. (2000) for temperature field. Wygnanski *et al.* (1992) measured the spatial distribution of mean velocity for various jet Reynolds numbers using Hot-Wire-Anemometer. Schneider and Goldstein (1994) employed LDA technique for the measurement of flow characteristics and provided data for the self-similar velocity profile and jet spread rate. In this paper, the performance of several turbulence models implemented in to a commercial CFD solver is critically assessed for the wall jet flow configuration by comparison with the experimental data in the literature mentioned above.

2. PROBLEM DESCRIPTION AND SOLUTION METHODOLOGY

2.1 Computational domain and boundary conditions

Experimental data of Wygnanski et al. (1992) and Schneider and Goldstein (1994) are used to validate turbulence models for flow field character and that of AbdulNour et al. (2000) for thermal field. Figure 1 shows the typical wall-jet flow configuration and computational domains corresponding to these three cases used in the present study. For the cold wall-jet configuration, the jet inlet Reynolds number based on the nozzle-exit width and the nozzle-exit velocity is varied from 5000 to 19000. This is achieved by applying suitable mass flow at the plenum inlet. The rightmost boundary is treated as exit and gradient of all variables is set to zero. As indicated in the Figure 1, constant atmospheric pressure is applied on free boundaries to ensure smooth entrainment of flow from outside. Experimental data of AbdulNour et al. (2000) is used to validate turbulence models for the heated wall-jet case.

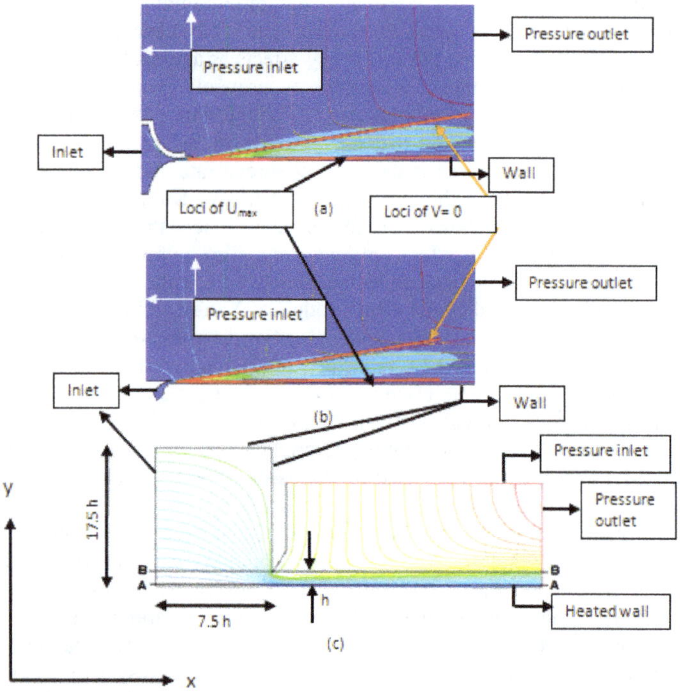

Fig 1. Wall-jet flow configuration for (a) Schneider and Goldstein (1994) (b) Wygnanski (1992) (c) AbdulNour et al. (2000).

The computational domain includes plenum for which dimensions are taken as 7.5h×17.5h in the axial and transverse directions, respectively. Here h is the slot width. Mass flow inlet boundary condition was employed at the left most boundary in such a way that the jet inlet Reynolds number based on the nozzle-exit width and the nozzle-exit velocity is 7700. For heated plate, two different thermal boundary conditions were employed, one as uniform heat flux of 935 W/m2 and another one as constant temperature of 318.15K. The fluid inside the plenum was maintained at temperature of 295.15K.

2.2 Governing equations

For incompressible flows, the Reynolds Averaged Navier Stokes (RANS) equation governing the mean flow along with the equations for κ-ε turbulence model can be written in Cartesian tensor form as:

$$\frac{DU_i}{Dx_i} = 0 \tag{1}$$

$$\frac{DU_i}{Dx_i} = -\frac{1}{\rho}\frac{\partial P}{\partial x_i} + \upsilon\frac{\partial^2 U_i}{\partial x_j \partial x_j} - \frac{\partial \overline{u_i u_j}}{\partial x_j} \tag{2}$$

$$\frac{D\kappa}{Dt} = \frac{\partial}{\partial x_j}\left[\left(\upsilon + \frac{\upsilon_t}{\sigma_\kappa}\right)\frac{\partial \kappa}{\partial x_j}\right] + P_\kappa - \varepsilon + D \tag{3}$$

$$\frac{D\varepsilon}{Dt} = \frac{\partial}{\partial x_j}\left[\left(\upsilon + \frac{\upsilon_t}{\sigma_\kappa}\right)\frac{\partial \varepsilon}{\partial x_j}\right] + C_{\varepsilon 1} f_1 \frac{1}{T_t} P_\kappa - C_{\varepsilon 2} f_2 \frac{\tilde{\varepsilon}}{T_t} + E \tag{4}$$

where

$$-\overline{u_i u_j} = 2\upsilon_t S_{ij} - \frac{2}{3}\kappa\delta_{ij} \tag{5}$$

$$S_{ij} = \left(\partial U_i / \partial x_j + \partial U_j / \partial x_i\right)/2 \tag{6}$$

$$P_\kappa = -\overline{u_i u_j}\left(\frac{\partial U_i}{\partial x_j}\right) \tag{7}$$

The commercial solver ANSYS FLUENT 6.3 has been used for the present study. Thirteen turbulence models, namely those of Standard k-ε (SKE, Launder and Spalding (1974)), Realizable k-ε (RKE, Shih et al. (1995)), Reynolds stress model (RSM, Launder et al. (1975)), Shear stress transport (SST, Menter (1994)), Abid (Abid (1991)), AKN (Abe et al. (1994)), CHC (Chang et al. (1995)), LB (Lam and Bremhorst (1981)), LS (Launder and Sharma (1974)), v^2-f (Durbin (1991)), SSA (Sarkar and So (1997)), YS (Yang and Shih (1993)), and SA (Spalart and Allmaras (1992)) were selected for validation. The SSA model was incorporated in Fluent using a user-defined-function (UDF), whereas all other models are available in the ANSYS FLUENT software. The details of SKE, RKE, RSM, SST, and v^2-f models are given in References 13, 19, 16, 17, and, 6 respectively and not repeated here for the sake of brevity. The constants for these models are taken as default value given in Reference 5. For Low-Reynolds-Number models the details of constants, additional terms, wall boundary conditions and damping functions are given Tables 1, 2 and 3 respectively. The simulations for high Reynolds number models (i.e. SKE, RKE and RSM) have been carried out using standard wall function of Launder and Spalding (1973) as well as "enhanced wall treatment" (ewt) approach, available in Fluent. For ewt the grid was refined in such a way that the maximum y+ ≈1. The ewt is a two-layer model where laminar-viscosity affected region (i.e. Rey < 200) is modeled using one equation approach of Wolfstein (1969). In one equation model, equation for turbulent kinetic energy is same as that in SKE model, but the turbulent viscosity is modeled as:

$$\mu_{t2Layer} = \rho C_{\mu} \ell_{\mu} \sqrt{k} \quad (8)$$

The length-scale appeared in equation is computed based on the approach proposed by Chen and Patel (1988).

$$\ell_{\mu} = yc_l \left(1 - e^{\mathrm{Re}_y / A_\mu}\right) \quad (9)$$

the ε field is calculated as

$$\varepsilon = \frac{k^{3/2}}{\ell_{\varepsilon}} \quad (10)$$

$$\ell_{\varepsilon} = yc_l \left(1 - e^{\mathrm{Re}_y / A_\varepsilon}\right) \quad (11)$$

Where,

$$c_l = \kappa C_{\mu}^{-3/4}, \; A_{\mu} = 70, \; A_{\varepsilon} = 2c_l \quad (12)$$

Further details are available in References 4 and 5.

Table 1. Low Reynolds number *k*-ε models and their model constants.

	Model	C_μ	$C_{\varepsilon 1}$	$C_{\varepsilon 2}$	σ_k	σ_ε
Abid (1991)	Abid	0.09	1.44	1.92	1.0	1.3
Lam-Bremhorst (1981)	LB	0.09	1.44	1.92	1.0	1.3
Launder-Sharma (1974)	LS	0.09	1.44	1.92	1.0	1.3
Yang-Shih (1993)	YS	0.09	1.44	1.92	1.0	1.3
Abe-Kondoh-Nagano (1994)	AKN	0.09	1.50	1.90	1.4	1.3
Chang-Hsieh-Chen (1995)	CHC	0.09	1.44	1.92	1.0	1.3
Sarkar-So (1997)	SSA	0.096	1.44	1.92	1.0	1.43

Table 2. Additional terms and wall boundary conditions for the Low Re. turbulence models.

Model	D	E	Wall B.C.
Abid	0	0	$k = 0; \varepsilon = 2\upsilon \left(\frac{\partial \sqrt{\kappa}}{\partial y}\right)^2$
LB	0	0	$k = 0; \varepsilon = \upsilon \frac{\partial^2 k}{\partial y^2}$
LS	$2\upsilon \left(\frac{\partial \sqrt{\kappa}}{\partial y}\right)^2$	$2\upsilon\upsilon_t \left(\frac{\partial^2 u}{\partial y^2}\right)$	$k = 0; \varepsilon^* = 0$
YS	0	$\upsilon\upsilon_t \left(\frac{\partial^2 u}{\partial y^2}\right)$	$k = 0; \varepsilon = 2\upsilon \left(\frac{\partial \sqrt{\kappa}}{\partial y}\right)^2$
AKN	0	0	$k = 0; \varepsilon = 2\upsilon \left(\frac{\partial \sqrt{\kappa}}{\partial y}\right)^2$
CHC	0	0	$k = 0; \varepsilon = \upsilon \frac{\partial^2 k}{\partial y^2}$
SSA	0	$\exp\left[-\left(\frac{\mathrm{Re}_t}{40}\right)^2\right]\left[-0.57\frac{\varepsilon\bar{\varepsilon}}{\kappa} + 0.5\frac{(\varepsilon^*)^2}{\kappa} - 2.25\frac{\varepsilon}{\kappa}P_\kappa\right]$	$k = 0; \varepsilon = 2\upsilon \left(\frac{\partial \sqrt{\kappa}}{\partial y}\right)^2$

2.3 Grid sensitivity study

A detailed grid independence study has been conducted to ensure good solution accuracy. Figure 2(a) and (b) show the results of the grid independence study on the velocity profile for the case of Wygnanski et al. (1992) for Reynolds number of 5000 and on the heat transfer coefficient for the case of AbdulNour et al. (2000) for Reynolds number of 7700, respectively. For the grid "m0", initially, total number of cells was taken as 52416. Subsequently, this grid was adapted in Fluent to see if results are independent of grid refinement. For all the cases, grid sensitivity study was done using RKE with enhanced wall treatment (ewt). The final grid was chosen in such a way that y+ for the first grid point from the plate wall is less than five, whereas for the entire computational domain sufficiently fine base-mesh was used to capture essential flow features.

For the case of cold wall jet, the sensitivity of self-similar velocity profile at X/s = 20 for the case of Re=5000 are compared with the data of Wygnanski et al. (1992) shown in Figure 2(a) for four different grids. It is seen from Figure 2(a) that results for all the four grids are similar. However, the local skin friction coefficient plotted in Figure 2(b) and compared with the data of Eriksson et al. (1998) is found to be sensitive to y+ value. It can be seen in Figure 2(b) that a considerable scatter and deviation in predicted values of skin friction coefficient from experimental data is due to higher value of y+. In order to eliminate such discrepancies, y+ value less than 1.5 is needed. Therefore, a typical grid, similar to "m3" has been used for the entire validation study. In order to ensure that are results are independent of grid refinement, the grids are generated in such a way that the refinement is done within region AA and BB, as shown in Figure 1(c). Figure 2(c) shows the effect of grid refinement on heat transfer coefficient using realizable k-ε model with enhanced wall treatment (ewt) for uniform heat flux B.C. for Re=7700 and compared with the experimental data of AbdulNour et al. (2000). In this figure 'm1' indicates grid distribution of 426×50 with y+>50, 'm2' indicates grid distribution of 426×65 with y+~30, 'm3' indicates grid distribution of 426×74 with y+~10 and the fourth grid, 'm4' is generated in such a way that the grid counts in the transverse direction remains same as 74, whereas y+ comes down to approximately value of 0.9. It is clear from the Figure 2(c) that except for results on coarser grid, 'm1', results for all the other three finer grids are the same. Therefore, grid 'm4', is taken for all high-Reynolds-number models with ewt, and for all other low-Reynolds-number model cases. For high Reynolds number models with standard wall-function (swf), grid 'm2', in which y+ was maintained close to 30 is used in the rest of the study.

The convergence criteria for all the models are set to 10-6 for continuity and momentum equations. The total number of iteration required for most of the models varies from 8000 to 9000 iterations. The number of iterations required for RSM is nearly 50000, which is substantially high. The second order upwind scheme is used for the spatial discretization of the convective terms in the governing equations.

Table 3. The damping functions used in the Low Re. models.

Model	f_1	f_2	f_μ
Abid	1	$\left[1-\frac{2}{3}\exp\left(-\frac{Re_T^2}{36}\right)\right]\left[1-\exp\left(\frac{Re_y}{12}\right)\right]$	$\tanh(0.008\,Re_y)(1+4Re_T^{-3/4})$
LB	$1+(0.05/f_\mu)$	$1-\exp(-Re_T^{\,2})$	$[1-\exp(-0.0165\,Re_T)^2(1+20.5/Re_T)]$
LS	1	$1-0.3\exp(-Re_T^{\,2})$	$\exp\left[-3.4/(1+Re_T/50)^2\right]$
YS	$\frac{1}{1+c_k/\sqrt{Re_T}}$	$\frac{1}{1+c_k/\sqrt{Re_T}}$	$\left(1+\frac{c_k}{\sqrt{Re_T}}\right)\times$ $\left[1-\exp\left(-1.5\times10^{-4}Re_y-5.0\times10^{-7}Re_y^{\,3}-1.0\times10^{-10}Re_y^{\,5}\right)\right]^{1/2}$
AKN	$\left[1-\exp\left(-y^k/3.1\right)\right]^2\left[1-0.3\exp\left\{-\left(\frac{Re_T}{6.5}\right)^2\right\}\right]$	1	$\left[1-\exp\left(-y^k/14\right)\right]^2\left[1+\frac{5}{Re_T^{\,3/4}}\exp\left\{-\left(\frac{Re_T}{200}\right)^2\right\}\right]$
CHC	1	$[1-0.01\exp(-Re_T^{\,2})][1-\exp(-0.0631\,Re_y)]$	$[1-\exp(-0.0215\,Re_y)]^2(1+31.66/Re_T^{\,5/4})$
SSA	1	1	$(1+3/Re_T^{\,3/4})\times[1+80\exp(-Re_\varepsilon)]$ $\times\left[1-\exp\left(-Re_\varepsilon/43-Re_\varepsilon^{\,2}/330\right)\right]^2$

3. RESULTS AND DISCUSSION

In this section, for the turbulence models considered, first, the self-similar velocity profile, jet spread rate, mean velocity profile in the inner region and applicability of general similarity, proposed by Wygnanski et al. (1992) have been compared with the experimental data. Subsequently, the temperature profiles, thermal boundary layer thicknesss and the heat transfer coefficient predicted by the different turbulence models are compared with that of AbdulNour et al. (2000). The better model is identified based on factors such as good agreement with experimental data and lower computational time.

3.1 Comparison of Velocity Field

Figure 3 show the comparison of self-similar velocity profile normalized by the maximum velocity at the same stream wise location for different models with the results of Wygnanski et al. (1992) and Schneider and Goldstein (1994) for Reynolds number of 10000 and 19000. The experimental uncertainty in the measurements for velocity is around 3% as reported by Schneider and Goldstein (1994). Seven different turbulence models namely Standard k- (SKE), Realizable k- (RKE), Reynolds stress model (RSM), Shear stress transport (SST), Sarkar and So (SSA), Spalart and Allmaras (SA) and v^2-f model (v^2-f) are chosen for comparison. The standard wall function (swf) is used for near wall treatment with SKE while the enhanced wall treatment (ewt) option is used with RKE and RSM. The results predicted by all the models closely follow the data of Schneider and Goldstein (1994) within the experimental uncertainty level. Results using all the models are in good agreement with the data with a slight deviation observed using RSM_ewt and SA models.

In Figures 4(a-c) the predicted normalized mean velocity profile in the inner or the near wall region of the wall jet is compared with the experimental data of Wygnanski et al. (1992). The results using SKE_swf, RKE_ewt and SST models have been compared against the experimental data for Re varying in the range of 5000 to 19000, in Figures 4 (a)-(c), respectively. From this figure, it is observed that the results obtained using the different models show a closer agreement to the experimental data for Re = 5000 while deviating significantly from the data for Re = 19000. The velocity profiles obtained using RKE_ewt gives a better agreement with the experimental data as compared to the other two models. For the sake of comparison, the log law has also been plotted in the figure. As pointed out by Gerodimos and So (1997), except SKE_swf, all other models show considerably small limited log region. Considerable under prediction in the near wall velocity profile has been observed using the RSM_ewt, SSA, and v^2-f models, which are not shown here.

Figure 5 presents the variation of the turbulent intensity with the wall normal distance along the downstream distance at different axial locations for Reynolds numbers of 5000 & 19000, respectively. In this figure results using only RKE_ewt model are presented here, the predictions of the other models are quite similar in nature to that of RKE_ewt. Unlike experimental data the results show a self-similar profile for the turbulent intensity and produces a good agreement with the data of Wygnanski et al. (1992) for Re of 19000.

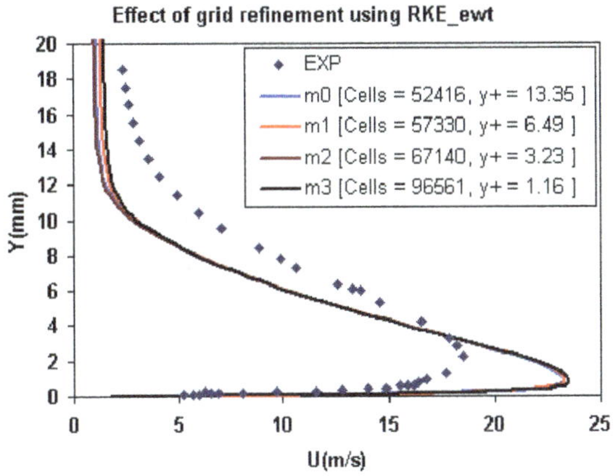

(a)

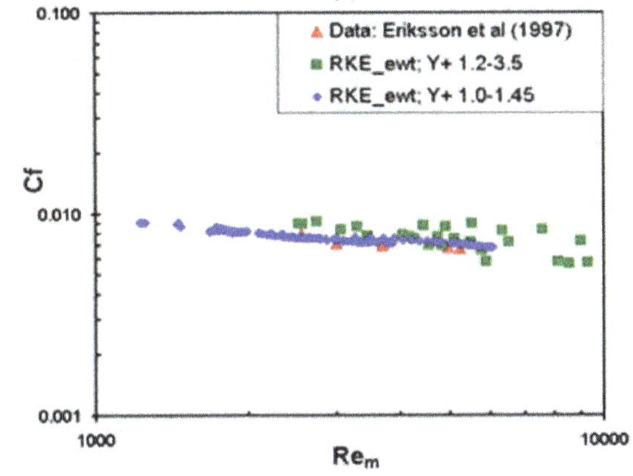

(b)

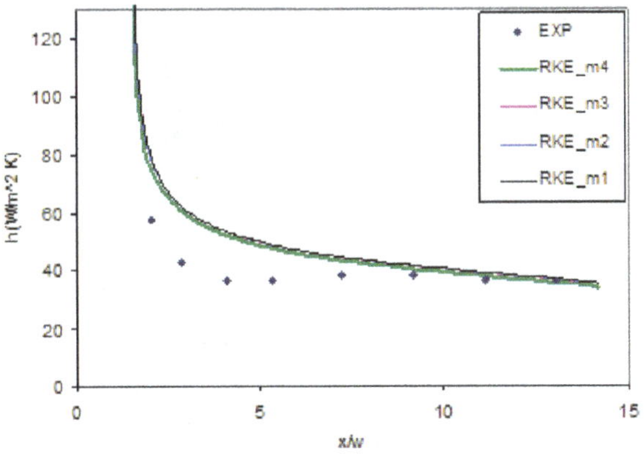

(c)

Fig 2. (a) Effect of grid refinement on self-similar velocity profile for Re = 5000 at X/s = 20 (b) Effect of y+ on skin friction coefficient Re = 5000 (c) Effect of grid refinement on heat transfer coefficient using RKE with uniform heat flux BC for Re = 7700.

Figure 6 shows the comparison of the maximum velocity with the downstream distance. The ratio of $(Uj/Um)^2$ is plotted against the normalized downstream distance (X – X0)/w in order to have a linear profile as the velocity scale is expected to decay approximately as 1/X1/2. The stream wise distance is measured from a virtual origin X0. It is seen that the results using RKE_ewt and SKE_swf produce a good match with the data whereas RSM_ewt and v^2-f models over predicts. All other models produce significant under prediction.

In Figure 7, the jet-spread rate predicted by various models is compared with the data of Schneider and Goldstein (1994) for Re of 14000. The RKE_ewt model shows good match, however SKE_swf and RSM_ewt produce significant over prediction in the downstream and rest of the models show under prediction.

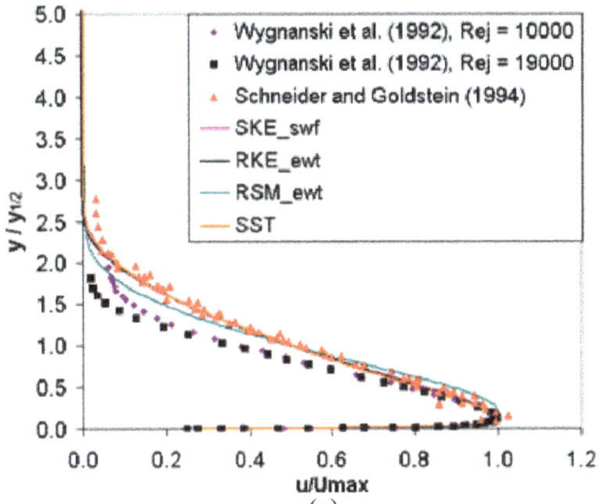

(a)

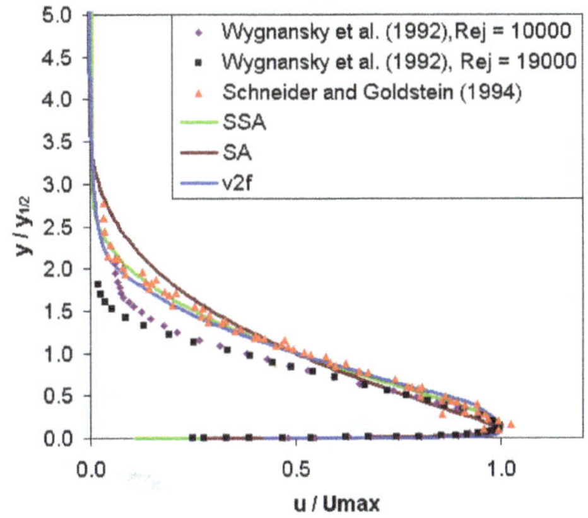

(b)

Fig 3. Comparison of self-similar velocity profiles for different models with the experimental data.

Figures 8 and 9 show the comparison of normalized velocity decay and jet growth, respectively. It has been observed that most of the models show similar trend, therefore in these figures predictions using only RKE_ewt are shown on two different length scales. As suggested by Wygnanski et al. (1992), the data for the Reynolds numbers in range of 5000-19000 have been correlated based on non-dimensional velocity decay (Umax **υ**/J) and jet spread (Ym/2 J/**υ**2) using equations 13 and 14 below and, in Table 4, the correlation parameters Ay, Au, m and n for different models are compared for Reynolds number 5000 and 19000.

$$Y_{m/2}\, J/\upsilon^2 = A_y \left[X\, J/\upsilon^2\right]^m \tag{13a}$$

$$U_{\max}\, \upsilon/J = A_u \left[X\, J/\upsilon^2\right]^n \tag{13b}$$

$$Y_{m/2}\, J/\upsilon^2 = A_y \left[(X - X_0) J/\upsilon^2\right]^m \tag{14a}$$

$$U_{\max}\, \upsilon/J = A_u \left[(X - X_0) J/\upsilon^2\right]^n \tag{14b}$$

As shown in Table 4, the exponents for the velocity decay and jet growth are same as estimated by Wygnanski et al. (1992), but constant coefficient differs considerably for different models. However, this deviation for length scale, based on virtual origin is more, specially at low Reynolds number. The agreement in coefficients estimated for SKE_swf, RKE_ewt and RSM_ewt is slightly better as compared to other models, with average value of coefficients using RSM_ewt being closest to data. However, as shown in Figures 6 and 7, jet spread, velocity decay and near wall velocity profile using RSM_ewt does not agree well with experiments.

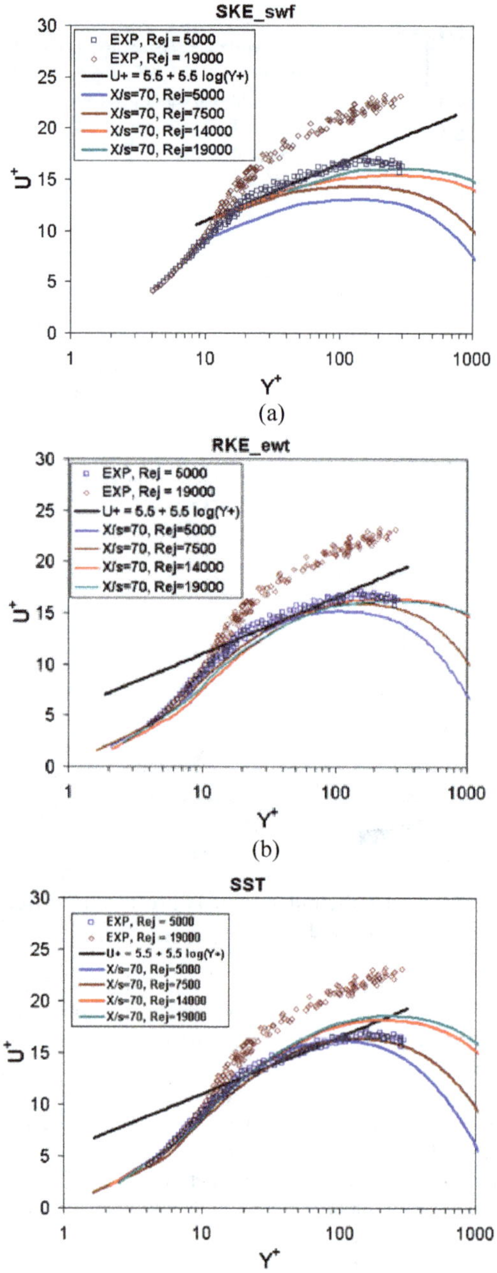

(c)

Fig 4. Comparison of the normalized mean velocity profile in the inner layer of the wall jet for different models.

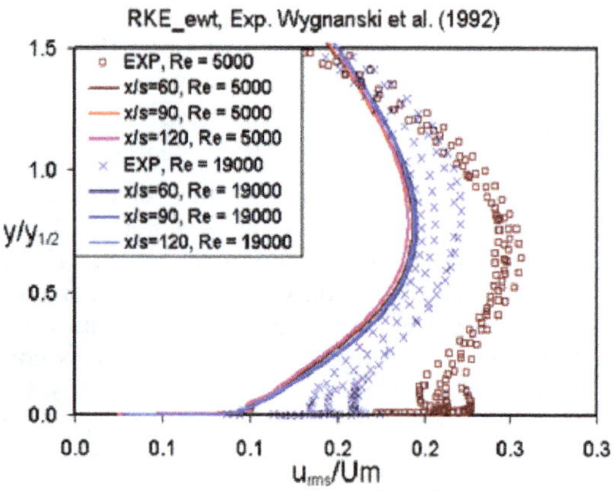

Fig 5. Comparison of the turbulent intensity with experiments.

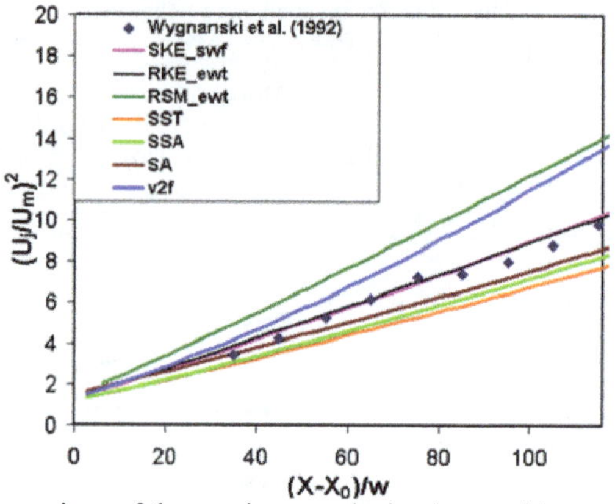

Fig 6. Comparison of the maximum velocity decay with measurements at Re = 19000.

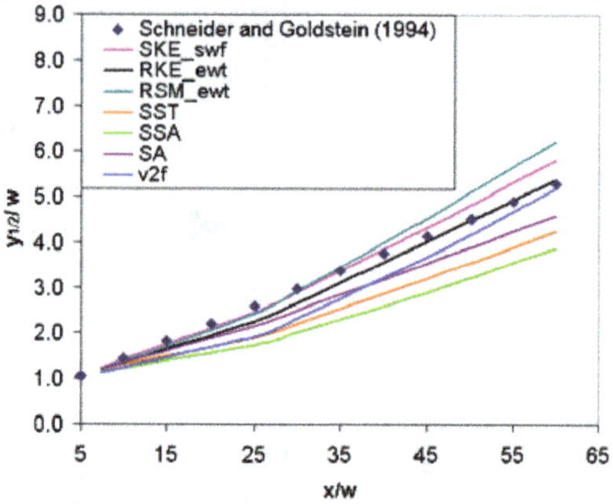

Fig 7. Comparison of jet spread rate for different models with experiment at Re = 19000.

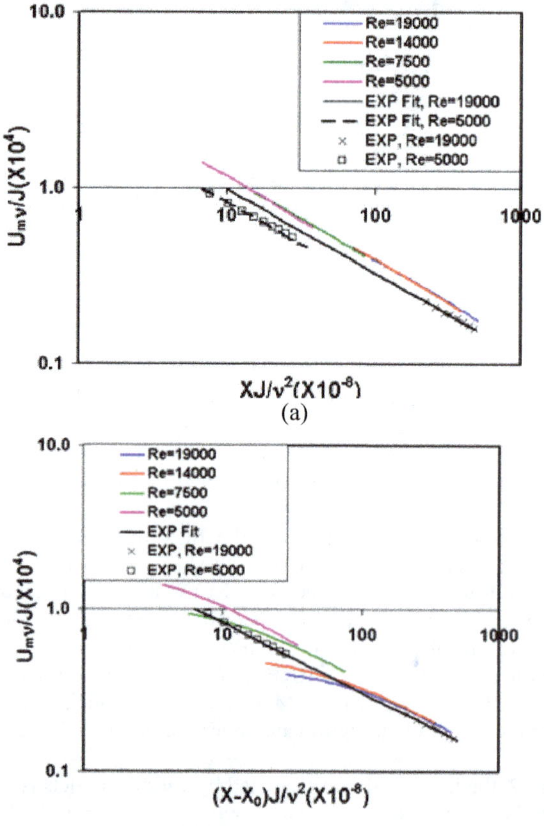

(b)

Fig 8. Comparison of normalized maximum velocity decay with data for two different length scales.

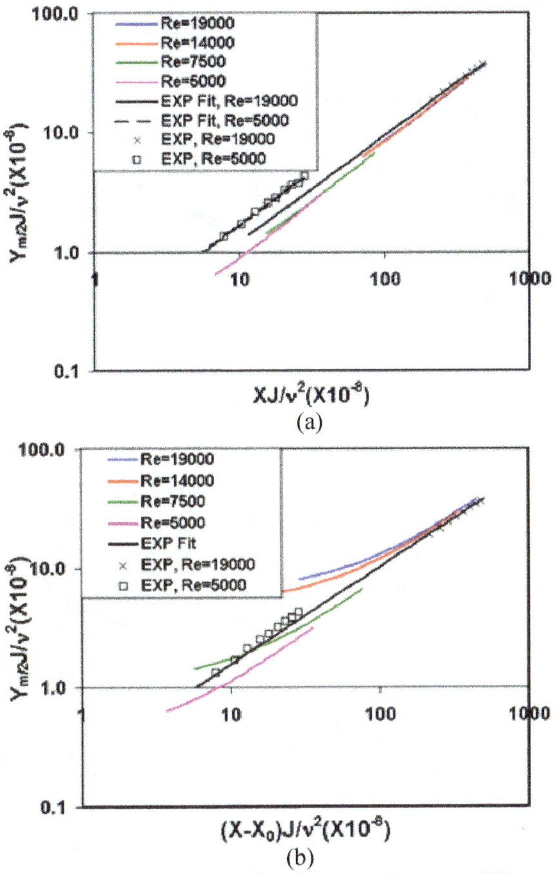

Fig 9. Comparison of jet-growth with data for two different length scales.

The difference in estimated coefficients is reflected in the Figures 8 and 9. As pointed out earlier, the virtual origin based scaling, shows a greater deviation in the maximum velocity as well as in the jet-spread rate. It has been noticed that unlike Wygnanski et al. (1992), all the models show dependency of length and velocity scales on Reynolds number with the scaling base on virtual origin.

In Table 5, the mean error in predicting the jet half widths are listed for different models. Also, in Table 5 the error in predicting the jet-spread rate is compared for different models using the two sets of experimental data. It is noticed that among the various turbulence models considered, SA shows least percentage error in predicting the jet-spread rate followed by SST and RKE_ewt. It is found that in case of SST, the RMS values in predicting the jet half width is as high as 46 as compared to that by RKE_ewt. Also, the applicability of Spalart–Allmaras model is limited to non-reacting flows only. However, RKE_ewt shows better versatility when it comes to a wide range of applications where turbulent combustion processes are involved.

3.2 Comparison of Thermal Field

In this section comparison between predicted results using various turbulence models and the experimental data of AbdulNour et al. (2000) is presented for the thermal characteristics of wall jet. It has been observed that in terms of near field and self-similar velocity profiles, temperature field and heat transfer coefficient, none of the low Reynolds number k-ε model give satisfactory results. Therefore, in order to avoid clutter, only for selected models, the temperature field is compared. Subsequently, the error in prediction of heat transfer coefficient for all of the thirteen models are plotted and compared. For the sake of comparison, results using SKE with standard wall function (swf) are also shown in figures. Finally, models are ranked based on their ability to predict the heat transfer coefficient accurately.

Table 4. Comparison of correlation parameters with experimental data of Wygnanski et al. (1992) for different models

Actual origin based															
Constants	EXP	SKE_swf		RKE_ewt		RSM_ewt		v2f		SST		SSA		SA	
Reynolds number ->		5000	19000	5000	19000	5000	19000	5000	19000	5000	19000	5000	19000	5000	19000
A_u	1.473	1.860	1.890	1.930	1.900	1.700	1.690	2.150	1.850	2.200	2.150	2.200	2.150	2.050	2.050
n	**-0.472**	-0.472	-0.472	-0.472	-0.472	-0.472	-0.472	-0.472	-0.472	-0.472	-0.472	-0.472	-0.472	-0.472	-0.472
A_y	1.445	1.340	1.670	1.200	1.600	1.450	1.920	1.060	1.700	1.000	1.250	0.930	1.200	1.110	1.370
m	**0.881**	0.881	0.881	0.881	0.881	0.881	0.881	0.881	0.881	0.881	0.881	0.881	0.881	0.881	0.881
Virtual origin based															
A_u	0.557	0.650	0.560	0.680	0.560	0.610	0.480	0.680	0.500	0.780	0.640	0.750	0.620	0.720	0.620
n	**-0.428**	-0.428	-0.428	-0.428	-0.428	-0.428	-0.428	-0.428	-0.428	-0.428	-0.428	-0.428	-0.428	-0.428	-0.428
A_y	9.246	8.000	13.200	7.300	12.500	8.700	15.000	7.000	14.250	6.000	10.000	5.750	9.800	6.500	10.500
m	**0.804**	0.804	0.804	0.804	0.804	0.804	0.804	0.804	0.804	0.804	0.804	0.804	0.804	0.804	0.804

Table 5. Comparison of error in jet spread rate prediction for different turbulent models.

Error in jet spread rate for different models				
Turbulence models	Wygnanski et al. (1992)	%Error	Schneider and Goldstein (1994)	%Error
Exp data	0.077	0.0	0.077	0.0
SKE_swf	0.097	-26.4	0.097	-26.3
RKE_ewt	0.090	-17.2	0.088	-14.0
RSM_ewt	0.109	-41.8	0.096	-24.2
SST	0.068	11.5	0.074	3.3
SSA	0.062	19.6	0.070	8.5
SA	0.070	8.7	0.075	2.9
v2f	0.094	-22.0	0.109	-40.9

The non-dimensional temperature profiles for the isothermal plate are compared for different turbulence models at various x/w locations in Figures 10(a) to (c). It is seen from the Figure 10(c) that at x/w=13.02 v^2-f model deviate maximum from the data. It has been observed that at this position with AKN (not shown in the figure), near wall region is captured well but beyond y/w=0.03 it shows considerable deviation from data. The SST, SA, and RKE with ewt models give equally good predictions, whereas slight over prediction has been observed using SKE with ewt. The near field region (i.e. x/w = 2.2) is captured very well by v^2-f model, followed by SA. At this position all other models show slight under prediction.

In Figure 11, the development of thermal boundary layer is compared for different turbulence models. Non-dimensional quantities δt/w and x/w are plotted for two different values of theta of 0.1 and 0.15 in Figures 11(a) and (b) respectively.

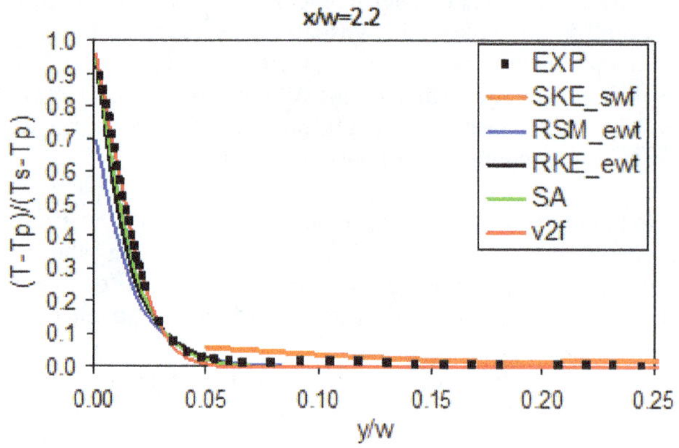

(a)

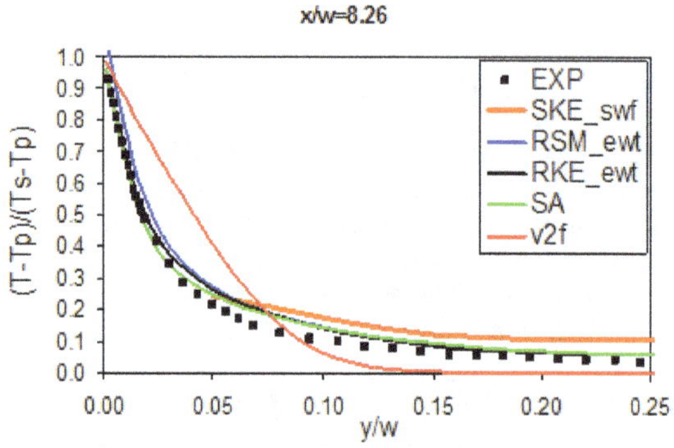

(b)

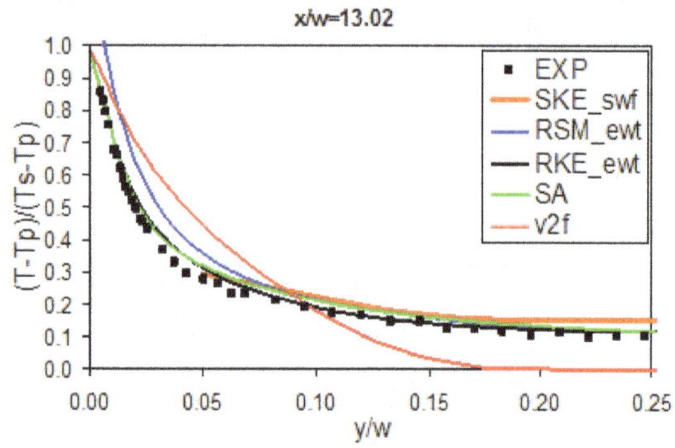

(c)

Fig 10. Comparison of temperature field for an isothermal plate with different turbulence models with the data of AbdulNour et al (2000) for Re = 7700.

The initial region (i.e. x/w < 8), where growth of the thermal boundary layer is nonlinear, could not be captured by most of the low-Reynolds-number turbulence models. The prediction using YS and v^2-f models show good agreement with data, however, both of these models fail to capture change in the slope at x/w ~ 8. In this regard, realizable k-ε model with ewt over predicts but shows correct trends in terms of initial development and subsequent change in thermal boundary layer growth. The models of SA, SST, and RSM with ewt also show similar trends but slope change x/w ~ 8 predicted by these models is smeared.

The comparisons of heat transfer coefficient predicted by the turbulence models for uniform temperature and for uniform heat flux BCs are shown in Figures 12(a) and (b) respectively. The initial region using SKE as well as RKE with ewt is captured in the same way by both the models. However, in the further downstream, results deviate with SKE model. It is seen that increasing heat-transfer-coefficient for x/w ≥ 5 and its reduction after x/w > 8 could be captured well using RSM with ewt for both the cases. In this aspect the model of SA also shows good agreement with data.

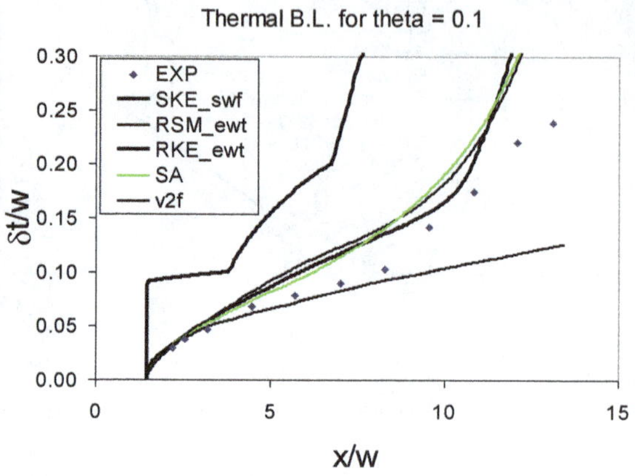

(a)

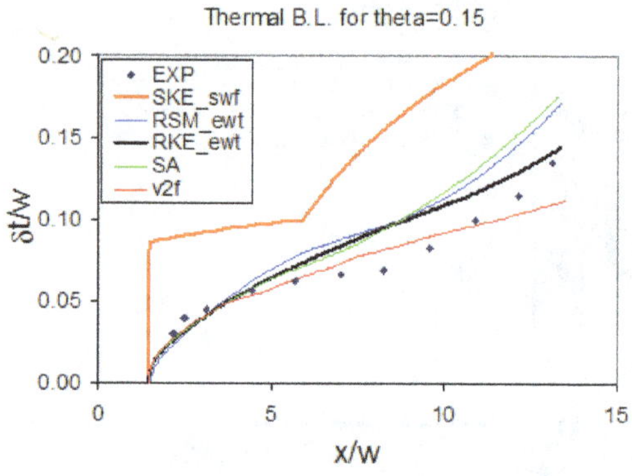

(b)

Fig 11. Comparison of thermal boundary layer for θ = 0.1 and θ = 0.15, for constant wall temperature case with the data of AbdulNour et al (2000) for Re = 7700.

The percentage errors in heat transfer coefficient for uniform temperature and for uniform heat flux BCs with different models are listed in Tables 6. The percentage error in heat transfer coefficient is calculated by taking the average value for the region x/w > 6 from Figure 12. It is seen from Table 6 that RSM with ewt gives least error in predicting the heat transfer coefficient. The error in heat transfer coefficient predicted by SSA is the second lowest, next to RSM with ewt followed by SA, RKE with ewt, and SST.

Table 6. Average errors, its standard deviation and rms errors in the heat transfer coefficient prediction by different models.

Turbulence models	Constant heat flux BC			Uniform temperature BC		
	Mean	σ	rms	Mean	σ	rms
SKE_swf	-22.0	1.1	42.3	-67.9	14.0	84.7
SKE_ewt	-0.6	16.0	19.6	16.7	16.2	26.6
RKE_ewt	-10.1	12.0	22.4	6.6	13.3	15.5
V2f	41.2	9.4	83.0	56.6	8.5	75.9
SST	-8.2	13.8	21.2	9.2	14.8	18.8
RSM_swf	39.9	4.4	76.0	37.6	4.3	4.3
RSM_ewt	3.5	2.7	7.7	14.5	5.5	5.5
SA	30.5	7.4	58.4	7.0	10.5	10.5
AKN	-4.1	15.9	20.4	7.7	24.0	24.0
Abid	-42.5	27.7	84.5	-21.0	28.4	28.4
LB	-41.2	26.9	81.7	-21.4	28.4	28.4
LS	-31.9	11.6	59.3	-59.5	36.7	36.7
YS	42.4	9.9	84.9	58.2	8.2	8.2
CHC	-38.2	27.1	76.8	-17.5	27.5	27.5
SSA	18.3	14.0	81.7	-6.9	22.4	22.4

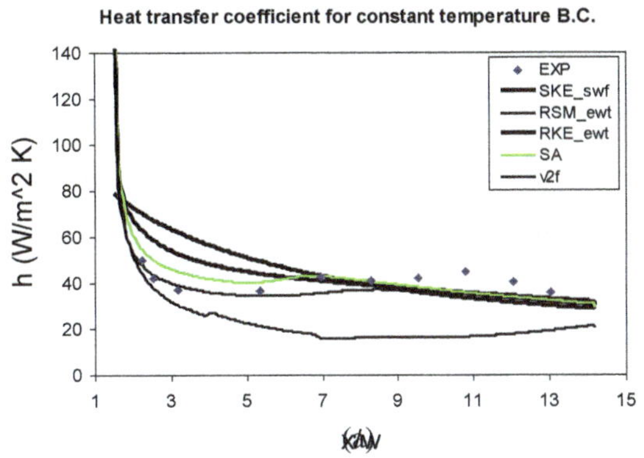

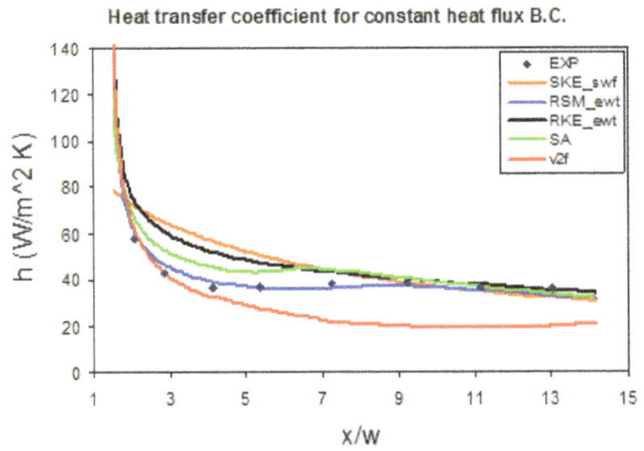

(b)

Fig 12. Comparison of local heat transfer coefficient for isothermal and isoflux boundary conditions with the data of AbdulNour et al (2000) for Re = 7700.

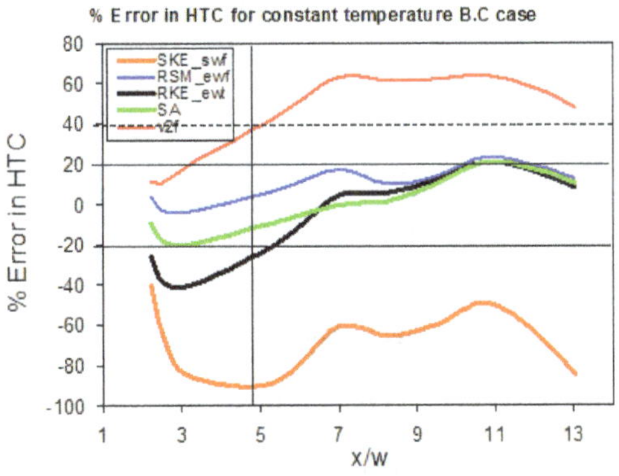

(a)

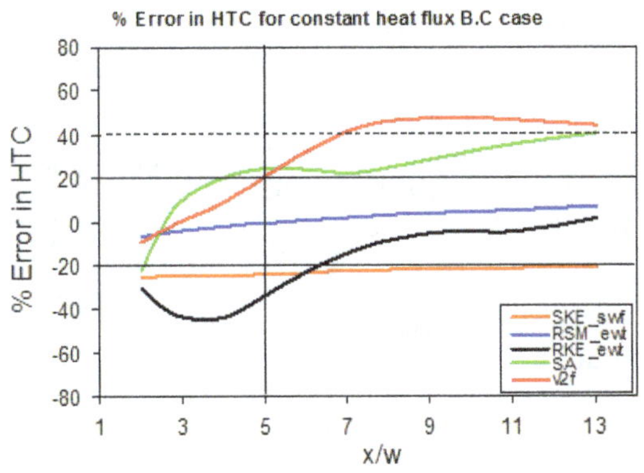

(b)

Fig 13. Comparison of error bands and selection criteria for the best turbulence models.

Figures 13(a) and (b) show the axial variation error in heat transfer coefficient plotted separately for uniform temperature and constant heat flux cases, respectively. For gas turbine combustor liners, accurate predictions beyond x/w > 5 are very important to correctly estimate film-cooling effectiveness. Therefore, models are categorized for ± 20 and 40 percent accuracy in predicting the heat transfer coefficient. It is seen from Figure 13(a) and (b) that only RSM with ewt could satisfy the requirement of ± 20 percent error band for constant heat flux as well as uniform temperature cases. On the other hand, for the allowable error band of ± 40 percent, apart from RSM with ewt there are four models, namely SA, SST, and RKE and SKE with ewt, which satisfy this requirement. Also, it is seen that beyond x/w > 5 the percentage error for RSM, SA and RKE with ewt falls between 0-10%, which is within the experimental uncertainty of around 10 % reported by AbdulNour et al. (2000). The computational time required for RSM is substantially higher than that of the other RANS models. Therefore, the models of SA and RKE with ewt offer a better option when both accuracy and computational time are taken into consideration. These two models show consistently accurate predictions for both the velocity and thermal field. In case of gas turbine combustor, since flow field is much more complex, the RKE with ewt would be the preferred choice over the SA model.

4. SUMMARY AND CONCLUSIONS

In the present study, emphasis is laid upon the accurate prediction of heat transfer and flow field characteristics of heated wall jet flows in order to earmark a suitable turbulence model. To achieve this, the results of thirteen turbulence models, implemented in ANSYS FLUENT are compared against the experimental data of Wygnanski et al. (1992), and Schneider and Goldstein (1994) for the cold wall jet configuration and against the data of AbdulNour et al. (2000) for heated wall jet configuration. It is seen that only a few of these models could accurately capture the complex flow features of the wall jet. The performance of the models for flow field characteristics like maximum velocity decay and jet spread rate been compared with the experimental measurements of Wygnanski et al. (1992) using the length scale as virtual origin. None of the models is found to show Reynolds number independence for length scale based on virtual origin. The near wall velocity profile captured using Realizable k-ε (RKE) with enhanced wall treatment (ewt) shows the best agreement with the experimental data as compared to the other models. The spread rate of the jet is an important parameter that singles out the model, which can predict the flow field of the heated wall jet correctly. The model of Spalart-Allmaras is found to capture most of the flow features well but its applicability is limited to only non-reacting flows. The shear stress transport (SST) and Sarkar & So (SSA) models do well in predicting jet spread rate but lack consistency in predicting other flow features like jet half width accurately. The Spalart Allmaras (SA), Reynolds Stress Model (RSM) and Realizable k-ε (RKE) with enhanced wall treatment (ewt) are the models, which also satisfy this requirement. For the heated wall jet configuration of AbdulNour et al. (2000), considerable deviation has been observed using standard k-ε (SKE) with standard wall function (swf) whereas that of v^2-f shows good prediction of temperature profiles in the near field region. However, the v^2-f model is found to deviate from the data in the downstream region where the velocity profiles exhibit similarity. In the prediction of heat transfer coefficient, RSM followed by SA and RKE with ewt, is found to be the closer to the experimental data compared to the rest of the models. The computational time required for RSM is substantially higher than that of the other RANS models. Therefore, RKE with ewt would be the best choice among various turbulence models to capture the complex flow physics of heated wall jets with least computational time required.

NOMENCLATURE

Ui	Velocity tensor
P	Pressure
$\overline{u_i u_j}$	Turbulent stress tensor
k	Kinetic energy
Pk	Production of kinetic energy
ε	Dissipation rate
f1, f2	Damping functions in the 'ε' equation.
D,E	Addition terms in the K- ε turbulence model.
$C\varepsilon_1$, $C\varepsilon_2$	Constants in the K- ε equation.
S_{ij}	Strain rate tensor
ν_t	Turbulent kinematic viscosity
μ_t	Turbulent dynamic viscosity
ℓ	Length scale
Re	Reynolds number
H	Slot height width
SA	Spalart Allmaras
SKE	Standard k-ε
RKE	Realizable k-ε
RSM	Reynolds stress mode
SST	Shear-stress transport
Ewt	Enhanced wall treatment
Swf	standard wall function

ACKNOWLEDGEMENTS

The authors would like to express their sincere gratitude to Professor Hukam C. Mongia at Purdue University for his collaboration and mentorship during the course of this work. The authors would like to appreciate the efforts of Mr. Subashish Battacharjee and Ms. Savithiri for their help in preparing the manuscript.

REFERENCES

AbdulNour, R.S., Willenborg, K., McGrath, J.J., Foss, J.F., and AbdulNour, B.S., 2000, "Measurements of the convective heat transfer coefficient for a planar wall jet: uniform temperature and uniform heat flux boundary conditions," Experimental Thermal and Fluid Science, **22**(3-4), 123-131.
http://dx.doi.org/10.1016/S0894-1777(00)00018-2

Abe, K., Kondoh, T., and Nagano, Y., 1994, "A new turbulence model for predicting fluid flow and heat transfer in separating and reattaching flows - I. flow field calculations," International Journal of Heat and Mass Transfer, **37**(1), 139-151.
http://dx.doi.org/10.1016/0017-9310(94)90168-6

Abid, R., 1991, "A two-equation turbulence model for compressible flows," Technical Report AIAA-91-1781, AIAA 22nd Fluid Dynamics, Plasma Dynamics and Lasers Conference, Honolulu, Hawaii.

ANSYS Fluent 6.3, documentations.

ANSWER user's manual, version 4.00, 1999, Rev. C, Analytic and Computational Research, Inc. (ACRi), 1931 Stradella Rd, Bel Air, CA, 90077.

Chang, K.C., Hsieh, W.D., and Chen, C.S., 1995, "A modified low-Reynolds-number turbulence model applicable to recirculating flow in pipe expansion," ASME Journal of Fluids Engineering, **117**(3), 417-423.
http://dx.doi.org/10.1115/1.2817278

Chen, H.C., and Patel, V.C., 1988, "Near-wall turbulence models for complex flows including separation," AIAA Journal, **26**(6), 641-648.
http://dx.doi.org/10.2514/3.9948

Durbin, P.A., 1991, "Near-wall turbulence closure modeling without damping functions," Theoretical Computational Fluid Dynamics, **3**(1), 1-13.
http://dx.doi.org/10.1007/BF00271513

Eriksson, J.G., Karlsson. R.I., and Persson, J., 1998, "An experimental study of a two-dimensional plane turbulent wall jet," Experiments in Fluids, **25**(1), 50-60.
http://dx.doi.org/10.1007/s003480050207

Gerodimos, G., and So, R.M.C., 1997, "Near-wall modeling of plane turbulent wall jets," Transactions of the ASME Journal of Fluids Engineering, **119**(2), 304-313.
http://dx.doi.org/10.1115/1.2819135

Kumar, G.N., and Mongia, H.C., 1999, "Validation of Turbulence models for wall jet computations as applied to combustor liners," AIAA-1999-2250, 35th AIAA/ASME/ASEE Joint Propulsion Conference, Los Angeles, California.

Kumar, G.N., and Mongia, H.C., 2000, "Validation of Near Wall Turbulence models for Film Cooling Applications in Combustors," AIAA-2000-0480, 38th AIAA Aerospace Sciences Meeting, Reno, NV.

Lam, C.K.G., and Bremhorst, K., 1981, "A modified form of the k-epsilon model for predicting wall turbulence," ASME Journal of Fluids Engineering, 103, 456-460.
http://dx.doi.org/10.1115/1.3240815

Launder, B.E., and Spalding, D.B., 1972, "Lectures in mathematical models of turbulence," Academic Press, London, England.

Launder, B.E., and Spalding, D.B., 1974, "The numerical computation of turbulent Flows," Computer Methods in Applied Mechanics and Engineering, **3**(2), 269-289.
http://dx.doi.org/10.1016/0045-7825(74)90029-2

Launder, B.E., and Sharma, B.I., 1974, "Application of the energy-dissipation model of turbulence to the calculation of flow near a spinning disc," Letters in Heat Mass Transfer, **1**(2), 131-138.
http://dx.doi.org/10.1016/0094-4548(74)90150-7

Launder, B.E., Reece, G.J., and Rodi, W., 1975, "Progress in the development of a Reynolds-Stress turbulence closure," Journal of Fluid Mechanics, **68**(3), 537-566.
http://dx.doi.org/10.1017/S0022112075001814

Launder, B.E., and Rodi, W., 1979, "The turbulent wall jet," Progress in Aerospace Sciences, **19**(2-4), 81-128.
http://dx.doi.org/10.1016/0376-0421(79)90002-2

Menter, F.R., 1994, "Two-equation eddy-viscosity turbulence models for engineering applications," AIAA Journal, **32**(8), 1598-1605.
http://dx.doi.org/10.2514/3.12149

Narasimha, R., Narayan, K.Y., and Parthasarathy, S.P., 1973, "Parametric analysis of turbulent wall jets in still air," Aeronautical Journal, **77**, pp 355-359.

Parneix, S., and Durbin, P., 1997, "Numerical simulation of 3D turbulent boundary layers using the V^2-F model," Annual Research Briefs, Center for Turbulence Research, NASA Ames/Stanford Univ. 135-148.
http://airex.tksc.jaxa.jp/pl/dr/19990063260/en

Sarkar, A. and So, R.M.C., 1997, "A critical evaluation of near-wall two-equation models against direct numerical simulation," International Journal of Heat and Fluid Flow, **18**(2), 197-208.
http://dx.doi.org/10.1016/S0142-727X(96)00088-4

Schneider, M.E., and Goldstein, R.J., 1994, "Laser Doppler measurement of turbulence parameters in a two-dimensional plane wall jet," Physics of Fluids, **6** (9), 3116-3129.
http://dx.doi.org/10.1063/1.868136

Shih, T.H., Liou, W.W., Shabbir, A., Yang, Z., and Zhu, J., 1994, "A new k-epsilon eddy viscosity model for high Reynolds number turbulent flows: Model development and validation," Computers and Fluids, **24**(3), 227-238.
http://dx.doi.org/10.1016/0045-7930(94)00032-T

Spalart, P.R., and Allmaras, S.R., 1992, "A one-equation turbulence model for aerodynamic flows," Proceedings of the 30th AIAA Aerospace Sciences Meeting and Exhibit, **92**(2).
http://dx.doi.org/10.1051/meca:2007025

Wolfshtein, M., 1969, "The velocity and temperature distribution in one-dimensional flow with turbulence augmentation and pressure gradient," International Journal of Heat and Mass Transfer, **12**(3), 301-318.
http://dx.doi.org/10.1016/0017-9310(69)90012-X

Wygnanski, I., Katz, Y., and Horev, E., 1992, "On the applicability of various scaling laws to the turbulent wall jet," Journal of Fluid Mechanics, **234**, 669-690.
http://dx.doi.org/10.1017/S002211209200096X

Yang, Z., and Shih, T.H., 1993, "New Time Scale Based k-epsilon model for near wall turbulence," AIAA Journal, **31**(7), 1191-1198.
http://dx.doi.org/10.2514/3.11752

DETERMINING HEAT TRANSFER COEFFICIENT OF HUMAN BODY

A. Najjaran*, Ak. Najjaran, A. Fotoohabadi, A.R. Shiri

Islamic Azad University Branch of Shiraz, Shiraz, Fars, Iran, 7154845589, Iran

ABSTRACT

In this paper, the aim is obtaining convection coefficient of human body. This field of study is essential in study of ventilation systems, astronauts' clothes and any other fields in which human body is the main concern. At first a 3D human body has been designed by unstructured grids. Feet and hands are stretched completely in considered sample. Two postures (standing and supine) are considered for body. Soles and the back of entire body are considered in contact with the ground respectively in these postures. Other parts of human body are exposed to surrounding air. The heat transfer and the body temperature are assumed steady and constant. The results are obtained by applying finite volume method for each grid and extracted by the weighted area method in Fluent®. Then the attained results are validated with the recent experimental results. Good agreement is observed between the obtained results and the previous experimental results. Finally two formulas are derived for natural convection coefficient of human body.

Keywords: *convective heat transfer coefficient, standing, supine, human body, numerical simulation.*

1. INTRODUCTION

Natural convection is a process which doesn't need any external force to complete the thermal cycle. For instance, if a heat source is assumed in a cooler space, the surrounding air which is close to the heat source is warmed and expanded. So because of the decrease of its density the air goes upward. Then in consequence of vicinity with cooler air, its temperature decreases and it compresses again and therefore it goes back to its initial position in the cycle. This process takes place only because of the temperature difference between the heat source and surrounding air. This paper is aimed at finding the natural convection coefficient of human body. This coefficient is essential for determining the exchanged convectional heat between human body and surrounding air.

In previous papers convection coefficient of human body is determined by different methods. The reported convection coefficients are different and contain questionable accuracies. The reasons of differences in reported coefficients are various. Firstly, the natural convection coefficient is affected by gravity force. The effect of gravity force direction is different in various body postures and would cause different results. Secondly, human body sample shape causes an important effect. Complicated body models increase airflow disturbance and so the flow disturbance around the body should be analyzed. Thirdly, as Kurazumi et al. mentioned, 10 to 20 % of body surface doesn't participate in convectional heat transfer (for example because of ground contact) but in the majority of recent researches convectional heat transfer is considered in all entire body surface (area ratio 1) for simplicity. Fourthly, in experimental studies unlike analytical methods, measuring quantities has some errors. Fifthly, experimental studies are with some limitations which decrease the accuracy of results. For example obtaining a specific thermal condition in laboratory is difficult or convectional heat should be analyzed in plastically mannequins. Furthermore numbers of sensors which are used in experimental researches are considered as disturbing factor and should be limited too. For example, the convection coefficient in different members is different; so this parameter should be measured for each member separately. But the applied sensors are considered as disturbing structures in heat transfer and should be limited. In some studies, the radiated heat is calculated and then is subtracted from the total heat for determining convection coefficient. The radiation heat depends on angle factor.

By considering the 3D body of human, the angle factor has various amounts and so it causes different heat radiation. Therefore the calculated quantities for convection heat and convection coefficient don't have proper accuracy. In this paper, natural convection coefficient of human body is calculated for two usual postures (supine & standing). Of course these postures are samples and other postures can be analyzed too. Despite of previous studies skin and meat is considered for the body. The areas which have no convection heat transfer with surroundings are calculated and subtracted from the total area. The assumed area of the body is optional and so it doesn't have the limitations of the previous methods. Although the area of the actual heat transfer surfaces might be less or more than the considered amount, but because the convection heat transfer coefficient is calculated per unit area so the errors can be neglected. Unknown parameters are determined from the weighted area method. The obtained results are validated with recent experimental results. Finally two formulas are obtained from the results.

2. GOVERNING EQUATIONS

Heat transfer of body with surrounding air can be divided into four main methods of Convection, Conduction, Radiation and Evaporation. Convection heat flux of human body follows conjugate equations because of having a specific structure similar to fins. Human body is the same as six cylindrical fins and for each part the formula of the cylindrical fins is used for calculating its heat flux. Of course in some studies human body is considered as a collection of one spherical and five cylindrical fins. By considering a steady state heat transfer and

* Corresponding Author Email: anajaran@iaushiraz.ac.ir

applying these equations for human body, the natural convection coefficient and also convection heat flux are determined.

Equations (1) to (3) show the continuity, momentum and energy governing equations. These equations can be used in the conditions in which the viscosity is constant or zero.

$$\frac{D\rho}{Dt}=\frac{\partial\rho}{\partial t}+V.\nabla\rho=-\rho\nabla.V \tag{1}$$

$$\rho\frac{DV}{Dt}=\rho(\frac{DV}{Dt}+V.\nabla V)=F-\nabla P+\mu\nabla^2 V+\frac{\mu}{3}3\nabla(\nabla.V) \tag{2}$$

$$\rho C_p\frac{DT}{Dt}=\rho C_p(\frac{DT}{Dt}+V.\nabla T)=\nabla.(k\nabla T)+q'''+\beta T\frac{DT}{Dt}+\mu\phi_v \tag{3}$$

where V is the velocity vector, T is the surrounding temperature, t is the time, F is volumetric force per unit volume, C_p is specific heat capacity in constant pressure, P is static pressure, ρ is fluid density, β is thermal expanding coefficient of fluid, ϕ_v is viscosity losses (which is irreversible part of heat transfer because of the viscosity forces) and q''' is the heat generation per unit volume. The thermal expanding coefficient of fluid is in $\beta=-(1/\rho)(\partial\rho/\partial T)_P$ which the symbol of P denotes the constant pressure. For a Noble gas this parameter is equal 5 to $\beta=1/T$ which T is the absolute temperature in terms of Kelvin. D/Dt is the total or partial differential which might become apparent as $(\delta(\nabla V)/\delta T)$. The main factor for natural convection heat transfer is temperature differences. Warming and cooling processes of body by environmental air (or in fact temperature differences) causes environmental air density changes. Density variations and buoyancy forces result in fluid movement which causes natural convection. For example, during cooling process of body, the vicinity air gets warm and so its volume increases and its density decreases and so it goes upward.

3. SOLUTION PROCEDURE

At first a sample of human body is designed by Solid works software and then its grids are generated by Gambit software. The final designed model is transferred to fluent software. Boundary conditions (such as body temperature, air density, surrounding temperature and pressure) are set as inputs. The software uses Boussinesq and boundary layer approximations for solving the problem. The initial guesses are obtained by software and the desired accuracies of results of governing equations (Equations (1) To (3)) are defined by users. By convergence of the problem the considered parameters are determined in each grid. Finally the outputs are extracted by use of weighted area method.

4. BOUSSINESQ AND BOUNDARY LAYER APPROXIMATIONS

The governing equations of natural convection are dependent on each other and have elliptical and partial differential forms. Unwished viscosity deviations which are caused by the temperature changes also cause calculations to be in a conjugate form. Generally some methods are used to make these equations simple which in these methods Boussinesq and boundary layer approximations are very important. These approximations are widely used for natural convection. An important condition for validation of these approximations is β (T-T∞) << 1. When the temperature differences are low this simplification is valid. In Boussinesq approximation the density deviations are neglected. Moreover density differences of the fluid usually are estimated by temperature effect (the effect of pressure is neglected). The density differences are estimated for the thermal buoyancy by Equation (4).

$$\rho_\infty-\rho=\rho\beta(T-T_\infty) \tag{4}$$

The other method is boundary layer approximation. In natural convection, the velocity out of the boundary layer is very low and is affected with the surrounding pressure. The surrounding pressure is the hydrostatic pressure. In addition increasing of heat or mass of the fluid is assumed to be limited to thin layer close to the body surface. So it is assumed that the gradients in the tangent direction of the surface are very smaller than the ones that are in the normal (perpendicular) direction. The main result of boundary layer approximation is that the flow terms (except the vertical ones) are neglected in momentum and heat equations. These assumptions are losing their validity in unsteady state heat transfer. Therefore in this paper the ground temperature is assumed the same as surrounding temperature.

5. INITIAL CONDITION AND RESULTS' CONVERGENCY

There is no general law for convergence. So each problem has its specific technique. As is said in the natural convection problems the flow velocity is low. Viscosity deviations cause considerable conjunction too. So for convergence of the results, at the beginning of the solving velocities are assumed close to zero and In addition heat transfer is assumed steady. Then for several iterations the energy equations are solved for determining a better initial guess. After that, the continuum and momentum equations are solved with energy equation simultaneously. Then the flow velocity is increased to determine the flow treatment. This performance is done by increasing the gravity acceleration. After obtaining the magnified flow treatment, the gravity acceleration is decreased in several consecutive processes until real amounts are attained.

6. MATHEMATICAL DEFINITION OF WEIGHTED AREA METHOD

The weighted area average (or in fact weighted mean) of a non-empty set of convection coefficients

$$\{h_{conv,1},h_{conv,2},...,h_{conv,n}\}$$

With non-negative weights

$$\{w_1,w_2,...,w_n\}$$

Is the quantity

$$\bar{h}_{conv}=(\Sigma_{i=1}^n w_i h_{conv,i})/(\Sigma_{i=1}^n w_i)$$

Therefore convection coefficients of grids with a high weight contribute more to the weighted area than do convection coefficients of grids with a low weight. The weights cannot be negative. Some may be zero, but not all of them because division by zero is not allowed. The formulas are simplified when the weights are normalized such that they sum up to 1, i.e. $\Sigma_{i=1}^n w_i=1$. For such normalized weights the weighted area is simply $\bar{h}_{conv}=\Sigma_{i=1}^n w_i h_{conv,i}$.

7. RESULTS AND DISCUSSIONS

In the previous experimental studies air temperature, relative humidity and flow velocity are measured using an Assmann aspiration Psychrometer and a non-directional hot bulb-type anemometer. The vertical air temperature profile and the temperatures of all room wall surfaces are also taken with T-type thermocouples. Considering the effects of relative humidity and the other parameters on the natural convection coefficient, in this paper these parameters are set as inputs and so it is no need of using these devices. Skin temperature, conduction coefficient, specific heat capacity and average density of human body are considered 308 K, 0.209 W/mK , 3470J/KgK and 1000 kg/m^3, respectively. In addition, the air pressure is considered $101325 Pa$. The surrounding is considered uniform and the heat transfer is assumed steady. The remaining fluid characteristics in natural convection are considered constant. The considered human body exchanges heat just with the air. The local differences of the flow velocity in lower and upper part of body are neglected and the whole body is exposed to natural convection with the same characteristic. As is said in the introduction the calculations based on total heat have some problems. Therefore in this paper, calculations are done based on the convection total heat.

One of the differences between the present work and the previous ones is that the body is organized from a large number of approximating surfaces. For each part surface area or area ratio is determined. So the area ratios of recent researches that were related to limited number of surfaces are not used. A human body sample includes 1874427 approximating surfaces, 191524 boundary nodes and 896726 cells in standing posture. Also it contains 1572162 approximating surfaces, 143830 boundary nodes and 771108 cells in supine posture (see Figure 1). Grids are considered unstructured and the cells are considered tetrahedron.

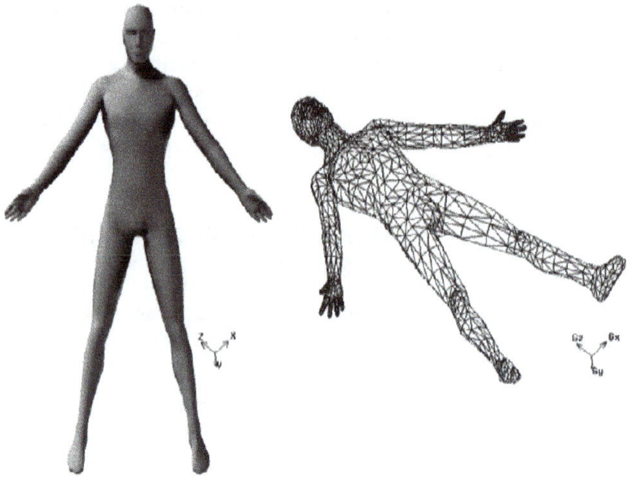

Fig. 1 Human Body model (standing and spine postures)

As shown in Figs. 2 and 3 concentration of grids is more around the body. This concentration leads to more accurate results and decreases the number of grids. The reasons of the excess number of grids and concentration of grids around the body are the complex structure of body and conjugation of thermal heat exchange around the body. Body grid generation is done in such a way that different parts of body have no contact with each other. For example fingers don't have any contact with each other. Figure 2 shows the grid test of analyzed samples.

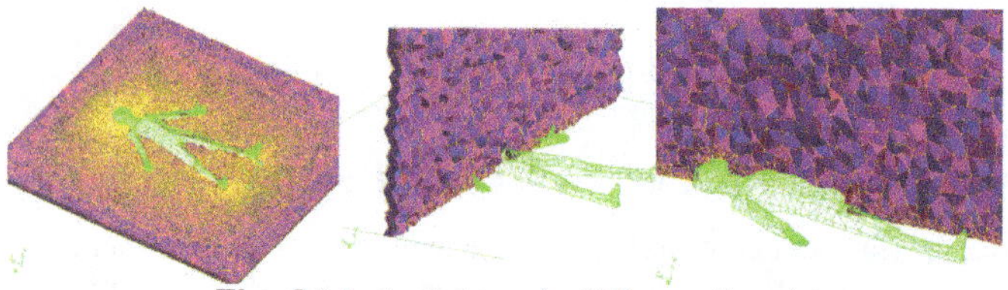

Fig. 2 Mesh gird test in different directions

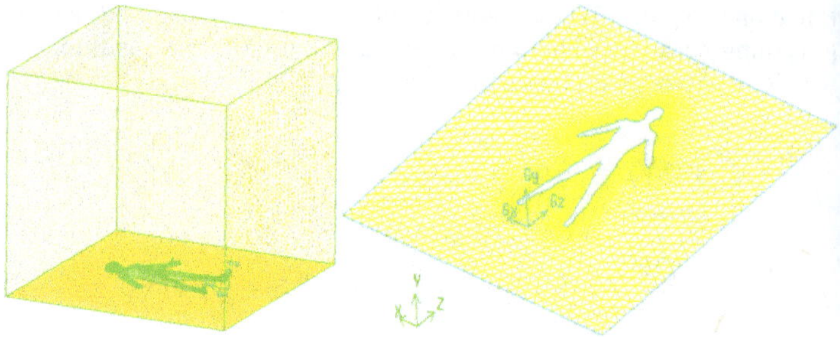

Fig. 3 Analyzed domains and meshed model in supine posture

A domain with dimension of 2 × 2 × 2 meter is considered around the body (see Figure 3). This domain is produced in such a way that in standing posture soles and in supine posture the whole back of the body are in contact with the ground. It is limited from the ground and from the other side is exposed to the air. The ground and the air are in a thermal balance and are considered to have the same temperature.

After calculating the convection coefficient for each grid, the average convection coefficient is determined for the entire body. For determining the average of obtained results several computational methods exist such as actual area average, actual vertex average, standard deviation and the weighted area average. In this paper, the results are determined based on the last two methods. By comparing the results it is concluded that the weighted area average method has a better performance in determining average between unbalanced parameters. Due to size differences of surfaces using standard deviation method faces with problem. Table 1 shows some obtained results for standing posture by the use of standard deviation method (the temperatures are in Kelvin and the convection coefficient unit is W/m^{2}K.

Table 1

Case	*Sur. Tem.*	*Body Tem.*	*Conv. Coef.*
1	295	304	3.563
2	293	298	3.715

The histogram diagram of the results obtained by standard deviation method (see Fig. 4) shows the wide dispersal of the natural convection coefficient.

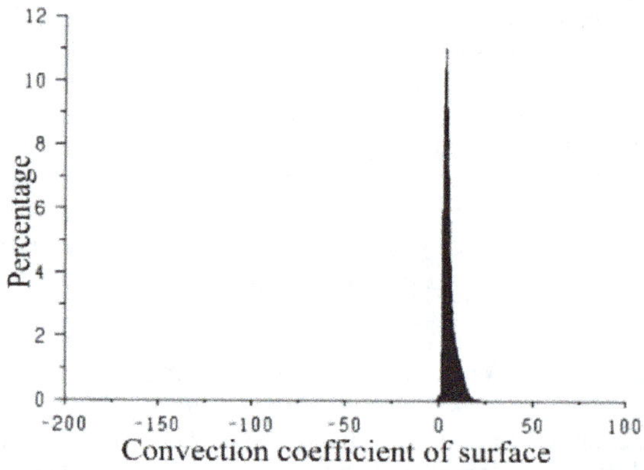

Fig. 4 Histogram diagram of convection coefficient obtained by standard division method

According to the statistical data, the most percentage of the outputs is close to the coefficients which are expressed in Table 1. The obtained results by weighted area method are shown in Table 2 It shows six cases which are considered for two standing and supine postures. In Table 2 Sur. Tem. denotes surrounding temperature and Re. Er. denotes relative error. The temperatures are in Kelvin and the convection coefficients are in W/m^2K.

Table 2

Sur. Tem.	*Skin Tem.*	*Obtained Standing Coef.*	*Standing Coef. Eq. (6)*	*Re Er %*	*Obtained Supine Coef.*	*Supine Coef Eq. (7).*	*Re %*
289	308	3.3	3.29	0.41	2.03	2.60	22.02
291	308	3.18	3.16	0.57	2.54	2.50	1.63
293	308	3.09	3.03	2.06	2.4	2.39	0.56
295	308	2.90	2.88	0.66	2.28	2.26	0.69
297	308	2.65	2.72	2.52	2.03	2.13	4.65
299	308	2.24	2.54	11.66	1.78	1.98	10

In Kurazumi et al. formulas are derived for convection coefficients of supine and standing postures from experimental results. Equations (6) and (7), represent the obtained formulas for standing and supine respectively.

$$h_{supine} = 1.183\Delta T^{0.347} \quad (6)$$

$$h_{supine} = 0.881\Delta T^{0.368} \quad (7)$$

Table 2 also represents the relative errors for each case. These errors are calculated by comparing the obtained results with the results which are attained by Equations (6) and (7). The mean relative errors of the results related to standing and supine postures are, 2.98% and 6.59% respectively.

By using the obtained results of numerical simulation, two formulas are achieved for standing and supine postures. Equation (8) represents the obtained formula for standing.

$$h_{standing}^{0.5} = 1.951 - 16535(Ln(\Delta T)) / \Delta T^2 \quad (8)$$

where h (W/m^2K) is the convection coefficient and ΔT is the temperature difference of the skin and surrounding. Equation (9) represents the convection coefficient for supine posture.

$$h_{supine}^{0.5} = 1.734 - 15.016(Ln(\Delta T)) / \Delta T^2 \quad (9)$$

Equations (8) and (9) are valid when the temperature difference is not high (e.g. no bigger than 40 degree). Figure 5 shows the results which are reported by Hard and Dubois, Nielsen and Pedersen, Ohmoto and Mochida, Kurazumi et al., Nishi and Gagge and Najjaran et al. for convection coefficient of standing and supine postures. As is observed in Figure 5 By decreasing of the temperature difference, the convection coefficient also decreases. When the air temperature decreases, body decreases the skin temperature to prevent the extreme thermal losses. By decreasing the temperature difference between bodies and surrounding, the flow velocity also decreases. Consequently the coefficient of convection heat transfer decreases. Also it is observed that the convection coefficient in the standing posture is continuously more than the sitting posture (see Figure 5). Good agreement is seen between the obtained curves and Kurazumi et al. results.

By comparing the obtained results of all cases it is conducted that by increasing the temperature difference between body and surrounding air, the velocities in both postures increases layer by layer.

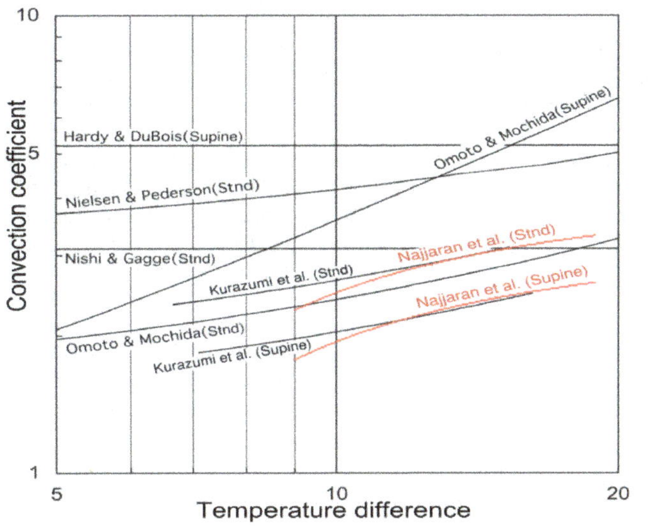

Fig. 5 Comparison of the obtained natural convection coefficient with the ones which are obtained by experimental methods in recent studies

In Fig. 5 the convection coefficients of postures which don't have any contact with the ground are not analyzed. These postures are unusual and are considered only because of experimental limitations (such as preventing the contact of body surfaces with each other [52]). Velocity path lines are factors which show the velocity gradient of air around human. Figures 6 and 7 show velocity contours around the body in standing and supine postures in two cases.

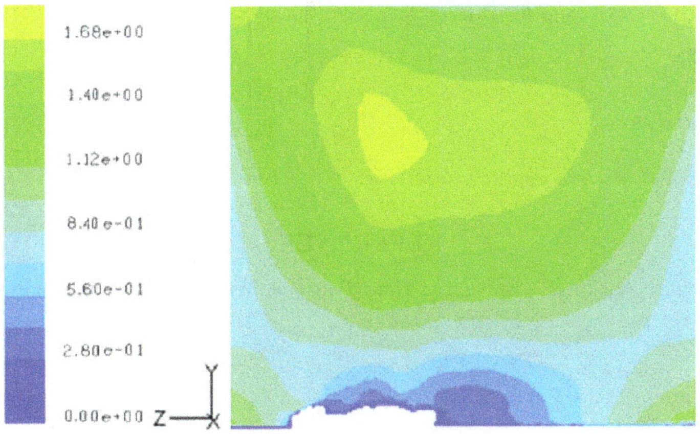

Fig. 6 Contour velocities in supine posture

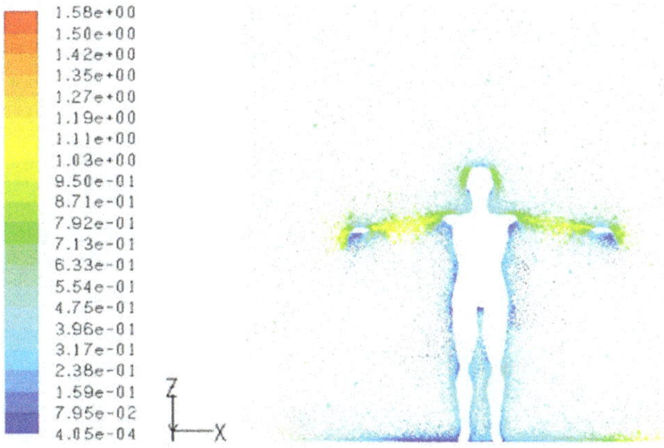

Fig. 7 Contour velocities in standing posture

8. CONCLUSIONS

The aim of this research is determining of natural convection coefficient of human body which has many applications in different fields. A sample of human body was designed by unstructured grids. Two different postures (standing and supine) were considered for sample. Some parts of the body were in contact with the ground in these postures. The rest of surfaces of the body were exposed to the surrounding air. For determining results the weighted area method was used. Two formulas were reported for the considered postures for natural convection coefficient. The obtained results were compared with the experimental results of previous researches and their errors were calculated. Good agreement was observed between the obtained results and the previous experimental results. No need of experimental setup, fast determination of the coefficient in various conditions and accuracy of the results are the benefits of applying this method.

NOMENCLATURE

C	specific heat
F	volumetric force (per unit volume)
h	convection coefficient
P	static pressure
q	Heat generation
t	time
T	surrounding temperature
V	velocity vector
w	surface weight

Greek Symbols

β	Thermal fluid expanding coefficient
ρ	fluid density (kg/m^3)
ϕ_v	Viscosity losses

Subscripts

p	constant pressure

REFERENCES

Bejan, A., 2002, *Convection Heat Transfer*, Wiley & Sons, Hoboken, NJ.

Hardy, JD., Dubois, EF., 1938, "The Technic Of Measuring Radiation And Convection," *Journal of Nutrition*, **15**(5), 461–475.

Kurazumi, Y., Tsuchikawa, T., Ishiia, J., Fukagawaa, K., Yamatoc, Y., Matsubarad, N., 2008, "Radiative and Convective Heat Transfer Coefficients of the Human Body in Natural Convection," *Building and Environment*, **43**, 2142–2153. http://dx.doi.org/10.1016/j.buildenv.2007.12.012

Kurazumi, Y, Tsuchikawa, T, Yamato, Y, Kakutani, K, Matsubara, N, Horikoshi, T, 2003, "The Posture and Effective Thermal Convection Area Factor of the Human Body," *Japanese Journal of Biometeology*, **40**(1), 3–13.

Kurazumi, Y, Tsuchikawa, T, Matsubara, N, Horikoshi, T, 2004, "Convective Heat Transfer Area of the Human Body," *European Journal of Applied Physiology*, **93**(3), 273–85. http://dx.doi.org/10.1007/s00421-004-1207-1

Mochida, T., 1977, "Convective And Radiative Heat Transfer Coefficients For Human Body," *Transactions of AIJ*, **258**, 63–69.

Mochida, T., 1977, "Mean Skin Temperature Weighted with Both Heat Transfer Coefficient and Skin Area," *Transactions of AIJ*; **259**,67– 73.

Mochida, T, Nagano, K, Shimakura, K, Kuwabara, K, Nakatani, T, Matsunaga, K, 1999, "Characteristics Of Convective Heat Transfer Of Thermal Manikin Exposed In Air Flow From Front," Journal *of Human and Living Environment*, **6**(2), 98–103.

Nielsen, M, Pedersen, L., 1952, "Studies on the Heat Loss by Radiation and Convection from the Clothed Human Body," *Acta Physiologica Scandinavica*, **27**, 272–94. http://dx.doi.org/10.1111/j.1748-1716.1953.tb00943.x

Nishi, Y, Gagge, AP., 1970, "Direct Evaluation of Convective Heat Transfer Coefficient by Naphthalene Sublimation," *Journal of Applied Physiology*, **29**(6), 830–8.

Ohmoto, Y., Mochida, T., 1995, "Influence of Posture upon Convective Heat Transfer Coefficient for the Human Body," *Proc. of the 19th symposium on human–environment system*, Kyoto, Japan; 25–34.

12

HALL AND ION SLIP EFFECTS ON FREE CONVECTION HEAT AND MASS TRANSFER OF CHEMICALLY REACTING COUPLE STRESS FLUID IN A POROUS EXPANDING OR CONTRACTING WALLS WITH SORET AND DUFOUR EFFECTS

Odelu Ojjela*, N. Naresh Kumar

Department of Applied Mathematics
Defence Institute of Advanced Technology, Deemed University, Pune - 411025, India.

ABSTRACT

This article deals the Hall and ion slip currents on free convection flow, heat and mass transfer of an electrically conducting couple stress fluid through porous channels with chemical reaction, Soret and Dufour effects. Assume that there is symmetric suction or injection along the expanding or contracting walls, which are maintained at different constant temperatures and concentrations. The governing partial differential equations are reduced to nonlinear dimensionless ordinary differential equations using the similarity transformations and solved numerically by the method of quasilinearization. The effects of various parameters on non-dimensional velocity components, temperature distribution and concentration are discussed in detail and shown in the form of graphs.

Keywords: *MHD; Couple stress fluid; Free convection; Chemical reaction; Porous medium; Soret and Dufour.*

1. INTRODUCTION

The MHD flow through porous channels has attracted the attention of several researchers due to their applications in science and engineering fields. Examples of these were found in the transport of biological flows between expanding or contracting vessels, the synchronous pulsation of porous diaphragms, MHD power generators, magnetic filtration and separation, jet printers, micro fluidic devices, polymeric liquids, boundary layer control, etc. The couple stress fluid theory developed by Stokes (1966) presents the generalization of the classical viscous fluid theory that sustains couple stresses and the body couples. The concept of couple stresses takes place due to the mechanical interactions in the fluid medium and in this theory the velocity field was defined in place of rotational field. Further Stokes (1968) discussed the effect of couple stresses through hydromagnetic channels in a fluid medium.

Many researchers studied the fluid flow in porous channels with expanding or contracting walls. Srinivasacharya et al. (2009) considered the problem on couple stress fluid flow and heat transfer in a porous channel with expanding and contracting walls and the solution was discussed using quasilinearization method. Alam et al. (2013) studied the couple stress fluid flow between expanding or contracting channel and an analytical approximate solution obtained for reduced governing equations. Analytical solutions were obtained for two dimensional incompressible symmetric and asymmetric flow of viscous fluid through porous channels between expanding or contracting walls discussed by Si et al. (2010 and 2011). Uchida and Akoi (1977) modeled the unsteady incompressible viscous fluid flows inside expanding or contracting pipe. When heat and mass transfer occur simultaneously in a moving fluid, it can be observed that an energy flux can be created not only by temperature gradients but also by concentration gradients. The energy flux caused by a concentration gradient is termed the Dufour effect. On the other hand, mass fluxes can also be generated by temperature gradients are known as the Soret effect. These effects are studied as second order phenomena and had applications in many areas such as petrology, geosciences, hydrology, etc. An incompressible flow, heat and mass transfer of a viscous fluid between expanding or contracting walls is considered by Srinivas et al. (2012) with weak permeability, Soret and Dufour effects. Srinivasacharya and Kaladhar (2011) examined the steady mixed convection heat and mass transfer of couple stress fluid past a vertical plate. An incompressible flow of couple stress liquids with Soret and Dufour effects studied by Malasetty et al. (2006).

Dursunkaya and Worek (1992) considered the diffusion-thermo and thermal-diffusion effects in transient and steady free convection flow over a vertical surface. The effects of thermal diffusion and diffusion thermo on the mixed convection heat and mass transfer of a non-Newtonian fluid in a porous medium with thermal radiation effect was investigated by Mahmoud and Megahed (2013). Nithyadevi and Yang (2009) examined numerically the Soret and Dufour effects on the free convection flow of water over a partially heated enclosure. Postelnicu (2004) considered the problem of MHD free convection flow, heat and mass transfer over a vertical surface with Soret and Dufour effects. The steady laminar incompressible mixed convection heat and mass transfer of viscous fluid embedded in the non Darcy porous medium over a stretching sheet with chemical reaction, Soret and Dufour effects analyzed by Dulal and Mondal (2011). Ganeshan et al. (2012) investigated the chemical reaction effect on natural convection flow over a vertical plate with Soret and Dufour effects. The steady free convection heat and mass transfer from the vertical cone with Soret and Dufour effects considered by Cheng (2010) and taking variable temperature and concentration.

In most of the electrically conducting fluid flow problems, the Hall and Ion-slip terms in Ohms law were neglected, then the influence of Hall current and Ion-slip are important under the presence of strong magnetic field. The effects of Soret and Dufour on the mixed convection heat and mass transfer of second grade fluid over a

* Corresponding author Email: odelu3@yahoo.co.in ; odelu@diat.ac.in

stretching sheet is studied with Hall and ion slip currents by Hayat and Nawaz (2011). Elgazery (2009) examined the flow, heat and mass transfer of an electrically conducting viscous fluid with chemical reaction, Hall and ion slip by considering temperature dependent thermal diffusivity and viscosity. An unsteady magnetohydrodynamic natural convection heat and mass transfer from vertical surface with chemical reaction and radiation effects investigated by Chamkha et al. (2011). Patil and Kulkarni (2008) discussed the effect of chemical reaction on convective flow and heat transfer of polar fluid in a porous medium with internal heat generation. The application for quasilinearization technique to the flow of viscoelastic fluid in a porous annulus presented by Bhatnagar et al. (1994).

This paper investigates the influence of Hall and ion slip on unsteady two dimensional MHD natural convection heat and mass transfer of couple stress fluid in a porous medium between expanding or contracting walls with chemical reaction, Soret and Dufour effects. The reduced flow field equations are solved by using the quasilinearization method. The results are obtained in the form of graphs and discussed in detail for the velocity components, temperature distribution and concentration with respect to various fluid and geometric parameters.

2. FORMULATION OF THE PROBLEM

Consider an unsteady laminar incompressible free convection flow of couple stress fluid in a porous medium through two parallel horizontal walls. Assume that both upper and lower walls have equal permeability and expand or contract at a time dependent rate. Therefore their separation is a function of time a(t). The lower and upper walls are maintained at constant temperatures T_1, T_2 and concentrations C_1, C_2 respectively. Let the fluid be injected or aspirated uniformly and orthogonally through the channel walls at an absolute velocity V_1. The flow is subjected to a uniform magnetic field perpendicular to the flow direction with the Hall and ion-slip effects.

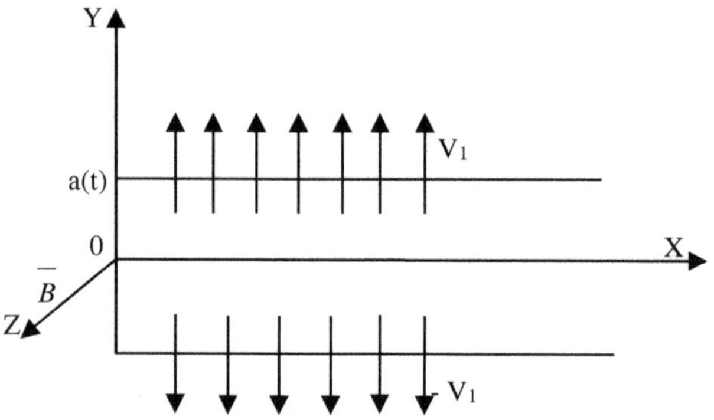

Fig. 1 Physical model of the problem

The equations governing the couple stress fluid flow, heat and mass transfer in the presence of magnetic field, Soret and Dufour effects and in the absence of body forces and body couples are given by

$$\nabla \cdot \bar{Q} = 0 \tag{1}$$

$$\rho\left[\frac{\partial \bar{Q}}{\partial t} + (\bar{Q}.\nabla)\bar{Q}\right] = -\,grad\,p - \mu\ curl\,(curl\,(\bar{Q}))$$

$$-\eta\ curl\,(curl\,(curl\,(curl\,(\bar{Q})))) - \frac{\mu}{k_1}\bar{Q} + \bar{J}\times\bar{B} + \overline{F_B} \tag{2}$$

$$\rho c\left[\frac{\partial T}{\partial t} + (\bar{Q}.\nabla)T\right] = k\nabla^2 T + 2\mu D{:}\,D + 4\eta\,[(\nabla\,\bar{\omega}){:}\,(\nabla\bar{\omega})^T] +$$

$$4\eta'\,[(\nabla\,\bar{\omega}){:}(\nabla\,\bar{\omega})] + \frac{\mu}{k_1}\left(\bar{Q}\right)^2 + \frac{\left(\bar{J}\right)^2}{\sigma} + \frac{\rho D_1 k_T}{c_s}\nabla^2 C \tag{3}$$

$$\left[\frac{\partial C}{\partial t} + (\bar{Q}.\nabla)C\right] = D_1\nabla^2 C - k_2\,(C - C_1) + \frac{D_1 k_T}{T_m}\nabla^2 T \tag{4}$$

where $\overline{F_B}$ is the buoyancy force and it is defined as $\left(\rho g \beta_T\,(T - T_1) + \rho\beta_C\,(C - C_1)\right)\hat{i}$, $\bar{Q}$ is velocity, p is pressure, k_1 is permeability parameter, σ is electrical conductivity, ρ is density, $\bar{B}$ is magnetic field, $\bar{E}$ is electric field, T is the temperature distribution, $\bar{\omega}$ is the rotation vector, k is the thermal conductivity, c is the specific heat at constant temperature, k_T is the thermal diffusion ratio, T_m is the mean temperature, c_s is the concentration susceptibility, D_1 is the coefficient of mass diffusivity, η and η' are the couple stress fluid parameters, C is the concentration, k_2 is the chemical reaction rate and $\bar{J}$ is current density.

The force stress tensor τ and the couple stress tensor M that arises in the theory of couple stress fluids are given by

$$\tau = (-p + \lambda\,\mathrm{div}\,\bar{Q})I + \mu\,[\,\mathrm{grad}\,\bar{Q} + (\mathrm{grad}\,\bar{Q})^T] +$$

$$\tfrac{1}{2} I\ [\mathrm{div}\,M + \rho C] \tag{5}$$

and

$$M = mI + 2\eta\ \mathrm{grad}(\mathrm{curl}\,\bar{Q}) + 2\eta'(\mathrm{grad}\,(\mathrm{curl}\,\bar{Q}))^T \tag{6}$$

where m is 1/3rd Trace of M and ρC is the body couple tensor. The quantity λ is the material constant and η' is the constant associated with couple stresses. The dimensions of the material constant λ is that of viscosity where as the dimensions of η and η' are those of momentum. These material constants are considered by the inequalities,

$$\mu \geq 0,\quad 3\lambda + 2\mu \geq 0,\ \eta \geq 0,\ \eta' \leq \eta \tag{7}$$

Neglecting the displacement currents, the Maxwell equations and the generalized Ohm's law are

$$\nabla.\bar{B} = 0,\quad \nabla\times\bar{B} = \mu'\bar{J},\quad \nabla\times\bar{E} = \frac{\partial \bar{B}}{\partial t},$$

$$\bar{J} = \frac{\sigma B_0}{\alpha e^2 + \beta e^2}\left((\alpha e\ v - \beta e\ u)\,\hat{i} - (\alpha e\ u + \beta e\ v)\,\hat{j}\right) \tag{8}$$

Where $\bar{B} = B_0\,\hat{k} + \bar{b}$, $\bar{b}$ is the induced magnetic field β_e is the Hall parameter β_i is the ion slip parameter and μ' is magnetic permeability. Let the induced magnetic field be negligible compared to the applied magnetic field so that magnetic Reynolds number is small, the electric field is zero and magnetic permeability is constant throughout the flow field.

The velocity is $\bar{Q} = u\hat{i} + v\hat{j}$,

Following Si et al. (2010 and 2011), we take the velocity components as,

$$u(x,\xi,t) = -\frac{\upsilon x}{a^2} F^{/}(\xi,t),\quad v(x,\xi,t) = \frac{\upsilon}{a} F(\xi,t),$$

Following Srinivas et al. (2012), we take the temperature distribution as

$$T(x,\xi,t) = T_1 + \frac{\mu V_1}{\rho a c}\left(\varphi_1(\xi) + \frac{x^2}{a^2}\varphi_2(\xi)\right) \text{ and}$$

$$C(x,\xi,t) = C_1 + \frac{\dot{n}_A}{a\upsilon}\left(g_1(\xi) + \frac{x^2}{a^2} g_2(\xi)\right) \tag{9}$$

Where $\xi = \frac{y}{a(t)}$ and $F(\xi,t)$, $\varphi_1(\xi)$, $\varphi_2(\xi)$, $g_1(\xi)$, $g_2(\xi)$ are to be determined.

The boundary conditions on the velocity, temperature and concentration are

$u(x,\xi,t) = 0,\ v(x,\xi,t) = 0,\ \nabla\times\overline{Q} = 0,\ T(x,\xi,t) = T_1,$

$C(x,\xi,t) = C_1$ at $\xi = 0$

$u(x,\xi,t) = 0,\ v(x,\xi,t) = V_1,\ \nabla\times\overline{Q} = 0,\ T(x,\xi,t) = T_2,$

$C(x,\xi,t) = C_2$ at $\xi = 1$ (10)

Substituting (9) in (2), (3), and (4) then

$$f^{VI} = \frac{1}{\alpha^2}(\xi\beta f^{///} + 3\beta f^{//} + \mathrm{Re}(f^{/} f^{//} - f f^{///}) + f^{IV} - D^{-1} f^{//}$$

$$-\frac{Ha^2\,\alpha e}{\alpha e^2 + \beta e^2} f^{//} - \frac{EGr}{\mathrm{Re}\,\zeta}(\varphi_1^{/} + \zeta^2\varphi_2^{/}) - \frac{ShGm}{\mathrm{Re}\,\zeta}(g_1^{/} + \zeta^2 g_2^{/}) \tag{11}$$

$$\varphi_1^{//} = -2\varphi_2 - \mathrm{Re}\,\mathrm{Pr}(4 f^{/2} + \alpha^2 f^{//2} + D^{-1} f^2 + \frac{Ha^2}{\alpha e^2 + \beta e^2} f^2 +$$

$$\frac{\beta\xi\varphi_1^{/}}{\mathrm{Re}} + \frac{\beta\varphi_1}{\mathrm{Re}} - f\varphi_1^{/}) - 2\,\mathrm{Du}\, g_2 - \mathrm{Du}\, g_1^{//} \tag{12}$$

$$\varphi_2^{//} = -\mathrm{Re}\,\mathrm{Pr}(f^{//2} + \alpha^2 f^{///2} + D^{-1} f^{/2} + \frac{Ha^2}{\alpha e^2 + \beta e^2} f^{/2} +$$

$$\frac{\xi\beta g_2^{/}}{\mathrm{Re}} + \frac{3\beta g_2}{\mathrm{Re}} + 2 f^{/}\phi_2 - f\phi_2^{/}) - \mathrm{Du}\, g_2^{//} \tag{13}$$

$$g_1^{//} = -2 g_2 + Kr\, g_1 + Sc(-\beta g_1 - \xi\beta g_1^{/} + \mathrm{Re}\, f g_1^{/})$$

$$-ScSr(\varphi_1^{//} + 2\varphi_2) \tag{14}$$

$$g_2^{//} = Kr\, g_1 + Sc(-3\beta g_2 - \xi\beta g_2^{/} + \mathrm{Re}(f g_2^{/} - 2 f^{/} g_2))$$

$$-ScSr\varphi_2^{//} \tag{15}$$

Where prime denotes the differentiation with respect to ξ, $f(\xi) = \frac{F(\xi,t)}{\mathrm{Re}}$, $\alpha e = 1 + \beta e\beta i$.

The dimensionless form of temperature and concentration from (9) can be written as

$$T^* = \frac{T - T_1}{T_2 - T_1} = Ec(\varphi_1 + \zeta^2\varphi_2)$$

$$C = \frac{C - C_1}{C_2 - C_1} = Sh(g_1 + \zeta^2 g_2) \tag{16}$$

Where $Ec = \frac{\mu V_1}{\rho a c(T_2 - T_1)}$ is the Eckert number,

$Sh = \frac{\dot{n}_A}{a\upsilon(C_2 - C_1)}$ is the Sherwood number and $\zeta = \frac{x}{a}$ is the dimensionless axial variable.

The boundary conditions (10) in terms of f, φ_1, φ_2, g_1 and g_2 are

$$f(0) = 0,\quad f(1) = 1,$$
$$f^{/}(0) = 0,\quad f^{/}(1) = 0,$$
$$f^{//}(0) = 0,\quad f^{//}(1) = 0,$$
$$\varphi_1(0) = 0,\quad \varphi_1(1) = 1/Ec$$
$$\varphi_2(0) = 0,\quad \varphi_2(1) = 0,$$
$$g_1(0) = 0,\quad g_1(1) = 1/Sh$$
$$g_2(0) = 0,\quad g_2(1) = 0 \tag{17}$$

3. SOLUTION OF THE PROBLEM

The nonlinear equations (11), (12), (13), (14) and (15) are converted into the following system of first order differential equations by the substitution

$$(f, f^{/}, f^{//}, f^{///}, f^{IV}, f^{V}, \varphi_1, \varphi_1^{/}, \varphi_2, \varphi_2^{/}, g_1, g_1^{/}, g_2, g_2^{/}) =$$
$$(x_1, x_2, x_3, x_4, x_5, x_6, x_7, x_8, x_9, x_{10}, x_{11}, x_{12}, x_{13}, x_{14})$$

$$\frac{dx_1}{d\xi} = x_2,\ \frac{dx_2}{d\xi} = x_3,\ \frac{dx_3}{d\xi} = x_4,\ \frac{dx_4}{d\xi} = x_5,\ \frac{dx_5}{d\xi} = x_6,$$

$$\frac{dx_6}{d\xi} = \frac{1}{\alpha^2}(\xi\beta x_4 + 3\beta x_3 + \mathrm{Re}(x_2 x_3 - x_1 x_4) + x_5 - D^{-1} x_3$$

$$-\frac{Ha^2\,\alpha e}{\alpha e^2 + \beta e^2} x_3 - \frac{EGr}{\mathrm{Re}\,\zeta}(x_8 + \zeta^2 x_{12}) - \frac{ShGm}{\mathrm{Re}\,\zeta}(x_{12} + \zeta^2 x_{14})),$$

$$\frac{dx_7}{d\xi} = x_8,$$

$$\frac{dx_8}{d\xi} = -2x_9 - \frac{\mathrm{Re}\,\mathrm{Pr}}{1-DuScSr}(4x_2^2 + \alpha^2 x_3^2 + D^{-1}x_1^2 +$$

$$\frac{Ha^2}{\alpha e^2 + \beta e^2}x_1^2 + \frac{\beta\xi x_8}{\mathrm{Re}} + \frac{\beta x_7}{\mathrm{Re}} - x_1 x_8\ ) - \frac{Kr\,Du}{1-DuScSr}x_{11} +$$

$$\frac{Du\,Sc}{1-DuScSr}(-\beta x_{11} - \xi\beta x_{12} + \mathrm{Re}\,x_1 x_{12}\ ),$$

$$\frac{dx_9}{d\xi} = x_{10},$$

$$\frac{dx_{10}}{d\xi} = -\frac{\mathrm{Re}\,\mathrm{Pr}}{1-DuScSr}(x_3^2 + \alpha^2 x_4^2 + D^{-1}x_2^2 + \frac{Ha^2}{\alpha e^2 + \beta e^2}x_2^2$$

$$+\frac{\xi\beta x_{10}}{\mathrm{Re}} + \frac{3\beta x_9}{\mathrm{Re}} + 2x_2 x_9 - x_1 x_{10}\ ) -$$

$$\frac{Kr\,Du}{1-DuScSr}x_{13} + \frac{Du\,Sc}{1-DuScSr}(-3\beta x_{13} - \xi\beta x_{14} + \mathrm{Re}(x_1 x_{14} - 2x_2 x_{13}\ ))$$

$$\frac{dx_{11}}{d\xi} = x_{12},$$

$$\frac{dx_{12}}{d\xi} = -2x_{13} + \frac{ScSr\,\mathrm{Re}\,\mathrm{Pr}}{1-DuScSr}(4x_2^2 + \alpha^2 x_3^2 + D^{-1}x_1^2 +$$

$$\frac{Ha^2}{\alpha e^2 + \beta e^2}x_1^2 + \frac{\beta\xi x_8}{\mathrm{Re}} + \frac{\beta x_7}{\mathrm{Re}} - x_1 x_8\ ) + \frac{Kr}{1-DuScSr}x_{11} +$$

$$\frac{Sc}{1-DuScSr}(-\beta x_{11} - \xi\beta x_{12} + \mathrm{Re}\,x_1 x_{12}\ ), \quad \frac{dx_{13}}{d\xi} = x_{14},$$

$$\frac{dx_{14}}{d\xi} = \frac{ScSr\,\mathrm{Re}\,\mathrm{Pr}}{1-DuScSr}(x_3^2 + \alpha^2 x_4^2 + D^{-1}x_2^2 + \frac{Ha^2}{\alpha e^2 + \beta e^2}x_2^2 + \frac{\xi\beta x_{10}}{\mathrm{Re}}$$

$$+\frac{3\beta x_9}{\mathrm{Re}} + 2x_2 x_9 - x_1 x_{10}\ ) + \frac{Kr}{1-DuScSr}x_{13} +$$

$$\frac{Sc}{1-DuScSr}(-3\beta x_{13} - \xi\beta x_{14} + \mathrm{Re}(x_1 x_{14} - 2x_2 x_{13}\ )) \quad (18)$$

The boundary conditions in terms of $x_1, x_2, x_3, x_4, x_5, x_6, x_7, x_8, x_9, x_{10}, x_{11}, x_{12}, x_{13}, x_{14}$ are

$$x_1(0) = 0,\ x_2(0) = 0,\ x_3(0) = 0,\ x_7(0) = 0,\ x_9(0) = 0,$$

$$x_{11}(0) = 0,\ x_{13}(0) = 0,$$

$$x_1(1) = 1,\ x_2(1) = 0,\ x_3(1) = 0,\ x_7(1) = 1/Ec,\ x_9(1) = 0,$$

$$x_{11}(1) = 1/Sh,\ x_{13}(1) = 0 \quad (19)$$

The system of equations (18) is solved numerically subject to the boundary conditions (19) using quasilinearization method given by Bellman and Kalaba (1965).

4. SKIN FRICTION

Using equation (5), the shear stress τ is

$$\tau = \mu\frac{\partial u}{\partial y} \quad (20)$$

Then the coefficient of skin friction on the lower plate is given by

$$S_f = \left(\frac{2\tau}{\rho V_1^2}\right)_{\xi=0} = \frac{2}{\mathrm{Re}}\frac{x}{a}f''(0) \quad (21)$$

5. RESULTS AND DISCUSSION

The reduced flow field equations (11) to (15) are fully coupled nonlinear ordinary differential equations that is why we cannot solve analytically. Therefore, solving the above nonlinear equations using quasilinearization method, we obtained the solutions of non dimensional velocity components, temperature distribution and concentration for various fluid and geometric parameters such as Soret number Sr, Dufour number Du, Hall parameter βe, ion slip parameter βi, chemical reaction parameter Kr, thermal Grashof number Gr and solutal Grashof number Gm and presented in the form of graphs.

The effects of Gr and Gm are shown in the Figs. 2 and 3. The Gr and Gm have the relative effect of the thermal and solutal buoyancy force to the viscous hydrodynamic force at the boundary layer. Hence their effects on velocity components and temperature are one and same. i.e., When Gr and Gm are increasing the radial velocity is also increasing, whereas the temperature distribution is decreasing and the axial velocity increases towards the center of the walls then decrease.

Figures 4 and 5 show the effects of βe and βi on velocity components, temperature distribution and concentration and from these it is analyzed that the concentration is increasing, whereas the radial velocity and temperature distribution are decreasing with the increasing of βe and βi. However, the axial velocity decreases towards the center of the plane then increases. It is because of the increase in the Hall and ion slip parameters reducing the effect of Lorentz force.

Figures 6 and 7 depict the variations of Sr and Du on temperature distribution and concentration. When Sr and Du increase the temperature distribution also increases, whereas the concentration decreases. This is because of an increase in Sr and Du increase the difference between the temperature of the fluid and the surface temperature and also increase in the difference between the surface concentration and the concentration of the fluid.

The effect of Kr on temperature distribution and concentration are presented in the Fig. 8. It is observed that when Kr is increasing the temperature distribution is also increasing, whereas concentration is decreasing towards the upper wall. It is clear that the increase in the Kr produces a decrease in the species concentration. This causes the concentration buoyancy effects to decrease as Kr increases. In addition the concentration boundary layer thickness decreases as Kr increases.

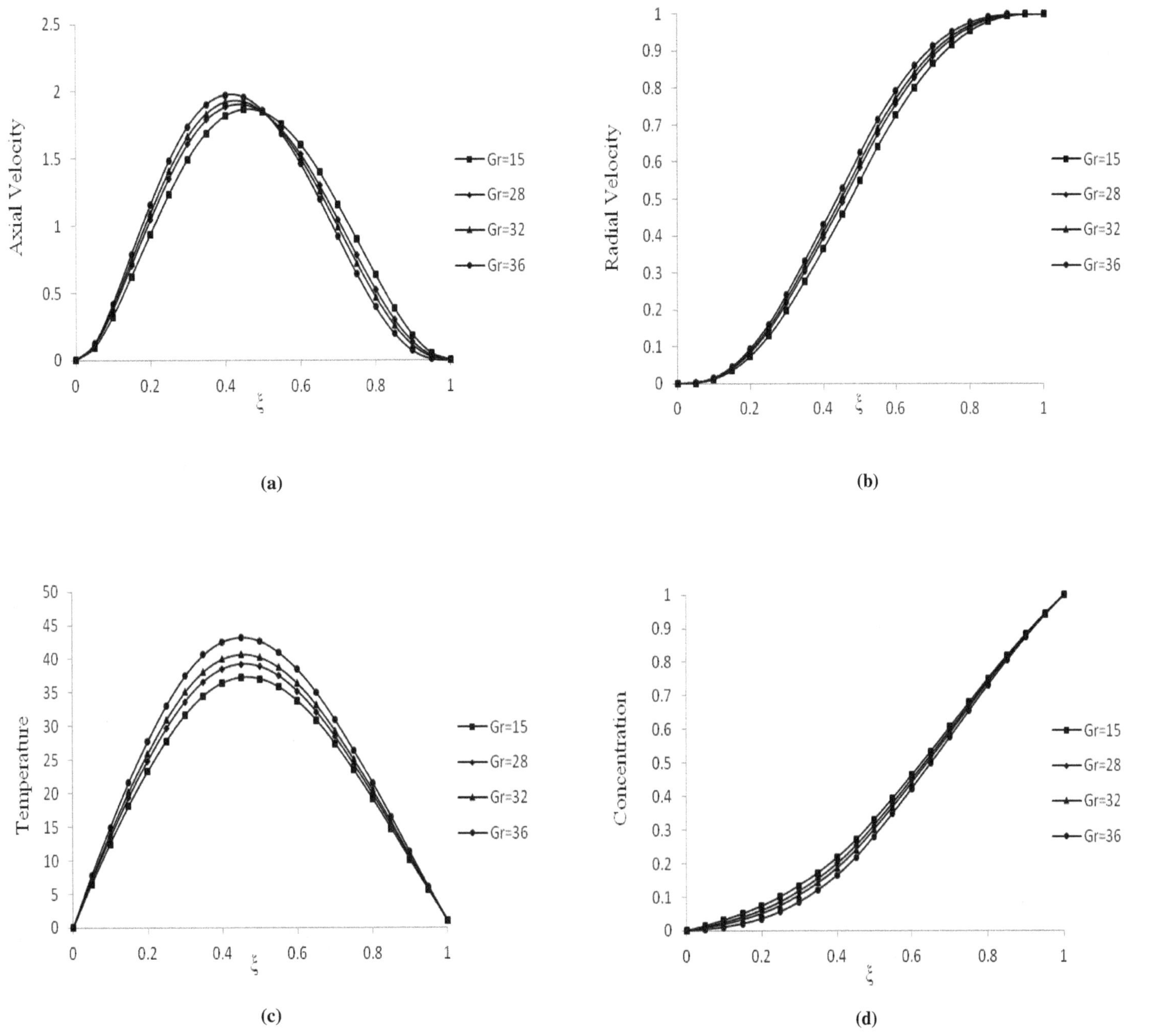

Fig. 2 Effect of Gr on **(a)** Axial Velocity, **(b)** Radial Velocity, **(c)** Temperature and **(d)** Concentration for Kr=2, βi=0.2, Gm=20, Sr=0.02, Du=0.2, βe=0.2, Sc=0.22, Pr=0.2, Re=2, D^{-1}=2, β=10, α=0.5, Ha=10.

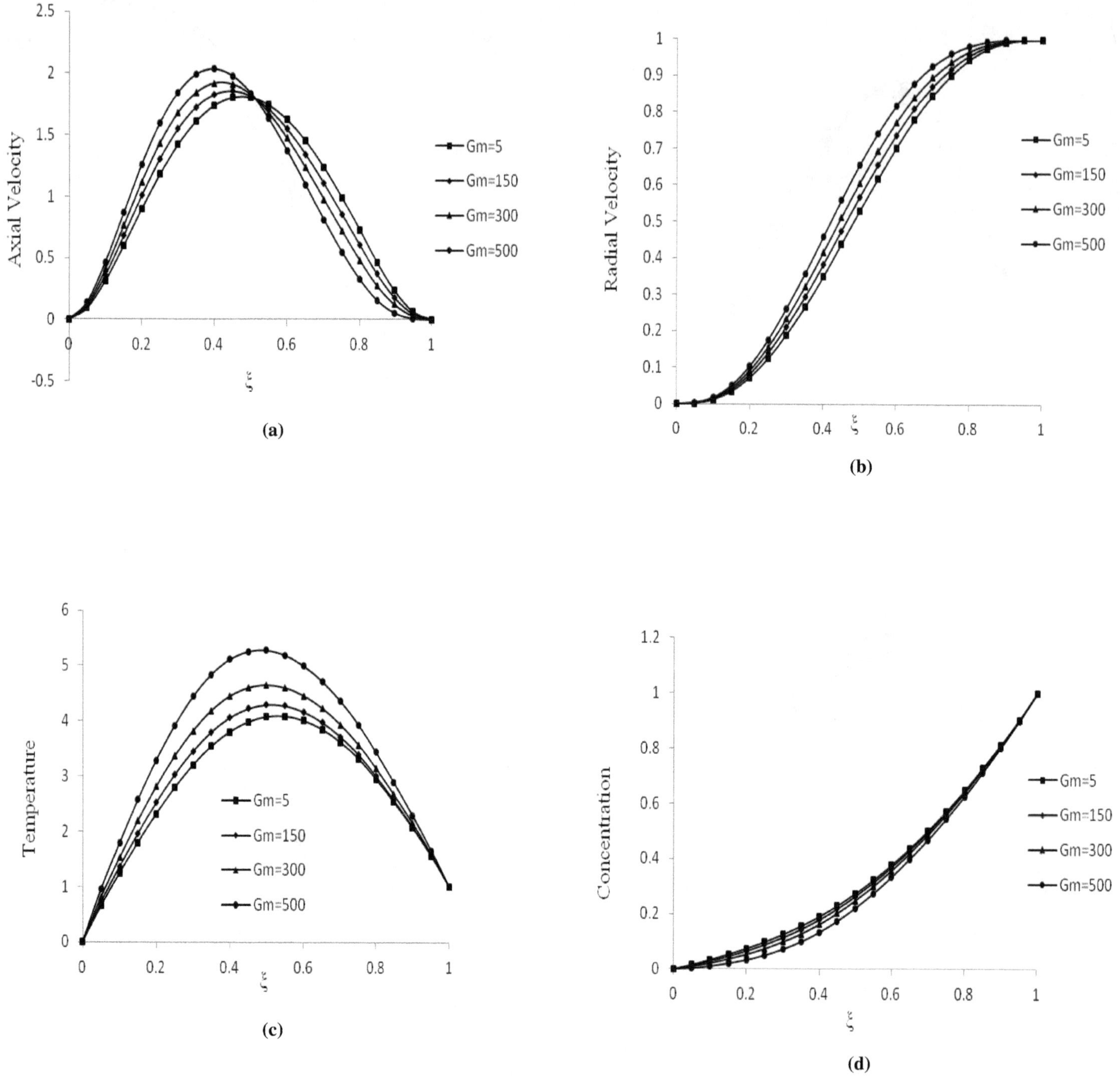

Fig. 3 Effect of Gm on (**a**) Axial Velocity, (**b**) Radial Velocity, (**c**) Temperature and (**d**) Concentration for Kr=2, βi=0.2, Gr=25, Sr=0.2, Du=0.2, βe=0.2, Sc=0.2, Pr=0.2, Re=2, D^{-1}=10, β=2, α=0.2, Ha=2.

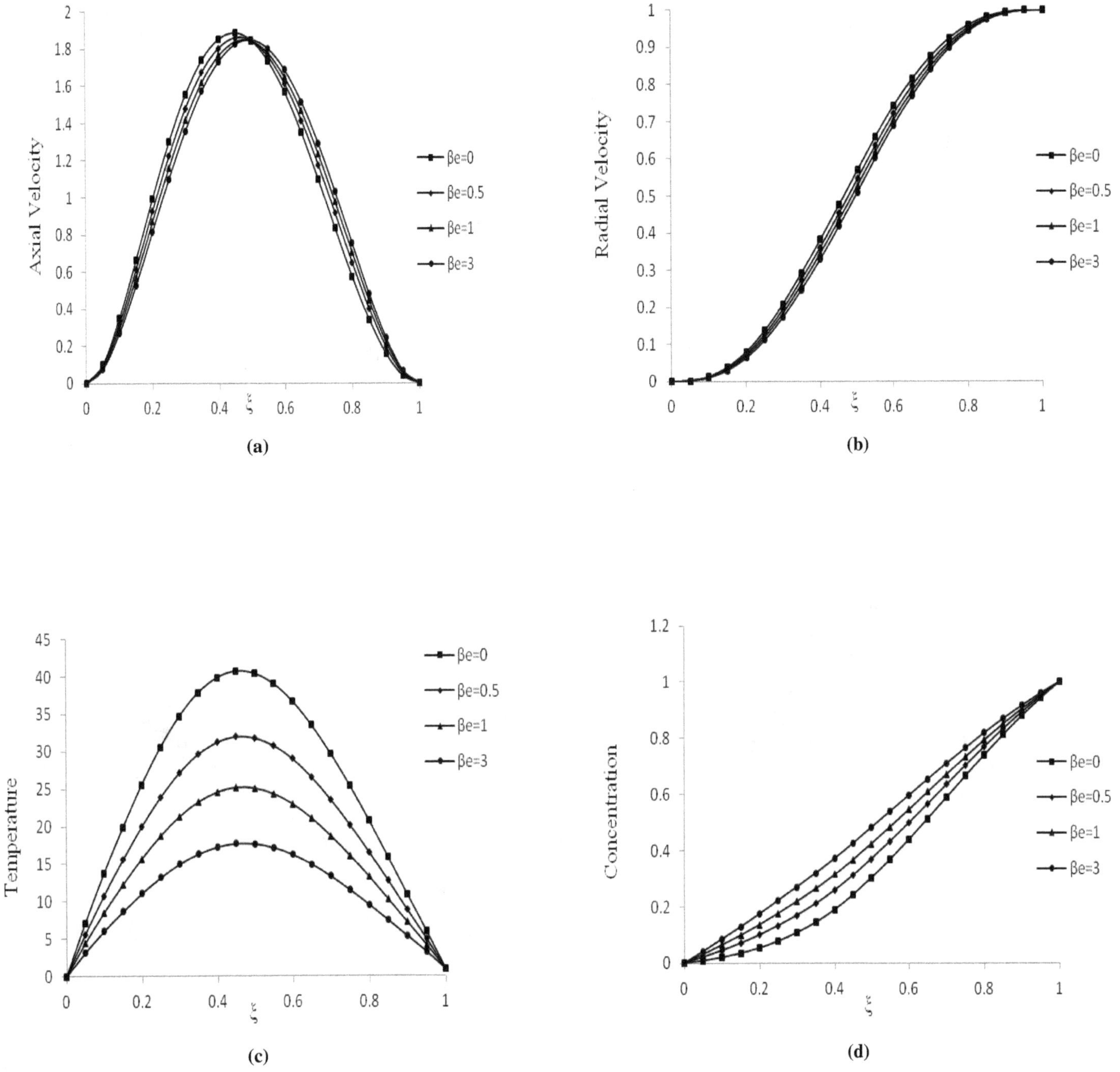

Fig. 4 Effect of βe on **(a)** Axial Velocity, **(b)** Radial Velocity, **(c)** Temperature and **(d)** Concentration for Kr=2, βi=0.2, Gr=20, Sr=0.02, Du=0.2, Gm=20, Sc=0.22, Pr=0.2, Re=2, D^{-1}=2, β=10, α=0.5, Ha=10.

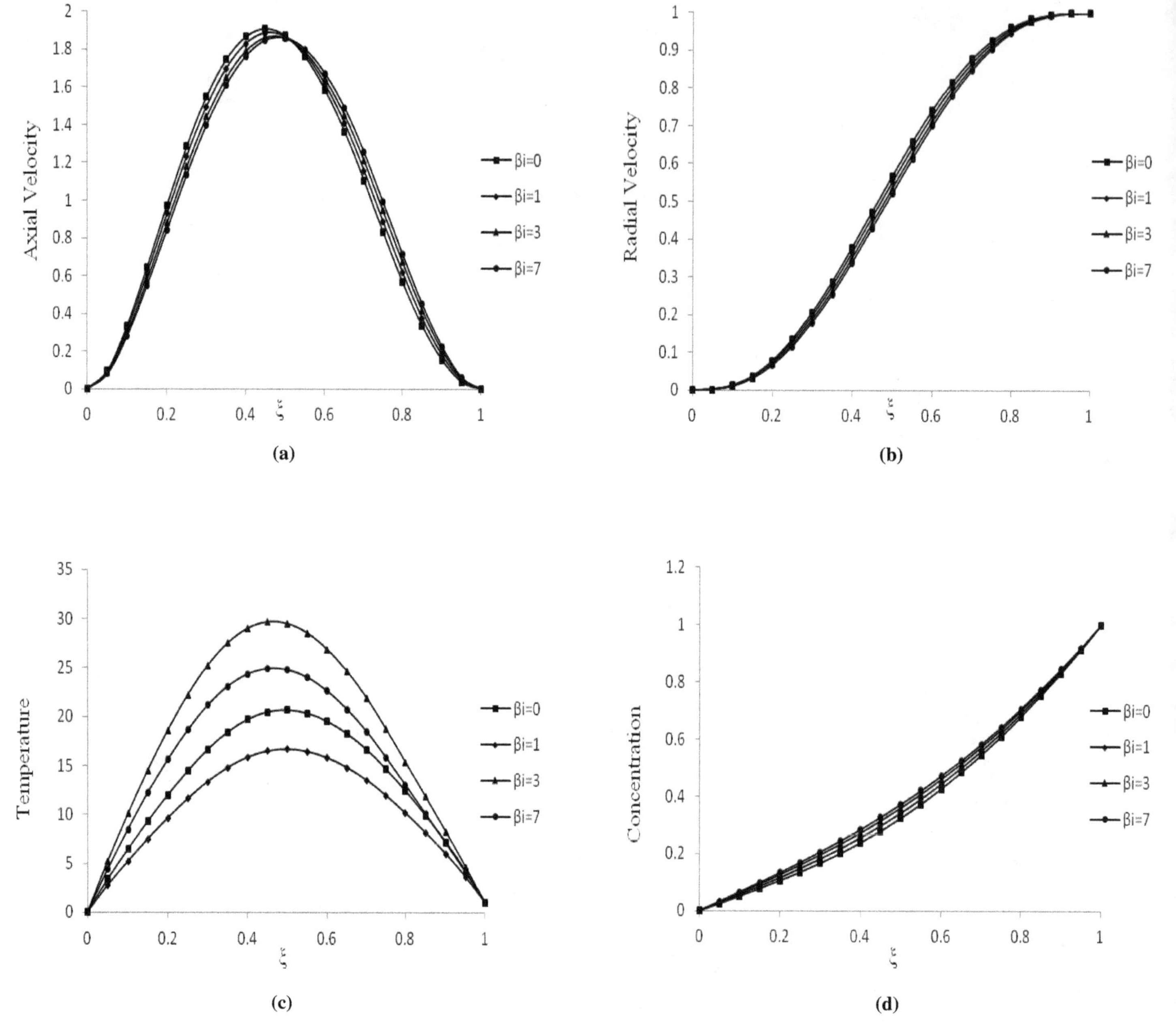

Fig. 5 Effect of βi on (**a**) Axial Velocity, (**b**) Radial Velocity, (**c**) Temperature and (**d**) Concentration for Kr=2, βe=0.2, Gr=20, Sr=0.02, Du=0.2, Gm=20, Sc=0.22, Pr=0.2, Re=2, D^{-1}=2, β=10, α=0.5, Ha=10

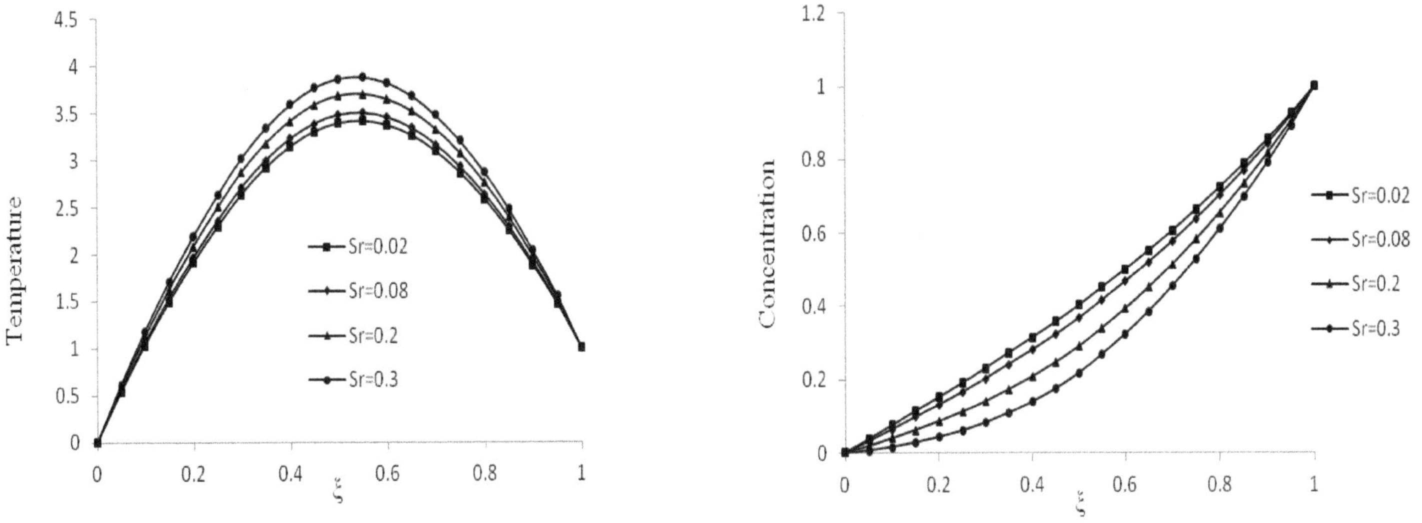

(a) (b)

Fig. 6 Effect of Sr on **(a)** Temperature and **(b)** Concentration for Kr=2, βe=0.2, Gr=10, βi =0.2, Du=2, Gm=10, Sc=0.2, Pr=0.2, Re=2, D^{-1}=2, β=2, α=0.2, Ha=2.

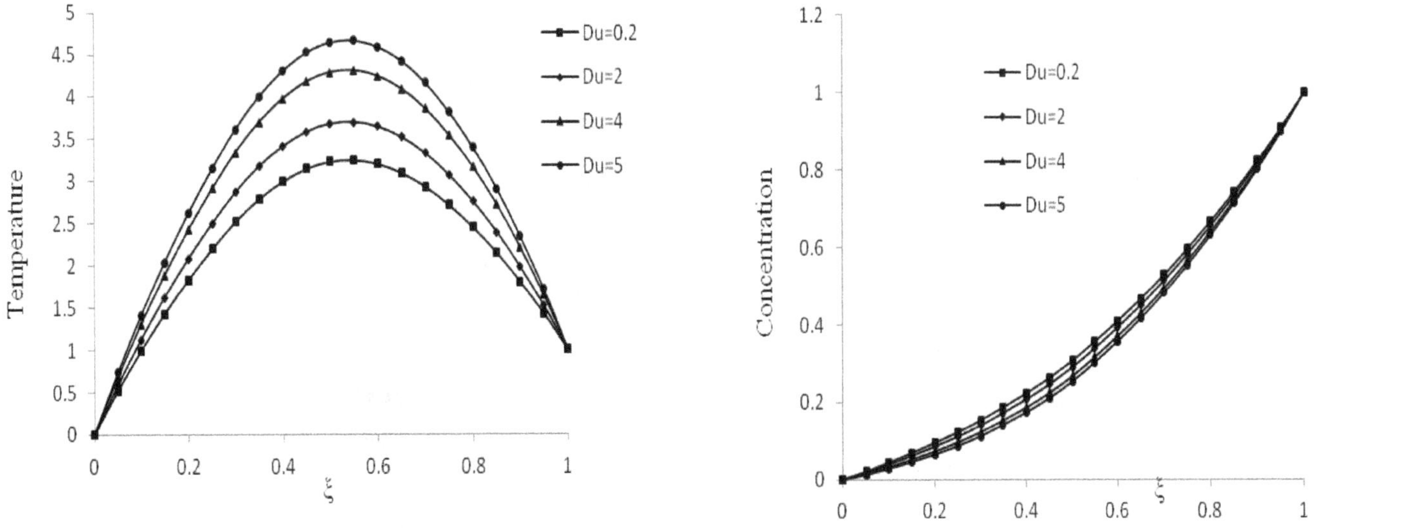

(a) (b)

Fig. 7 Effect of Du on **(a)** Temperature and **(b)** Concentrationfor Kr=2, βe=0.2, Gr=10, βi =0.2, Sr=0.2, Gm=10, Sc=0.2, Pr=0.2, Re=2, D^{-1}=2, β=2, α=0.2, Ha=2

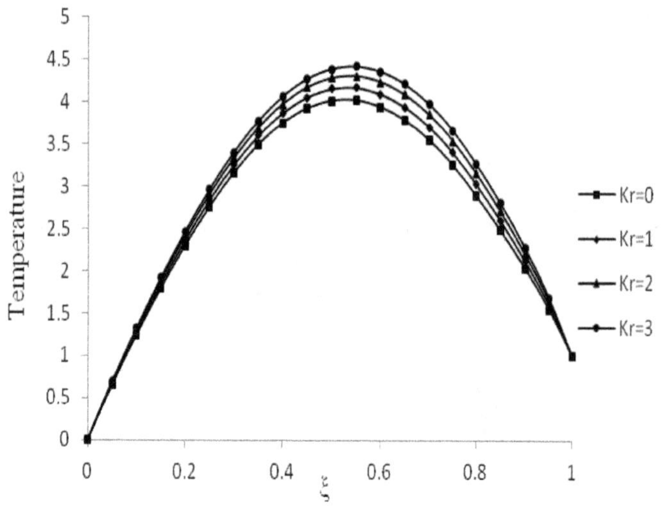

(a)

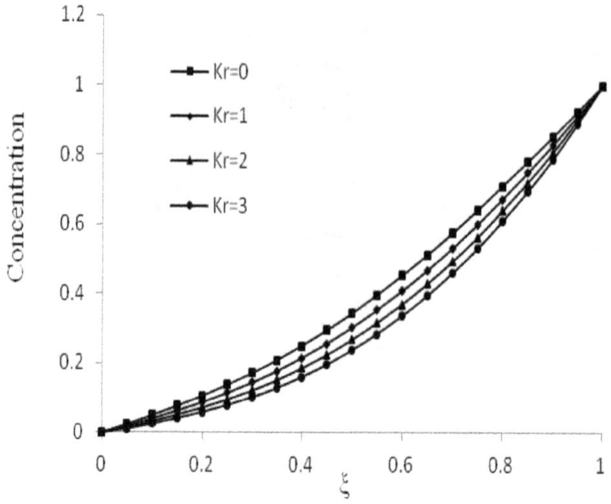

(b)

Fig. 8 Effect of Kr on **(a)** Temperature and **(b)** Concentration for Du=4, βe=0.2, Gr=10, βi =0.2, Sr=0.2, Gm=10, Sc=0.2, Pr=0.2, Re=2, D^{-1}=2, β=2, α=0.2, Ha=2.

6. CONCLUSIONS

The influence of the Hall and ion slip on an unsteady free convection heat and mass transfer of an electrically conducting couple stress fluid in a porous medium between expanding or contracting walls with chemical reaction, Soret and Dufour effects is considered. The reduced governing equations are solved numerically by the method of quasilinearization. The results are presented in the form of graphs for different values of fluid and geometric parameters and from these it is concluded that

- The influence of Sr and Du on temperature and concentration are similar.
- Kr reduces the concentration and enhances the temperature of the fluid.
- The velocity components and temperature distribution are exhibiting the same effects for Gr and Gm.
- The parameters Sr and Kr show the opposite effect for temperature and concentration.

NOMENCLATURE

t	Time
a(t)	Distance between parallel plates
V_1	Suction velocity
Pr	Prandtl number, $\mu c / k$
Re	Reynolds number, $\rho V_1 a / \mu$
$\bar{J}$	Current density
$\bar{B}$	Total magnetic field
$\bar{b}$	Induced magnetic field
B_0	Magnetic flux density
$\bar{E}$	Electric field
Ha	Hartmann number, $B_0 a \sqrt{\sigma / \mu}$
D^{-1}	Inverse Darcy parameter, a^2 / k_1
p	Fluid pressure
$\bar{Q}$	Velocity vector
c	Specific heat at constant temperature
Ec	Eckert number, $\mu V_1 / [\rho a c (T_2 - T_1)]$
k	Thermal conductivity
k_1	Permeability Parameter
u	Velocity component in x-direction
v	Velocity component in y-direction
T	Temperature
T_1	Temperature of the lower plate
T_2	Temperature of the upper plate
T^*	Dimensionless temperature, $(T - T_1)/(T_2 - T_1)$
D	Rate of deformation tensor
C	Concentration
D_1	Mass diffusivity
k_2	Chemical reaction rate
Kr	Non dimensional chemical reaction parameter, $k_2 a^2 / D_1$
Sc	Schmidt number, υ / D_1
Gr	Thermal Grashof number, $\rho g \beta_T (T_2 - T_1) a^3 / \upsilon^2$
Gm	Solutal Grashof number, $g \beta_C (C_1 - C_0) a^3 / \upsilon^2$
Sh	Sherwood number, $\dot{n}_A / [a \upsilon (C_1 - C_0)]$
$\dot{n}_A$	Mass transfer rate
Sr	Soret number, $D_1 k_T \upsilon V_1 / (c T_m \dot{n}_A)$
Du	Dufour number, $D_1 k_T \dot{n}_A \rho c / (\upsilon^2 V_1 c_s k)$
T_m	Mean temperature

k_T Thermal-diffusion ratio

c_s Concentration susceptibility

$\overline{F_B}$ Buoyancy force

$\hat{i}, \hat{j}, \hat{k}$ Unit vectors

Greek Symbols

ξ Dimensionless y coordinate, y / a

ζ Dimensionless axial variable, x / a

ρ Fluid density

μ Fluid viscosity

μ' Magnetic permeability

σ Electric conductivity

α Couple stress parameter, $\sqrt{\eta / (\mu a^2)}$

β_T Coefficient of thermal expansion

β_C Coefficient of solutal expansion

βi Ion slip parameter

βe Hall parameter

αe Hall and ion-slip parameter, 1+ βi βe

β Wall expansion ratios, $\dot{a}\, a / \upsilon$

REFERENCES

Vijay Kumar Stokes, 1966, "Couple Stresses in Fluids," *Phys of Fluids,* **9**, 1709-1715.
http://dx.doi.org/10.1063/1.1761925

Stokes V.K., 1968, "Effects of Couple Stress in Fluid on Hydromagnetic Channel Flow," *Phys.of Fluids*, **11**(5)**,** 1131-1133.
http://dx.doi.org/10.1063/1.1692056

Srinivasacharya D., Srinivasacharyulu N., Odelu Ojjela, 2009, "Flow and Heat Transfer of Couple Stress Fluid in a Porous Channel with Expanding and Contracting walls," *International Communications in Heat and Mass Transfer*, **36** (2), 180-185.
http://dx.doi.org/10.1016/j.icheatmasstransfer.2008.10.005

Najeeb Alam Khan, Amir Mahmood, Asmat Ara, 2013, "Approximate Solution of Couple Stress Fluid with Expanding or Contracting Porous Channel," *International Journal for Computer-Aided Engineering and Software,* **30**(3), 399-408.
http://dx.doi.org/10.1108/02644401311314358

Xin-hui SI, Lian-cun zheng, Xin-xin zhang, Ying chao, 2010, "Perturbation Solution to Unsteady Flow in a Porous Channel with Expanding or Contracting Walls in the Presence of a Transverse Magnetic Field," *Applied Mathematics and Mechanics -English Edition,* **31**(2), 151-158.
http://dx.doi.org/10.1007/s10483-010-0203-z

Xin-Hui Si, Lian-cun Zheng, Xin-Xin Zhang, Ying Chao, 2011, "Homotopy Analysis Solution for the Asymmetric Laminar Flow in a Porous Channel with Expanding or Contracting Walls," *Acta Mechanica Sinica,* **27** (2), 208–214.
http://dx.doi.org/10.1007/s10409-011-0430-3

Uchida S., Aoki H., 1977, "Unsteady Flows in a Semi-infinite Contracting or Expanding Pipe," *Journal of Fluid Mechanics,* **82,** 371–381.
http://dx.doi.org/10.1017/S0022112077000718

Srinivas S., Subramanyam Reddy A., Ramamohan T.R., 2012, "A Study on Thermal-diffusion and Diffusion-thermo Effects in a Two-dimensional Viscous Flow between Slowly Expanding or Contracting Walls with Weak Permeability," *International Journal of Heat and Mass Transfer,* **55**, 3008–3020.
http://dx.doi.org/10.1016/j.ijheatmasstransfer.2012.01.050

Srinivasacharya D., Kaladhar K., 2011, "Mixed Convection in a Couple Stress Fluid with Soret And Dufour Effects," *International Journal of Applied Mathematics and Mechanics,* **7**(20), 59-71.

Malasetty M.S., Gaikwad S.N., Swamy M., 2006, "An Analytical Study of Linear and Non-linear Double Diffusive Convection with Soret Effect in Couple Stress Liquids," *International Journal of Thermal Sciences,* **45**, 897–907.
http://dx.doi.org/10.1016/j.ijthermalsci.2005.12.005

Dursunkaya Z., Worek W.M., 1992, "Diffusion-thermo and Thermal-diffusion Effects in Transient and Steady Natural Convection from Vertical Surface," *International Journal of Heat and Mass Transfer,* **35**, 2060–2065.
http://dx.doi.org/10.1016/0017-9310(92)90208-A

Mahmoud M.A.A., Megahed A.M., 2013, "Thermal Radiation Effect on Mixed Convection Heat and Mass Transfer of a Non-Newtonian Fluid over a Vertical Surface Embedded in a Porous Medium in the Presence of Thermal-diffusion and Diffusion-thermo Effects, *Journal of Applied Mechanics and Technical Physics*, **54**(1), 90–99.
http://dx.doi.org/10.1134/S0021894413010112

Nithyadevi N., Ruey- Jen Yang, 2009, "Double Diffusive Natural Convection in a partially Heated Enclosure with Soret and Dufour Effects," *International Journal of Heat and Fluid Flow,* **30**, 902–910.
http://dx.doi.org/10.1016/j.ijheatfluidflow.2009.04.001

Adrian Postelnicu, 2004, "Influence of a Magnetic Field on Heat and Mass Transfer by Natural Convection from Vertical Surfaces in Porous Media considering Soret and Dufour Effects," *International Journal of Heat and Mass Transfer,* **47**, 1467–1472.
http://dx.doi.org/10.1016/j.ijheatmasstransfer.2003.09.017

Dulal Pal, Hiranmoy Mondal, 2011, "MHD Non-Darcian Mixed Convection Heat and Mass Transfer over a Non-linear Stretching Sheet with Soret–Dufour Effects and Chemical Reaction," *International Communications in Heat and Mass Transfer*, **38**, 463-467.
http://dx.doi.org/10.1016/j.icheatmasstransfer.2010.12.039

Ganesan P., Suganthi R.K., Loganathan P., 2012, "Soret and Dufour Effects in a Free Convective Doubly Stratified Flow over a Vertical Plate with Chemical Reaction,"*Chemical Engineering Communications,* **200**, 514–531.
http://dx.doi.org/10.1080/00986445.2012.712580

Ching-Yang Cheng, 2010, "Soret and Dufour Effects on Heat and Mass Transfer by Natural Convection from a Vertical Truncated Cone in a Fluid-saturated Porous Medium with Variable Wall Temperature and Concentration," *International Communications in Heat and Mass Transfer,* **37**, 1031–1035.
http://dx.doi.org/10.1016/j.icheatmasstransfer.2010.06.008

Hayat T., Nawaz M., 2011, "Soret and Dufour Effects on the Mixed Convection Flow of a Second Grade Fluid subject to Hall and Ion-slip Currents," *International Journal of Numerical Methods in Fluids,* **67**, 1073–1099.
http://dx.doi.org/10.1002/fld.2405

Nasser S. Elgazery, 2009, "The Effects of Chemical Reaction, Hall and Ion-slip Currents on MHD Flow with Temperature Dependent Viscosity and Thermal Diffusivity," *Communications in Nonlinear Science and Numerical Simulation,* **14** (4), 1267-1283.
http://dx.doi.org/10.1016/j.cnsns.2007.12.009

Ali Chamkha, Mansour M.A., Abdelraheem Aly, 2011, "Unsteady MHD Free Convective Heat and Mass Transfer from a Vertical Porous Plate with Hall Current, Thermal Radiation and Chemical Reaction Effects," *International Journal of Numerical Methods in Fluids,* **65**, 432–447.
http://dx.doi.org/10.1002/fld.2190

Patil P.M., Kulkarni P.S., 2008, "Effects of Chemical Reaction on Free Convective Flow of a Polar Fluid through a Porous Medium in the Presence of Internal Heat Generation," *International Journal of Thermal Sciences,* **47**, 1043–1054.
http://dx.doi.org/10.1016/j.ijthermalsci.2007.07.013

Bhatnagar R., Vayo H.W., Okunbor D., 1994, "Application of Quasilinearization to Viscoelastic Flow through a Porous Annulus," *International Journal of Non-Linear Mechanics,* **29**(1), 13-22.
http://dx.doi.org/10.1016/0020-7462(94)90048-5

Bellman R.E., Kalaba R.E., 1965, *Quasilinearization and Boundary-value Problems,* Elsevier publishing Co. Inc., New York.

Eckert E.R.G., Drake R.M., 1972, *Analysis of Heat and Mass Transfer,* McGraw-Hill, New York.

Sutton G.W., Sherman A., 1965, *Engineering Magnetohydrodynamics,* McGrawhill, New York.

13

CONTACT ANGLE MEASUREMENTS FOR ADVANCED THERMAL MANAGEMENT TECHNOLOGIES

Sally M. Smith* Brenton S. Taft†, Jacob Moulton

Air Force Research Laboratory Space Vehicles Directorate, Kirtland AFB, NM 87117, USA

ABSTRACT

This study investigates the wettability of fluid-solid interactions of interest for oscillating heat pipe (OHP) applications. Measurements were taken using two techniques: the sessile drop method and capillary rise at a vertical plate. Tested surface materials include copper, aluminum, and Teflon PFA. The working fluids tested were water, acetone, R-134a, and HFO-1234yf. A novel low-pressure experimental setup was developed for refrigerant testing. Results show that the refrigerants have significantly lower hysteresis than the water and acetone-based systems, which is thought to lead to better heat transfer in OHP design.

Keywords: *Wettability, Hysteresis, Interfacial Phenomena, Refrigerants, Oscillating Heat Pipe.*

1. INTRODUCTION

1.1 Problem Description

This study investigates the wettability of fluid-solid interactions for oscillating heat pipe (OHP) applications. Presented here are the methods and procedures for the experiments conducted, and a discussion of the results. Materials studied were substrates of mill-finish aluminum (alloy 6061), copper (alloy 101), and Teflon PFA. Working fluids used were distilled water, acetone, R-134a (1,1,1,2-Tetrafluoroethane), and HFO-1234yf (2,3,3,3-Tetrafluoropropene).

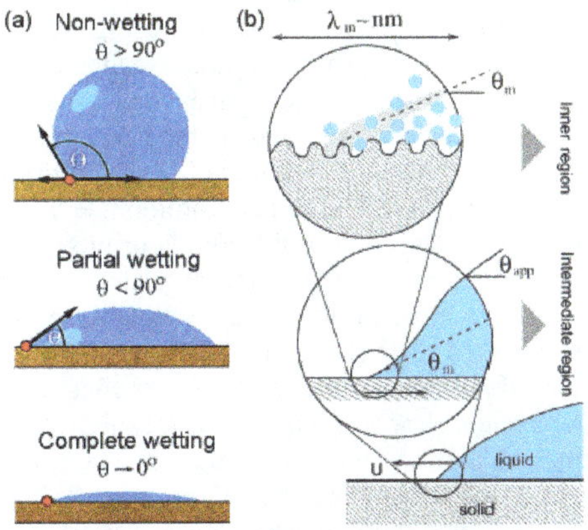

Fig. 1 (a) Contact angle of a fluid on solid substrate. (b) The moving meniscus seen on three different length scales (Khandekar *et al.*, 2010).

1.2 Relevance of Contact Angle to OHPs

Contact angle hysteresis is an important parameter in OHP performance. During OHP operation, the advancing and receding contact angles of liquid plugs change with filling ratio, working fluid, capillarity diameter, transfer power, and capillary length. Some results (Taft *et al.*, 2012) have demonstrated that increasing contact angle hysteresis negatively affects the heat transfer of the OHP, while other theories (Qu and Wu, 2011) suggest a decreasing Young contact angle, θ_0, leads to decreased active nucleation site density and deteriorated boiling heat transfer at the evaporator. It is not clear whether contact angle hysteresis or the Young contact angle has greater influence on OHP performance, or if they work collectively to reduce heat transfer. More research is needed before these phenomena can be well understood.

Unfortunately, the dynamic contact angle of varying working fluid/substrate combinations is not possible to estimate, and the database of measured contact angles is nearly an empty set. Dynamic contact angles are, however, worth considering for working fluid/substrate selection if the dynamic contact angle hysteresis is known, or can be measured (Qu *et al.*, 2003).

Table 1 Contact angle and strength of interactions (Khandekar *et al.*, 2010).

Contact Angle	Degree of Wetting
0°	Perfect wetting
$0° < \theta < 90°$	High wettability
$90° \leq \theta < 180°$	Low wettability
180°	Perfectly non-wetting

2. BACKGROUND

2.1 Contact Angles

The contact angle of a fluid-solid interaction characterizes the wettability of a solid surface by a liquid. Liquid with a small contact angle has high wettability, and will spread on the solid surface. The

* Currently at NASA Wallops Flight Facility, Wallops Island, VA, 23337, USA
† *Corresponding author. Email: afrl.rvsv@kirtland.af.mil*

contact angle of a liquid on a solid substrate depends on the roughness and the chemical homogeneity of the surface. The three-phase (solid/liquid/vapor) contact line is deformed due to physical and chemical heterogeneities (Zhang *et al.*, 2004).

The surface contact angle, θ, is a generic term that can describe a variety of angles a drop can make with a surface (Tadmor and Yadav, 2008). In this study, we will discuss four contact angles of interest: equilibrium Young contact angle, θ_0, the as-placed contact angle, θ_{AP}, the advancing contact angle, θ_A, and the receding contact angle, θ_R.

2.2 Young Equilibrium and As-Placed Contact Angles

The Young angle, θ_0, is the equilibrium contact angle of an ideal solid surface. In his original publication, Young (Schwartz, 1980) described static contact equilibrium as a balance of forces at the three-phase contact line. The current interpretation of Young's equation (Khandekar *et al.*, 2010; Diaz *et al.*., 2010; Benner *et al.*, 1982) is of a macroscopic relationship between interfacial tensions

$$\gamma_{SL} + \gamma_{LG} \cos(\theta_Y) = \gamma_{SG} \quad (1)$$

where γ_{SL} is the interfacial tension between liquid and solid states, γ_{LG} is the interfacial tension between liquid and gaseous states, and γ_{SG} is the interfacial tension between solid and gaseous states.

The determination of the Young angle is important for characterizing solid-liquid interfacial systems, because it is closely related to material properties (Marmur, 2006). The Young angle represents true thermodynamic equilibrium: mechanical, chemical and thermal. Therefore to satisfy Young's equation, one must have a surface that is chemically homogenous and perfectly smooth (Khandekar *et al.*, 2010). In practice, these requirements mean that the Young angle cannot be directly determined experimentally.

Therefore, in practice, the "static" contact angle of a non-ideal surface is reported as the as-placed contact angle, θ_{AP}, in which a drop of fluid is gently placed on the surface. This angle is highly dependent on the thermophysical properties of the liquid and the vapor, the physico-chemical structure of the solid substrate, and ambient conditions – particularly temperature and humidity (Khandekar *et al.*, 2010). There is experimental evidence that when a sessile drop is placed on a solid surface, the apparent contact angle can vary by several degrees, and often tends toward the advancing contact angle value (Butt *et al.*, 2007).

It has been shown that multiple values of contact angles can be measured on the same surface, even when that surface is smooth and homogenous down to the atomic level (Torrigiani, 2005). This variation in the as-placed contact angle value can be explained by the presence and varying thickness of an adsorbed film that develops next to the triple-phase contact line (Butt *et al.*, 2007). There is a need for careful humidity control, particularly when using water as the liquid phase (Holmes-Farley, 1985). Ambient temperature, relative humidity, vapor pressure, adsorption constants, and evaporation rates all play significant roles in the value of the as-placed contact angle, θ_{AP} (Diaz *et al.*, 2010). As such, the as-placed angle is non-unique, but is often reported as an auxilary measurement.

2.3 Hysteresis

A better method for characterizing a solid surface is to report the maximal advancing, θ_A, and minimal receding, θ_R, contact angles (Rodriquez-Valverde *et al.*, 2010). This is because θ_A and θ_R are extreme values and are considered means of obtaining thermodynamic properties (Tadmor and Yadav, 2008). These dynamic angles can also be used to estimate the Young angle, θ_0, as seen in this analysis and in earlier studies (Della Volpe *et al.*, 2002; Tadmor, 2004). If the contact angle is measured while the volume of the drop is increasing, this is called the advancing angle, θ_A, as seen in Fig. 2(a). Practically, this is done just before the three-phase contact line starts to advance. Similarly, if the angle is measured while the volume is decreasing, this is called the receding angle, θ_R, as in Fig. 2(b).

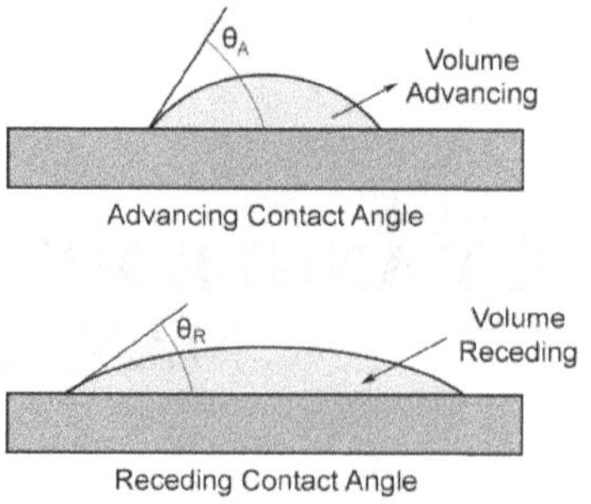

Fig. 2 (a) Advancing contact line and (b) receding contact line of a drop on a horizontal surface.

The difference between the advancing and receding angles is known as the dynamic contact angle hysteresis, $\Delta\theta$. The size of the [θ_A, θ_R] domain is usually attributed to the surface roughness of the solid substrate, and the Young angle lies somewhere within this domain (Van Mourik, 2013). Contact angle hysteresis is useful for characterizing surface roughness, heterogeneity, and mobility.

In addition to surface roughness, hysteresis is also influenced by microscopic chemical heterogeneity, drop size relative to physical topography, molecular reorientation, impurities on the surface, and the penetration of liquid molecules into the solid surfaces (Khandekar *et al.*, 2010; Erbil *et al.*, 1999). These defects change the value of the hysteresis, which is reported as an absolute value. However, the absolute values of the angular deviations of θ_A and θ_R from the Young angle θ_0 are typically different, i.e. $| \theta_A - \theta_0 | \neq | \theta_R - \theta_0 |$ (Tadmor, 2004). That is to say, the Young angle does not necessarily fall in the center of the contact angle hysteresis.

2.4 Measurement Techniques

There is a wide range of techniques used to measure the wettability of a fluid-solid interfacial interaction (Pappas *et al.*., 2013). Even for simple microscopic examination, equipment can involve a goniometer (Sklodowaka *et al.*, 1999; Gajewski, 2008), tensiometer (Extrand, 2003; Shirtcliffe *et al.*, 2004; Tang *et al.*, 2004), and CCD (charge-couple device) or digital camera (Bernardin *et al.*, 1997; Lamour and Hamraoui, 2010). For experimental setups using separate software analysis, the angle is typically measured using either a custom MATLAB script or by using one of several Java plugins.

The two most frequently reported methods of measuring contact angle through microscopic examination are the sessile drop method and the Wilhelmy plate method, but other common techniques use the mutual displacement of two immiscible fluids through a capillary, the spreading of a liquid between two parallel plates, and rotation of a cylinder partially submerged in liquid, and the capillary rise of a liquid on a partially submerged plate (Dussan, 1979).

The choice of contact angle method depends directly on the geometry of the system. For this study, two techniques were used: the sessile drop method and capillary rise at a flat plate. Because of the physical limitations of the refrigerants of interest, R-134a and HFO-1234yf, the capillary rise method was chosen for ease of integration into a vacuum setup. The sessile drop method was also performed for comparison with the capillary method, as well as an assessment of our experimental set-up using existing literature values.

2.5 Sessile Drop Method

To measure the as-placed contact angle, θ_{AP}, a liquid drop is placed on a horizontal solid surface. Fig. 3 demonstrates the basic configuration. In this study, the drop was photographed using a digital camera and measured with separate software analysis. Because drop size can vary between tests, and θ_{AP}, θ_A and θ_R are functions of drop size (Tadmor and Yadav, 2008; Amirfazli *et al.*, 1998; Yekta-Fard and Ponter, 1988; Herzberg and Marian, 1970; Good and Koo, 1979), this process was repeated several times in order to calculate an average.

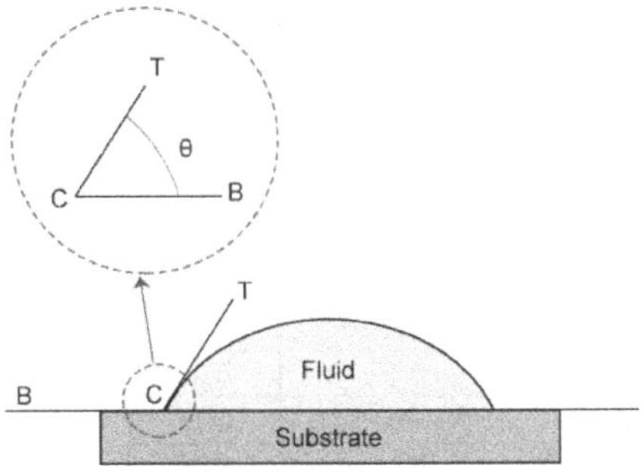

Fig. 3 Sessile drop technique of contact angle measurement. Line T denotes the tangent to a drop's profile at the triple-phase contact point, or point C. Line B denotes the baseline tangent to the substrate surface. The contact angle θ is between Line B and Line T at point C.

The sessile drop method can also be used to measure the advancing and receding contact angles and thus contact angle hysteresis. The dynamic angles can be measured by placing a liquid drop on a horizontal surface and then slowly tilting the surface; the measurement is taken just before the wetting line begins to advance, when the angles of the leading and receding edges provide the advancing and receding angles, respectively. This method has an accuracy of approximately 1 to 5° (Erbil *et al.*, 1999). However, it has been shown that θ_A and θ_R obtained by tilting the surface are functions of the tilt angle and differ from those of planar surfaces.

A second, more accurate way to obtain the dynamic angles is to use the tip of a needle or fine wire to add or remove liquid from a static sessile drop. As liquid is slowly added to the drop, the angle is repeatedly measured until the maximum advancing angle , θ_A, is obtained, or just before the wetting line begins to advance. Similarly, as liquid is removed from the drop, the minimum receding angle, θ_R, is obtained just before the wetting line begins to recede. This technique typically has a higher accuracy than the tilting plate approach (Erbil, 1999), and therefore was chosen for this study.

It should be noted that on non-ideal surfaces, wetting lines tend to continuously attach to and detach from the surface, creating an unsteady movement. This causes difficulty in both the measurement and interpretation of the contact angle (Tripathi *et al.*, 2010). To account for this variation, we repeated each individual measurement ten times to acquire an average value.

2.6 Capillary Rise Method

For a vertical, flat plate brought into contact with a pool of liquid, the liquid will rise on the plate to a height h (Fig. 4). The height of this capillary rise can be obtained from a straightforward integration of the Laplace equation of capillarity (Budziak and Neumann, 1990), as detailed in Section 3.4.

Although it is not clear if a "static" contact angle can be obtained using this technique, the dynamic contact angles are clearly accessible. The advancing angle, θ_A, can be achieved by lowering the vertical plate into the liquid, reducing the height of the capillary rise. Similarly, the receding angle, θ_R, can be found by withdrawing the plate from the liquid, raising the height of the capillary rise (Budziak and Neumann, 1990). This technique is illustrated in Fig. 5.

Typically, plate movement is achieved by attaching the plate to a motor-driven mechanism and then raising and lowering the plate into the liquid. However, for this experiment, we used a stationary plate setup and increased or decreased the volume of the liquid to raise or lower the liquid level. That is, by adding liquid we simulated a dropping plate and obtained the advancing contact angle; by removing liquid we simulated a rising plate to find the receding contact angle.

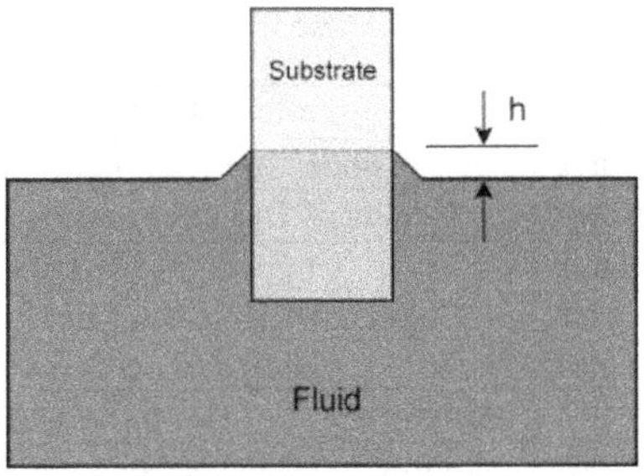

Fig. 4 Schematic of capillary rise at a vertical plate. The height h is the height of the capillary rise.

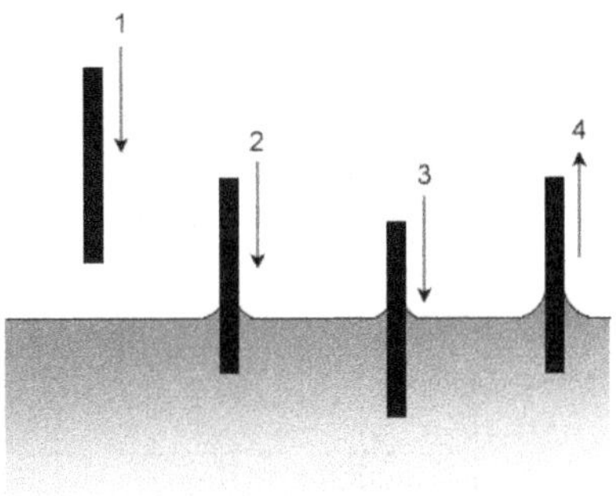

Fig. 5 A submersion cycle: (1) the sample approaches the liquid, (2) the sample is in contact with the liquid surface, (3) the liquid rises up, creating an advancing contact angle, (4) the sample is pulled up, creating a receding contact angle.

2.7 Literature Values

Relevant to this experiment is existing data on the interaction of potential working fluids (water, acetone, HFO-1234yf, and R-134a) with various OHP materials (PFA, copper, and aluminum). While there exists a significant amount of data for the interaction of water with the solid materials, the other working fluids have a nearly empty data set.

In the following tables, we report literature values for contact angle measurement with acetone and with R134a. There is no existing data available for either acetone or HFO-1234yf with any of the substrates of interest (PFA, copper, and aluminum). In Table 2, the reported values were measured using the sessile drop technique. In Table 3, measurements were taken by direct observation of the capillary rise on a vertical plate.

3. MATERIALS AND METHODS

3.1 Test Matrix

Because this experiment was intended to investigate properties of working fluids and materials for OHP applications, the following materials were chosen for contact angle measurement: substrate materials of mill-finish aluminum (alloy 6061), copper (alloy 101), and Teflon PFA; working fluids of distilled water, acetone (Univar 100%), DuPont™ Suva® R-134a (1,1,1,2-Tetrafluoroethane), and Honeywell HFO-1234yf (2,3,3,3-Tetrafluoropropene). Although water is increasingly less popular as an OHP working fluid, it was chosen as a reference liquid for comparison to literature values.

The test matrix, Table 4, details which techniques used both the sessile drop and capillary rise techniques, or only capillary rise, based on physical restrictions of the working fluids (i.e. the refrigerants are not liquid at typical room pressure and temperature).

Table 2 Literature values for as-placed contact angle (θ_{AP}) of water using sessile drop technique.

Fluid	Surface	Temp (°C)	Humidity	θ_{AP}	θ_A	θ_R
Water	Teflon PFA (Goswami *et al.*, 2008)	20 - 25	70%*	110°	110°	95°
Water	Teflon PFA (Hung *et al.*, 1999)	*N/A*	72%*	115°	*N/A*	*N/A*
Water	Teflon PFA (Extrand, 2003)	*N/A*	*N/A*	*N/A*	109°	84°
Water	Copper 101 (Shoji and Zhang, 1984)	20	71%*	71°	92°	48°
Water	Copper 101 (Yekta-Fard and Ponter, 1985)	20	100%	78°	*N/A*	*N/A*
Water	Copper 101 (Extrand, 2003)	*N/A*	80%*	69°	*N/A*	*N/A*
Water	Copper 101 (Larmour and Hamraoui, 2010)	20 – 100	72%*	9-74° **	*N/A*	*N/A*
Water	Copper 101 (Li *et al.*, 2008)	*N/A*	65%*	74°	*N/A*	*N/A*
Water	Aluminum 6061 (Larmour and Hamraoui, 2010)	50 – 150	72%*	60-90° **	*N/A*	*N/A*
Water	Aluminum 6061 (Larmour and Hamraoui, 2010)	< 120	72%*	*N/A*	90°	*N/A*
Water	Aluminum (unknown alloy) (Extrand, 2003)	*N/A*	80%*	83°	*N/A*	*N/A*
Water	Aluminum 6061 (Cayabyab *et al.*, 2013)	*N/A*	80%*	69°	*N/A*	*N/A*

* *Assumed relative humidity based on location of test*

** *Results from tests of varying temperature*

Table 3 Literature values for contact angle of R-134a using capillary rise technique

Fluid	Surface	Temp (°C)	Humidity	θ
R-134a	Copper 101 (Vadgama and Harris, 2007)	20°	72%*	6.5°
R-134a	Aluminum 3003 (Vadgama and Harris, 2007)	20°	72%*	8.1°

* *assumed relative humidity base on location of test*

Table 4 Test Matrix

	Water	Acetone	R-134a	HFO-1234yf
Aluminum (Alloy 6061)	Sessile Drop Capillary Rise	Sessile Drop Capillary Rise	Capillary Rise*	Capillary Rise*
Copper (Alloy 101)	Sessile Drop Capillary Rise	Sessile Drop Capillary Rise	Capillary Rise*	Capillary Rise*
Teflon PFA	Sessile Drop Capillary Rise	Sessile Drop Capillary Rise	Capillary Rise*	Capillary Rise*

* *performed under vacuum*

3.2 Sessile Drop Procedure

The experimental facility for the sessile drop technique consisted of a camera-based setup, as seen in Fig. 6. Images were captured using a Canon 30D digital camera with an EFS 60 mm macro lens (1:2.8 USM). A diffuser was placed between the lamp and the sample to minimize heat input and to provide a uniformly bright background of light.

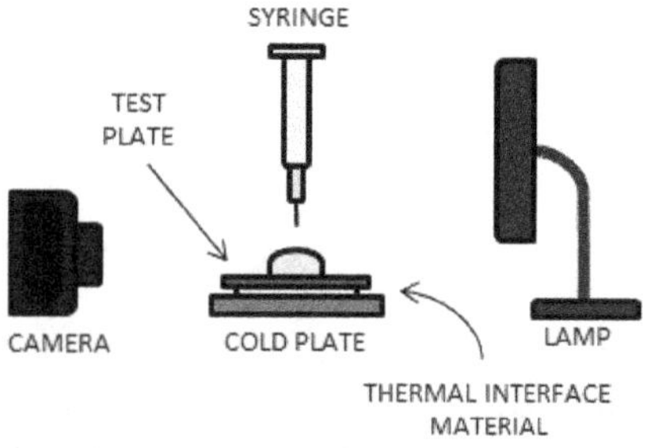

Fig. 6 Experimental setup for sessile drop capture.

The temperature of the solid surface was controlled using a cold plate and thermal pad (Parker Chomerics CHO-THERM 1671), with water circulating through the cold plate at 20°C. The solid substrate, thermal pad, and cold plate were clamped together to reach the minimum pressure required for the thermal pad to be effective.

Drops were gently placed on the sample using a 30 mL Luer-Lok™ syringe with a 25G x 1" Turemo® needle. Drop size was measured to be an average 0.040 mL for water, and 0.022 mL for acetone. Between tests, the samples were cleaned with isopropyl and distilled water. This prevented the build-up of residue, particularly from the acetone drops.

For advancing and receding contact angles, liquid was either added or removed from the drop using the syringe and needle. The camera continuously took photos during this process at an average rate of 4.2 photos per second.

3.3 Capillary Rise Procedure

The experimental setup for the capillary rise procedure used the same camera and lens as the sessile drop procedure, and included the lamp with diffuser for a high contrast background. However, plate alignment was changed to vertical and the plate was suspended in a clear beaker of fluid, as in Fig. 7. The syringe attached to a stand and a long plastic tube was attached to the syringe needle and then secured to the interior surface of the beaker for stability. This kept fluid movement at the base of the beaker to minimize interference with the fluid surface. Liquid was added to the beaker for an advancing contact angle, and removed from the beaker for a receding contact angle. The camera continuously took photos at a rate of approximately 4.2 photos per second.

It should be noted that the speed at which the plate moves (i.e. the speed of the contact line) has an effect on dynamic angle measurements. Both the contact line velocity and acceleration influence the dynamic contact angle; in particular, the dynamic contact angle is larger for higher contact line acceleration (Xu *et al.*, 2011). Our speed was chosen based on physical limitations of the camera and syringe system. The resultant average volumetric flow rate of the syringe was 5 cm^3/s, or a surface level change of 0.6 mm/s.

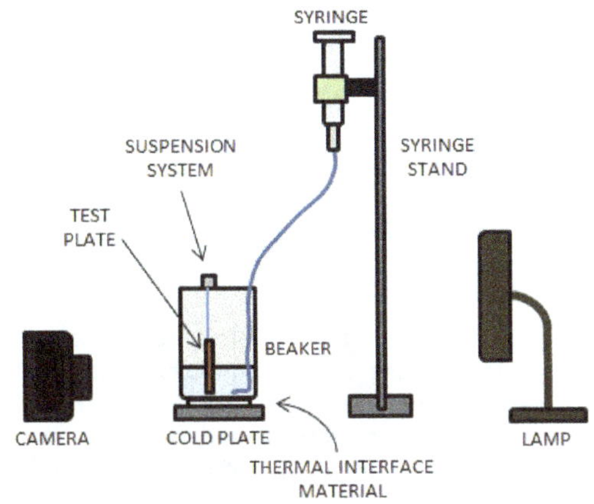

Fig. 7 Experimental setup for capillary rise at a vertical plate under ambient conditions.

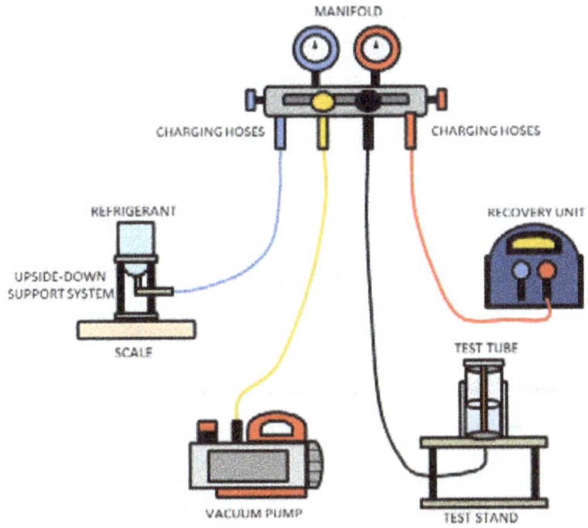

(a) Overall view of experimental setup for capillary rise at a vertical plate under vacuum conditions.

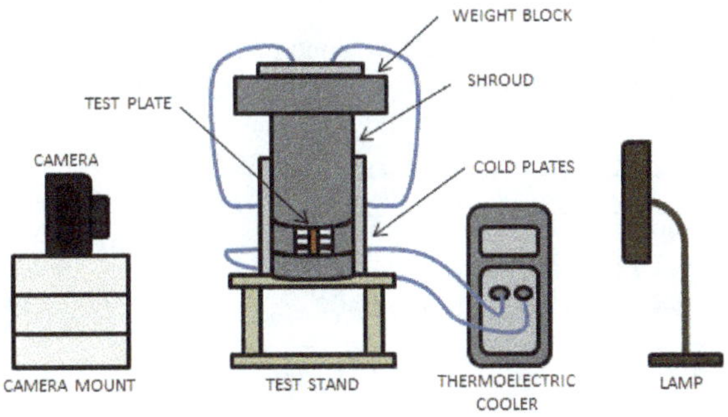

Fig. 8 (b) Test stand detail.

Because the two refrigerants of interest to this study, R-134a and HFO-1234yf, are not liquid at room temperature and pressure, we built a separate system for measuring the dynamic contact angles under vacuum. We constructed a small vacuum apparatus using a clear PETG (polyethylene terephtalate glycol-modified) tube and two custom-built aluminum caps. For safety, an aluminum shroud surrounded the tube, with small viewing windows for light and camera access. Two cold plates were clamped next to the aluminum shroud and connected in series with a NESLAB RTE7 chiller. Water was circulated through this system at 5°C. The experimental setup is shown in Fig. 8.

Initially, rough vacuum is pulled on the entire system to a range of approximately 40 to 70 torr. The refrigerant can is tapped, and then opened to fill the test tube. As refrigerant flows into the tube, the internal pressure of the tube rises to arrange of 35 to 50 psi (1810 to 2585 torr). Once a sufficient amount of liquid refrigerant is in the test tube, the manifold lines are closed. After measurements are taken, the recovery unit is used to remove the refrigerant from the test tube, manifold, and the rest of the system.

Because of physical restrictions of the system, raising and lowering the liquid level and/or the plate was not experimentally practical. Thus to find the advancing angle of the fluid/solid interface, the tube was tilted toward the camera, moving the liquid surface higher on the plate on the edge closest to the camera. Similarly, the receding angle was measured by tilting the tube away from the camera, lowering the liquid surface level on the camera side. This replicated the motion of the traditional fluid setup, performed for water and acetone by the capillary rise technique.

3.4 Analysis

For the sessile drop technique, the advancing, receding, and as-placed angles were measured using the DropSnake Java plugin, available from the National Institute of health, which applies active contours to an image after the user defines points along the drop outline (Stalder *et al.*, 2010). The raw images were initially post-processed for sharpness and clarity and converted to a black and white format. The DropSnake plugin was chosen over the other ImageJ options (Kwok *et al.*, 1995) because it allows for separate angle measurements of each side of the drop, and because the contour placement method facilitates faster and easier processing.

For the capillary rise technique, the advancing and receding angles captured were measured using a modified form of the Laplace equation. Assuming the vertical plate is sufficiently wide, the Laplace equation (Pogorzelski, *et al.*, 2012) integrates into:

$$\sin\theta = 1 - \frac{\Delta\rho g h}{2\gamma_{LV}} \qquad (2)$$

where $\Delta\rho$ is the difference in density between the fluid and the substrate, g is the acceleration due to gravity, γ_{LG} is the liquid-gas surface tension, h is the capillary rise, and θ is the contact angle. Some fluid/solid systems formed a capillary depression instead of a rise, which created an advancing angle greater than 90°. For these cases, a modified Laplace equation was used that subtracted the angle from 180°.

For both methods, the equilibrium Young contact angle, θ_0, was calculated from the advancing and receding angles using Tadmor's equation (2004):

$$\theta_0 = \arccos\left(\frac{\Gamma_A \cos(\theta_A) + \Gamma_R \cos(\theta_R)}{\Gamma_A + \Gamma_R}\right) \qquad (3)$$

where

$$\Gamma_R = \left(\frac{\sin^3\theta_R}{(2 - 3\cos\theta_R + \cos^3\theta_R)}\right)^{1/3} \qquad (4)$$

and

$$\Gamma_A = \left(\frac{\sin^3 \theta_A}{2 - 3\cos\theta_A + \cos^3 \theta_A} \right)^{1/3} \tag{5}$$

4. RESULTS

Table 5 reports the contact angles obtained through the sessile drop technique. The receding, advancing, and as-placed angles were captured in images and processed through software. The Young's equilibrium angle was determined using Tadmor's equation (Eq. 2). Each case was repeated at least ten times to obtain an average value. The standard deviation is included in the table.

As noted earlier, wetting is affected by a large number of factors – not only liquid properties but also substrate properties and system conditions. (Lewis, 2006; Kumar and Prabhu, 2007). For this reason, we report the temperature and relative humidity for each system during testing. Laboratory conditions varied across different days, so the table indicates the conditions for each system.

Measurements were not attainable for acetone-aluminum and acetone-copper systems because the angle was too small to measure with the available equipment. Thus, the Young's equilibrium angle, θ_0, could not be calculated for these systems.

Table 6 reports the contact angles obtained through capillary rise at a vertical plate. The receding and advancing angles were determined using the height of the capillary rise and Eq. 1. The Young's equilibrium angle was calculated from Tadmor's equation (Eq. 2). Each measurement was repeated at least ten times to obtain a value, and standard deviation is included in the table.

Figure 9 presents results for the sessile drop technique (traditional fluids) and makes a comparison to some available literature values. The water-based systems correlate relatively well with previous research. Published values for acetone-based systems were not available, but our results matched the predicted trend, based on related research, of a contact angle lower than in water-based systems. These results are reported in Table 5, and literature values are summarized in Table 2.

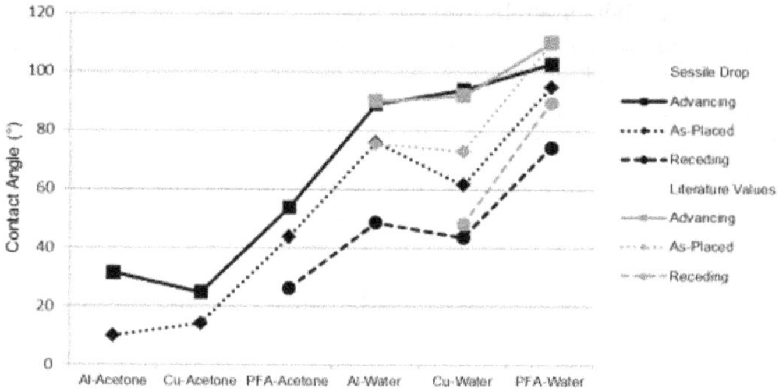

Fig. 9 Contact angle measurements of water and acetone using sessile drop technique.

Figure 10 presents the results for both the sessile drop technique and the capillary rise technique. The figure includes both traditional fluids (water, acetone) and refrigerants (R-134a, HFO-1234yf). The sessile drop technique was performed only for the traditional fluids because of physical limitations of the refrigerants, while the capillary rise technique was used on all four fluids. It should be noted that the capillary rise measurements employed different experimental setups for the refrigerants and the traditional fluids. These results are reported in Table 5 and Table 6.

Figure 11 summarizes the measured dynamic contact angle hysteresis values, or the absolute value of the difference between the advancing contact angle, θ_A, and the receding contact angle, θ_R. The figure includes results for all four working fluids, and compares the results from the sessile drop technique and the capillary rise technique.

Table 5 Results and standard deviation of experimental advancing, receding and as-placed angles, and calculated Young's equilibrium angle, using the sessile drop technique. Tests were performed under the following atmospheric conditions: [1] 19°C and 33% humidity; [2] 22°C and 21% humidity; [3] 22°C and 45% humidity; [4] 22°C and 18% humidity; [5] 23°C and 32% humidity; [6] 19°C and 41% humidity.

System	θ_R (°)	θ_{AP} (°)	θ_A (°)	θ_0 (°)
Water-Aluminum	48.68 ± 1.2 [1]	76.08 ± 2.3 [1]	89.15 ± 3.4 [1]	67.02
Water-Copper	43.50 ± 3.6 [2]	61.50 ± 3.7 [4]	93.98 ± 2.7 [2]	65.78
Water-PFA	74.39 ± 3.2 [2]	94.86 ± 1.9 [4]	102.84 ± 4.0 [2]	86.91
Acetone-Aluminum	*N/A*	9.93 ± 1.1 [1]	31.30 ± 1.5 [1]	*N/A*
Acetone-Copper	*N/A*	14.18 ± 3.0 [5]	24.71 ± 3.6 [2]	*N/A*
Acetone-PFA	26.08 ± 3.6 [3]	43.50 ± 1.2 [6]	53.75 ± 2.8 [3]	38.98

Table 6 Results and standard deviation of experimental advancing, receding and as-placed angles, and calculated Young's equilibrium angle, using the capillary rise at a vertical plate technique. Tests were performed under the following atmospheric conditions: [1] 19°C and 33% humidity; [2] 22°C and 21% humidity; [3] 20°C and 34% humidity; [4] 21°C and 54% humidity; [5] 21°C and 58% humidity; [6] 21°C and 55% humidity.

System	θ_R (°)	θ_A (°)	θ_0 (°)
Water-Aluminum	46.99 ± 1.4 [1]	99.33 ± 2.9 [1]	69.44
Water-Copper	53.24 ± 2.6 [2]	94.14 ± 0.7 [2]	71.40
Water-PFA	76.08 ± 2.6 [2]	94.15 ± 0.9 [2]	84.49
Acetone-Aluminum	70.47 ± 1.4 [3]	73.56 ± 1.6 [3]	79.16
Acetone-Copper	77.37 ± 0.8 [2]	80.99 ± 1.3 [2]	72.00
Acetone-PFA	83.78 ± 1.3 [2]	85.52 ± 0.6 [2]	84.64
R134a-Aluminum	60.11 ± 1.7 [4]	60.88 ± 1.4 [4]	60.50
R134a-Copper	58.92 ± 1.2 [5]	59.58 ± 1.7 [5]	59.25
R134a-PFA	53.23 ± 0.9 [5]	62.16 ± 2.7 [5]	52.86
HFO1234yf-Copper	62.16 ± 2.2 [6]	64.22 ± 1.1 [6]	63.19

It can be seen from Fig. 9 that, within the sessile drop technique, the acetone-based systems had lower contact angles and lower dynamic contact angle hysteresis than the water-based systems. This is confirmed by the capillary rise technique results (Fig. 10) for water and acetone-based systems. This will likely impact OHP design, although further study is needed to determine the influence of hysteresis on heat transfer performance.

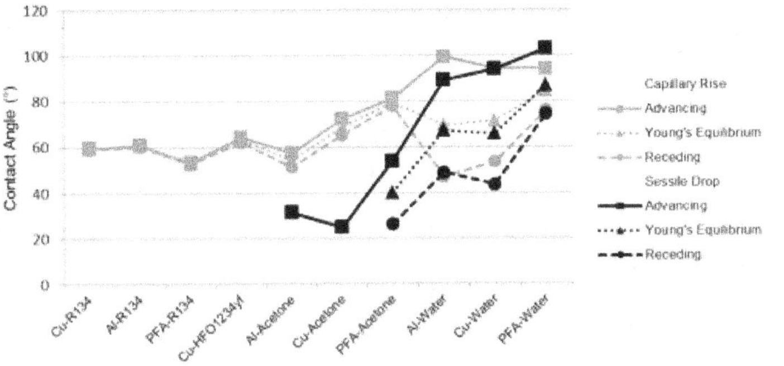

Fig. 10 Contact angle measurements of water, acetone, R-134a, and HFO-1234yf using capillary rise and sessile drop techniques.

Figure 10 demonstrates that the two techniques produce variations in contact angles, in particular for the acetone-based systems. It was noticed that advancing and receding angles obtained through capillary rise at a vertical plate, regardless of the fluid/solid system, are systematically greater than those obtained by the sessile drop technique. This reinforces the notion that a contact angle value is highly dependent on the physical conditions in which it is measured, among other factors. Therefore data comparison must use a relative ranking with consistent procedures and conditions. However, within data from the capillary rise technique, the refrigerants (R-134a and HFO-1234yf) exhibited the lowest dynamic contact angle hysteresis.

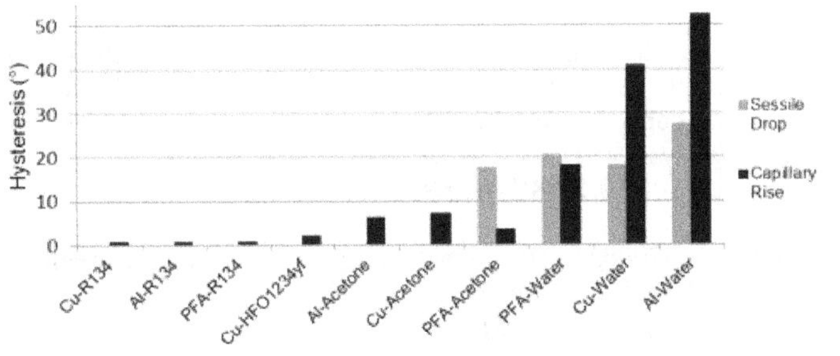

Fig. 11 Measured contact angle hysteresis of water, acetone, R-134a, and HFO-1234yf using sessile drop and capillary rise techniques.

We also noted that the low humidity conditions produced a smaller contact angle, i.e. larger drop radius, than the literature values. This is evident from comparing our sessile drop technique results with literature values (Fig. 9), where the literature values were generally measured in higher humidity environments (Table 2, 3). This corresponds to similar results that show the equilibrium fraction relative humidity increasing with decreasing drop radius (Lewis, 2006).

Not presented with the results is the study performed on an R-134a-copper system (Vadgama *et al.,* 2007), reported in Table 3. The reported angle was 6.5°, which is significantly lower than our results. This study included little information on the type of angle measured and conditions for measurement (e.g. relative humidity, temperature, pressure), so therefore a direct comparison cannot be made. Also, the analysis process used by the Vadgama *et al.* (2007) was a combination of visual observation and a polynomial fitting approach, in which the shape of the meniscus was estimated directly by drawing a tangent to this polynomial at the intersection of the surface and the edge of the drop, i.e. at the three-phase contact line. However, when we applied this technique to our acetone-based systems, we found that it significantly underestimated the advancing and receding angles when compared to our sessile drop technique measurements. Thus we decided that using the Laplace equation (Eq. 2) for analysis was a better choice because it more accurately captured the physical behavior of the system based on work done by Pogorzelski *et al.* (2012).

5. CONCLUSIONS

This study investigated the wettability of fluid-solid interactions for advanced heat transfer applications. Measurements were taken using two techniques: the sessile drop method and capillary rise at a vertical plate. The tested surface materials were copper, aluminum, and Teflon PFA. The working fluids tested were traditional fluids, water and acetone, and refrigerants, R-134a and HFO-1234yf. A novel low-pressure experimental setup was developed for refrigerant testing. Results show that the refrigerants have significantly lower hysteresis than the water and acetone-based systems, which is thought to lead to better heat transfer in an OHP design. This data contributes to the nearly empty set of dynamic contact angle data for the substrates and working fluids of interest.

To complete this data set, further study should be performed on the refrigerants to determine contact angle values with HFO-1234yf with the two additional substrates of interest, PFA and aluminum. A study of the two traditional fluids using the refrigerants' vacuum system for capillary rise technique would allow us to draw a more accurate comparison between traditional and non-traditional fluids.

Further testing and comparison of refrigerant-filled OHPs will be useful for comparing wettability data and OHP performance. The correlation between dynamic contact angle hysteresis and OHP performance is not yet known, and more research is needed before this phenomenon can be well understood.

ACKNOWLEDGEMENTS

This research was supported by the Air Force Office of Scientific Research, and the Universities Space Research Association. The authors would also like to thank Dr. Derek Hengeveld and Julio Alvarado for their advice and assistance in designing and running the system for capillary rise of the refrigerants.

NOMENCLATURE

g	Acceleration Due to Gravity (m/s^2)
h	Capillary Rise (m)

Greek Symbols

γ	Interfacial Tension (N/m)
Δ	*Hysteresis, Difference*
θ	Contact Angle (degrees)
ρ	Density (kg/m^3)

Subscripts

0	Young's Equilibrium
A	Advancing
AP	As Placed
R	Receding
SL	Solid to Liquid States
LG	Liquid to Gaseous States
SG	Solid to Gaseous States

REFERENCES

Amirfazli, A., Kwok, D.Y., Gaydos, J., 1998, A.W. Neumann. "Line tension measurements through drop size dependent of contact angle." *Journal of Colloid Interface Science*, **205,** pg. 1-11. http://dx.doi.org/10.1006/jcis.1998.5562

Benner, R.E., Scriven, L.E., Davis, H.T., 1982, “Structure and stress in the gas-liquid-solid contact region.” *Faraday Symposia of the Chemical Society*. Vol. 16, Royal Society London, pp. 169.

Bernardin, J.D., Mudawar, I., Walsh, C.B., Franses, E.I., 1997, "Contact Angle Temperature Dependence for Water Droplets on Practical Aluminum Surfaces," *International Journal of Heat and Mass Transfer*, **40**(5), pp. 1017-1033.
http://dx.doi.org/10.1016/0017-9310(96)00184-6

Budziak, C.J., Neumann, A.W., 1990, “Automation of the Capillary Rise Technique for Measuring Contact Angles,” *Colloids and Interfaces*, **43**, pp. 279-293.

Butt, H.J., Golovko, D.S., Bonaccurso, E., 2007, "On the derivation of Young's equation for sessile drops: Nonequilibrium effects due to evaporation." *Journal of Physical Chemistry B*, **111**, pp. 5277.
http://dx.doi.org/10.1021/jp065348g

Cayabyab, J., Cu, J.R.L., Leron, A.M.S., 2006, "Contact Angle Measurements," University of the Philippines Diliman. URL: http://www.scribd.com/doc/131275900/06-Contact-Angle-Measurements [cited 20 June 2013]

Della Volpe, C., Maniglio, D., Morra M., Siboni, S., 2002, "The determination of a 'stable-equilibrium' contact angle on heterogeneous and rough surfaces," *Colloids and Surfaces A: Physicochemical and Engineering Aspects*, **206**, pp. 47.
http://dx.doi.org/10.1016/S0927-7757(02)00072-9

Diaz, M.E., Fuentes, J., Cerro, R.L., Savage, M.D., 2010, "Hysteresis during contact angle measurement," *Journal of Colloid and Interface Science*, **343**, pp. 574-583
http://dx.doi.org/10.1016/j.jcis.2009.11.055

Dussan, E. B., 1979, "On the spreading of liquids on solid surfaces: static and dynamic contact lines." *Annual Review of Fluid Mechanics*. **11**, pp. 371-400.
http://dx.doi.org/10.1146/annurev.fl.11.010179.002103

Erbil, H.Y., Mc Hale, G., Rowan, S.M., Newton, M.I., 1999, "Determination of the receding contact angle of sessile drops on polymer surfaces by evaporation," *Langmuir*, **15**, pp. 7378-7385.
http://dx.doi.org/10.1021/la9900831

Extrand, C.W., 2003, “Contact Angles and Hysteresis on Surfaces with Chemically Heterogeneous Islands,” *Langmuir*, **19**, pp. 3793-3796.
http://dx.doi.org/10.1021/la0268350

Gajewski, A., 2008, "Contact Angle and Sessile Drop Diameter Hysteresis on Metal Surfaces," *International Journal of Heat and Mass Transfer*, **51**(19-20), pp. 4628-46
http://dx.doi.org/10.1016/j.ijheatmasstransfer.2008.01.027

Good, R.J., Koo, M.N., 1979, "The effect of drop size on contact angle." *Journal of Colloid and Interface Science*, **71**, pp. 283-292.
http://dx.doi.org/10.1016/0021-9797(79)90239-X

Goswami, S., Klaus, S., Benziger, J., 2008, “Wetting and Absorption of Water Drops on Nafion Films.” *American Chemical Society*, **24**, pp. 8627-8633.

Herzberg, W.J., Marian, J.E., 1970, "Relationship between contact angle and drop size." *Journal of Colloid and Interface Science*, **33**, pp. 161-163.
http://dx.doi.org/10.1016/0021-9797(70)90083-4

Holmes-Farley, S.R., Reamey, R.H., McCarthy, T.J., Deutch, J., Whitesides, G.M., 1985, "Acid-based behavior of carboxylic acid groups covalently attached at the surface of polyethylene: the usefulness of contact angle in following the ionization of surface functionality." *Langmuir*, **1**, pp. 725.
http://dx.doi.org/10.1021/la00066a016

Hung, M., Resnick, P.R., Smart, B.E., Buck, W.H., 1999, “Fluorinated Plastics, Amorphous.” *Concise Polymeric Materials Encyclopedia*, pp. 499-501.

Khandekar, S., Panigrahi, P.K., Lefevre, F., Bonjour, J., 2010, “Local Hydrodynamics of Flow in a Pulsating Heat Pipe: A Review,” *Frontiers in Heat Pipes*, **1**, 023003. http://dx.doi.org/10.5098/fhp.v1.2.3003

Kumar, G., Narayan Prabhu, K., 2007, "Review of non-reactive and reactive wetting of liquids on surfaces," *Advances in Colloid and Interface Science*, **133**, pp. 61-89.
http://dx.doi.org/10.1016/j.cis.2007.04.009

Kwok, D.Y., Budziak, C.J., Neumann, A.W., 1995, "Measurements of Static and Low Rate Dynamic Contact Angles by Means of an Automated Capillary Rise Technique," *Journal of Colloids and Interface Science*, **173**, pp. 143-150.
http://dx.doi.org/10.1006/jcis.1995.1307

Lamour, G., Hamraoui, A., 2010, "Contact Angle Measurements Using a Simplified Experimental Setup," *Journal of Chemical Education*, **87**(12), pp. 1403-1407.
http://dx.doi.org/10.1021/ed100468u

Lewis, E.R., 2006, "The effect of surface tension (Kelvin effect) on the equilibrium radius of a hygroscopic aqueous aerosol particle," *Journal of Aerosol Science*, **37**, pp. 1605-1617.
http://dx.doi.org/10.1016/j.jaerosci.2006.04.001

Li, G., Wang, B., Liu, Y., Tan, T., Song, X., Li, E., *et al.*, 2008, "Stable Superhydrophobic Surface: fabrication of interstitial cottonlike structure of copper nanocrystals by magnetron sputtering," *Science and Technology of Advanced Materials*, **9**, pp. 1-6.
http://dx.doi.org/10.1088/1468-6996/9/2/025006

Marmur. A., 2006, "Soft contact: measurement and interpretation of contact angles," *Soft Matte*, **2** (1), pp. 12-17.
http://dx.doi.org/10.1039/b514811c

Pappas, D., Copeland, C., Jensen, R., 2007, “Wettability Tests of Polymer Films and Fabrics and Determination of Their Surface Energy by Contact-Angle Methods”, Army Research Laboratory, ARL-TR-4052, March 2007, URL: http://www.dtic.mil/cgi-bin/GetTRDoc?AD=ADA466437 [cited 1 August 2013]

Pogorzelski, S.J., Berezowski, Z., Rochowski, P., Szurkowski, J., 2012, "A novel methodology based on contact angle hysteresis approach for surface changes monitoring in model PMMA-Corega Tabs system," *Applied Surface Science*, **258**, pp. 3652-2658.
http://dx.doi.org/10.1016/j.apsusc.2011.11.132

Qu, J., Wu, H., 2011, "Thermal Performance Comparison of Oscillating Heat Pipes with SiO2/water and Al2O3/water Nanofluids," *International Journal of Thermal Sciences*, **50**(10), pp.1954-1962.
http://dx.doi.org/10.1016/j.ijthermalsci.2011.04.004

Qu, W., Fan, C., Ma, T., 2003, “Contact angle hysteresis and capillary resistance of pulsating heat pipe,” *Journal of Engineering Thermophysics*.

Rodriguez-Valverde, M.A., Montes Ruiz-Cabello, F.J., Gea-Jodar, P.M., Kamusewitz, H., Cabrerizo-Vilchez, M.A., 2010, "A new model to estimate the Young contact angle from contact angle hysteresis measurements," *Colloids and Surfaces A: Physicochemical and Engineering Aspects,* **265**, pp. 21-27. http://dx.doi.org/10.1016/j.colsurfa.2010.01.055

Schwartz, A., 1980, "Contact angle hysteresis: a molecular interpretation," *Journal of Colloid and Interface Science*, **75**, pp. 404. http://dx.doi.org/10.1016/0021-9797(80)90465-8

Shirtcliffe, N.J., McHale, G., Newton, M.I., Chabrol, G., Perry, C.C., 2004, "Dual-Scale Roughness Produces Unusually Water-Repellent Surfaces," *Advanced Materials*, **16**(21), pp. 1929-1932. http://dx.doi.org/10.1002/adma.200400315

Shoji, M., Zhang. X.Y., 1984, "Study of Contact Angle Hysteresis (In Relation to Boiling Surface Wettability)." *Japanese Society of Mechanical Engineers International Journal*, **37**(3), pp. 560-567.

Sklodowaka, A., Wozniak, M., Matlakowska, R., 1999, "The Method of Contact Angle Measurements and Estimation of Work of Adhesion in Bioleaching of Metals," *Biological Procedures Online*, **1**, pp. 114-121. http://dx.doi.org/10.1251/bpo14

Stalder, A.F., Melchior, T., Muller, M., Sage, D., Blu, T., Unser, M., 2010, "Low-bond axisymmetric drop shape analysis for surface tension and contact angle measurements of sessile drops," *Colloids and Surfaces A: Physicochem. Eng. Aspects*, **364**, pp. 72-81. http://dx.doi.org/10.1016/j.colsurfa.2010.04.040

Tadmor, R., 2004, "Line Energy and the Relation between Advancing, Receding, and Young Contact Angles," *Langmuir,* **20**, pp. 7659-7664. http://dx.doi.org/10.1021/la049410h

Tadmor, R., Yadav, P.S., 2008, "As-placed contact angles for sessile drops," *Journal of Colloid and Interface Science*, **317**, pp. 241-246. http://dx.doi.org/10.1016/j.jcis.2007.09.029

Taft, B.S., Williams, A.D., Drolen, B.L., 2012, "Review of Pulsating Heat Pipe Working Fluid Selection," *Journal of Thermophysics and Heat Transfer,* **26**(4), pp. 651-656. http://dx.doi.org/10.2514/1.T3768

Tang, Z.G., Black, R.A., Curran, J.M., Hunt, J.A., Rhodes, N.P., Williams, D.F., 2004, "Surface Properties and Biocompatibility of Solvent-cast Poly[e-caprolactone] Films," *Biomaterials*, **25**, pp. 4741-4748. http://dx.doi.org/10.1016/j.biomaterials.2003.12.003

Torrigiani, M., 2005, "Wetting in the presence of Langmuir Films," MS Thesis, University of Alabama in Huntsville.

Tripathi, A., Khandekar, S., Panigraphi, P.K., 2010, "Oscillatory Contact Line Motion Inside Capillaries," *15th International Heat Pipe Conference*, April 25-30, Clemson, USA.

Vadgama, B., Harris, D.K., 2007, "Measurements of the contact angle between R134a and both aluminum and copper surfaces," *Environmental Thermal and Fluid Science.* **31**(8), pp. 979-984. http://dx.doi.org/10.1016/j.expthermflusci.2006.10.010

Van Mourik, S., 2002, "Numerical Modeling of the Dynamic Contact Angle," MS Thesis, University of Groningen, Department of Mathematics, URL: http://www.math.rug.nl/~veldman/Scripties/VanMourik-afstudeerverslag.pdf [cited 1 August 2013]

Xu, S.H., Wang, C.X., Sun, Z.W., Hu, W.R., 2011, "The influence of contact line velocity and acceleration on the dynamic contact angle: An experimental study in microgravity," *International Journal of Heat and Mass Transfer*, **54**, pp. 2222-2225. http://dx.doi.org/10.1016/j.ijheatmasstransfer.2011.01.018

Yekta-Fard, M., Ponter, A., 1985, "Surface Treatment and its Influence on Contact Angles of Water Drops Residing on Teflon and Copper," *The Journal of Adhesion*, **18**, pp. 197-206. http://dx.doi.org/10.1080/00218468508079683

Yekta-Fard, M., Ponter, A.B., 1988, "The influences of vapor environment and temperature on the contact angle-drop size relationship," *Journal of Colloid and Interface Science*, **126**, pp. 134-140. http://dx.doi.org/10.1016/0021-9797(88)90107-5

Zhang, J., Li, J., Han, Y., 2004, "Superhydrophobic PTFE Surfaces by Extension," *Macromolecular Rapid Communications*, **25**, pp. 1105–1108. http://dx.doi.org/10.1002/marc.200400065

14

EXPERIMENTAL STUDY OF COEFFICIENT OF THERMAL EXPANSION OF ALIGNED GRAPHITE THERMAL INTERFACE MATERIALS

Hsiu-Hung Chen[a], Yuan Zhao[b], Chung-Lung Chen[a,*]

[a] *Department of Mechanical and Aerospace Engineering, University of Missouri, Columbia, MO 65211*
[b] *Teledyne Scientific, Thousand Oaks, CA 91360*

ABSTRACT

Carbon-based materials draw more and more attention from both academia and industry: its allotropes, including graphene nanoplatlets, graphite nanoplatlets and carbon nanotubes, can readily enhance thermal conductivity of thermal interface products when served as fillers. Structural-optimization in micro/nano-scale has been investigated and expected to finely tune the coefficient of thermal expansion (CTE) of thermal interface materials (TIMs). The capability of adjusting CTE of materials greatly benefits the design of interface materials as CTE mismatch between materials may result in serious fatigue at the interface region that goes through thermal cycles. Recently, a novel nano-thermal-interface material has been developed, which is composed of tin (Sn) solder and graphite nanoplatlets. CTE of such sort of TIMs can be adjusted to match well with the substrate materials. A customized, optical CTE measuring system was built to measure CTEs of these thin and flexible samples. The averaged CTEs of samples made by this new approach range from -0.267×10^{-6}/°C to 5×10^{-6}/°C between 25°C and 137°C, which matches CTEs of typical semiconductor materials (the CTE of silicon is ~3×10^{-6}/°C in the same temperature range). This unique CTE-matching feature of a bonding material will have great potential to impact future development of high power microelectronics devices.

Keywords: *Graphene, Graphite, Thermal Interface, Thermal Expansion.*

1. INTRODUCTION

Thermal interface material (TIM) plays a critical role in microelectronics packaging. The main function of a TIM is to thermally connect various components in a microelectronics package. Thermal resistance and the ability to safely bond layers with vastly different coefficient of thermal expansion (CTE) are among the key factors for any TIMs in determining the overall performance and reliability of electronic devices in practical applications. As power density of microelectronics devices increase rapidly, the reliability and instability in the performance of TIMs is a growing concern in many applications (Prasher, 2006).

Conventional TIMs are mostly based on low electrically conductive materials (also low in thermal conductivity), *e.g.* polymers, greases, or adhesives, mixed with high thermal conductive particles, such as silver, silicon oxide or aluminum oxide. Such sort of TIMs (with K in the range of 0.1 to 4 W/m-°C) is a bottleneck in the whole thermal conductive path and highly restricts heat dissipation from the heat source. Recently, the outstanding thermal properties of carbon nanotubes (CNTs) that can reach up to 3,000 W/m-°C (Balandin *et al.*, 2008) draw extreme attention from both industry and academia. Single wall CNTs (SWCNTs) or multi-wall CNTs (MWCNTs) have been investigated and added to polymer resin materials as fillers (Choi *et al.*, 2001; Biercuk *et al.*, 2002; Yu *et al.*, 2006; Amama *et al.*, 2006; Tao Tong *et al.*, 2007). The enhancement on overall thermal conductivity of such TIMs is obvious; however, it is not as high as was expected. On the other hand, graphene and graphite nanoplatelets (GNPs), prepared from exfoliated natural graphite, provide excellent thermal conductivity on their own and in TIMs matrix (Balandin *et al.*, 2008). Impressively, the hybrid filler of GNPs and SWCNTs demonstrate a further enhancement of thermal conductivity and surpasses the performance of the individual SWCNT and GNP fillers, due to synergistic effect (Yu *et al.*, 2008). With the same loading ratio, TIM with GNPs fillers outperforms that with SWCNTs fillers by almost double the overall thermal conductivity. GNPs connected by SWCNTs in network qualitatively explain how synergetic effect happens. On the other hand, the introduction of SWCNTs into GNP fillers depresses the electrical conductivity. Such kind of trend actually favors the application of TIMs.

When filled in polymer matrix, GNPs demonstrates remarkable enhancement in thermal conductivity, even better than the case with SWCNTs. This phenomenon has been observed and studied, and is due to weak thermal couple at CNTs/base material interface and prohibitive cost (Shahil and Balandin, 2012). The decrease of K in higher temperatures at higher loading can be described by Umklapp process, where phonon scattering tends to happen more at in higher temperatures (Balandin, 2011).

Instead of introducing pristine carbon materials to polymer composite, chemically functionalized and exfoliated GNPs facilitate nanocomposites by enhancing the interaction between the epoxy and graphite filler. Both thermal conductivity and electrical resistivity have been improved (Ganguli *et al.*, 2008). The use of functionalized GNPs increases K by minimizing interfacial phonon scattering. The addition of GNPs as fillers also delays curing temperature of silicone polymer, because GNPs hinder the mobility of the polymer chains (Raza *et al.*, 2010).

Literature survey also indicates that the CTE of a bonding agent can be changed by infutrating it with fillers of different CTEs. Since

* *Corresponding author. Email:chencl@missouri.edu*

exfoliated graphite has a lower CTE value than the pure polymer, the mixing of graphite and polymer will result in a compromise between the two materials. Wang *et al.* verified this assumption by their experiments of epoxy mixed with carbon nano-materials (Wang *et al.*, 2009). In their study, obvious CTE reductions were observed on the epoxies when 1% SWCNT and 5% GO were added to them.

It can be noticed that (a) polymer resin, even though it can be enhanced in folds when expensive CNTs or graphene are added, performs relatively poor in thermal conductivity; (b) metal, silicon and carbon (including allotropes) all have high thermal conductivity. Silicon and carbon have small CTEs: some allotropes of carbon even have negative values. Therefore, the effect of CTE mismatch that may cause cracks and fatigues should be taken into consideration when new TIMs are designed. Among the carbon-based materials, graphite seems to be a more promising TIM candidate when both its natural property of thermal conductivity, simplicity and low-cost of fabrication are taken into account.

Zhao, el al. (Zhao *et al.*, 2012; Zhao *et al.*, 2011) prototyped a new TIM structure: vertically-aligned graphite laminated with solder layers (Fig. 1). With the laminated structure it is easy to form perfectly straight and highly conductive paths along the desired heat flow direction. The flexibility of the TIM could also be altered by adjusting the thicknesses of the solder layers and graphite sheets so that it could conform to surfaces with different roughness and hardness. They achieved an overall thermal resistivity of approximately 0.035°C/(W/cm^2) at an assembly pressure of 30 Psi with a 200-μm-thick TIM (graphite-to-solder ratio of 8), and the number decreased when the compression pressure increased. Achieving the outstanding thermal performance at a much larger bonding line thickness is an attractive and desirable characteristic in microelectronics packaging because thermal induced stress (mainly from the CTE mismatch among materials) is much smaller based on material mechanics principles.

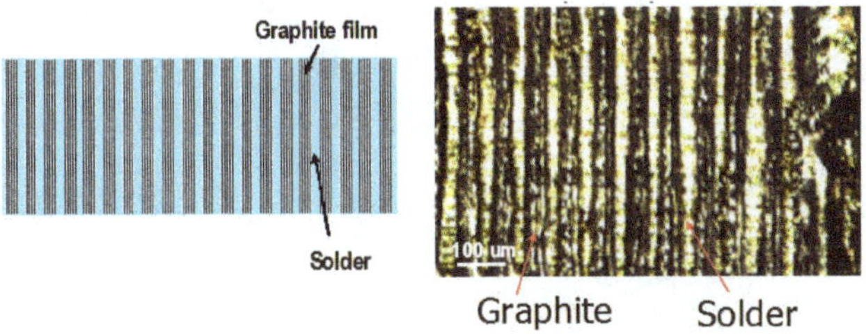

(a) Schematic of graphite TIM structures

(b) Sample microscopic image of a cross section

Fig. 1 Schematic and microscopic image of laminated TIM structure.

Applying thermally mismatched materials may result in serious induced stress, or even fatigue/cracks, at the interface region that go through thermal cycles. Therefore, obtaining CTE information for TIM of interests is necessary. There are various dilatometers available to study CTEs for solids (Boccaccini and Hamann, 1999; Neumeier *et al.*, 2008; Winkler *et al.*, 1993). The most commonly used technique is a contact-based expansion measurement during heating (Neumeier *et al.*, 2008; Winkler *et al.*, 1993): when the sample expands, a rod (or a plate) that rests slightly on the surface of the specimen is pushed and the expansion is then measured. This type of measurement requires that the sample is in the direct contact to the sensor, which may disturb the testing results when the sample is fragile or flexible. An optical dilatometer, on the other hand, is a non-contact device that is able to measure CTE of almost any kind of materials (Boccaccini and Hamann, 1999; Raether *et al.*, 2001).

In this research, an experimental study of CTEs of the vertically-aligned graphite/solder TIMs is presented. A series of tests were conducted to quantify impacts of adding metal layers with different thickness onto surfaces of a graphite thin film, on its CTE. A direct optical CTE measurement system was designed and built to execute the corresponding CTE measurements.

2. EXPERIMENTAL TEST SYSTEM AND APPROACH

2.1 CTE Measurement System

The CTE measurement system is based on an optical approach (Fig. 1), comprised of a microscope (Olympus BXFM, Tokyo, Japan), a CCD camera (UI-1240LE, IDS GmbH, Obersulm, Germany), a motorized XY stage with motorized actuators (8302 Picomotor, Newport Corporation), and a ceramic heater (Ultramic CER-1-01-00005, Watlow, Fenton, MO). Two metal tracks are first screwed onto the stage, which is customized for this test bed. The XY stage is fastened to one track, as shown in **II** of Fig. 1. Quartz slides sandwich the ceramic heater and a spacer, in order to form a sample holding site. The use of quartz minimizes the heating effect introduced from itself (CTE of 0.33×10^{-6}/°C) and allows visible light to pass through. A Teflon mounting gadget (with a viewing window) which can be utilized to fasten the sample holding site is fabricated, and allows expansion room for the ceramic heater.

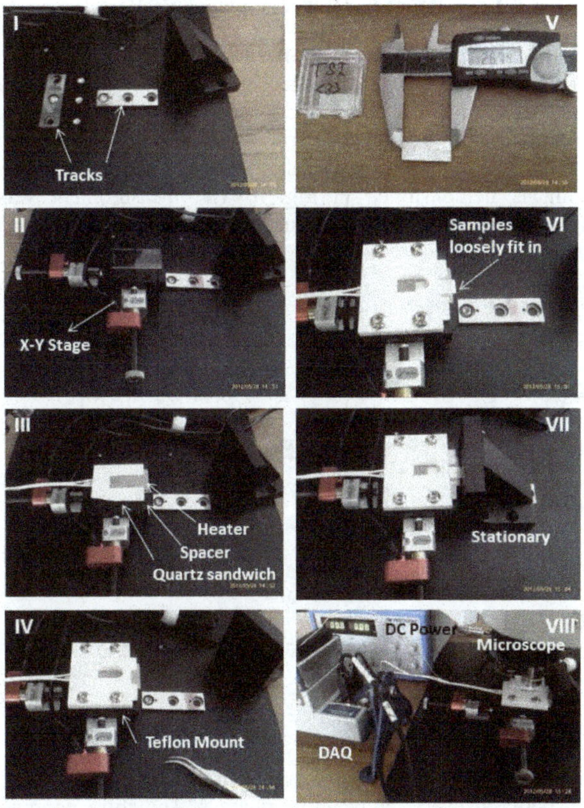

Fig. 2 Optical CTE test bed. From I to VIII: step-by-step assembly procedures.

The sample length along the measuring direction is taken by a caliper, shown in **V**. Then the sample, able to expand freely, is inserted into the sample holding site. A triangular metal block, serving as a stationary reference, is fastened on the other track (shown in **I** of Fig. 2). About one tenth of the sample will be extruded out (**VI** of fig. 2). By doing this, every time before the image is taken at its corresponding set temperature the sample can be "zeroed" by moving the motorized XY stage with an appropriate distance towards the stationary. Temperature is controlled by a DC power supply, providing electric power to the ceramic heater. The sample temperature is monitored by a K-type thermocouple, connected to a data acquisition (DAQ) system (NI-cDAQmx, National Instrument).

When a thin, long sample is heated, its total length usually increases. If the total length is recorded along with its corresponding set temperature, the CTE of a material can then be determined. This change in linear dimension is estimated as

$$\alpha = \frac{1}{L_{25^\circ C}} \frac{\Delta L}{\Delta T} \quad (1)$$

where α is the CTE, L is the total length of the sample at a reference temperature, ΔL is the difference of sample length before and after heating, and ΔT is the corresponding temperature difference.

2.2 TIM Sample Preparation

TIM samples are made out of graphite films and silver-tin solders (Indalloy 121: Ag 3.5% and Sn 96.5% with CTE: 30×10^{-6}/°C and reflow temperature of 221°C). The graphite film was processed from commercially available graphite sheets (GrafTech International) through a series of heat and compress procedures. Tin metallization was completed by electroplating technique. The detail fabrication processes were discussed in the previous publication (Zhao *et al.*, 2012; Zhao *et al.*, 2011). Our target temperature range is focused between 25°C and lower than 200°C.

3. TEST RESULTS AND ANALYSIS

3.1 Test Bed Validation with Silicon and Copper

Before testing TIM samples, we calibrated our test bed by measuring the CTEs of pure Silicon strip and Copper Alloy 101 (purity of 99.99%), and compared them with literature values. A Silicon strip (24.39mm × 1.2mm × 0.5mm) was tested and its linear expansion during heating is demonstrated in Fig. 3. The measurement was repeated twice on the same sample. A linear curve fit was applied on each data set.

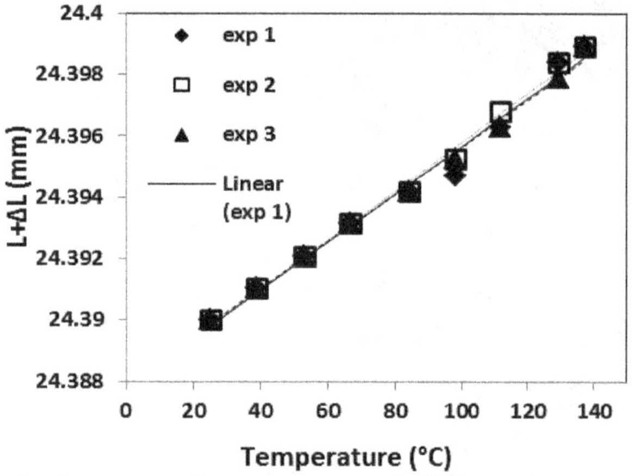

Fig. 3 Linear expansion measurement of a Silicon strip.

By using Eq. (1), α can be determined for each individual measurement between every temperature intervals. The corresponding results for each individual test, curve-fitted with least square method, are listed in Table 1. The measured CTE at 81°C can be treated as an averaged CTE since that temperature is at the middle of our temperature range. The measured CTEs and literature value are shown side-by-side in Fig. 4.

Table 1 Least square fit for each linear expansion measurements on Silicon strip.

	Least Square Fit	CTE (/°C)	Bias
exp 1	y1=0.000078x + 24.388 (R^2=0.98)	3.20×10^{-6}	5.3%
exp 2	y2=0.000080x + 24.388 (R^2=0.99)	3.28×10^{-6}	7.9%
exp 3	y3=0.000077x + 24.388 (R^2=0.99)	3.16×10^{-6}	3.9%

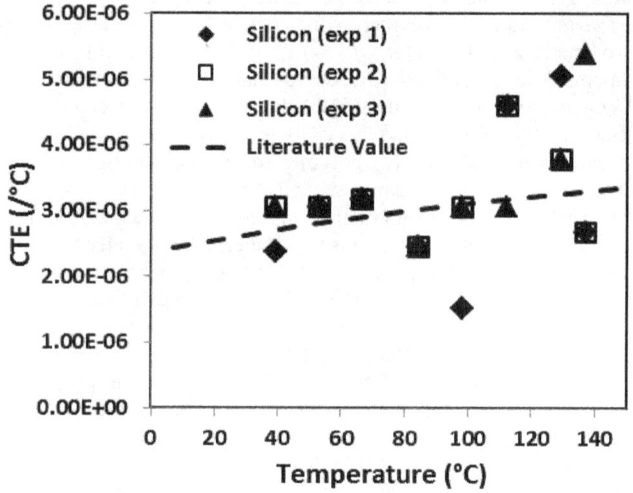

Fig. 4 CTE translated from linear expansion measurements and literature value (Roberts, 1981; Okada and Tokumaru, 1984).

A Copper Alloy 101 (23.33mm × 1mm × 0.25mm) strip was tested and its linear expansion during heating is displayed in Fig. 5. The measurement was repeated twice on the same sample. A linear curve-fit was applied on each data set.

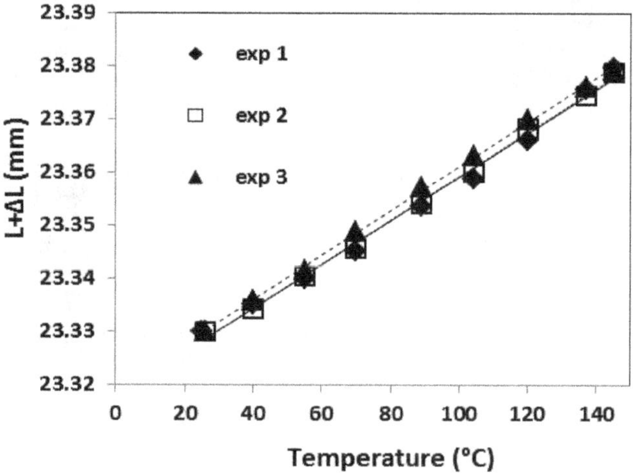

Fig. 5 Linear expansion measurement of a Copper Alloy 101 strip.

Similarly, by dividing the coefficient representing the slope for each first-order line with 23.33mm (original length), the averaged CTEs are obtained and listed in Table 2. The measured CTE at 81°C can be treated as an averaged CTE. The biggest bias among these measurements is 5% larger than the literature value (17.1×10^{-6} at 81°C), shown in Fig. 6 (Hahn, 1970).

Table 2 Least square fit for each linear expansion measurements on Copper Alloy 101.

	Least Square Fit	CTE (/°C)	Bias
exp 1	y1=0.00041x + 23.318 (R^2=0.99)	17.70×10^{-6}	3.5%
exp 2	y1=0.00041x + 23.318 (R^2=0.99)	17.79×10^{-6}	4%
exp 3	y3=0.00041x + 24.319 (R^2=0.99)	17.96×10^{-6}	5%

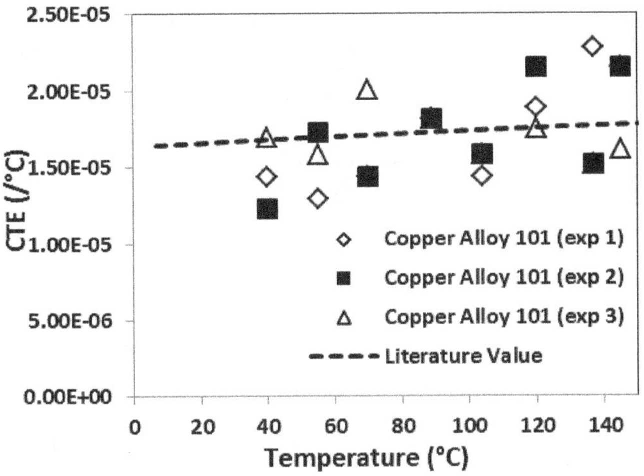

Fig. 6 CTE translated from linear expansion measurements and literature value (Hahn, 1970).

3.2 Uncertainty Analysis

The optical approach on CTE measurement demonstrated here is capable of recording visual information and translating it into linear CTEs. There is rich amount of natural markers visible under microscope that can be chosen as a reference. An example is demonstrated in Fig. 7 that one natural marker (a white dot pointed by a black arrow) was chosen and usedto trace thermal expansion at three different temperatures. By counting the pixel differences among different temperatures, one can easily determine ΔL when temperature varies.

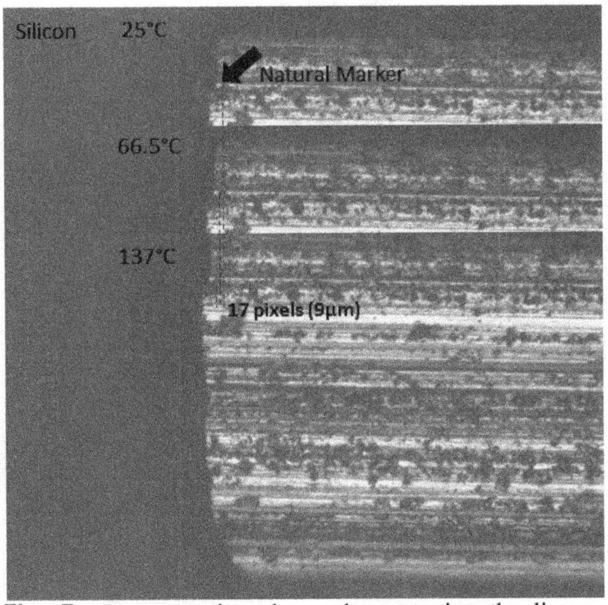

Fig. 7 Representative photos demonstrating the linear expansion traced from a natural marker (under 10× objective).

Error may occur in a variety of sources. One of the error sources comes from the resolution provided by digital imaging. Since the images taken in this research have the resolution of 0.265μm per pixel (20× used in TIM samples), the error caused due to miscounting one pixel for 10°C temperature for a 25mm-long sample will result in a CTE error of 1.05×10^{-6}/°C. This accounts for ±3% for materials with CTEs like copper, and ±16% for materials like silicon. It should be noted that this estimated error is a single-shot error. Repeats of experiments are able to minimize this sort of error.

We also realized that some samples are not uniform on their surface tomography. There are visible ripples on the uneven surface. Error estimation was conducted based on a simplified model (schematic shown in Fig 8). The samples were covered by a quartz slide with a space ranging from 0.2mm to 0.4mm, resulting in a gap roughly from 0.1mm to 0.3mm. If a sample evenly zigzags like a line segment (segment number N) shown in Fig. 8, its CTE from visible changes can be written as

$$CTE = \frac{\sqrt{(L+\Delta L)/N)^2 - d^2} - \sqrt{(L/N)^2 + d^2}}{\sqrt{(L/N)^2 + d^2 \Delta T}} \tag{2}$$

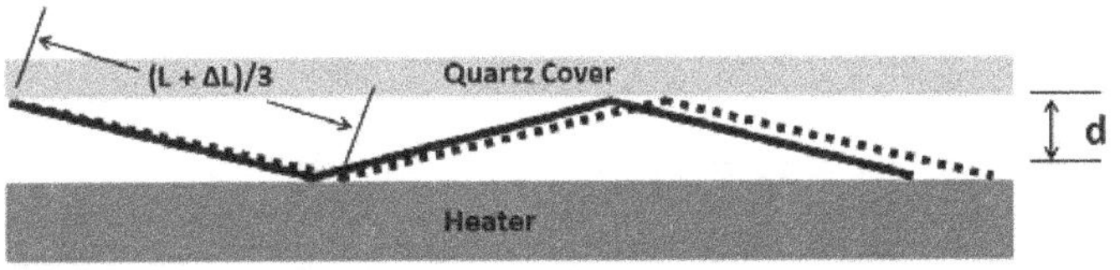

Fig. 8 Schematic of a sample with two zigzags and gap size d.

For a material with CTE = 4×10^{-6}/°C and with 25mm in length, the CTE sensitivity during measurements towards the chosen gap size and zigzag numbers are analyzed and shown in Fig. 9. First of all, such sort of error is a positive offset. Secondly, although more zigzags will result in bigger errors, there is only 1.45% increase when nine zigzags and larger gap size of 0.3mm are applied. This shows that CTEs are relatively insensitive to the geometry irregularities, like ripples, compared with errors caused from image resolution.

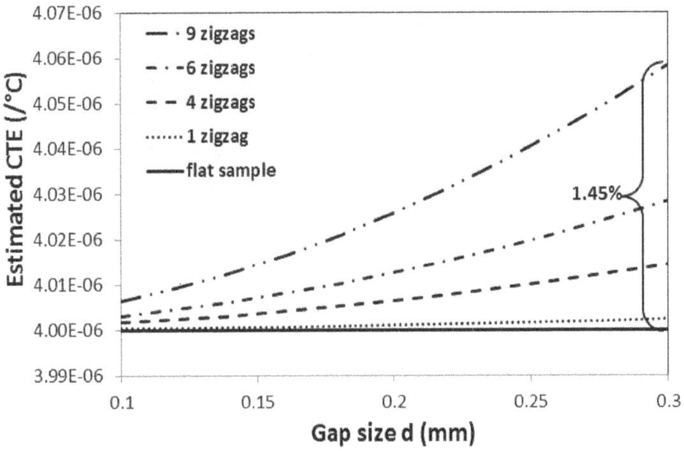

Fig. 9 Analysis of error generated due to surface irregularities and clamping gap size.

Although optical resolution and zigzag effect may affect the accuracy of the measurement, other traditional approaches (e.g. Dilatometer) have no position to replace the optical method, as traditional approaches are unable to handle thin and fragile samples demonstrated in this research.

We validate our test bed by comparing measured CTE of Silicon with the literature value. CTE for Silicon based on our measurement seems to jump up and down; however, the temperature-dependent trend can still be observed. The largest bias among these measurements is 7.9% larger than the literature value (~3.04×10^{-6} at 81°C), shown in Table 1 and Fig. 4. We assume that this percentage of difference is acceptable.

A more complete plot of copper CTE v.s. temperature can be found in the previous paper (Hahn, 1970). In some published literatures, such temperature-dependent CTE can be modeled by a fourth-order equation with the temperature range from -253°C to 527°C. In our application we are focusing on a narrower temperature range from 25°C to 150°C, which can be assumed as a first-order equation, listed in

Tables 1 and 2. These two tests validated our CTE measurement system.

3.3 CTE Measurement on TIM Samples

As the first step, we conducted a series of experimental measurements to evaluate impacts on CTE of a thin graphite film by adding Sn layers with different thickness on both sides of its surfaces. Four test coupons of tin-on-graphite were fabricated, numbered #1 to 4. The geometries of the four test coupons are listed in Table 3. Sample #1, 2, 3 and 4 are about 26mm in length.

Table 3 Tested samples from Teledyne Scientific.

Test Coupons	Graphite Thickness (μm)	Tin layer Thickness (μm)
1	120	0
2	120	0.5
3	120	1
4	120	2

Fig. 10 shows sample appearances. The thickness for all TIM samples ranges from 0.12mm to 0.2mm.

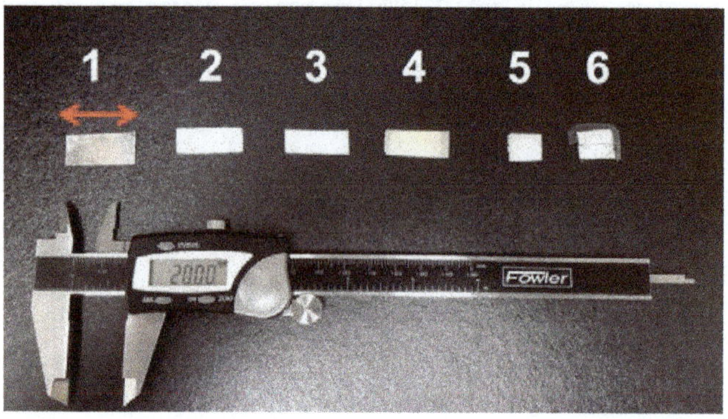

Fig. 10 TIM samples. The red arrow represents the CTE measuring direction.

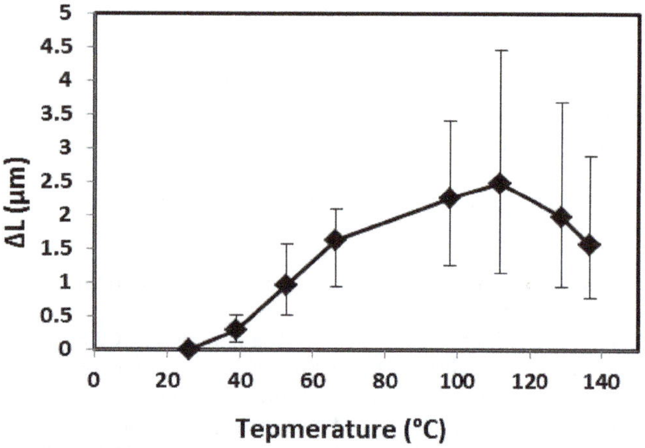

Fig. 11 Averaged linear expansion of TIM sample #2.

A typical measurement of TIM samples is shown in Fig. 11. Each point in the figure represents an average from three different experimental results. The upper and lower ends of each bar represent maximum and minimum linear expansion, respectively, at its corresponding temperature. All four samples were characterized three times and averaged. The corresponding thermal changes are shown in Fig. 12. The averaged CTE is calculated from each temperature segment. For example, there are seven CTEs for the adjacent temperature points in each independent experiment as any temperature differences can be served as a ΔT in Eq. 1. The temperature range during CTE measurement is from 25°C to 137°C.

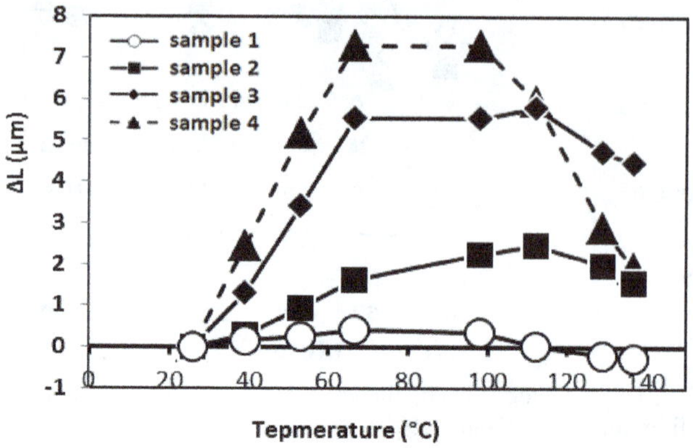

Fig. 12 Linear expansion of TIM samples #1, #2, #3 and #4.

Two thicker graphite-tin TIM samples (#5 and #6, both are ~0.2mm in thickness) were tested as well. Unlike samples #1 to 4, these TIM samples have a vertical-oriented multi-layered structure, in which graphite and tin solder layers are laminated (Fig. 13). The CTE measurement was executed along z-direction. The samples have a graphite-to-solder ratio of 10:1. The thermal expansion of samples #5 and #6 are shown in Figs 14 and 15, respectively. Each sample is tested three times and averaged.

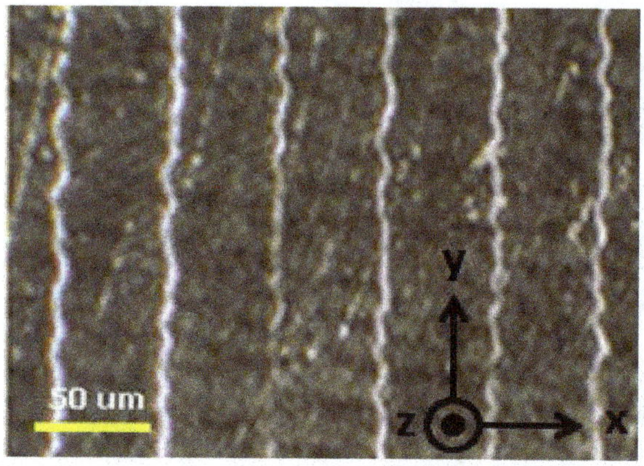

Fig. 13 TIM sample with 10:1 graphite-to-solder ratio.

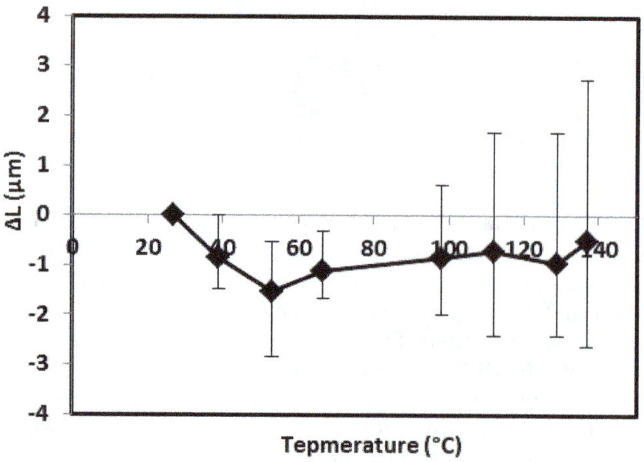

Fig. 14 Linear expansion measurement of sample #5.

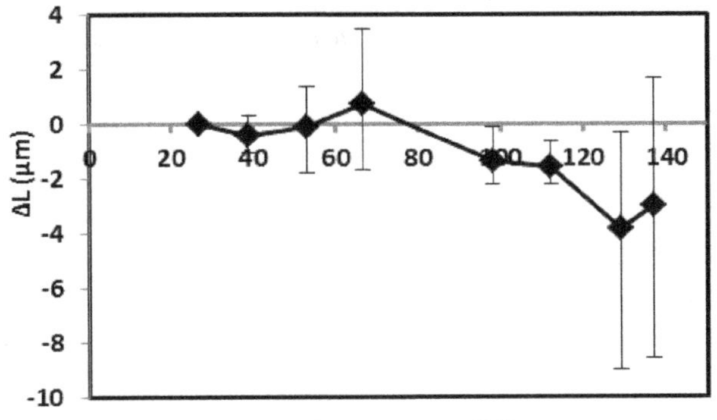

Fig. 15 Linear expansion measurement of sample #6.

Table 4 Summarizes the CTEs and the corresponding lengths of six TIM samples. For sample #1, #2, #3 and #4, only data below 100°C is calculated, in order to better demonstrate the effect of the tin layers added to the samples and avoid the unusual CTEs with negative trends.

Table 4 Summary of CTEs for TIM samples.

Sample #	Sample length (mm)	Averaged CTE (/°C)
1	**26.46**	**0.28×10^{-6}**
2	**24.87**	**0.62×10^{-6}**
3	**26.12**	**2.93×10^{-6}**
4	**26.9**	**5×10^{-6}**
5	**14.86**	**-0.267×10^{-6}**
6	**14.75**	**-0.225×10^{-6}**

When coming to TIM samples, they perform quite differently from ordinary materials (*e.g.* pure copper and pure silicon tested in this research). For samples #1 to #4, they expanded quite linearly during heating when temperatures are below 100°C. It can be observed that the decreasing trend of CTEs below 100°C is highly correlated with the ratios of composition of graphite films and metal layers. The averaged CTEs show a clear trend in Fig. 16 and follow the mechanics of composite and indicate that we can readily modify CTEs of regular solder by laminating solder layers with aligned graphite films, which will also significantly enhance thermal transport of the solder due to high in-plane thermal conductivity of graphite films. When temperatures are over 100°C, the samples begin to shrink. The phenomena of negative CTE is unusual, assumabily due to the special design on graphite micro-structures.

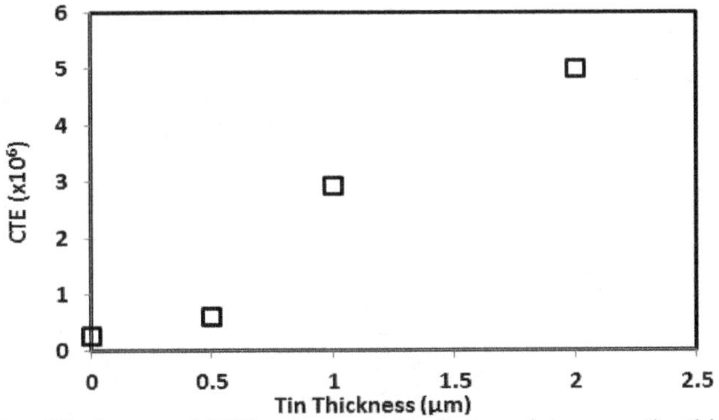

Fig. 16 Averaged CTEs of sample #1, 2, 3, and 4 versus the thickness of tin below 100°C.

Samples of #5 and #6 were from the same bigger sample and cut from different spots. The composition structures are vertically oriented, as demonstrated in Fig. 13, different from #1 to #4 with single graphite layer. Higher solder contents are expected to provide better comformability to surface roughness for the molten state of solder during the bonding process. Test results for #5 and #6, whose multiple tin layers are laminated among graphite films, show zero expansion. This indicates that CTE of a TIM is highly dependent on its structure besides its material composition and further investigation, especially from the microstructure and manufacture point of views, is needed.

4. CONCLUSIONS

An optical CTE measuring system was developed and validated by taking CTE measurements on Copper Alloy 101 and pure silicon. Such validation matches well with published literature when temperatures of the samples are within the room temperature and 150°C. Our CTE measurement stops at 150°C because the TIM samples start softening and melting over 200°C. This defines our measuring temperature range within ~120°C. Longer samples may improve the accuracy by increasing ΔL when temperature changes. On the other hand, future application of automatic focusing and computer-based image processing may eliminate the subjective judgment made by human eyes.

We successfully developed a new approach of fabricating CTE-adjustable TIMs and measured their corresponding CTEs. We observed unique characteristics on CTEs of the new TIMs. Reasons for (a) negative overall CTEs, and (b) positive CTEs below 100°C and negative CTEs over 100°C, observed in TIM samples are still unclear. However, it is proven that the new type of TIMs can be designed to match the CTE for targeted bonding materials. By plating metal layers on graphite films, CTEs for TIMs can be tuned from 0 to 5×10^{-6}/°C. Also, the laminated structures of the TIMs can significantly reduce the CTEs of the TIMs.

This unique CTE-matching feature of a bonding material will have great potential to impact the future development of high heat flux, high power microelectronics devices.

ACKNOWLEDGEMENTS

The authors gratefully acknowledge the contracts from DARPA (N66001-09-C-2015). The views expressed are those of the authors and do not reflect the official policy or position of the Department of Defense or the U.S. Government. In the event permission is required, DARPA is authorized to reproduce the copyrighted material for use as an exhibit or handout at DARPA-sponsored events and/or to post the material on the DARPA website. The authors also appreciate Tony Liao from Teledyne Scientific for preparing TIM samples.

REFERENCES

Amama, P. B., Ogebule, O., Maschmann, M. R., 2006, "Dendrimer-Assisted Low-Temperature Growth of Carbon Nanotubes by Plasma-Enhanced Chemical Vapor Deposition," *Chemical Communications (Cambridge England),* (27), 2899-2901. http://dx.doi.org/10.1039/B602623K

Balandin, A. A., 2011, "Thermal Properties of Graphene and Nanostructured Carbon Materials," *Nat Mater,* **10** (8), 569-581. http://dx.doi.org/10.1038/nmat3064

Balandin, A. A., Ghosh, S., Bao, W., 2008, "Superior Thermal Conductivity of Single-Layer Graphene," *Nano Letters,* **8** (3), 902-907. http://dx.doi.org/10.1021/nl0731872

Biercuk, M. J., Llaguno, M. C., Radosavljevic, M., 2002, "Carbon Nanotube Composites for Thermal Management," *Applied Physics Letters,* **80** (15), 2767. http://dx.doi.org/10.1063/1.1469696

Boccaccini, A. R., and Hamann, B., 1999, "Review in Situ High-Temperature Optical Microscopy," *Journal of Materials Science,* **34** (22), 5419-5436.
http://dx.doi.org/10.1023/A:1004706922530

Choi, S. U. S., Zhang, Z. G., Yu, W., 2001, "Anomalous Thermal Conductivity Enhancement in Nanotube Suspensions," *Applied Physics Letters,* **79** (14), 2252-2254.
http://dx.doi.org/10.1063/1.1408272

Ganguli, S., Roy, A., and Anderson, D., 2008, "Improved Thermal Conductivity for Chemically Functionalized Exfoliated graphite/epoxy Composites," *Carbon,* **46** (5), 806-817.
http://dx.doi.org/10.1016/j.carbon.2008.02.008

Hahn, T. A., 1970, "Thermal Expansion of Copper from 20 to 800 K - Standard Reference Material 736," *Journal of Applied Physics,* **41** (13), 5096-5096.
http://dx.doi.org/10.1063/1.1658614

Neumeier, J. J., Bollinger, R. K., Timmins, G. E., 2008, "Capacitive-Based Dilatometer Cell Constructed of Fused Quartz for Measuring the Thermal Expansion of Solids," *Review of Scientific Instruments,* **79** (3), 033903-033903.
http://dx.doi.org/10.1063/1.2884193

Okada, Y., and Tokumaru, Y., 1984, "Precise Determination of Lattice Parameter and Thermal Expansion Coefficient of Silicon between 300 and 1500 K," *Journal of Applied Physics,* **56** (2), 314-314.
http://dx.doi.org/10.1063/1.333965

Prasher, R., 2006, "Thermal Interface Materials: Historical Perspective, Status, and Future Directions," *Proceedings of the IEEE,* **94** (8), 1571-1586.
http://dx.doi.org/10.1109/JPROC.2006.879796

Raether, F., Springer, R., and Beyer, S., 2001, "Optical Dilatometry for the Control of Microstructure Development during Sintering," *Materials Research Innovations,* **4** (4), 245-250.
http://dx.doi.org/10.1007/s100190000101

Raza, M. A., Westwood, A. V. K., and Stirling, C., 2010, "Graphite nanoplatelet/silicone Composites for Thermal Interface Applications," *Advanced Packaging Materials: Microtech, 2010. APM '10. International Symposium on,* 34-48.
http://dx.doi.org/10.1109/ISAPM.2010.5441382

Roberts, R.,B., 1981, "Thermal Expansion Reference Data: Silicon 300-850 K," *Journal of Physics D: Applied Physics,* **14** (10), L163-L166; L163-L166.
http://dx.doi.org/10.1088/0022-3727/14/10/003

Shahil, K. M. F., and Balandin, A. A., 2012, "Graphene - Multilayer Graphene Nanocomposites as Highly Efficient Thermal Interface Materials," *Nano Letters,* , 120117152312004.
http://dx.doi.org/10.1021/nl203906r

Tao Tong, Yang Zhao, Delzeit, L., 2007, "Dense Vertically Aligned Multiwalled Carbon Nanotube Arrays as Thermal Interface Materials," *Components and Packaging Technologies, IEEE Transactions on,* **30** (1), 92-100.
http://dx.doi.org/10.1109/TCAPT.2007.892079

Wang, S., Tambraparni, M., Qiu, J., 2009, "Thermal Expansion of Graphene Composites," *Macromolecules,* **42** (14), 5251-5255.
http://dx.doi.org/10.1021/ma900631c

Winkler, S., Davies, P., and Janoschek, J., 1993, "High-Temperature Dilatometer with Pyrometer Measuring System and Rate-Controlled Sintering Capability," *Journal of Thermal Analysis and Calorimetry,* **40** (3), 999-1008.
http://dx.doi.org/10.1007/BF02546859

Yu, A., Ramesh, P., Sun, X., 2008, "Enhanced Thermal Conductivity in a Hybrid Graphite Nanoplatelet - Carbon Nanotube Filler for Epoxy Composites," *Advanced Materials,* **20** (24), 4740-4744.
http://dx.doi.org/10.1002/adma.200800401

Yu, A., Itkis, M. E., Bekyarova, E., 2006, "Effect of Single-Walled Carbon Nanotube Purity on the Thermal Conductivity of Carbon Nanotube-Based Composites," *Applied Physics Letters,* **89** (13), 133102.
http://dx.doi.org/10.1063/1.2357580

Zhao, Y., Strauss, D., Chen, Y. C., 2012, "Experimental Study of A High Performance Aligned Graphite Thermal Interface Material," *ASME Conference Proceedings.*

Zhao, Y., Strauss, D., Liao, T., 2011, "Development of a High Performance Thermal Interface Material with Vertically Aligned Graphite Platelets," *ASME Conference Proceedings,* **2011** (38921), T30010-T30010-7.
http://dx.doi.org/10.1115/AJTEC2011-44169

15

RADIATION EFFECTS ON MHD NATURAL CONVECTION FLOW ALONG A VERTICAL CYLINDER EMBEDDED IN A POROUS MEDIUM WITH VARIABLE SURFACE TEMPERATURE AND CONCENTRATION

Machireddy Gnaneswara Reddy*

Department of Mathematics, Acharya Nagarjuna University Campus, Ongole - 523 001, Andhra Pradesh, India

ABSTRACT

The numerical solution of transient natural convection MHD flow past a vertical cylinder embedded in a porous medium with surface temperature and concentration along with thermal radiation is presented. The temperature and concentration level at the cylinder surface are assumed to vary as power law type functions, with exponents m and n respectively in the stream wise co-ordinate. The governing boundary layer equations are converted into a non-dimensional form. A Crank-Nicolson type of implicit finite-difference method is used to solve the governing non-linear set of equations. Numerical results are obtained and presented with various thermal and mass Grashof numbers and power law variations. Transient effects of velocity, temperature and concentration are analyzed. Local and average skin-friction, Nusselt number and Sherwood number are shown graphically. The numerical predications have been compared with the existing information in the literature and good agreement is obtained.

Keywords: *Free convection, Heat transfer, Radiation, Finite-difference Scheme, vertical cylinder*

1. INTRODUCTION

Natural convection flows are frequently encountered in nature. They have wide applications in Science and Technology. These types of problems are commonly encountered in start-up of a chemical reactor and emergency cooling of a nuclear fuel element. In the case of power or pump failure, similar conditions may arise for devices cooled by forced circulation, as in the core of a nuclear reactor. In the glass and polymer industries, hot filaments, which are considered as a vertical cylinder, are cooled as they pass through the surrounding environment. The analytical methods fail to solve the problems of unsteady natural convection flows past a semi-infinite vertical cylinder. The advanced numerical methods and computer technology have shown the way in which such difficult problems can be solved. Finite difference methods play a major role in solving partial differential equations. Several investigators under different boundary conditions have analyzed steady free convection along vertical cylinders. Convective flow through porous media is a branch of research undergoing rapid growth in fluid mechanics and heat transfer. This is quite natural because of its important applications in environmental, geophysical and energy related engineering problems. Tien and Vafai (1990) have presented an excellent review of natural convection flow in porous media and have stressed the importance of the non-Darcy effects such as the inertia and boundary effects.

Unsteady natural convection flow of a viscous incompressible fluid over a heated vertical cylinder is an important problem relevant to many engineering applications. In the glass and polymer industries, hot filaments, which are considered as vertical cylinders, are cooled as they pass through the surrounding environment. For these types of non-linear problems, the exact solution is not possible. Sparrow and Gregg (1956) provided the first approximate solution for the laminar buoyant flow of air bathing a vertical cylinder heated with a prescribed surface temperature, by applying the similarity method and power series expansion. Lee et al. (1988) investigated the problem of natural convection in laminar boundary layer flow along slender vertical cylinders and needles for the power-law variation in wall temperature. Velusamy and Grag (1992) presented the numerical solution for transient natural convection over heat generating vertical cylinders of various thermal capacities and radii. The rate of propagation of the leading edge effect was given special consideration by them.

The effects of heat and mass transfer on the flow near vertical circular cylinder have been realized in many engineering and physical problems such as transport processes industry, ocean circulations due to heat current and difference in salinity etc. Combined buoyancy effects of thermal and mass diffusion along vertical cylinders have been given scant attention in the literature. Experimental results of pure and simultaneous heat and mass transfer by free convection along a vertical cylinder for $\Pr = 0.71$ and $Sc = 0.63$ are given by Bottemanne (1972). Rani (2003) studied the transient natural convection along a vertical cylinder with variable temperature and mass diffusion, by employing an implicit finite-difference method of Crank-Nicolson type. Chen and Yuh (1980) considered the effects of heat and mass transfer on natural convective flow along a vertical cylinder, where the surface of the cylinder was either maintained at a uniform temperature/concentration or subjected to uniform heat/mass flux. They concluded that the combined buoyancy force from thermal and species diffusion provide larger Nusselt and Sherwood numbers for uniform surface heat/mass flux. Ganesan and Rani (1999) analyzed the unsteady free convection on vertical cylinder with variable heat and mass flux.

The study of flow problems, which involve the interaction of several phenomena, has a wide range of applications in the field of science and technology. One such study is related to the effects of free convection MHD flow, which plays an important role in agriculture, engineering and petroleum industries. The problem of free convection under the influence of a magnetic field has attracted the interest of many researchers in view of its application in geophysics and in

* Email: mgrmaths@gmail.com.

astrophysics. The problem under consideration has important applications in the study of geological formations; in the exploration and thermal recovery of oil; and in the assessment of aquifers, geothermal reservoirs and underground nuclear waste storage sites. Results obtained from this study will be helpful in the prediction of flow, heat transfer and solute or contaminant dispersion about intrusive bodies such as salt domes, magnetic intrusions, piping and casting systems. Ganesan and Rani (2003) analyzed the magnetic field effect on a moving vertical cylinder with constant heat flux. Elbashbeshy (1997) studied heat and mass transfer along a vertical plate with variable surface temperature and concentration in the presence of magnetic field. Agarwal et al. (1989) considered the effect of MHD free convection and mass transfer flow past a vibrating infinite vertical circular cylinder. Combined heat and mass transfer effects on moving vertical cylinder that of steady and unsteady flow were investigated by Takhar et al. (2000), by using an implicit finite-difference scheme of Crank-Nicolson type. A numerical solution for the transient natural convection flow over a vertical cylinder under the combined buoyancy effect of heat and mass transfer was obtained by Ganesan and Rani (1998), by means of an implicit finite-difference scheme. Gnaneswara Reddy (2012) have analyzed unsteady free convective flow past a semi-infinite vertical plate with uniform heat and mass flux.

Heat transfer by simultaneous radiation and convection has applications in numerous technological problems, including combustion, glass production, furnace design, the design of high temperature gas cooled nuclear reactors, nuclear reactor safety, fluidized bed heat exchanger, fire spreads, advanced energy conversion devices such as open cycle coal and natural gas fired MHD, solar fans, solar collectors natural convection in cavities, turbid water bodies, photo chemical reactors and many others when heat transfer radiation is of the same order of magnitude as by convection, a separate calculation of radiation and their superposition without considering the interaction between them can lead to significant errors in the results, because of the presence of the radiation in the medium, which alters the temperature distribution within the fluid. Therefore, in such situation heat transfer by convection and radiation should be solved for simultaneously. In this context, Abd El-Naby et al. (2003) considered the effects of the radiation on unsteady free convective flow past a semi-infinite vertical plate with variable surface temperature using Crank-Nicolson finite difference method. The combined radiation and free convection flow over a vertical cylinder was presented by Yih (1999). Radiation and mass transfer effects on flow of an incompressible viscous fluid past a moving vertical cylinder was studied by Ganesan and Loganathan (2002). Gnaneswara Reddy and Bhaskar Reddy (2009) presented the radiation and mass transfer effects on unsteady MHD free convection flow past a moving vertical cylinder. The chemically reactive species and radiation effects on MHD convective flow past a moving vertical cylinder studied by Gnaneswara Reddy (2013).

No analytical or numerical work on transient natural convection along a vertical cylinder embedded in a porous medium of thermal and mass diffusion with power law variation in wall temperature and concentration with thermal radiation has been reported. This type of problem has non-similar boundary layers, governed by unsteady, non-linear, coupled equations. Hence, it is proposed to study the effects of variable surface temperature and concentration along a vertical cylinder by an implicit finite-difference scheme of Crank-Nicolson type. The behavior of the velocity, temperature, concentration, skin-friction, Nusselt number and Sherwood number has been discussed for variations in the governing thermophysical and hydrodynamical parameters.

2. MATHEMATICAL ANALYSIS

Consider an unsteady two-dimensional laminar free convective heat and mass transfer flow of a viscous incompressible electrically conducting and radiating optically thick fluid past a vertical cylinder of radius r_0 embedded in a porous medium. The x-axis is taken along the axis of the cylinder and the radial coordinate r is taken normal to the cylinder. Initially, the fluid and the cylinder are at the same temperature T'_∞ and the concentration C'_∞. At time $t'>0$, the temperature and concentration of the cylinder are raised to $T'=T'_\infty+(T'_w-T'_\infty)x^m$, $C'=C'_\infty+(C'_w-C'_\infty)x^n$ respectively and are maintained constantly thereafter. A uniform magnetic field is applied in the direction perpendicular to the cylinder. The fluid is assumed to be slightly conducting, and hence the magnetic Reynolds number is much less than unity and the induced magnetic field is negligible in comparison with the applied magnetic field. It is further assumed that there is no applied voltage, so that electric field is absent. It is also assumed that the radiative heat flux in the x-direction is negligible as compared to that in the radial direction. The viscous dissipation is also assumed to be negligible in the energy equation due to slow motion of the cylinder. Also, it is assumed that there is no chemical reaction between the species and the fluid. The foreign mass present in the flow is assumed to be at low level and hence Soret and Dufour effects are negligible. It is also assumed that all the fluid properties are constant except that of the influence of the density variation with temperature and concentration in the body force term (Boussinesq's approximation). Then, under the above assumptions the governing boundary layer equations are

Continuity equation

$$\frac{\partial(ru)}{\partial x}+\frac{\partial(rv)}{\partial r}=0 \tag{1}$$

Momentum equation

$$\begin{aligned}&\frac{\partial u}{\partial t'}+u\frac{\partial u}{\partial x}+v\frac{\partial u}{\partial r}\\&=g\beta(T'-T'_\infty)+g\beta^*(C'-C'_\infty)+\frac{\nu}{r}\frac{\partial}{\partial r}\left(r\frac{\partial u}{\partial r}\right)-\frac{\sigma B_0^2}{\rho}u-\frac{\nu}{K'}u\end{aligned} \tag{2}$$

Energy equation

$$\frac{\partial T'}{\partial t'}+u\frac{\partial T'}{\partial x}+v\frac{\partial T'}{\partial r}=\frac{\alpha}{r}\frac{\partial}{\partial r}\left(r\frac{\partial T'}{\partial r}\right)-\frac{1}{\rho c_p}\frac{1}{r}\frac{\partial}{\partial r}(rq_r) \tag{3}$$

Mass diffusion equation

$$\frac{\partial C'}{\partial t'}+u\frac{\partial C'}{\partial x}+v\frac{\partial C'}{\partial r}=\frac{D}{r}\frac{\partial}{\partial r}\left(r\frac{\partial C'}{\partial r}\right) \tag{4}$$

The initial and boundary conditions are

$t'\le 0$: $u=0, v=0, T'=T'_\infty, C'=C'_\infty$ for all $x\ge 0$ and $r\ge 0$

$$\begin{aligned}&t'>0:\ u=0, v=0,\ T'=T'_\infty+(T'_w-T'_\infty)x^m,\quad at\ r=r_0\\&C'=C'_\infty+(C'_w-C'_\infty)x^n\\&u=0,\ T'=T'_\infty,\ C'=C'_\infty\quad at\quad x=0 \text{ and } r\ge r_0\\&u\to 0,\ T'\to T'_\infty,\ C'\to C'_\infty\quad as\quad r\to\infty\end{aligned} \tag{5}$$

By using the Rosseland approximation (Brewster (1992)), the radiative heat flux q_r is given by

$$q_r=-\frac{4\sigma_s}{3k_e}\frac{\partial T'^4}{\partial r} \tag{6}$$

where σ_s is the Stefan-Boltzmann constant and k_e - the mean absorption coefficient. It should be noted that by using the Rosseland approximation, the present analysis is limited to optically thick fluids. If the temperature differences within the flow are sufficiently small, then Equation (6) can be linearized by expanding T'^4 into the Taylor series about T'_∞, which after neglecting higher order terms takes the form

$$T'^4 \cong 4T'^3_\infty T' - 3T'^4_\infty \tag{7}$$

In view of Equations (6) and (7), Equation (3) reduces to

$$\frac{\partial T'}{\partial t'} + u\frac{\partial T'}{\partial x} + v\frac{\partial T'}{\partial r} = \frac{\alpha}{r}\frac{\partial}{\partial r}\left(r\frac{\partial T'}{\partial r}\right) + \frac{16\sigma_s T'^3_\infty}{3k_e \rho c_p}\frac{1}{r}\frac{\partial}{\partial r}\left(r\frac{\partial T'}{\partial r}\right) \tag{8}$$

In order to write the governing equations and the boundary conditions in dimensionless form, the following non-dimensional quantities are introduced.

$$X = \frac{x}{r_0},\ R = \frac{r}{r_0}, U = \frac{ur_0}{\nu},\ V = \frac{vr_0}{\nu},\ t = \frac{\nu t'}{r_0^2}, T = \frac{T' - T'_\infty}{T'_w - T'_\infty}$$
$$K = \frac{K'\nu}{r_0^3}, M = \frac{\sigma B_0^2 r_0^3}{\rho \nu^2},\ C = \frac{C' - C'_\infty}{C'_w - C'_\infty}, Gr = \frac{g\beta r_0^3 (T'_w - T'_\infty)}{\nu^2} \tag{9}$$
$$Gc = \frac{g\beta^* r_0^3 (C'_w - C'_\infty)}{\nu^2},\ Sc = \frac{\nu}{D},\quad N = \frac{k^* k}{4\sigma T'^3_\infty}$$

In view of the Equation (9), the Equations (1), (2), (8) and (4) reduce to the following non-dimensional form

$$\frac{\partial (RU)}{\partial X} + \frac{\partial (RV)}{\partial R} = 0 \tag{10}$$

$$\frac{\partial U}{\partial t} + U\frac{\partial U}{\partial X} + V\frac{\partial U}{\partial R} = Gr\,T + Gc\,C + \frac{1}{R}\frac{\partial}{\partial R}\left(R\frac{\partial U}{\partial R}\right) - \left(M + \frac{1}{K}\right)U \tag{11}$$

$$\frac{\partial T}{\partial t} + U\frac{\partial T}{\partial X} + V\frac{\partial T}{\partial R} = \frac{1}{\Pr}\left(1 + \frac{4}{3N}\right)\frac{1}{R}\frac{\partial}{\partial R}\left(R\frac{\partial T}{\partial R}\right) \tag{12}$$

$$\frac{\partial C}{\partial t} + U\frac{\partial C}{\partial X} + V\frac{\partial C}{\partial R} = \frac{1}{Sc}\frac{1}{R}\frac{\partial}{\partial R}\left(R\frac{\partial C}{\partial R}\right) \tag{13}$$

The corresponding initial and boundary conditions are

$$t \le 0:\ U = 0,\ V = 0,\ T = 0,\ C = 0 \ \text{ for all } X \ge 0 \text{ and } R \ge 0$$
$$t > 0: U = 0,\ V = 0,\ T = X^m, C = X^n \ \ at\ \ R = 1$$
$$U = 0,\ T = 0, C = 0 \ \ at\ \ X = 0 \text{ and } R \ge 1 \tag{14}$$
$$U \to 0,\ T \to 0,\ C \to 0 \ \ as\ \ R \to \infty$$

Knowing the velocity, temperature and concentration fields, it is interesting to find the skin-friction, Nusselt number and Sherwood numbers are defined as follows.

Local and average skin-frictions in non-dimensional form are

$$\tau_X = -\left(\frac{\partial U}{\partial R}\right)_{R=1} \tag{15}$$

$$\bar{\tau} = -\int_0^1 \left(\frac{\partial U}{\partial R}\right)_{R=1} dX \tag{16}$$

Local and average Nusselt numbers in non-dimensional form are

$$Nu_X = -X\left[\frac{\left(\frac{\partial T}{\partial R}\right)}{T_{R=1}}\right] \tag{17}$$

$$\overline{Nu} = -\int_0^1 \left[\frac{\left(\frac{\partial T}{\partial R}\right)_{R=1}}{T_{R=1}}\right] dX \tag{18}$$

Local and average Sherwood numbers in non-dimensional form are

$$Sh_X = -X\left[\frac{\left(\frac{\partial C}{\partial R}\right)_{R=1}}{C_{R=1}}\right] \tag{19}$$

$$\overline{Sh} = -\int_0^1 \left[\frac{\left(\frac{\partial C}{\partial R}\right)_{R=1}}{C_{R=1}}\right] dX \tag{20}$$

3. NUMERICAL TECHNIQUE

In order to solve the unsteady, non-linear, coupled Equations (10) - (13) under the conditions (14), an implicit finite difference scheme of Crank-Nicolson type has been employed.

The finite difference equations corresponding to Equations (10) - (13) are as follows:

$$\frac{\left[U_{i,j}^{n+1} - U_{i-1,j}^{n+1} + U_{i,j}^{n} - U_{i-1,j}^{n} + U_{i,j-1}^{n+1} - U_{i-1,j-1}^{n+1} + U_{i,j-1}^{n} - U_{i-1,j-1}^{n}\right]}{4\Delta X}$$
$$+ \frac{\left[V_{i,j}^{n+1} - V_{i,j-1}^{n+1} + V_{i,j}^{n} - V_{i,j-1}^{n}\right]}{2\Delta R} + \frac{V_{i,j}^{n+1}}{1 + (j-1)\Delta R} = 0 \tag{21}$$

$$\frac{\left[U_{i,j}^{n+1} - U_{i,j}^{n}\right]}{\Delta t} + U_{i,j}^{n}\frac{\left[U_{i,j}^{n+1} - U_{i-1,j}^{n+1} + U_{i,j}^{n} - U_{i-1,j}^{n}\right]}{2\Delta X}$$
$$+ V_{i,j}^{n}\frac{\left[U_{i,j+1}^{n+1} - U_{i,j-1}^{n+1} + U_{i,j+1}^{n} - U_{i,j-1}^{n}\right]}{4\Delta R}$$
$$= Gr\frac{\left[T_{i,j}^{n+1} + T_{i,j}^{n}\right]}{2} + Gc\frac{\left[C_{i,j}^{n+1} + C_{i,j}^{n}\right]}{2}$$
$$+ \frac{\left[U_{i,j-1}^{n+1} - 2U_{i,j}^{n+1} + U_{i,j+1}^{n+1} + U_{i,j+1}^{n+1} - 2U_{i,j}^{n} + U_{i,j+1}^{n}\right]}{2(\Delta R)^2}$$
$$+ \frac{\left[U_{i,j+1}^{n+1} - U_{i,j-1}^{n+1} + U_{i,j+1}^{n} - U_{i,j-1}^{n}\right]}{4[1 + (j-1)\Delta R]\Delta R} - \left(M + \frac{1}{K}\right)\frac{\left[U_{i,j}^{n+1} + U_{i,j}^{n}\right]}{2} \tag{22}$$

$$\frac{[T_{i,j}^{n+1} - T_{i,j}^{n}]}{\Delta t} + U_{i,j}^{n}\frac{\left[T_{i,j}^{n+1} - T_{i-1,j}^{n+1} + T_{i,j}^{n} - T_{i-1,j}^{n}\right]}{2\Delta X}$$
$$+ V_{i,j}^{n}\frac{\left[T_{i,j+1}^{n+1} - T_{i,j-1}^{n+1} + T_{i,j+1}^{n} - T_{i,j-1}^{n}\right]}{4\Delta R}$$

$$= \frac{1}{\Pr}\left(1+\frac{4}{3N}\right)\left(\frac{\left[T_{i,j-1}^{n+1}-2T_{i,j}^{n+1}+T_{i,j+1}^{n+1}+T_{i,j-1}^{n}-2T_{i,j}^{n}+T_{i,j+1}^{n}\right]}{2(\Delta R)^2} + \frac{\left[T_{i,j+1}^{n+1}-T_{i,j-1}^{n+1}+T_{i,j+1}^{n}-T_{i,j-1}^{n}\right]}{4\Pr[1+(j-1)\Delta R]\Delta R}\right) \qquad (23)$$

$$\frac{[C_{i,j}^{n+1}-C_{i,j}^{n}]}{\Delta t}+U_{i,j}^{n}\frac{\left[C_{i,j}^{n+1}-C_{i-1,j}^{n+1}+C_{i,j}^{n}-C_{i-1,j}^{n}\right]}{2\Delta X} + V_{i,j}^{n}\frac{\left[C_{i,j+1}^{n+1}-C_{i,j-1}^{n+1}+C_{i,j+1}^{n}-C_{i,j-1}^{n}\right]}{4\Delta R}$$
$$= \frac{1}{Sc}\frac{\left[C_{i,j-1}^{n+1}-2C_{i,j}^{n+1}+C_{i,j+1}^{n+1}+C_{i,j-1}^{n}-2C_{i,j}^{n}+C_{i,j+1}^{n}\right]}{2(\Delta R)^2} + \frac{\left[C_{i,j+1}^{n+1}-C_{i,j-1}^{n+1}+C_{i,j+1}^{n}-C_{i,j-1}^{n}\right]}{4Sc[1+(j-1)\Delta R]\Delta R} \qquad (24)$$

The region of integration is considered as a rectangle with sides $X_{\max}$ (=1) and $R_{\max}$ (=14), where $R_{\max}$ corresponds to $R=\infty$, which lies very well outside the momentum, energy and concentration boundary layers. The maximum of R was chosen as 14 after some preliminary investigations, so that the last two of the boundary conditions (14) are satisfied. Here, the subscript i - designates the grid point along the X - direction, j - along the R - direction and the superscript n along the t - direction. The appropriate mesh sizes considered for the calculation are ΔX = 0.05, ΔR = 0.25, and time step Δt = 0.01. During any one-time step, the coefficients $U_{i,j}^{n}$ and $V_{i,j}^{n}$ appearing in the difference equations are treated as constants. The values of U, V, T and C are known at all grid points at t = 0 from the initial conditions. The computations of U, V, T and C at time level (*n+1*) using the known values at previous time level (*n*) are calculated as follows.

The finite difference Equation (24) at every internal nodal point on a particular i - level constitute a tri-diagonal system of equations. Such a system of equations is solved by Thomas algorithm as described in Carnahan et al. (1969). Thus, the values of C are found at every nodal point on a particular i at $(n+1)^{\text{th}}$ time level. Similarly, the values of T are calculated from the Equation (23). Using the values of C and T at $(n+1)^{\text{th}}$ time level in the Equation (22), the values of U at $(n+1)^{\text{th}}$ time level are found in a similar manner. Thus the values of C, T and U are known on a particular i - level. The values of V are calculated explicitly using the Equation (21) at every nodal point on a particular i - level at $(n+1)^{\text{th}}$ time level. This process is repeated for various i - levels. Thus, the values of C, T, U and V are known at all grid points in the rectangular region at $(n+1)^{\text{th}}$ time level. Computations are carried out till the steady state is reached. The steady state solution is assumed to have been reached, when the absolute difference between the values of U as well as temperature T and concentration C at two consecutive time steps are less than 10^{-5} at all grid points.

After experimenting with few sets of mesh sizes, they have been fixed at the level ΔX =0.05, ΔR =0.25, and the time step Δt =0.01.In this case, spatial mesh sizes are reduced by 50% in one direction, and then in both directions, and the results are compared. It is observed that, when the mesh size is reduced by 50% in X - direction and R - direction, the results differ in the fourth decimal places. The computer takes more time to compute, if the size of the time-step is small. Hence, the above mentioned sizes have been considered as appropriate mesh sizes for calculation.

The local truncation error in the finite-difference approximation is $O\left(\Delta t^2+\Delta R^2+\Delta X\right)$ and it tends to zero as ΔX, ΔR and Δt tend to zero. Hence the scheme is compatible. Stability and compatibility ensures convergence (1967). The derivatives involved in the Equations (15) - (20) are evaluated using five-point approximation formula and the integrals are evaluated using Newton-Cotes closed integration formula.

4. RESULTS AND DISCUSSION

A representative set of numerical results is shown graphically in Figs. 1-12, to illustrate the influence of governing parameters viz., radiation parameter N, thermal Grashof number Gr, solutal Grashof number Gc, magnetic parameter M, permeability parameter K, Prandtl number $\Pr$, Schmidt number Sc, exponent in the power law variation of the wall temperature m, exponent in the power law variation of the wall concentration n on the velocity, temperature and concentration, skin-friction, Nusselt number and Sherwood number. Here the value of Pr is chosen as 0.71, which corresponds air. The values of Sc are chosen such that they represent water vapour (0.6) and Ammonia (0.78).

In order to ascertain the accuracy of the numerical results, the present study is compared with the previous study. The velocity profiles for Gc = 2.0, N = 0.0, M = 0.0, Pr = 0.7, X = 1.0, Sc =0.6 are compared with the available solution of Rani (2003), in Fig.1. It is observed that the present results are in good agreement with that of Rani (2003).

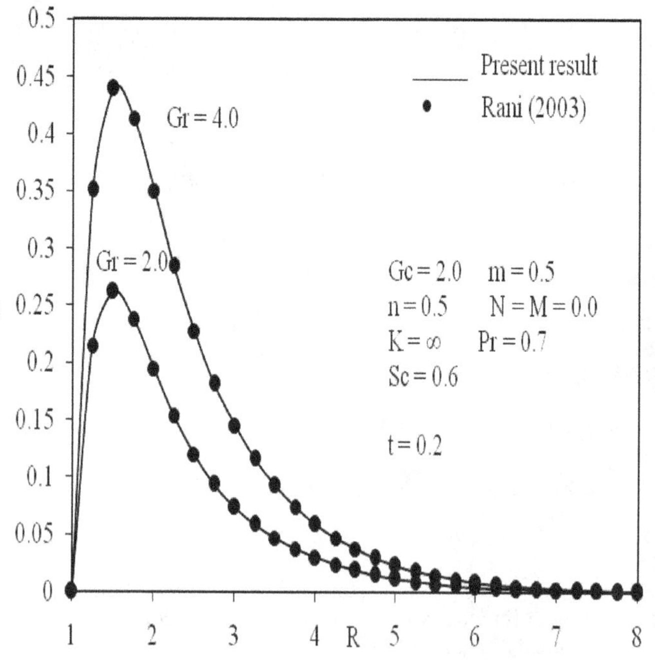

Fig.1 Comparison of velocity profiles.

The transient and steady state velocity profiles at X=1.0 for different values Gr, Gc, M, K and Sc are shown in Fig. 2. The thermal Grashof number signifies the relative effect of the thermal buoyancy (due to density differences) force to the viscous hydrodynamic force in the boundary layer flow. Here the positive values of Gr correspond to cooling of the cylinder. As expected, it is noticed that an increase in Gr leads to a rise in the values of velocity due to enhancement in the buoyancy force. The solutal Grashof number Gc defines the ratio of the species buoyancy force to the viscous hydrodynamic force. It is found that the fluid velocity increases and the peak value become more distinctive due to an increase in the concentration buoyancy force represented by Gc (Fig.2.). The contribution of mass diffusion to the buoyancy force increases the maximum velocity significantly. The time required to reach the steady state velocity increases as Gr or Gc increases. It is observed that an increase in the magnetic field parameter leads to a decrease in the velocity field. It is because that the application of transverse magnetic field will result in a resistive type force (Lorentz force) similar to drag force which tends to resist the fluid flow and thus reducing its velocity. Also, the boundary layer thickness decreases with an increase in the magnetic parameter. A rise in permeability parameter accompanying a decrease in porous resistance in the momentum equation (11) i.e. $-\left[\frac{1}{K}\right]U$, induces a substantial rise in transient velocity, U. The Schmidt number Sc signifies the ratio of the momentum diffusivity to the mass (species) diffusivity. It physically relates the relative thickness of the hydrodynamic boundary layer and mass-transfer (concentration) boundary layer. It is noticed that as the Schmidt number increases the velocity decreases. It is observed that the time required to reach the steady state velocity increases with an increase in M or Sc .

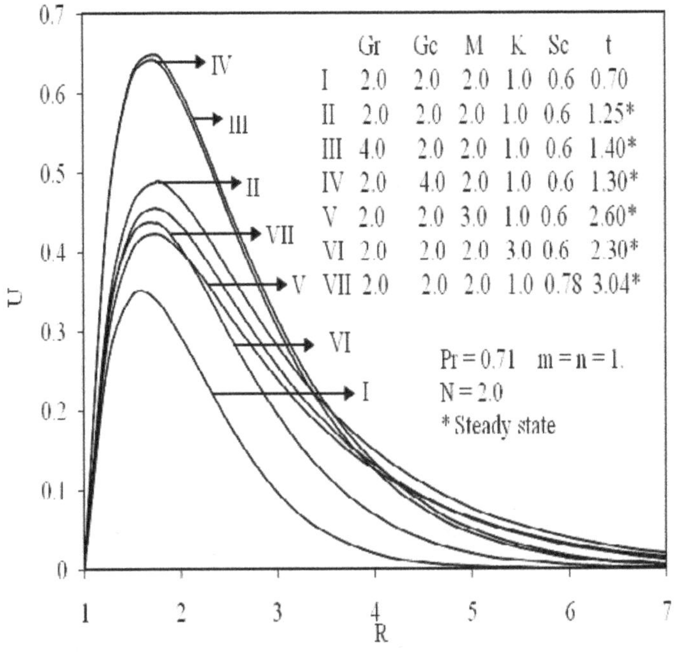

Fig.2 Transient velocity profiles at X=1.0 for different Gr, Gc, M , K and Sc.

Figure 3 depicts the transient and steady state velocity profiles for different values of radiation parameter N , m and n . The radiation parameter N (i.e., Stark number) defines the relative contribution of conduction heat transfer to thermal radiation transfer. It can be found that an increase in N leads to a decrease in the velocity within the boundary layer as well as decreased thickness of the hydrodynamic boundary layer. It is also noticed that the velocity decreases with an increase in m or n . The time required to reach the steady state velocity increases as N or m or n increases.

The transient and steady-state temperature profiles at X=1.0 for different values of m , n and N are shown in Fig.4. It is found that as N increases from 2.0 to 5.0, the temperature decreases markedly throughout the length of the cylinder. As a result the thermal boundary layer thickness decreases due to a rise in N values. Also, it is noticed that the temperature decreases as m or n increases. The effect of m is more important on temperature than n. It is also seen that the time taken to reach the steady state temperature increases with an increase in m or n or N.

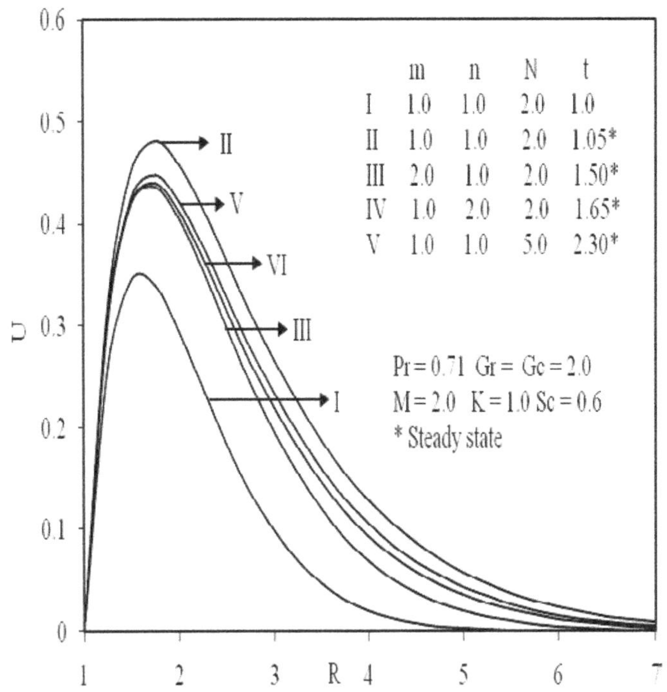

Fig. 3 Transient velocity profiles at X=1.0 for different m, n and N.

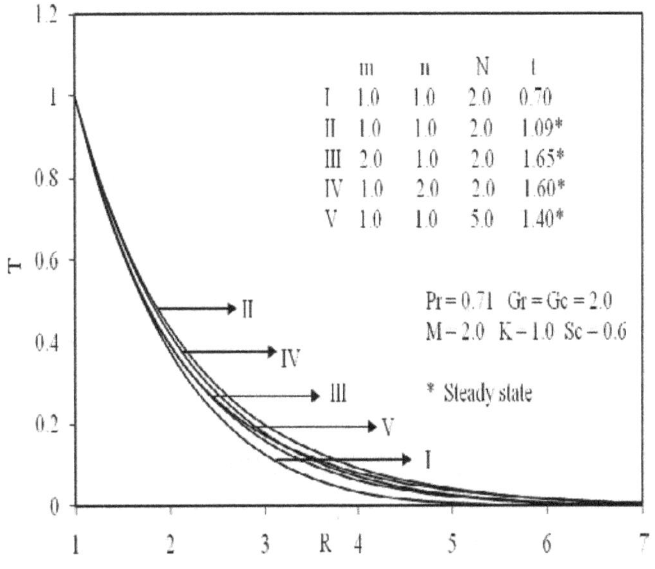

Fig.4 Transient temperature profiles at X=1.0 for different m, n and N.

The concentration profiles for different values of m and n are shown in Fig. 5. It is found that the concentration decreases with an increase in m or n . Here, the effect of n is more important than the effect of m . The time required to reach the steady state concentration increases with an increase in m or n . The transient and steady-state concentration profiles at X=1.0 for different values of N and Sc are shown in Fig.6. It is seen that the concentration increases with an increase in N . As Sc increases the mass transfer rate increases. Hence, the concentration decreases as Sc increases. The time required to reach the steady state concentration increases with an increase in Sc or N .

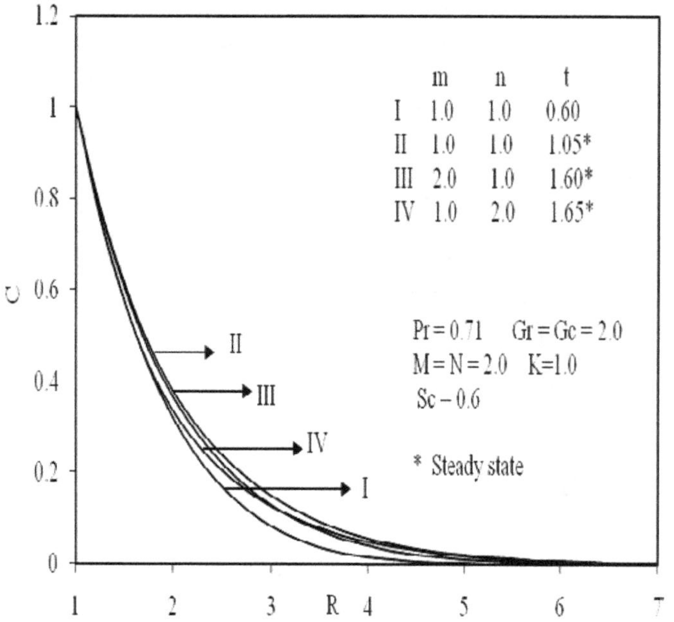

Fig.5 Transient concentration profiles at X = 1.0 for different m and n.

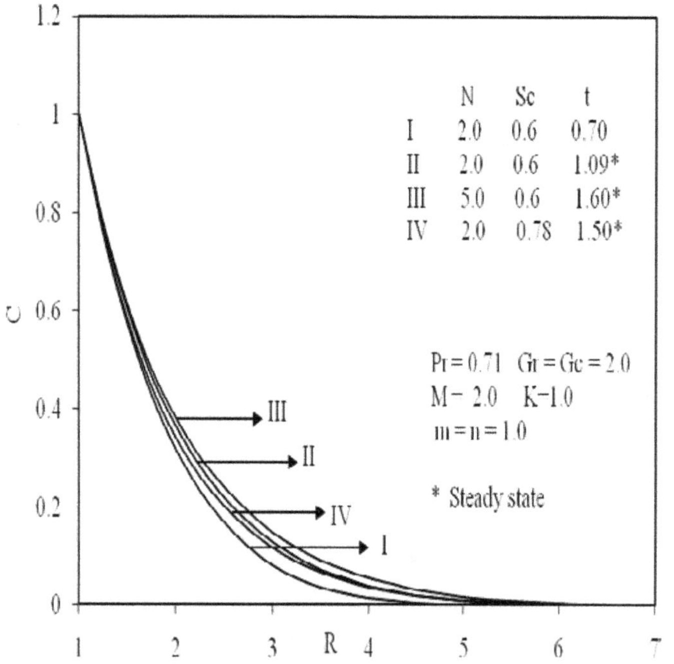

Fig.6 Transient profiles at X= 1.0 for different N and Sc.

The effects of m, n, N and Sc on the local skin-friction (τ_x) are shown in Fig. 7. It is observed that, the local skin-friction decreases with an increase in n or N or Sc, while it increases with an increase in m. Fig. 8 displays the effects of m, n, N and Sc on the average skin-friction ($\overline{\tau}$). The average skin-friction decreases with an increase in n or N or Sc, while it increases with an increase in m. The local Nusselt number (Nu_X) for different values of m, n, N and Sc are shown in Fig. 9. It is noticed that, the local Nusselt number decreases with an increase in m or n or Sc, while it increases with an increase in N. Fig.10 shows the effects of m, n, N and Sc on the average Nusselt number ($\overline{Nu}$). The average Nusselt number decreases with an increase in m or n or Sc, while it increases with an increase in N. The effects of m, n, N and Sc on the local and average Sherwood numbers are shown in Figs. 11 and 12 respectively. It is seen that both the local and average Sherwood numbers increase with an increase in n or Sc, while they decrease with an increase in m or N.

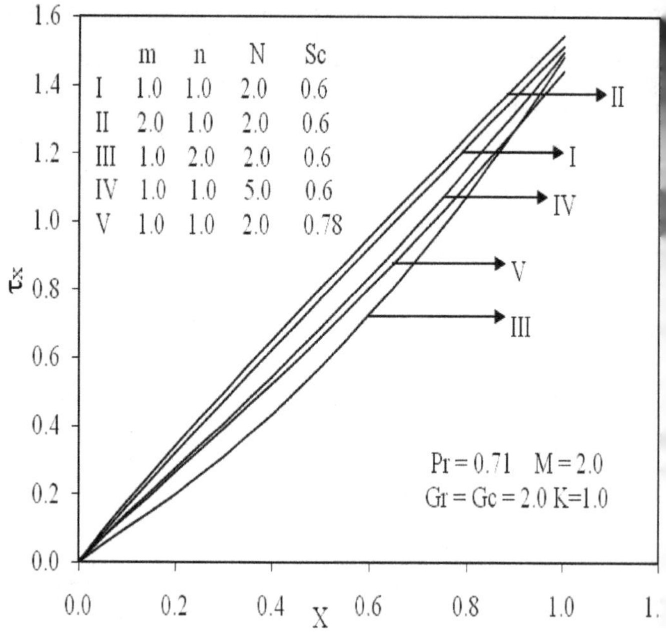

Fig.7 Local skin-friction.

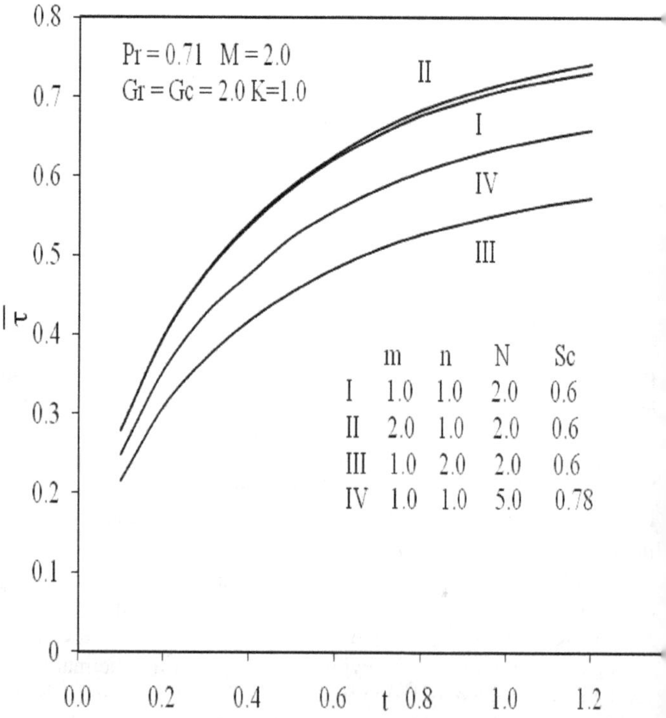

Fig.8 Average skin-friction.

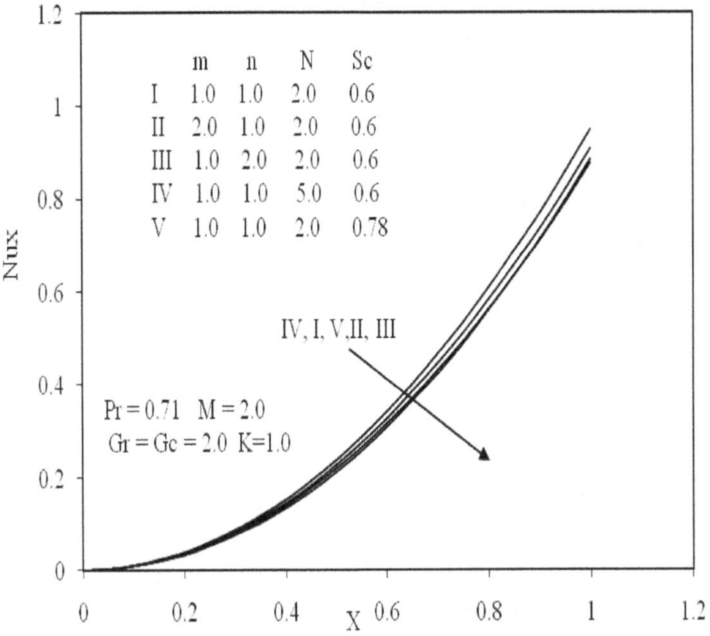

Fig. 9 Local Nusselt number.

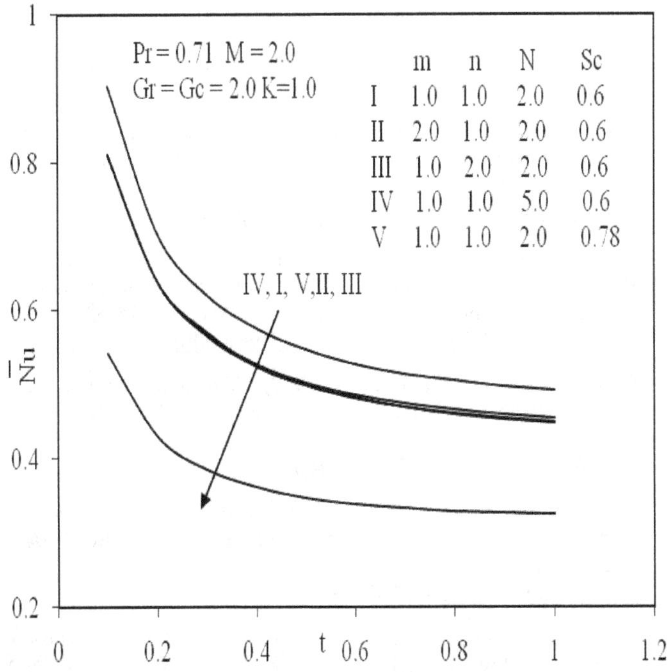

Fig. 10 Average Nusselt number.

5. CONCLUSIONS

A numerical study has been carried for the thermal radiation unsteady MHD natural convection past a vertical cylinder with variable surface temperature and mass diffusion. The fluid is gray, absorbing-emitting but non-scattering medium and the Rosseland approximation is used to describe the radiative heat flux in the energy equation. The dimensionless governing equation is derived. A Crank-Nicolson type of implicit method is used to solve these equations. This study is compared with the available solution in the literature and good agreement is found to exist.

1. As the magnetic field parameter M increases, the transient velocity decreases.
2. A rise in permeability parameter induces a substantial rise in transient velocity U .
3. The transient velocity increases with an increase in Gr or Gc.
4. At small values of the radiation parameter N , the velocity and temperature of the fluid increases sharply near the cylinder as the time t increase.
5. The concentration reduces with an increase in m or n or Sc.
6. The local and average skin-friction $\overline{\tau}$ decreases with an increase M and increases with increasing value of N and Sc .
7. The average Nusselt number $\overline{Nu}$ increases with increasing value of radiation parameter N and decreasing values of Sc .
8. The Sherwood number $\overline{Sh}$ increases as Sc increases.

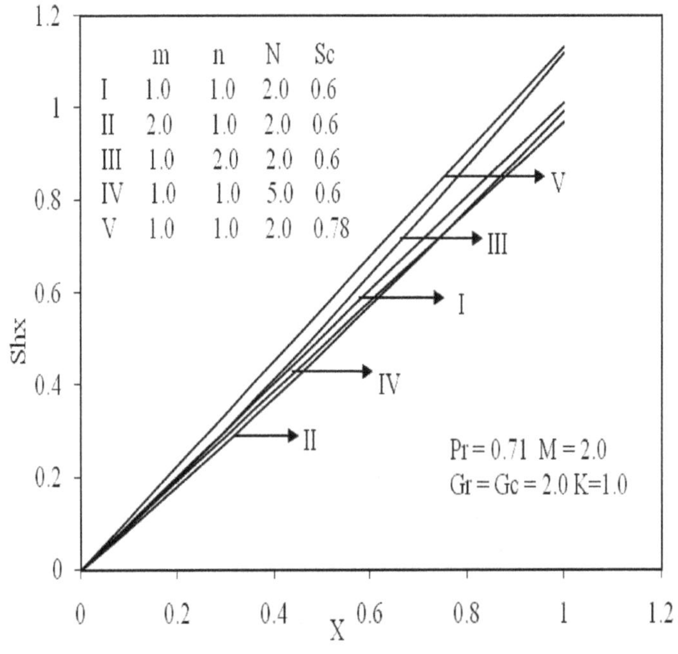

Fig.11 Local Sherwood number.

ACKNOWLEDGEMENTS

The author is very thankful to the editor and reviewers for their encouraging comments and constructive suggestions to improve the presentation of this manuscript.

NOMENCLATURE

B_0	magnetic field strength
C'	species concentration
C	dimensionless species concentration
D	the species diffusion coefficient
Gr	thermal Grashof number
Gc	modified Grashof number

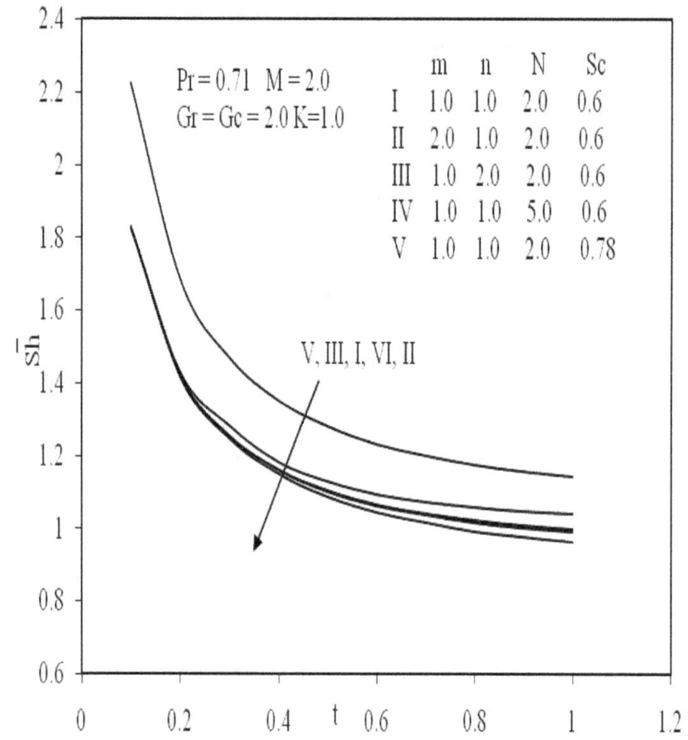

Fig.12 Average Sherwood number.

g	acceleration due to gravity
i	grid point along the X - direction
j	grid point along the R - direction
m	exponent in power law variation of wall temperature
n	exponent in power law variation of wall concentration
K	permeability parameter
M	magnetic parameter
N	radiation parameter
$\overline{Nu}$	average Nusselt number
Nu_x	local Nusselt number
Pr	Prandtl number
q_r	radiative heat flux
R	dimensionless radial co-ordinate
r	radical co-ordinate
r_0	radius of cylinder
Sc	Schmidt number
$\overline{Sh}$	average Sherwood number
Sh_X	local Sherwood number
T	temperature
x	axial co-ordinate measured vertically upward direction
t	time
X	dimensionless axial co-ordinate
U,V	dimensionless velocity components in X,R directions

Greek symbols

α	thermal diffusivity
β	volumetric coefficient of thermal expansion
$\beta*$	volumetric coefficient of expansion with concentration
Δt	grid size in time
ΔR	grid size in radical direction
ΔX	grid size in axial direction
k_e	mean absorption coefficient
ν	kinematic viscosity
ρ	density
σ_s	Stefan-Boltzmann constant
τ_x	local skin-friction
$\overline{\tau}$	average skin-friction

Subscripts

w	condition on the wall
∞	ree-stream condition

REFERENCES

Abd El-Naby, M.A., Elsayed, M.E., Elbarbary, M.E., and Nader, Y.A., 2003, "Finite Difference Solution of Radiation Effects on MHD Free Convection Flow over a Vertical Plate with Variable Surface Temperature," *J. Appl. Math.*, **2**, 65-86. http://dx.doi.org/10.1155/S1110757X0320509X

Agarwal, A.K., Kishor, B., and Raptis, A., 1989, "Effect of MHD Free Convection and Mass Transfer on the Flow Past a Vibrating Infinite Vertical Circular Cylinder," *Heat Mass Transfer*, **24**, 243-250.

Bottemanne, F.A., 1972, "Experimental Results of Pure and Simultaneous Heat and Mass Transfer by Free Convection about a Vertical Cylinder for $\Pr = 0.71$ and $Sc = 0.63$," *Appl. Sci. Res.*, **25**, 372-381. http://dx.doi.org/10.1007/BF00382310

Brewster, M.Q., 1992, *Thermal Radiative Transfer and Properties*, John Wiley & Sons, New York.

Carnahan, B., Luther, H.A. and Wilkes, J.O. 1969, *Applied Numerical Methods*, John Wiley & Sons, New York.

Chen, T.S., and Yuh, C.F., 1980, "Combined Heat and Mass Transfer in Natural Convection along a Vertical Cylinder," *Int. J. Heat Mass transfer*, **23**, 451-461. http://dx.doi.org/10.1016/0017-9310(80)90087-3

Elbashbeshy, E.M.A., 1997, "Heat and Mass Transfer along a Vertical Plate with Variable Surface Temperature and Concentration In The Presence Of Magnetic Field," *Int. J. Eng. Sci.*, **34**,515-522. http://dx.doi.org/10.1016/S0020-7225(96)00089-4

Ganesan, P., and Rani, H.P., 1998, "Transient Natural Convection Cylinder With Heat And Mass Transfer," *Heat Mass Transfer*, **33**,449-455. http://dx.doi.org/10.1007/s002310050214

Ganesan, P., and Rani, H.P., 1999, "Unsteady Free Convection on A Vertical Cylinder with Variable Heat and Mass Flux," *Heat Mass Transfer*, **35**, 259-265. http://dx.doi.org/10.1007/s002310050322

Ganesan, P. and Loganathan, P., 2002, "Radiation and Mass Transfer Effects on Flow of an Incompressible Viscous Fluid Past a Moving Vertical Cylinder," *Int. J. Heat Mass Transfer*, **45**, 4281-4288. http://dx.doi.org/10.1016/S0017-9310(02)00140-0

Ganesan, P., and Rani, H.P., 2003, "Magnetic Field Effect on a Moving Vertical Cylinder with Constant Heat Flux," *Heat Mass Transfer*, **39**, 381-386.

Gnaneswara Reddy, M., and Bhaskar Reddy, N., 2009, "Radiation and Mass Transfer Effects On Unsteady MHD Free Convection Flow Past A Moving Vertical Cylinder," *Theoret. Appl. Mech.*, **36**(3), 239-260. http://dx.doi.org/10.2298/TAM0903239G

Gnaneswara Reddy, M., 2012, "Unsteady Free Convective Flow Past a Semi-infinite Vertical Plate with Uniform Heat and Mass Flux," *Journal of Petroleum and Gas Exploration Research*, **2**(3), 052-056.

Gnaneswara Reddy, M., 2013, "Chemically Reactive Species and Radiation Effects on MHD Convective Flow Past a Moving Vertical Cylinder," *Ain Shams Engineering Journal*, **4**, 879–888. http://dx.doi.org/10.1016/j.asej.2013.04.003

Lee, H.R., Chen, T.S. and Armaly, B.F. 1988, "Natural Convection along Slender Vertical Cylinders with Variable Surface Temperature," *J. Heat transfer*, **100**, 103-108. http://dx.doi.org/10.1115/1.3250439

Rani, H.P., 2003, "Transient Natural Convection along a Vertical Cylinder with Variable Surface Temperature and Mass Diffusion," *Heat Mass Transfer*, **40**, 67-73. http://dx.doi.org/10.1007/s00231-002-0372-1

Richtmyer Robert, D., and Morton, K.W., 1967, *Difference Methods for Initial-Value Problems*, John Wiley & Sons, New York.

Sparrow, E.M., and Gregg, J.L., 1956, "Laminar Free Convection Heat Transfer From the Outer Surface of a Vertical Circular Cylinder," *Trans. ASME*, **78**, 1823-1830.

Takhar, H.S., Chamkha, A.J., and Nath, G., 2000, "Combined Heat and Mass Transfer along a Vertical Cylinder with Free Stream," *Heat Mass Transfer*, **36**, 237-246. http://dx.doi.org/10.1007/s002310050391

Tien, C.L., and Vafai, K., 1990, "Convective and Radiative Heat Transfer in Porous Media," *Adv. Appl. Mech.* **33**, 225–281.

THERMAL CHARACTERIZATION OF AS4/3501-6 CARBON-EPOXY COMPOSITE

Bradley Doleman[a], Messiha Saad[a,*]

[a] *North Carolina A&T State University, Greensboro, NC, 27411, USA*

ABSTRACT

Thermal diffusivity, specific heat, and thermal conductivity are important thermophysical properties of composite materials. These properties play a significant role in the engineering design process of space systems, aerospace vehicles, transportation, energy storage devices, and power generation including fuel cells. This paper examines these thermophysical properties of the AS4/3501-6 composite using the xenon flash method to measure the thermal diffusivity in accordance with ASTM E1461 and differential scanning calorimetry to measure the specific heat in accordance with ASTM E1269. The thermal conductivity was then calculated using a proportional relationship between the density, specific heat, and thermal diffusivity.

Keywords: *Thermal Diffusivity; Specific Heat; Thermal Conductivity; Xenon Flash Method*

1. INTRODUCTION

As today's technology continues to develop at a rate that was once unimaginable, the demand for new materials that will outperform traditional materials also increases dramatically. To meet these challenges, monolithic materials are being combined to develop new unique materials called composites. The formation of composites provides properties unobtainable separately with either constituent. Besides improvements in the mechanical properties such as tensile strength, stiffness, and fatigue endurance, materials must retain functionality at much higher operating temperatures than before. Due to extreme temperatures, material properties may alter in operation resulting in severely reduced properties which may lead to catastrophic failures during usage. Thermal properties play a significant role in design applications, determining safe operating temperatures, process control characteristics, and quality assurance of these materials.

The objective of this paper is to develop a thermal properties database for the carbon-epoxy AS4/3501-6 composites. The AS4 carbon fiber used is a unidirectional continuous PAN based fiber. The 3501-6 epoxy resin is amine cured and provides low shrinkage during the curing process while maintaining excellent resistance to chemicals and solvents. The 3501-6 was developed to operate in a temperature environment up to 350°F (177°C). The AS4/3501-6 carbon-epoxy composite used in the investigation is an 8-ply laminate compiled of laminas alternating between 0° and 90° orientations. The composite material has a high gloss and smooth black finish that was surface treated to improve the fiber-to-resin interfacial bond strength which met the Hexcel aerospace specification, HS-CP-5000.

The thermophysical properties of AS4/3501-6 carbon-epoxy have been investigated using experimental methods. The flash method was used to measure the thermal diffusivity of the composite. This method is based on the American Society for Testing and Materials standard, ASTM E1461. In addition, the Differential Scanning Calorimeter was used in accordance with the ASTM E1269 standard to measure the specific heat. The measured thermal diffusivity, specific heat, and density data were used to compute the thermal conductivity of the AS4/3501-6 carbon-epoxy composite. Thermal conductivity is the property that determines the working temperature levels of the material; it plays a critical role in the performance of materials in high temperature applications, and it is an important parameter in problems involving heat transfer and thermal structures.

The materials used in the investigation were developed and fabricated at the Center for Composite Materials Research (CCMR) at North Carolina A&T State University. The mechanical properties of the AS4 composite have been measured (Akangah and Shivakumar, 2013). The AS4/3501-6 autoclave processed carbon/epoxy composite was made from a unidirectional [non-woven] carbon fiber/epoxy tape pre-preg molded in its particular [8-ply 0/90-degree] stacking sequence of the following specification: fiber area weight of 150 g/m^2 with a resin content of 33% [that is 33% weight portion of the uncured pre-preg tape, which is the area weight of the fiber plus the area weight of the resin = 150 g/m^2 (fiber) + 74 g/m^2 (resin) = 224 g/m^2 (total area weight of the uncured pre-preg)]. AS4/3501-6 is a carbon composite commonly used and recommended for general purpose structural applications (United States, 1999). Manufactured by Hexcel Corporation, the AS4's epoxy resin has properties that allows for the composite to withstand temperatures up to 200°C as well as provide a high strength to weight ratio. In short, it can withstand some high temperature applications and also provide the light weight desired in industries to reduce fuel consumption. For example, this material can be used for wind power turbine blade applications (3501-6, 1998; HexTow, 2010).

2. THERMAL DIFFUSIVITY MEASUREMENTS

In order to determine the working temperature levels of a material, it is expedient to measure how rapidly heat will pass through that material. This measurement, or property, is called thermal diffusivity. Thermal diffusivity plays a critical role in the performance of materials in high temperature applications, and this material property can be measured in several different ways. There are steady-state methods as well as transient techniques. Available procedures include Thermal Wave Interferometry (TWI), Thermographic methods, the Flash Method, the

* *Corresponding Author Email: saad@ncat.edu*

Hot-wire method, and others. Transient techniques have been preferred in measuring thermal properties of materials more recently; however, the Flash Method is the most common of these methods (Nunes dos Santos, 2007; Patrick and Saad, 2012).

2.1 The Flash Method

In 1961, W.J. Parker founded the Flash method, and it is the most frequently used transient photothermal technique and has the versatility of using a xenon lamp or laser as the energy source. In many countries, it is considered a standard for thermal diffusivity measurements of solid materials (Cernuschi *et al.*, 2004). As adopted by the United States, the laser flash method is a standard test method and is defined by the American Society for Testing and Materials E1461 standard. It involves a small cylindrical, thin disk specimen being heated to a desired temperature, usually between 20 and 500°C. Once the disk has reached the specified temperature the front face is subjected to a quick radiant energy pulse as shown in Fig. 1. The energy source can be a laser or a lamp. A detector measures the resulting temperature change with respect to time on the rear face of the sample. The data acquisition system then records the temperature change of the rear face of the specimen versus time. A graphical representation of this data is called the thermogram of the flash. Figure 2 displays the theoretical model thermogram. The time in which it takes the rear face of the specimen to reach half the maximum temperature rise is called the halftime, $t_{1/2}$.

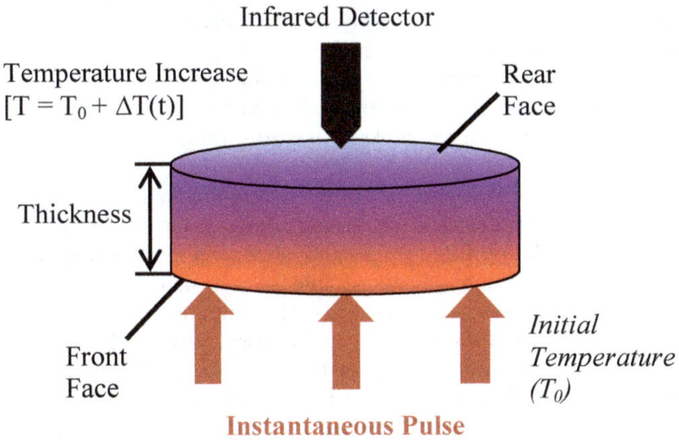

Fig. 1 Schematic of the Flash Method

Employing Carslaw and Jeager's equation of the temperature distribution within a thermally insulated solid of uniform thickness L, Parker *et al.* (1961) was able to derive a mathematical expression to calculate thermal diffusivity (Carslaw *et al.*, (1959). An abbreviated version of this derivation is given as:

$$T(x,t) = \frac{1}{L}\int_0^L T(x,0)dx + \frac{2}{L}\sum_{n=1}^{\infty} exp\left(\frac{-n^2\pi^2 at}{L^2}\right) cos\frac{n\pi x}{L}\int_0^L T(x,0)cos\frac{n\pi x}{L}dx \quad (1)$$

where α is the thermal diffusivity of the solid material. If a pulse of radiant energy, *Q*, is instantaneously and uniformly absorbed into a small depth referred as *g*, at the front face (x=0) of the thermally insulated solid material, the temperature distribution at that initial point is given by:

$$T(x,0) = \frac{Q}{\rho \cdot c_p \cdot g} \quad \text{for} \quad 0 < x < g \quad (2)$$

$$T(x,0) = 0 \quad \text{for} \quad g < x < L \quad (3)$$

These initial conditions are substituted into the temperature distribution equation above. It is considered that the very small depth g of an opaque solid will yield a small angle. Additionally, it is known that for a very small angle, θ, sin(θ) ≈ θ and cos(nπx/L) = $(-1)^n$. Once this is applied, the temperature distribution at the rear face (x=L) is expressed as:

$$T(L,t) = \frac{Q}{\rho \cdot C \cdot L}\left[1 + 2\sum_{n=1}^{\infty}(-1)^n exp\left(\frac{-n^2\pi^2}{L^2}\alpha t\right)\right] \quad (4)$$

Parker *et al.* (1961) then defined two dimensionless parameters, V and ω as:

$$V(L,t) = \frac{T(L,t)}{T_m} \quad (5)$$

$$\omega = \frac{\pi^2 \alpha t}{L^2} \quad (6)$$

where T_m is the maximum temperature at the rear face. Combining equations 4, 5, and 6 results in (Parker *et al.*, 1961):

$$V = 1 + 2\sum_{n=1}^{\infty}(-1)^n(e^{-\omega n^2}) \quad (7)$$

Setting V = 0.5 allows for the determination of ω, and then substitution into equation 7 allows for a mathematical equation for thermal diffusivity to be stated as:

$$\alpha = 0.1388\frac{L^2}{t_{1/2}} \quad (8)$$

This derivation by W.J. Parker is a theoretical model of the flash method and is the ideal case. It assumes that the specimen is mostly homogeneous and isotropic, that there is one dimensional heat flow, and that there are no heat losses from the specimen (Cernuschi *et al.*, 2002). It also assumes that energy pulse is uniformly subjected across the front face of the specimen and that the pulse is instantaneous. Because of this assumption, many researchers have developed correction factors since the time of Parker's original derivation. These included but are not limited to Cowan, Clark and Taylor, Koski, and Heckman. Each of these correction factors use different methods or a combination of methods to reanalyze the theoretical model and impose additional parameters. The Clark and Taylor correction factor accounts for radiation heat losses and is used in the research conducted in this experiment. Clark and Taylor (1975) examined the thermogram at different points before the maximum temperature rise was reached and developed a correction factor. The correction factor is calculated using the following equation:

$$K_r = -0.3461467 + 0.361578\left(\frac{t_{3/4}}{t_{1/4}}\right) - 0.06520543\left(\frac{t_{3/4}}{t_{1/4}}\right)^2 \quad (9)$$

Specifically, they analyzed the time to reach 25 percent and 75 percent of the maximum temperature change. The corrected thermal diffusivity equation as defined by Clark and Taylor is

$$\alpha_{corrected} = \frac{\alpha K_R}{0.13885} \quad (10)$$

2.2 Experimental Apparatus

In general, the ASTM E1461 standard (2007) provides the minimum requirements for the apparatus. The key components are the flash source, specimen holder, temperature response detector, recording device, and an environmental enclosure is needed when testing above and below room temperature (ASTM, 2007). The flash source can be any device able to emit a quick energy pulse, usually a lamp or laser. The apparatus used in this facility was commercialized and purchased from Anter Corporation. The name of the apparatus is the FlashLine ™ 2000, and it utilizes a high intensity xenon lamp as the pulse source. The pulse duration time should be less than 2% halftime of the specimen to be measured in order to keep the error due to finite pulse less than 0.5%. The apparatus is automated and capable of testing up to four specimens in each run.

The thermal property analyzer also contains a vacuum capable environmental enclosure in which nitrogen gas is used to evacuate the chamber. The detector should be any sensor that can measure a linear electrical output proportional to a small temperature rise, and it, along with its amplifier, must have a response time of no more than 2% of the half-time. The Indium Antimonide (InSb) Infrared Detector outputs a linear electrical signal proportional to a small temperature change experienced by the rear face of the specimen after the energy pulse has been initiated. The data acquisition system can be pre-programmed within one time period for the acceptable resolution of at least 1% for the quickest thermogram the system can deliver (ASTM, 2007). Figure 2 shows the normalized temperature versus the normalized time.

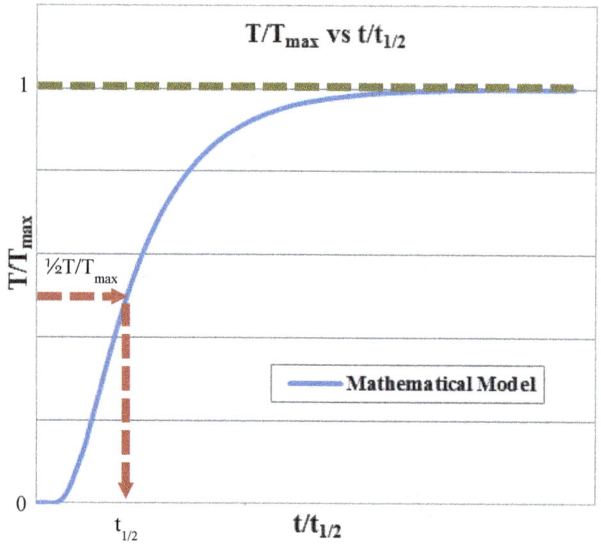

Fig. 2 Characteristic Thermogram of the Flash Method

2.3 Test Specimen Preparation

The test specimens were prepared to be thin circular disks of 10 to 30 mm in diameter, whose front face surface area is less than that of the energy pulse beam. According to ASTM E1461 standard (2007) each specimen should be thick enough to be representative of the test material but remain within 1 to 6 mm range. Overall, the optimum thickness depends upon the magnitude of the estimated thermal diffusivity and should be chosen so that the time to reach half of the maximum temperature falls within the 10 to 1000 ms range. In order to accomplish these specified dimensions, a drill press equipped with a diamond plated drill bit was used to cut the material to the appropriate diameter. When necessary, the specimens were milled to achieve the preferred thickness.

Both the rear and front faces were flat and parallel within 0.5% of their thickness to maintain pulse uniformity. The standard suggests that a thin, uniform layer of graphite be applied to both faces of the specimens to improve the capability of absorbing the applied energy flash by reducing the reflection from the specimen. This was not necessary for the experiments performed in this work due to the material nature of the AS4 composite.

2.4 Experimental Procedure

The experiments were conducted following the testing standard, ASTM E1461. 12.7 mm (0.5 inch) and 25.4 mm (1 inch) diameter samples were used depending on the availability or size limitations of the material. This will not account for any changes in the thermal diffusivity. The diameter, thickness, mass, and density were documented. Each sample was placed in the specimen holder housed inside a vacuum sealed environmental enclosure. The environmental enclosure was purged using nitrogen gas to form an inert environment for the samples.

Approximately 1 L of liquid nitrogen was manually poured in the receptacle of the IR detector. The sample thickness, diameter, and mass were input into the FlashLine™ 2000 System, and the test was initiated at ambient temperature. Each sample was tested to a maximum temperature of 175°C which is the service temperature of the material for general purpose structural applications. At each designated temperature, a minimum of three flashes were performed at a time. The results were compiled, analyzed, and necessary corrections were made. The time required for each experimental run varies depending on the range of temperatures tested and the temperature increment.

3. SPECIFIC HEAT MEASUREMENTS

Specific heat is a measurement of the amount of heat per unit mass that is required to raise the temperature of a material one degree Celsius. The differential scanning calorimetry (DSC) technique was used to measure the specific heat of the materials. This technique is based on the measurement of the change of the difference in the heat flow rate to the sample and to a reference sample while they are subjected to a controlled temperature program. Using the measured heat flow rate of the sample, differential scanning calorimetry can determine how a material's heat capacity varies with respect to temperature.

3.1 Differential Scanning Calorimetry (DSC)

Widely used for the measurement of specific heat, differential scanning calorimetry (DSC) is a thermoanalytical technique. Its methodology is defined by the ASTM E1269 standard. When performing a differential scanning calorimetry measurement, a test specimen and reference are enclosed in the same furnace together on a metallic block with high thermal conductivity within the calorimeter. The metallic block ensures a good heat-flow path between the specimen and reference. The sample and the reference are subjected to an identical temperature program. The heat capacity change in the specimen differs from that of the reference. The calorimeter measures the temperature difference and calculates heat flow from calibration data. As a result, the specific heat of the sample can be calculated using the heat flow results. Differential scanning calorimetry is an ASTM test method standard for determining specific heat capacity (ASTM, 2005).

To calculate the specific heat of an unknown material, the heat flux of the unknown and a reference must be measured using the differential scanning calorimeter. Using the measured heat flux and the known specific heat of the reference, the specific heat of the unknown material can be calculated by using the ratio method technique. Since the differential scanning calorimeter is at constant pressure, the change in enthalpy of the reference is equal to the heat absorbed or released by the reference (ASTM, 2005). The depiction mathematically is:

$$dQ = (m)dh \tag{11}$$

Equation (11) leads to the following relationship:

$$\dot{Q} = m\frac{dh}{dt} = m\frac{dq}{dt} \tag{12}$$

where *dq/dt* is the specific heat rate, and *dh/dt* is the change of enthalpy with respect to time. At constant pressure, the relationship for specific heat can be written as:

$$c_p = \frac{1}{m}\frac{dH}{dT} \tag{13}$$

Using the chain rule, the equation can be rewritten as:

$$c_p = \frac{1}{m}\frac{dt}{dT}\frac{dH}{dt} = \frac{1}{m}\frac{dQ}{dt}\frac{dt}{dT} \tag{14}$$

From equation (14), the specific heat can be written as:

$$c_p = \frac{\left(\frac{dQ}{dt}\right)\left(\frac{dt}{dT}\right)}{m} \tag{15}$$

where *dt/dT* is the inverse temperature distribution over time. Using equation (15) and the ratio method, the calibration constant, E, is multiplied through the specific heat equation for the reference:

$$c_{p_{ref}} = E\frac{\left(\frac{dQ}{dt}\right)_{ref}\left(\frac{dt}{dT}\right)}{m_{ref}} \tag{16}$$

Solving for the calibration constant:

$$E = \frac{c_{p_{ref}} m_{ref}}{\left(\frac{dQ}{dt}\right)_{ref}\left(\frac{dt}{dT}\right)} \tag{17}$$

To determine the specific heat of the unknown material, equation (16) is used again but in terms of the sample:

$$c_p = \frac{E}{m}\left(\frac{dQ}{dt}\right)\left(\frac{dt}{dT}\right) \tag{18}$$

Substituting equation (17) into equation (18):

$$c_p = \left[\frac{c_{p_{ref}} m_{ref}}{\left(\frac{dQ}{dt}\right)_{ref}\left(\frac{dt}{dT}\right)}\right]\left(\frac{1}{m}\right)\left(\frac{dQ}{dt}\right)\left(\frac{dt}{dT}\right) \tag{19}$$

The above equation can be reduced to:

$$c_p = c_{p_{ref}}\frac{m_{ref}}{m}\frac{\left(\frac{dQ}{dt}\right)}{\left(\frac{dQ}{dt}\right)_{ref}} \tag{20}$$

3.2 Experimental Apparatus

The calorimeter used in this research is the DSC 200 F3 Maia®, Differential Scanning Calorimeter manufactured by NETZSCH. It is a heat flux system that combines high stability, high resolution, and fast response time throughout a substantial temperature range. With the addition of the Intracooler 40, the temperature range extends from ambient temperature to cryostatic temperatures covering a larger temperature spectrum (-40˚C to 600˚C). The heating rate is adjustable from as low as 0.001K/min to as high as 100K/min while keeping a temperature accuracy of 0.1 K.

The DSC 200 F3 Maia® Differential Scanning Calorimeter consists of a furnace block, sample chamber, cooling system, heat flux sensor, and purge gas. The furnace block contains a miniature jacketed heater that provides the source of heat during the experiment. The furnace temperature is measured by a thermocouple integrated into the furnace walls. The sample chamber is sealed within the instrument's lid and has two additional lids to prevent a contamination from outside sources. The system's temperature can be reduced by using an Intracooler device. The calorimeter uses a high sensitivity type E heat flux sensor for its measurements (NETZSCH, 2008).

3.3 Test Specimen Preparation

Good thermal contact between the heat flux sensor and sample is vital for optimum results. To achieve this, the sample should lay as flush as possible with the bottom of the aluminum crucible. A 4-mm or 6-mm diameter and 1-mm thick sample can be used with this equipment using the corresponding crucible size. Then each sample was weighed three times, and the average mass was documented. Each sample was placed into the crucible, and a lid was positioned on top of the crucible to fully enclose the sample. Using tweezers, the crucible was then carefully placed on the heat flux sensor making sure the crucible was centered on the sensor.

3.4 Experimental Procedure

The differential scanning calorimetry experiment was performed following the ASTM E1269 testing standard for determining specific heat. The differential scanning calorimeter and data acquisition system were initialized and were allowed to reach thermal equilibrium. During this period the apparatus was purged with argon gas at a rate of 50 mL/min to produce an inert testing atmosphere. To measure the specific heat of a sample, a minimum of three runs must be performed.

Before the specific heat of the AS4 composite was determined, baseline and reference tests were performed. Since the samples were placed inside an aluminum crucible for testing, the crucible will add a contact resistance to the samples. The baseline corrects for this contact resistance increasing the accuracy of the results. The initial baseline run was performed by placing two empty crucibles in the designated location on heat flux sensors as seen in Fig. 3. The furnace was heated to the designated initial temperature of the program and held there isothermally at least four minutes while the calorimeter recorded the thermal curve. The crucibles were heated to the final temperature at a rate of 20°C/min and held isothermally again while the calorimeter recorded the thermal curve.

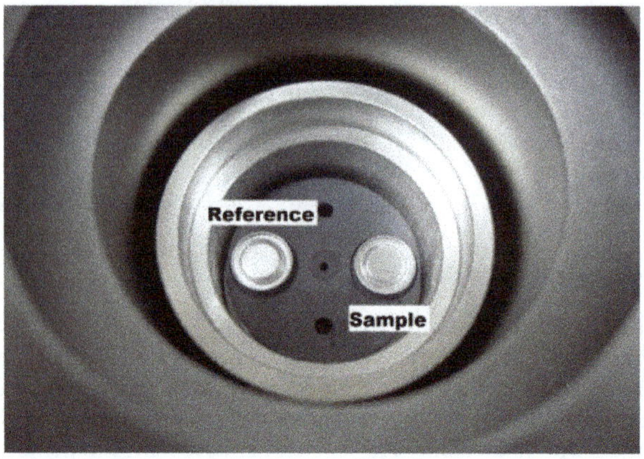

Fig. 3 Crucibles Placed on the Heat Flux Sensor

Following the baseline run, the calorimeter testing chamber was cooled to ambient temperature. The crucible on reference location in the testing chamber was replaced with a sapphire reference. After applying the previous baseline to correct for the aluminum crucible, the same temperature program used for the baseline was executed for the sapphire reference. The measured specific heat of the sapphire was compared to the known specific heat value for sapphire to determine the error. The test was repeated for the AS4/3501-6 samples. To verify that the baseline did not alter, a baseline was established after every fourth test. Using the measured sapphire as a reference, the ratio method was used to determine the specific heat of the AS4 carbon-epoxy.

4. RESULTS AND DISCUSSIONS

The flash method was used to measure the thermal diffusivity of the AS4/3501-6 composite. The thermal diffusivity was measured between room temperature and 175°C which is the temperature limitation for this material. Thermal diffusivity decreases over the temperature range by approximately 15% as shown in Fig. 4. This is apparent because the energy pulse will travel through the material faster at room temperature. In other words, the specific heat will be higher because it will take more heat to raise the temperature of the material by one degree Celsius thus signifying that the material has a lower threshold for heat at higher temperatures.

Three shots were recorded at each temperature during the experiment for each test. The data points represent the shots, or flashes, and the error at two standard deviations is displayed in Fig. 5. While there is more variation at the temperature range of 50° to 100°C, which is depicted by larger error bars, all of the data still falls within two standard deviations. This indicates that the error is minimal.

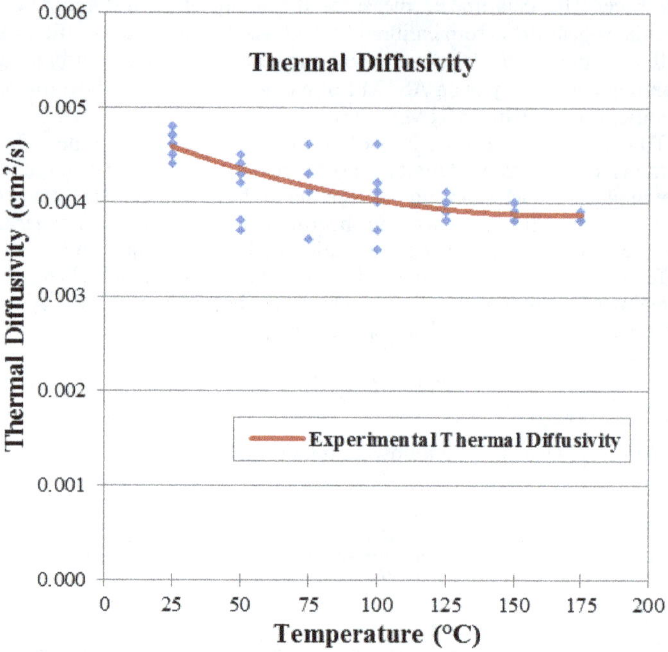

Fig. 4 Thermal Diffusivity of the AS4/3501-6 Composite

The error was calculated at each temperature, and the error bars shows the interval where 95% of the data collected at each shot should lie. Population data sets provide the statistical estimates known as the population mean value and the population standard deviation defined by (Figliola and Beasley, 2011):

$$\sigma = \sqrt{\frac{1}{N}\sum_{i=1}^{N}(a_i - \bar{a})^2} \tag{21}$$

The results from each temperature tested are considered as a population where N is the total number of measurements, a_i represents a single i^{th} measurement, and ā is the mean value of the data at each temperature. Additionally, the margins of error were calculated for each temperature using standard error of the mean as shown in equation (22).

$$SE = \frac{\sigma}{\sqrt{N}} \tag{22}$$

The critical value from the normal distribution chart based on 95% confidence is 1.96, and the margin of error, *e*, is determined by the following:

$$e = (\pm 1.96)SE \tag{23}$$

The percent error can now be found using:

$$error\ \% = \frac{e}{mean} \times 100\ \% \quad (24)$$

The results for the thermal diffusivity percent error at the corresponding temperatures have been calculated and are displayed in Table 1.

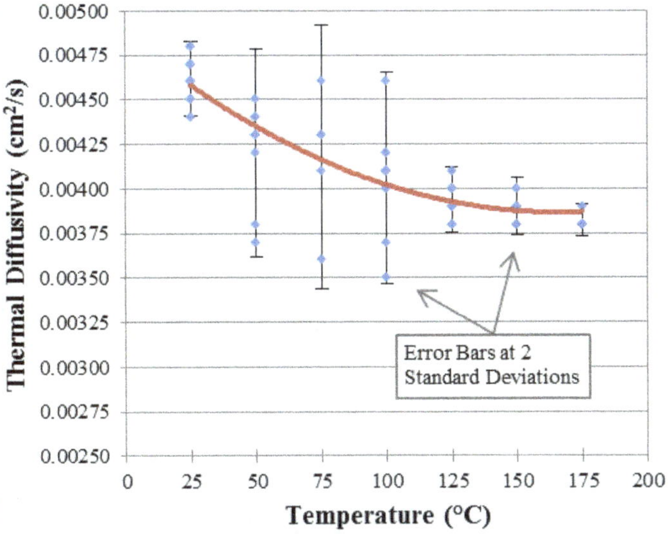

Fig. 5 Magnified View of the Thermal Diffusivity Error Bars

Table 1 Percent Error of the Thermal Diffusivity

Temperature (°C)	*Mean* (cm^2/s)	*Standard Deviation* (cm^2/s)	*Margin of Error* ($\pm$ cm^2/s)	*Percent Error* (%)
25	4.615E-03	1.01366E-04	4.44255E-05	0.96
50	4.200E-03	2.73861E-04	1.20025E-04	2.86
75	4.180E-03	3.31059E-04	1.45093E-04	3.47
100	4.073E-03	2.69986E-04	1.18327E-04	2.91
125	3.936E-03	8.81396E-05	3.86289E-05	0.98
150	3.900E-03	7.55929E-05	3.31300E-05	0.85
175	3.813E-03	3.30719E-05	1.44944E-05	0.38

According to the testing standard, the optimum thickness of the tested samples should be chosen so that the time to reach half of the maximum temperatures (half-time), $t_{1/2}$, falls within the 10 to 1000 ms (0.01 to 1 s) range (ASTM, 2007). To verify that the samples were the proper thickness, an initial test was performed to check the half-times of the tested material. The half-times attained at each temperature were recorded. The corresponding half-times at different temperatures are shown in Fig. 6. The half-time for each sample fell within the acceptable range according to the testing standard verifying that the proper thickness was chosen. According to the ASTM standard, the thickness may vary (1 to 6 mm) based on the thermal conductivity of the material (i.e., the more conductive a material is, the larger the thickness can be). The optimum thickness of the AS4 composite used in this research was 1.140 mm.

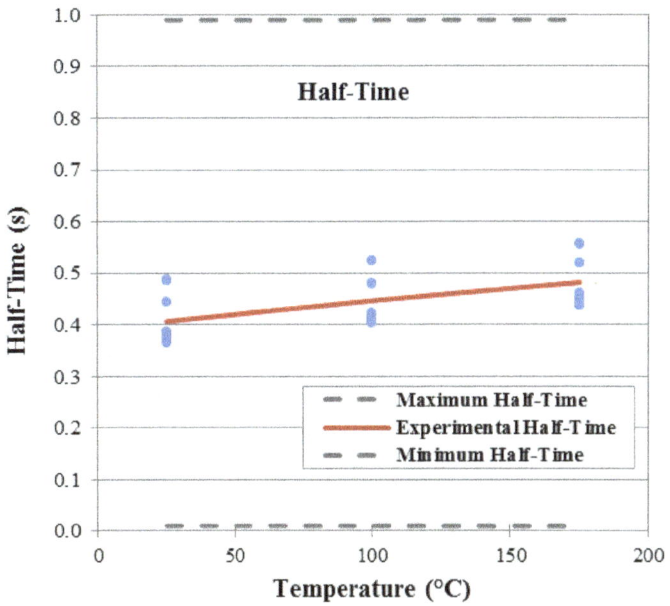

Fig. 6 Half-Times of the AS4/3501-6 Composite

The material has also been tested at the Oak Ridge National Laboratory (ORNL) in Oak Ridge, Tennessee. The apparatus at ORNL is capable of attaining cryogenic thermal diffusivity measurements. The subzero results for the AS4 composite are displayed and compared with the results from North Carolina A&T's Thermal Characterization Research Laboratory (TCL) as shown in Fig. 7. The TCL apparatus uses an infrared detector to measure the temperature changes; on the other hand, ORNL uses thermocouples instead. The results of the thermal diffusivity at both labs were consistent and within a 5% margin of error.

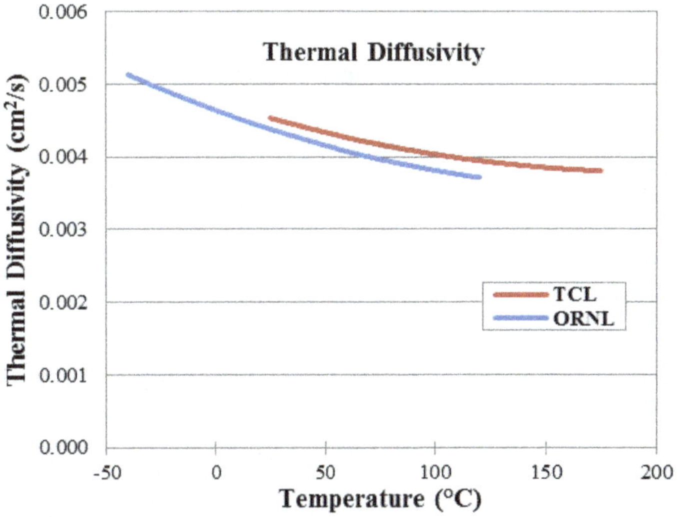

Fig. 7 Thermal Diffusivity of the AS4/3501-6 Composite

The specific heat of the AS4/3501-6 composite consistently increases from approximately 0.9 J/g °C at room temperature to 1.4 J/g °C at 175°C as shown in Fig. 8. The material has been tested at the TCL and also at ORNL. An agreement has been achieved at both labs. Figure 8 displays the specific heat measurements of TCL versus ORNL. The material has been tested at NETZSCH, and the data is also displayed in Fig. 8. All three specific heat results overlap indicating accurate and precise measurements. The heating curve was used in the analysis.

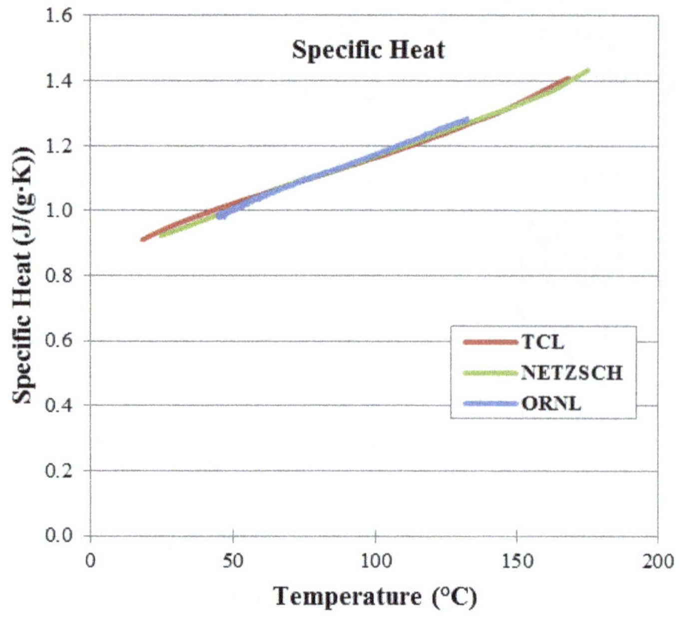

Fig. 8 Specific Heat of the AS4/3501-6 Composite

Using the density (the density for the tested AS4/3501-6 composite was 1.46 g/cm^3), specific heat, and thermal diffusivity, the thermal conductivity of the AS4/3501-6 composite was determined using the following relationship:

$$k = \rho c_p \alpha \tag{25}$$

The results for the thermal conductivity of the AS4/3501-6 composite are shown in Fig. 9. There is a 20% increase in the thermal conductivity of the composite over the service temperature range.

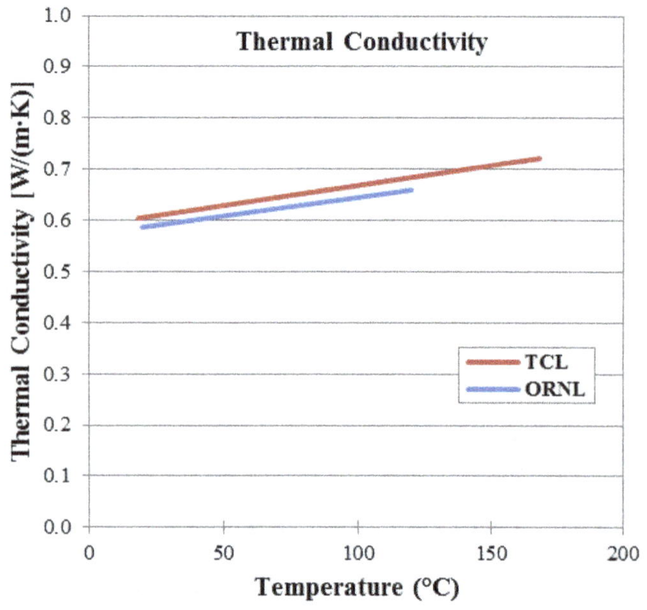

Fig. 9 Thermal Conductivity of the AS4/3501-6 Composite

Figure 10 shows the thermogram of the AS4/3501-6 carbon-epoxy composite as it compares with the mathematical model. The thermogram displays the relationship between the temperature divided by the maximum temperature versus the time divided by the half-time. The data acquisition system can be pre-programmed within one time period for the acceptable resolution of at least 1% for the quickest thermogram the system can deliver which is represented by the mathematical model.

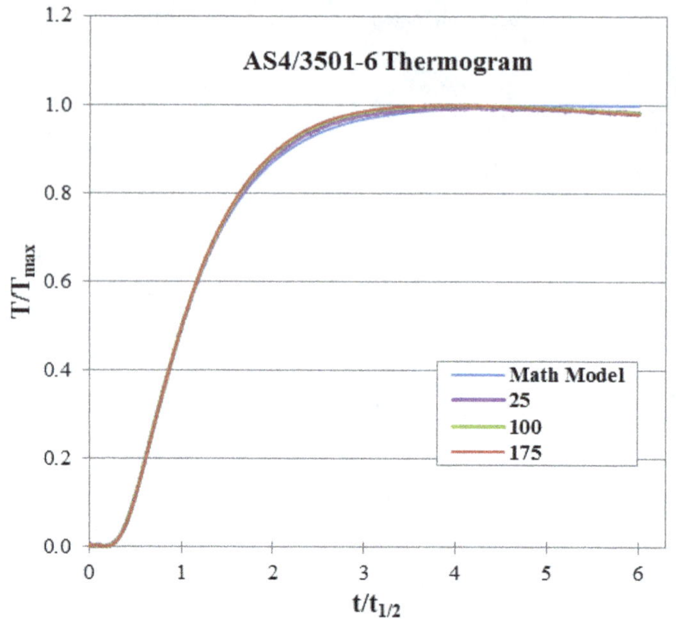

Fig. 10 Thermogram of the AS4/3501-6 Composite

As shown in Figure 10, immediately before the temperature reaches its maximum, it is noticed that there is slightly more radiation heat loss at room temperature than the 100°C and 175°C measurements. After the flash occurs, the heat will radiate more from the sample in a lower temperature environment than in a higher temperature environment. The AS4/3501-6 composite shows relatively no heat loss with respect to temperature. Table 2 shows the sample specifications of the AS4 composite while table 3 shows the measured thermal properties.

Table 2 Specifications of AS4/3501-6 Testing Specimens

Test	*Diameter* mm (inch)	*Thickness* mm (inch)	*Mass* (g)
Specific Heat	4.00 (0.157)	0.960 (0.038)	0.02013
Thermal Diffusivity	12.69 (0.50)	1.135 (0.447)	0.21000

Table 3 Thermal Properties of AS4/3501-6

Temperature (°C)	*Density* (g/cm^3)	*Thermal Diffusivity* (cm^2/s)	*Specific Heat* (J/g*K)	*Thermal Conductivity* (W/m*K)
25	1.4629	0.004615	0.94171	0.6127445
50	1.4629	0.004200	1.02508	0.6347415
75	1.4629	0.004180	1.09773	0.6492697
100	1.4629	0.004073	1.16040	0.6622167
125	1.4629	0.003936	1.23705	0.6803271
150	1.4629	0.003900	1.32987	0.7088875
175	1.4629	0.003813	1.40960	0.7376534

5. CONCLUSIONS

The thermophysical properties database of the AS4/3501-6 composite has been developed over the service temperature range of the material. Using the flash method, the thermal diffusivity was measured through the thickness. Specific heat was measured using the differential scanning calorimeter. The thermal conductivity was determined using the density, specific heat, and thermal diffusivity of the composite. The thermophysical properties have been validated using different apparatuses at ORNL and TCL from 25°C to 175°C. This composite material is suitable for general purpose structural applications.

ACKNOWLEDGEMENTS

The authors would like to acknowledge NASA-URC-Center for Aviation Safety (CAS) (Grant# NNX09AV08A), the Center for Composite Materials Research (CCMR) at North Carolina A&T State University, and the Oak Ridge National Laboratory's High Temperature Materials Laboratory (HTML) sponsored by the U.S. Department of Energy, Office of Energy Efficiency and Renewable Energy, and Vehicle Technologies Program.

NOMENCLATURE

$\bar{a}$	measurement mean
a_i	i^{th} measurement of the sample
c_p	specific heat (J/kg·K)
E	calibration constant
e	margin of error
g	small depth into sample (cm)
h	specific enthalpy (J/kg)
k	thermal conductivity (W/m·K)
K_r	correction factor
L	sample thickness (cm)
m	mass (kg)
N	number of measurements
Q	heat (J)
t	time (s)
T	temperature (K)
V	dimensionless quantity
x	distance (cm)
Greek Symbols	
α	thermal diffusivity (cm^2/s)
Δ	differential quantity
ρ	density (kg/m^3)
σ	population mean
ω	dimensionless quantity
Subscripts	
0	initial time step
½	half
max	maximum
ref	reference

REFERENCES

"3501-6 Epoxy Matrix," *Hexcel Product Data* (1998). http://www.hexcel.com/Resources/DataSheets/Prepreg-Data-Sheets/3501-6_eu.pdf.

Akangah, P. and Shivakumar, K., 2013, Assessment of Impact Damage Resistance and Tolerance of Polymer Nanofiber Interleaved Composite Laminates. *Journal of Chemical Science and Technology*, 2(2), 39-52.

Anter, "FlashLine™ Thermal Diffusivity Measuring System Operation and Maintenance Manual Part 1 Flashline 2000".

ASTM Standard E1269, 2005, "Standard Test Method for Determining Specific Heat Capacity by Differential Scanning Calorimetry," ASTM International, West Conshohocken, PA.

ASTM Standard E1461, 2007, "Standard Test Method for Thermal Diffusivity by the Flash Method," ASTM International, West Conshohocken, PA.

Baker, D., Reaves, R., and Saad, M., 2011, "Thermal Characterization of Carbon-Carbon Composites," IMECE2011-64061, *Proceedings of 2011 ASME International Mechanical Engineering Congress and Exposition,* Denver, CO. http://dx.doi.org/10.1115/IMECE2011-64061

Carslaw, H. S. and Jaeger, J. C., 1959, Conduction of Heat in Solids, Oxford University Press, New York.

Cernuschi, F., Bison, P.G., Figari, A., Marinetti, S., Grinzato, E., and Lorenzoni, L., 2002, "Comparison of Thermal Diffusivity Measurement Techniques," *Quantitative InfraRed Thermography Journal*, 28, 211-221.

Cernuschi, F., Bison, P.G., Figari, A., Marinetti, S., and Grinzato, E., 2004, "Thermal Diffusivity Measurements by Photothermal and Thermographic Techniques," *International Journal of Thermophysics* 25(2), 439-457. http://dx.doi.org/10.1023/B:IJOT.0000028480.27206.cb

Clark, L. M. and Taylor, R. E., 1975, "Radiation Loss in the Flash Method for Thermal Diffusivity," *Journal of Applied Physics*, 46(2), 714-719. http://dx.doi.org/10.1063/1.321635

Figliola, R. and Beasley, D., Theory and Design for Mechanical Measurements, John Wiley & Sons, 2011.

"HexTow® AS4," *Hexcel Product Data* (2010). http://www.hexcel.com/resources/datasheets/carbon-fiber-data-sheets/as4c.pdf.

Höhne, G., Hemminger, W. F., and Flammersheim, H. J., 2003, *Differential Scanning Calorimetry*, Springer, Germany.

NETZSCH, 2008, "Technical Data Sheet DSC 200 F3 Maia".

Nunes dos Santos, W., 2007, "Thermal Properties of Polymers by Non-Steady-State Techniques," Polymer Testing, 26(4), 556-566. http://dx.doi.org/10.1016/j.polymertesting.2007.02.005

Parker, W. J., Jenkins, R. J., Butler, C. P., and Abbott, G. L., 1961, "Flash Method of Determining Thermal Diffusivity Heat Capacity and Thermal Conductivity," *Journal of Applied Physics*, 32(9), 1679-1685. http://dx.doi.org/10.1063/1.1728417

Patrick, M. and Saad, M., 2012, "Examination of Thermal Properties of Carbon-Carbon and Graphitized Carbon-Carbon Composites," *Frontiers in Heat and Mass Transfer (FHMT)* 3 - 043007 (2012). http://dx.doi.org/10.5098/hmt.v3.4.3007

United States of America. Department of Defense. *Composite Materials Handbook Vol 2. Polymer Matrix Composites Materials Properties*. 1999. http://www.lib.ucdavis.edu/dept/pse/resources/fulltext/HDBK17-2F.pdf

LAMINAR NATURAL CONVECTION STUDY IN A QUADRANTAL CAVITY USING HEATER ON ADJACENT WALLS

Dipak Sen*, Probir Kumar Bose, Rajsekhar Panua, Ajoy Kumar Das, Pulak Sen

Mechanical Engineering Department, National Institute of Technology, Agartala , Tripura, 799055, India

ABSTRACT

A numerical analysis of laminar natural convection in a quadrantal cavity filled with water having variable length heaters attached on the adjacent walls have been made to examine heat and fluid flow. Numerical solutions are obtained using a commercial computational fluid dynamics package, FLUENT, using the finite volume method. Effects of the Rayleigh number, Ra, on the Nusselt number, Nu, as well as velocity and temperature fields are investigated for the range of Ra from 10^3 to 10^7. Computations were carried out for the non-dimensional heater lengths on the vertical wall (m=0.2, 0.4 and 0.6) and horizontal wall (n=0.2, 0.4 and 0.6). It is observed that heat transfer increases with increase in Rayleigh number and the flow strength increases with increase in size of heater on the vertical wall compared to the bottom wall and temperature fields are also affected. In contrast, with increase in size of heater on both side of adjacent walls flow strength does not changes significantly.

Keywords: *Quadrantal cavity, heater, Rayleigh number, Nusselt number*

1. INTRODUCTION

Natural convection in complex enclosures is an area of interest for several researchers. Simple square, rectangular or triangular cavities have been studied elaborately, but quadrantal enclosure has evoked considerable interest. This fact is amply reflected by the size of the research efforts during the past few decades dedicated to this topic. A detailed study of flow and heat transfer phenomena in quadrantal enclosures is useful in understanding the processes that occur in natural convection flows in building insulation materials, cooling of electrical equipments, solar collectors, geothermal applications, oil destruction etc.

The existing literature presents a vast number of studies on natural convection in enclosures. However, most of these studies have been related to either a vertically or a horizontally imposed heat flux or temperature difference. There is little work regarding natural convection in enclosures with heater used neighboring horizontal and vertical walls. Chu et al. (1976) investigated the effect of heater size, location, aspect ratio and boundary condition on two-dimensional, laminar, natural convection in rectangular channels both experimentally and numerically. They found that the maximum Nusselt number is obtained almost for all Rayleigh numbers when heater is located on the middle of the wall. In another study, Turkoglu and Yucel (1995) made a numerical study using control volume approach for the effect of heater and cooler locations on natural convection in cavities. They indicated that for a given cooler position; mean Nusselt number increases as the heater ismoved closer to the bottom horizontal wall. An experimental and numerical study of natural convection in a quadrantal cavity heated and cooled on adjacent walls was reported by Aydin and Yesiloz (2011a) and for inclined quadrantal cavity was reported by Aydin and Yesiloz (2011b). Chu and Hickox (1990) investigated the thermal convection with viscosity variation in a cavity with localized heating both experimentally and numerically. Besides these studies sometimes natural convection can be seen in cavities with discrete wall heat sources as indicated in Deng et al. (2002). Aydin et al. (1999) conducted a numerical study on buoyancy-driven laminar flow in an inclined square enclosure heated from one side and cooled from the adjacent side by using finite difference methods. In all of these studies, solution domain was chosen as square enclosure. Aydin and Yang (2000) studied the natural convection in an enclosure partially heated from the bottom wall and symmetrically cooled from the side walls. They observed that symmetrical flow fields are obtained when heater located at the centre of the bottom wall. Triangular shaped enclosures were investigated by some authors due to its shape is useful especially in the roof design or some of electronical devices. In the study of Asan and Namli (2001), the laminar natural convection heat transfer in triangular shaped roofs with different inclination angle and Rayleigh number in winter day conditions is investigated numerically using the finite volume method. They indicated that both aspect ratio and Rayleigh number affect the temperature and flow field. They also found that heat transfer decreases with the increasing of aspect ratio. Akinsete and Coleman (1982) illustrated the natural convection heat transfer in a triangular enclosure in steady-state regime. Moukalled andAcharya (2001) solved the governing equations of natural convection heat transfer inside a trapezoidal shaped geometry with baffles for building roofs in the conditions of summerlike and winter-like. They observed that in winter-like conditions, convection starts to dominate at a Rayleigh number much lower than that in summerlike conditions. Recently, Tzeng et al. (2005) proposed the Numerical Simulation Aided Parametric Analysis method to solve natural convection equations in streamline-vorticity form.

The aim of the present study is to investigate numerically the buoyancy-induced flow and heat transfer mechanisms in a water-filled quadrantal cavity from the heaters attached on both the vertical and horizontal walls, while curved wall is cold. To the best knowledge of the authors, this is the first natural convection study on this geometry with these boundary conditions.

* Corresponding Author Email: dipak_sen@ymail.com

2. PROBLEM DESCRIPTION

A schematic diagram of the physical domain is shown in Fig.1. A portion of both the vertical and the bottom wall is composed of heaters, the rest portion comprises of the adiabatic walls. The length of the heaters in the vertical and the horizontal wall is represented by non-dimensional number m, the ratio of length of heater and length of vertical wall and n, the ratio of length of heater and length of horizontal wall respectively. The curved wall is considered as the cold wall. The flow is assumed to be steady and laminar. Constant fluid properties are assumed, except for the density changes with temperature that induce buoyancy forces, so the Boussinesq approximation is adopted.

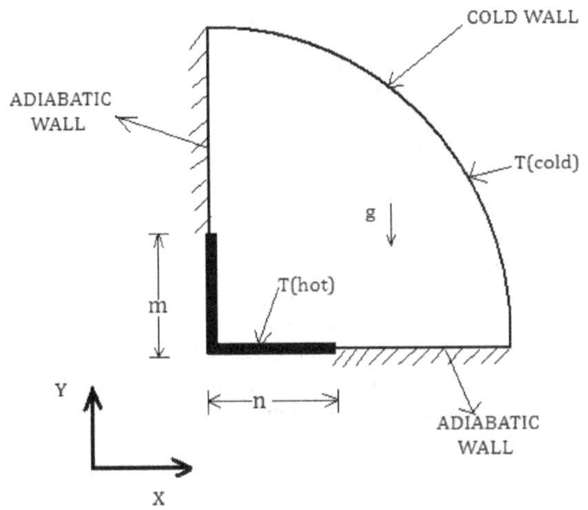

Fig.1 Schematic diagram of the physical domain

3. GOVERNING EQUATIONS

The continuity, momentum and energy equations for a two dimensional laminar flow of an incompressible Newtonian fluid is considered. Following assumptions are made: there is no viscous dissipation, the cavity walls are impermeable, the gravity acts in negative y-direction, fluid properties are constant and fluid density variations are neglected except in the buoyancy term (the Boussinesq approximation) and radiation heat exchange is negligible. Using non-dimensional variables defined in the nomenclature, the non-dimensional governing equations are obtained as:

$$\frac{\partial U}{\partial X}+\frac{\partial V}{\partial Y}=0 \tag{1}$$

$$U\frac{\partial U}{\partial X}+V\frac{\partial U}{\partial Y}=-\frac{\partial P}{\partial X}+\Pr\left(\frac{\partial^2 U}{\partial X^2}+\frac{\partial^2 U}{\partial Y^2}\right) \tag{2}$$

$$U\frac{\partial V}{\partial X}+V\frac{\partial V}{\partial Y}=-\frac{\partial P}{\partial Y}+\Pr\left(\frac{\partial^2 V}{\partial X^2}+\frac{\partial^2 V}{\partial Y^2}+Ra\Pr\theta\right) \tag{3}$$

$$U\frac{\partial \theta}{\partial X}+V\frac{\partial \theta}{\partial Y}=\frac{\partial^2 \theta}{\partial X^2}+\frac{\partial^2 \theta}{\partial Y^2} \tag{4}$$

Appearing in Eqs. (2) and (3), Pr and Ra are the Prandtl and Rayleigh numbers, respectively, which are defined as

$$\Pr=\frac{\nu}{\alpha},\ Ra=\frac{g\beta H^3(T_h-T_C)}{\nu\alpha} \tag{5}$$

where β and υ are the thermal expansion coefficient and the kinematic viscosity of the fluid, respectively.

The non-dimensional parameters are listed as

$$X=\frac{x}{H},Y=\frac{y}{H},U=\frac{u}{\alpha/H},V=\frac{v}{\alpha/H},P=\frac{pH^2}{\rho\alpha^2}$$
$$\theta=\frac{T-T_C}{T_h-T_C} \tag{6}$$

3.1. Boundary conditions

Through the introduction of the non-dimensional parameters into the physical boundary conditions illustrated in Fig. 1, the following non-dimensional boundary conditions are obtained:

On the curved wall

$$\theta=0,\ U=V=0 \text{ at } 0<X<1 \text{ and } 0<Y<1 \tag{7}$$

On the bottom wall,

For the heater part:

$$\theta=1,\ U=V=0 \text{ at } Y=0 \text{ and } 0<X<n \tag{8}$$

For the adiabatic part:

$$\frac{\partial\theta}{\partial Y}=0,\ U=V=0 \text{ at } Y=0 \text{ and } n<X<1 \tag{9}$$

On the vertical wall,

For the heater part:

$$\theta=1,\ U=V=0 \text{ at } X=0 \text{ and } 0<Y<m \tag{10}$$

For the adiabatic part:

$$\frac{\partial\theta}{\partial X}=0,\ U=V=0 \text{ at } X=0 \text{ and } m<X<1 \tag{11}$$

4. NUMERICAL APPROACH

The continuity, momentum and the energy equations are solved using commercially available software FLUENT 6.3. Discretization of the momentum and energy equations is performed by a second order upwind scheme and pressure interpolation is provided by PRESTO scheme. Convergence criterion considered as residuals is admitted 10^{-3} for momentum and continuity equations and for the energy equation it is 10^{-6}. In this study, the mesh is structured in such a way that the path of the heat lines, which intersect the isotherms spanning orthogonally from the isothermal hot bottom wall to the isothermal cold curved wall for the conduction solution, is considered.

The Nusselt number along the hot wall can be defined as

$$Nu_r=\frac{qr}{k(T_{hot}-T_{cold})}\frac{\pi}{2} \tag{12}$$

Thus, we calculated Nusselt number manually using the Eq. (12).

4.1 Grid independency test

In the study, four different mesh sizes (40×40, 60×60, 80×80 and 100×100) are adopted in order to check the mesh independence. A detailed grid independence study has been performed, and results are obtained for the average Nusselt number, and the maximum values of the stream function, but any considerable changes were not obtained. Thus, a grid size of 80×80 is found to meet the requirements of both the grid independency study and the computational time limits.

Table 1 Relative error analysis with different grid sizes (T_h = 279.2 K, T_c=278.8 K, Ra = 10^4)

MESH SIZE	40x40	60x60	80x80	100x100
Ψ_{MAX}	2.01	2.0106	2.0263	2.0296
RELATIVE ERROR(%)	0.0298	0.7748	0.33	
Nu (avg)	5.9812	6.5737	7.1706	7.607
RELATIVE ERROR(%)	9.13	8.3243	5.737	

4.2 Validation

Due to lack of suitable results in the literature pertaining to the present configuration, the result obtained have been validated against the existing results for a quadrantal cavity filled with water medium when the bottom wall is heated and the vertical wall is cold (Aydin and Yesiloz, 2011a). Figures 2 and 3 indicate that the result showed good agreement with the literature. However, the published experimental results are not available for the enclosure configuration similar to that undertaken in this study with similar boundary conditions.

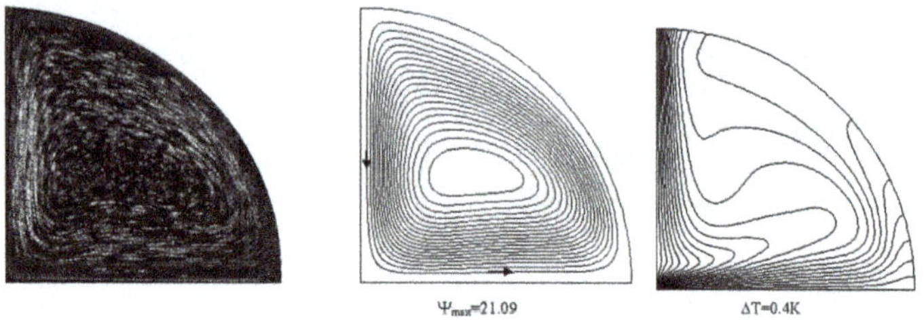

Fig. 2 Experimental (left) and numerical (right) streamline and isotherm for Ra=1.7×10^5 from Aydin and Yesiloz (2011a).

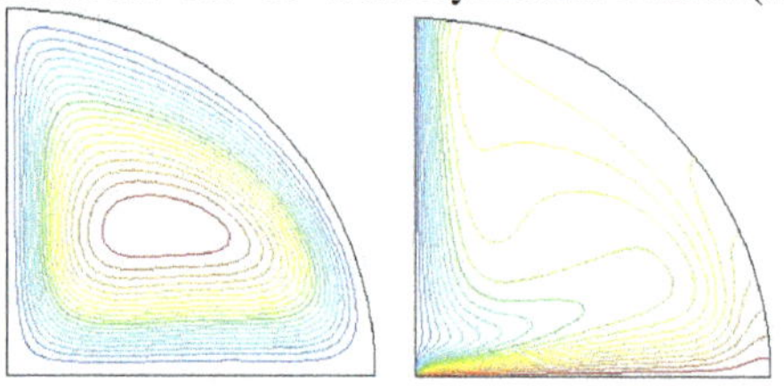

Ψ_{max}=21.07 ΔT=0.4K

Fig. 3 Numerical results from present study: streamline and isotherm for Ra=1.7×10^5

5. RESULTS AND DISCUSSIONS

Numerical analysis of laminar natural convection heat transfer and fluid flow is performed to obtain effects of Rayleigh number in an enclosure which is heated by the heaters attached in both the horizontal and vertical walls. The curved wall is assumed as the cold wall i.e. it has temperature less than that of the heater. The portion of the heaters in both the bottom and the vertical wall has higher temperature. The part on the bottom and the vertical wall other than the heater is considered as adiabatic. Results of flow fields and temperature distribution for different Rayleigh numbers, *Ra*, dimensionless length of heater in the vertical wall m, dimensionless length of fin on the bottom wall n, were plotted in this part of the study. Dimensionless length of the heater on the vertical wall varied as 0.2, 0.4 and 0.6. Also, dimensionless length of the thin fin on the bottom wall varied as 0.2, 0.4 and 0.6. The range of the Rayleigh number, *Ra*, was taken as 10^4 to 10^7.

Figures 4 and 5 show the streamlines and isotherms for different Rayleigh numbers at m = 0.2 and n = 0.2 to 0.6. As the value of Ra number is increased to 5x10^6 or more, two cells are formed on the top and mid-bottom part of the enclosure in anti-clock and clockwise direction respectively. No marked difference in streamline pattern takes place for a particular Rayleigh number having different geometries in heater size. As there is no obstruction of flow inside the quadrantal cavity, the heat transfer from the heaters to the cold curved wall occurs by buoyancy phenomena. With higher values of Rayleigh number there is crowding of streamlines on the vertical walls indicating boundary layer formation. The static temperature diagram shows concentration of isotherms near the cool curved wall, suggesting that heat transfer rate is high near the vicinity of the curved wall. Less packed isotherms are formed in the central part of the enclosure that indicates lower heat transfer. At the lowest Rayleigh number (*Ra* = 10^4), a relatively weak convective flow exists in the quadrantal cavity. As the Rayleigh number increases, the elliptical centre shifts towards the vertical wall and also the size of the elliptical centre increases along with the deformation of the elliptical centre. With Rayleigh numbers of 5x10^6 and more, there are two cell generated inside the closed enclosure. Isotherms show almost the same pattern for all Rayleigh numbers. Effects of the presence of heater on isotherms become stronger as the Rayleigh number increases. For lower Rayleigh numbers, the convection intensity in the enclosure is very weak as evident from the values of the stream functions. It means that the viscous forces are more dominant than the buoyancy forces at lower *Ra* numbers.

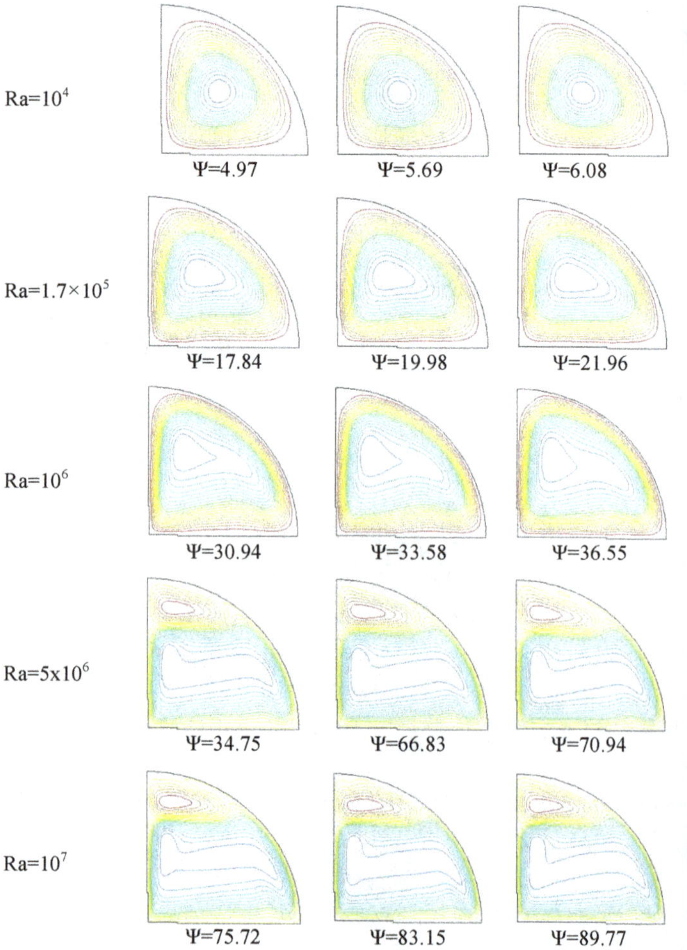

Fig. 4 Streamlines for dimensionless heater length, m= 0.2, column-wise different dimensionless heater lengths, n=0.2, n=0.4, n=0.6.

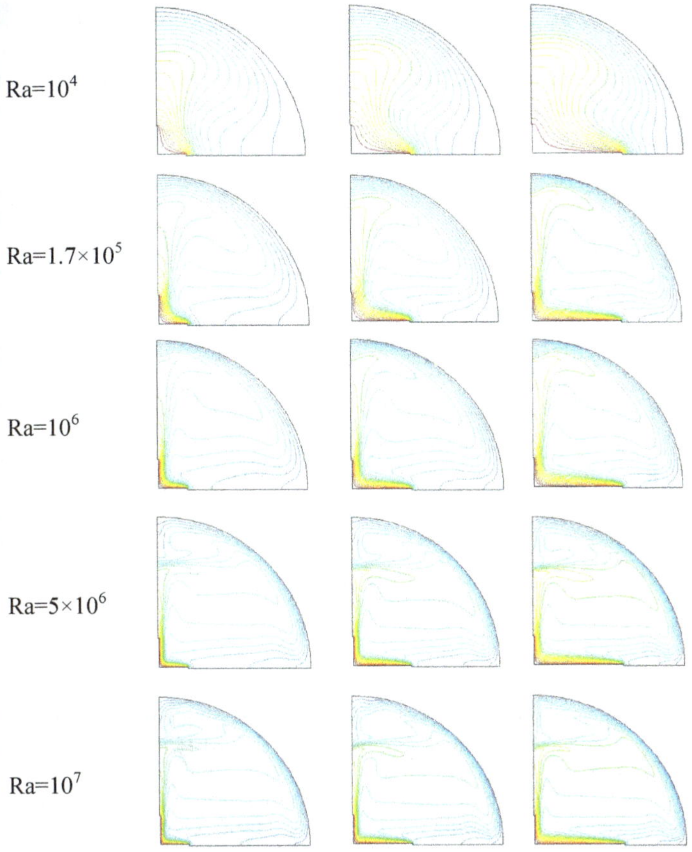

Fig. 5 Isotherms for dimensionless heater length, m= 0.2, column-wise different dimensionless heater lengths, n=0.2, n=0.4, n=0.6.

Different length of heaters are given in Figs. 6 and 7 for $Ra = 10^4$ to 10^7, n = 0.2 and m = 0.2, 0.4 and 0.6. At the lowest Rayleigh number Ra=10^4, streamlines form a nearly centrally located single cell forming an elliptical shape, and corresponding isotherms exhibit the characteristics of quasi-conduction. The hot fluid layer heated around the heater ascends upward by reason of decreasing density and the weak circulation in the enclosure. On the other hand, nearby the curve wall, the fluid layer with increased density moving downward sweeps the hot fluid layer forward. Since the curved wall is the cool wall, the hot fluid here exchanges heat with the cool curved wall and the cooler fluid here forms a cold fluid layer. These mutual effects of the hot and cold fluid layers cause isotherms to widen. Further increases in Rayleigh number, recirculation intensity increases to a degree that boundary layer formations are observed adjacent to the heater surface and cooled walls. The heated flow impinges to the curved wall to exchange heat. For the all configurations of the heater i.e. n=0.2 and m = 0.2, 0.4 and 0.6, an elliptical shape single cell has formed at the lowest Rayleigh number Ra = 10^4. As increases in the Rayleigh numbers two cells are formed and also boundary layer formation occurs near the top portion of the enclosure and the cavity is divided into two parts. The isotherm plots are concentrated towards the curved wall.

Figures 8 and 9 shows the streamlines and isotherms for different Rayleigh numbers at m=0.2, n=0.2; m=0.4, n= 0.4 and m=0.6, n=0.6. For the case of m=0.6 and n=0.6, the stream lines gets crowded near the bottom and the vertical surface with the increase in Rayleigh number thus causing boundary layer formation near the horizontal and the vertical walls. The isotherms form the above case is also crowded near the heater surface and the curved wall. This indicates that maximum heat transfer takes place in the vicinity of the heater surface and also the near the cold curved wall surface.

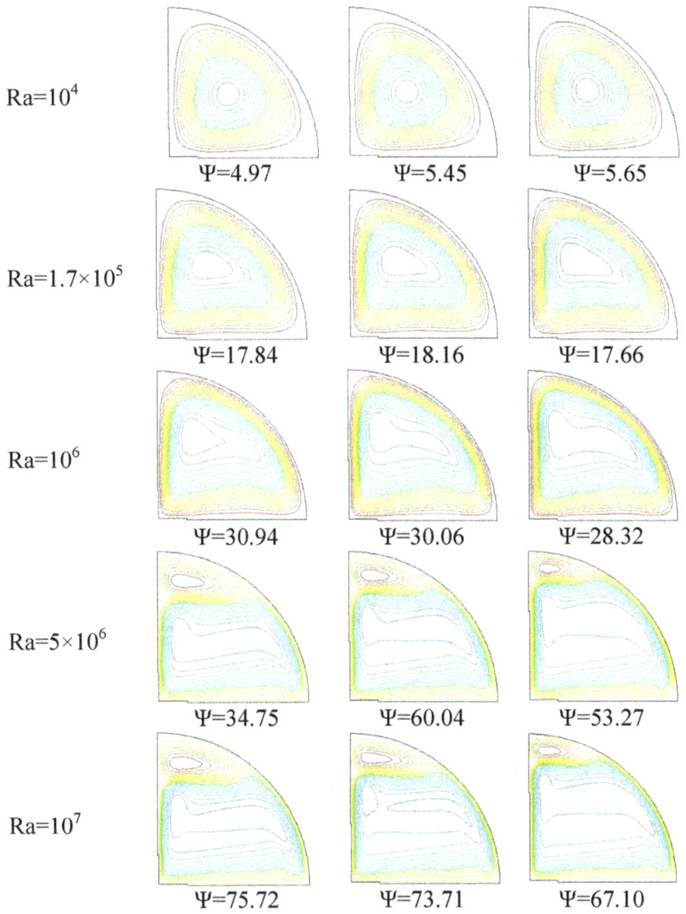

Fig. 6 Streamlines for dimensionless heater length, n= 0.2, column wise different dimensionless heater lengths, m=0.2, m=0.4, m=0.6

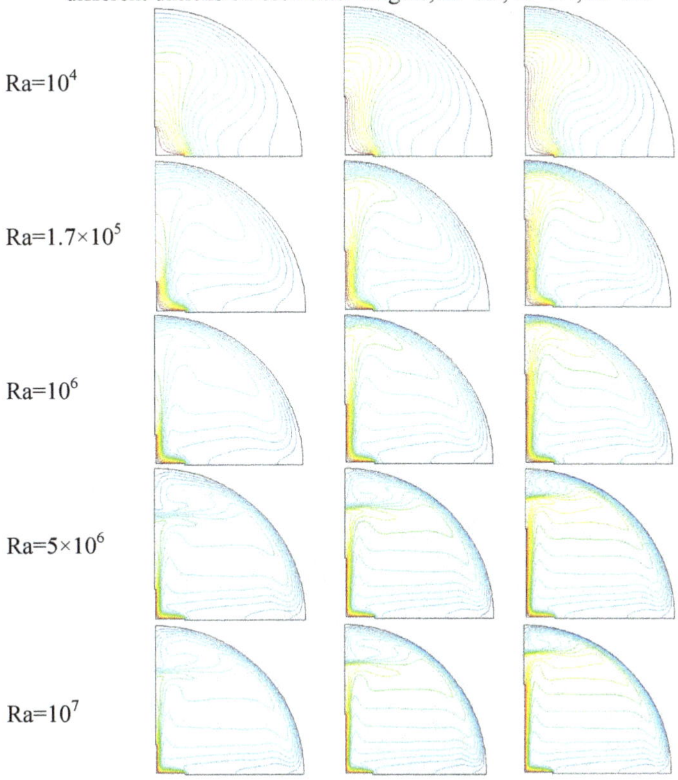

Fig. 7 Isotherms for dimensionless heater length, n= 0.2, column wise different dimensionless heater lengths, m=0.2, m=0.4, m=0.6.

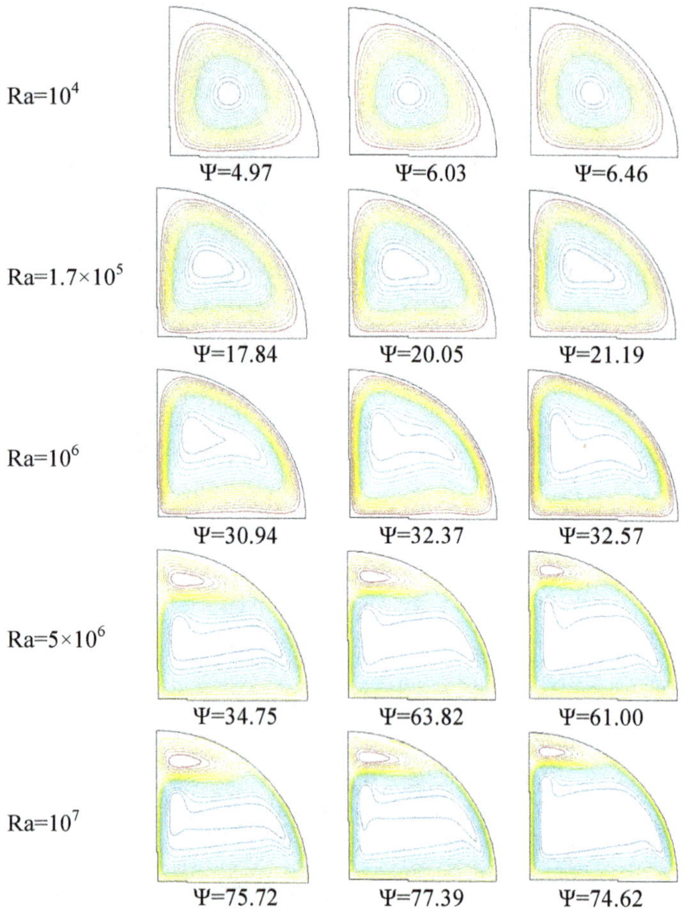

Fig. 8 Streamlines, column wise different dimensionless heater lengths, m=0.2 ,n=0.2; m=0.4, n=0.4; m=0.6, n=0.6.

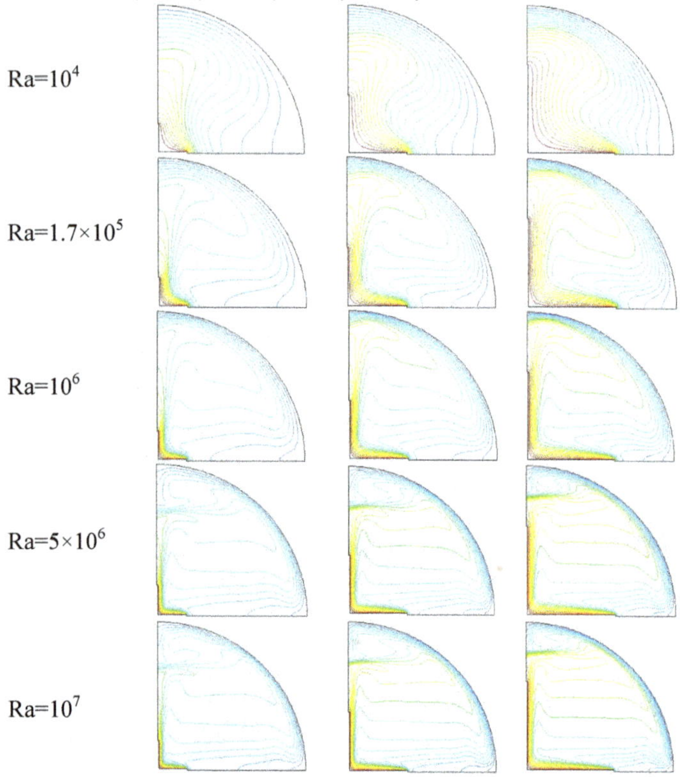

Fig. 9 Isotherms, column wise different dimensionless heater lengths, m=0.2 ,n=0.2; m=0.4, n=0.4; m=0.6, n=0.6.

Figure 10 shows the effect of dimensionless lengths of heater on the Nusselt number. The dimensionless heater length on the vertical wall is kept constant while that of the bottom horizontal wall is varied from 0.2, 0.4 and 0.6. It can be observed that for low Ra, the Nu for the three cases is very close. But as Ra increases, there is a large variation in the Nu. And when the heater length is the smallest, the value of Nu is minimum.

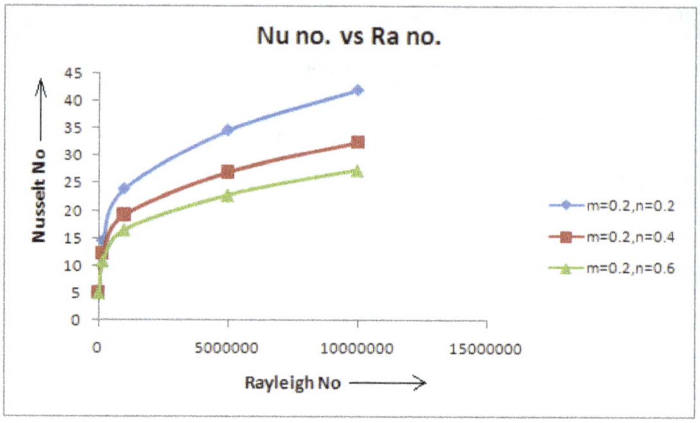

Fig. 10 Variation of mean Nu no. with Rayleigh number (10^4 to 10^7) for different heater lengths (n=0.2, 0.4, 0.6) for m=0.2.

Figure 11 is the case when the non-dimensional heater length on the bottom horizontal wall is kept constant at 0.2 and that of the vertical wall is varied from 0.2, 0.4 and 0.6. Similar to the above case the value of Nu for low Ra is almost same for the three cases due to the quasi-static regime, but Nu becomes stronger as the Rayleigh number increases.

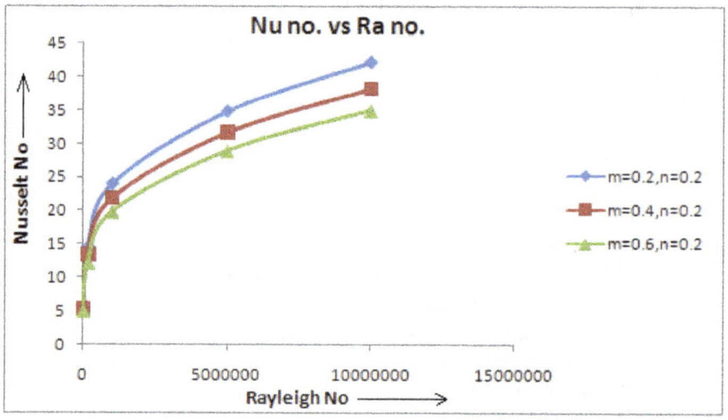

Fig. 11 Variation of mean Nu no. with Rayleigh number (10^4 to 10^7) for different heater lengths (m=0.2, 0.4, 0.6) for n=0.2.

Figure 12 represents our third study case where the increment in the non-dimensional heater lengths in both the horizontal and the vertical wall is same. From the Fig. 10, 11 and 12, it is observed that the deviation of Nusselt number depends on the non-dimensional lengths of the heaters. If the non-dimensional length of the heater on the vertical wall is more than or that on the bottom wall then the Nusselt number is high compared to cases where the length of heater on the vertical wall is less or equal to that of the bottom wall. The Nusselt number is also high for the cases when the non-dimensional lengths of the heater on both the wall is small, and it decreases with the simultaneous increase in the heater lengths.

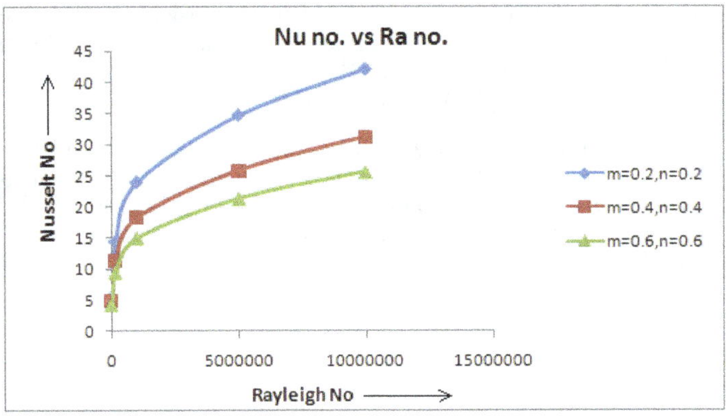

Fig. 12 Variation of mean Nu no. with Rayleigh number (10^4 to 10^7) for different heater lengths (m=0.2, n=0.2; m=0.4, n=0.4; m=0.6, n=0.6)

6. CONCLUSIONS

Numerical study has been performed to analyze the flow and temperature fields as well as the heat transfer rate on laminar natural convection in quadrentral cavity filled with water having heaters of varying lengths on both the vertical wall and the bottom wall. The curved wall is taken as the cold wall. The effects of Rayleigh number (10^4 to 10^7), non-dimensional heater lengths on the vertical wall (m=0.2, 0.4 and 0.6) and horizontal wall (n=0.2, 0.4 and 0.6) is studied.

The result of the numerical analysis lead to the following conclusions:-

1) Heat transfer increases with increase in Rayleigh number.
2) With smaller heater lengths, higher heat transfer is achieved at higher Rayleigh numbers. At small Rayleigh numbers the heat transfer is almost same irrespective of heater lengths.
3) With increase in the non-dimensional heater length on the vertical wall, the heat transfer decreases comparative with small heater lengths. This decrease in heat transfer is less if the heater length on the bottom wall is small and becomes drastic for larger lengths of the bottom heater.
4) As the non-dimensional length of the heater in the bottom wall is increased, the heat transfer decreases comparatively with lower heater lengths for the same Rayleigh number irrespective of the change in the heater lengths of the vertical wall.

NOMENCLATURE

Ra	Rayleigh number
Nu	Nusselt number
T	Temperature
Pr	Prandtl number
H	enclosure height
p	pressure
P	non-dimensional pressure
k	thermal conductivity
q	heat flux
c_p	specific heat
u	velocity component in the x-direction
v	velocity component in the v-direction
U	non-dimensional velocity component in the x-direction
V	non-dimensional velocity component in the y-direction
x,y	Cartesian coordinate system
X,Y	non-dimensional coordinates
r, ø	cylindrical coordinate system
m	non-dimensional heater length on vertical wall, m'/L
n	non-dimensional heater length on horizontal wall, n'/L

Greek Symbols

ψ	stream function
Ψ	non-dimensional stream function, ψ/α
ρ	density
β	co-efficient of thermal expansion
υ	kinematic viscosity
α	Thermal diffusivity
μ	dynamic viscosity
θ	dimensionless temperature

Subscripts

h	hot wall
c	cold wall
max	maximum
min	minimum

REFERENCES

Akinsete, V.A., and Coleman, T.A., 1982, "Heat Transfer by Steady Laminar Free Convection in Triangular Enclosures," International Journal of Heat and Mass Transfer, **25**, 991–998. http://dx.doi.org/10.1016/0017-9310(82)90074-6

Asan, H., and Namli, L., 2001, "Numerical Simulation Of Buoyant Flow In A Roof Of Triangular Cross Section Under Winter Day Boundary Conditions," *Energy and Buildings*, **33**, 753–757. http://dx.doi.org/10.1016/S0378-7788(01)00063-9

Aydin, O., Unal, A., and Ayhan, T., 1999, "A Numerical Study On Buoyancy-driven Flow in an Inclined Enclosure Heated and Cooled on Adjacent Walls," *Numerical Heat Transfer Part A: Applications*, **36**, 585–589. http://dx.doi.org/10.1080/104077899274589

Aydin, O., and Yesiloz, G., 2011a, "Natural Convection in a Quadrantal Cavity Heated and Cooled on Adjacent Walls," ASME *Journal of Heat Transfer*, **133**, 052501–7. http://dx.doi.org/10.1115/1.4003044

Aydin, O., and Yesiloz, G., 2011b, "Natural Convection in an inclined Quadrantal Cavity Heated and Cooled on Adjacent Walls," *Experimental Thermal and Fluid Science*, **35**, 1169–1176. http://dx.doi.org/10.1016/j.expthermflusci.2011.04.002

Aydin, O., and Yang, W.J., 2000, "Natural Convection in Enclosures with Localized Heating from Below and Symmetrical Cooling From Sides," *International Journal of Numerical Methods for Heat and Fluid Flow*, **10**, 518–529. http://dx.doi.org/10.1108/09615530010338196

Chu, H.H.S., Churchill, S.W., and Patterson, C.V.S., 1976, "The Effect of Heater Size, Location, Aspect Ratio, and Boundary Conditions on Two-Dimensional Laminar, Natural Convection in Rectangular Channels," ASME *Journal of Heat Transfer*, **98**, 1194–1201. http://dx.doi.org/10.1115/1.3450518

Chu, T.Y., and Hickox, C.E., 1990, "Thermal Convection with Large Viscosity Variation in an Enclosure with Localized Heating," ASME *Journal of Heat Transfer*, **112,** 388–395. http://dx.doi.org/10.1115/1.2910389

Deng, Q.H., and Tang, G.F., 2002, "A Combined Temperature Scale for Analyzing Natural Convection in Rectangular Enclosures with Discrete Wall Heat Sources," *International Journal of Heat and Mass Transfer*, **45,** 3437–3446. http://dx.doi.org/10.1016/S0017-9310(02)00060-1

Deng, Q.H., Tang, G.F., Li, Y., and Ha, M.Y., 2002, "Interaction between Discrete Heat Sources in Horizontal Natural Convection Enclosures," *International Journal of Heat and Mass Transfer*, **45,** 5117–5132. http://dx.doi.org/10.1016/S0017-9310(02)00221-1

FLUENT User's Guide, Release 6.3.26, Fluent Incorporated (2005-01-06).

Moukalled, F., and Acharya, S., 2001, "Natural Convection in Trapezoidal Enclosure with Offset Baffles," *Journal of Thermophysics and Heat Transfer*, **15**, 212–218. http://dx.doi.org/10.2514/2.6596

Turkoglu, H., and Yucel, N., 1995, "Effect of Heater and Cooler Locations on Natural Convection in Square Cavities," *Numerical Heat Transfer Part A: Applications*, **27,** 351–358. http://dx.doi.org/10.1080/10407789508913705

Tzeng, S.C., Liou, J.H., and Jou, R.Y., 2005, "Numerical Simulation-Aided Parametric Analysis of Natural Convection in a Roof of Triangular Enclosures," *Heat Transfer Engineering*, **26**, 69–79. http://dx.doi.org/10.1080/01457630591003899

18

INVESTIGATION OF PARTICULAR FEATURES OF THE NUMERICAL SOLUTION OF AN EVAPORATING THIN FILM IN A CHANNEL

Greg Ball, John Polansky, Tarik Kaya*

Department of Mechanical and Aerospace Engineering, Carleton University, Ottawa, Ontario, K1S 5B6, Canada

ABSTRACT

The fluid flow and heat transfer in an evaporating extended meniscus are numerically studied. Continuity, momentum, energy equations and the Kelvin-Clapeyron model are used to develop a third order, non-linear ordinary differential equation which governs the evaporating thin film. It is shown that the numerical results strongly depend on the choice of the accommodation coefficient and Hamaker constant as well as the initial perturbations. Therefore, in the absence of experimentally verified values, the numerical solutions should be considered as qualitative at best. It is found that the numerical results produce negative liquid pressures under certain specific conditions. This result may suggest that the thin film can be in an unstable state of tension; however, this finding remains speculative without experimental validation. Although similar thin-film models proved to be very useful in gaining qualitative insight into the characteristics of evaporating thin films, the results shown in this study indicate that careful experimental investigations are needed to verify the mathematical models.

Keywords: *Thin film, evaporating meniscus, capillary force, disjoining pressure*

1. INTRODUCTION

The study of thin films is important in many technological applications, including the cooling of electronics, evaporation, condensation, boiling, drying etc. The film dynamics are governed by various complex physical mechanisms such as surface tension, disjoining pressure, thermal conduction and phase change. Because of its importance, thin films have been widely studied.

The problem of modelling an evaporating thin film has been investigated by many authors using various techniques. Solutions governed by 3rd and 4th order Ordinary Differential Equations (ODE) have been proposed. In addition to the problem formulation (3rd versus 4th order), the existing models used different boundary conditions, and different techniques in the calculation of the mass transport across the interface. More recently, simulations based on the molecular dynamics were also attempted. Because of the very small length scales involved, few experimental works have been completed and majority of the models do not have the necessary validation required for providing further insight into whether the models produce accurate physical solutions.

Several authors have expanded upon current works by imposing different boundary conditions and observing the corresponding effect. Table 1 provides a sample of common variations undertaken by various authors. Note that the varying thermophysical properties, refers to whether the numerical model updates the fluid properties as a function of temperature in generating a solution.

Table 1 Comparison of some earlier works on modelling an evaporating meniscus.

Authors	Non-isothermal Interfacial Condition	Varying Thermophysical Properties	Slip Boundary condition	Polarity Effect	Superheat Effect (at least 5 K range)
Potash and Wayner (1972)	✓	-	-	-	-
Wayner et al. (1976)	✓	-	-	-	-
Moosman and Homsy (1980)	✓	-	-	-	-
Hallinan et al. (1994)	✓	-	-	-	-
Park et al. (2003)	-	-	✓	-	-
Qu et al. (2002)	✓	-	-	✓	-
Zhao et al. (2011)	✓	✓	-	-	✓
Wee et al. (2005)	✓	✓	✓	✓	-
Wang et al. (2007)	✓	-	-	-	✓
Present Study	✓	✓	✓	-	✓

* *Corresponding author. E-mail: tkaya@mae.carleton.ca.*

Of the parameters listed on Table 1, the following general trends were observed. Potash and Wayner (1972) were the first to show that both pressure gradient and evaporative mass flux reach a maximum within the thin-film region. Moosman and Homsy (1980) implemented a mathematical model to describe the thin-film characteristics using perturbation analysis. The effect of including an interfacial temperature gradient term, as well as varying thermophysical properties allowed for much more pronounced effects to be seen when superheat is varied (Zhao et al., 2011). An increased superheat serves to lessen the adsorbed film thickness, and creates a much more aggressive curvature increase as the film transitions from the adsorbed region through the thin-film region. In general, for higher superheats the film length is decreased, creating a much 'steeper' thin-film profile (Zhao et al., 2011). When polar effects were considered on the disjoining pressure model, the thin-film length was extended, all the while reducing the evaporative heat transfer (Wee et al., 2005). The wall slip boundary condition as introduced by Park et al. (2003), serves to elongate the thin film while yielding a lower pressure gradient (Wee et al., 2005).

In addition to the above trends, variance in channel width has also been studied, such as in Wang et al. (2007). The effect of an increasing channel width, was an increased thin-film length (Qu et al. 2002; Zhao et al., 2011). Du and Zhao (2012) attempted to quantify the effects of using altered evaporation models. They concluded that neglecting disjoining pressure terms have less of an effect than neglecting capillary pressure terms. They also concluded that the substrate thickness does have a minor effect on the total heat transfer.

Kou and Bai (2011) related wall slip to a temperature jump at the solid-liquid interface. They concluded that the presence of a temperature jump at the interface can reduce heat and mass transport characteristics.

It is important to mention that in the above works, solution methodology is similar. As such, the governing ODE is found to have a highly non-linear characteristic behaviour, regardless of order. If the initial conditions, such as slope and curvature, are assigned values of zero in the beginning of the thin film as one would assume, a constant thickness thin-film solution is obtained. As a result, perturbations must be applied in order to obtain a physical solution of interest. Considering the highly non-linear behaviour, it is difficult to identify a set of suitable perturbations that will yield a satisfactory solution. Most published works did not include their applied perturbations. Du and Zhao (2012) stated that the thickness perturbation, ε_1, should be set sufficiently close to zero and the slope perturbation, ε_2, should be set slightly larger than zero. Furthermore, altering parameters of the problem alters boundary conditions, which in turn have a profound effect on the perturbations. The solutions are also extremely sensitive to boundary conditions as noted in DasGupta et al. (1993).

Of note is the lack of relationship between the derivative perturbations and the magnitude of the applied superheat or channel width (Wang et al., 2007 and 2008; Park et al., 2003). These are not the only parameters that have an effect on applied perturbation. This research also notes extreme sensitivities to the slope perturbation, ε_2, as accommodation coefficient is altered.

Panchamgam et al. (2008) is the only work where the high-resolution data for thickness, slope and curvature were experimentally obtained and used as inputs to a macro-micro numerical model. Thus, the thin-film characteristics of a constrained vapour bubble - such as interfacial temperature profile, evaporative heat and mass flux – were directly calculated.

Morris (2003) presented a different and more rigorous approach to the same problem by matching the asymptotic solutions. He pointed out that the above methods divide the meniscus at an arbitrarily chosen point: nonlinear small region close to the apparent contact angle and relatively undisturbed region determined from the hydrostatics. In this work, it was shown that the above methods implicitly assume that the capillary number ($Ca = \mu V/\sigma$) of the induced flow is small.

In addition to the points raised by Morris (2003), we also identified that effects of certain variables often overlooked in these solutions, including the accommodation coefficient and Hamaker constant. It was found that these variables play a large role in the resulting thin-film characteristics.

In this paper, a mathematical model based on the augmented Young-Laplace and Kelvin-Clapeyron equations is studied. The main goal of our study is to investigate the effects of various parameters and identify the limits of the mathematical model in order to better understand the thin-film dynamics. A deliberate attempt is made to list the various parameters of the solution such as perturbations imposed and discuss in detail the validity limit of the solutions.

2. MATHEMATICAL MODEL

The meniscus is confined by a narrow channel with immiscible walls and of infinite depth, as illustrated in Fig. 1. The model developed herein, describes a thin film of non-polar fluid, evaporating in a steady-state condition, for a given superheat. The model proposed, follows a similar approach used in previous related works, e.g. Wee et al. (2005) and Panchamgam et al. (2008).

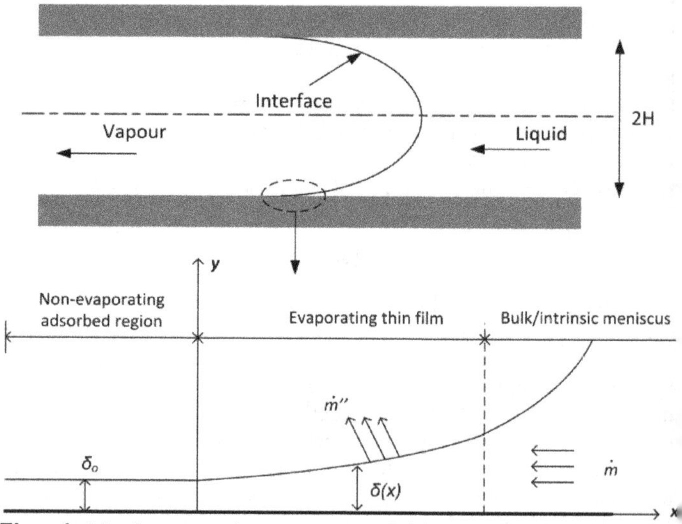

Fig. 1 Meniscus as it contacts a rigid and planar immiscible substrate.

The evaporating thin-film region of a meniscus is governed by the augmented Young-Laplace equation, as proposed by Wayner Jr. (1991), where the vapour pressure field is assumed uniform and constant.

$$P_v = P_l + P_d + P_c \tag{1}$$

The vapour pressure is balanced by the liquid, disjoining and capillary pressures found in the thin-film region. Of these pressures, the disjoining pressure for a pure, non-polar fluid is expressed as a function of the dispersion constant, A, and film thickness, δ, as follows

$$P_d = \frac{A}{\delta^3} \tag{2}$$

The dispersion constant, A, is related to the Hamaker constant, A_H, by the following relation: $A = A_H/6\pi$. Note that in some works both terms were alternatively used for the same value, leading to confusion in reproducing the results. The importance of the Hamaker constant and its effect on the results will be discussed later in the paper.

The capillary pressure is expressed as the product of the surface tension and the local curvature. The surface tension is assumed to vary linearly with the local interfacial temperature.

$$P_c = \sigma \kappa \tag{3}$$

where,

$$\sigma = a + bT_{lv} \tag{4}$$

$$\kappa = \frac{\delta''}{\left(1+\delta'^2\right)^{3/2}} = \frac{\delta''}{\alpha^{3/2}} \tag{5}$$

Combining Eqs. (1-5) and differentiating with respect to *x*, the following third-order differential equation is obtained.

$$\delta''' = \frac{3\delta'\delta''^2}{\alpha} + \frac{\alpha^{3/2}}{\sigma}\left(\frac{3A}{\delta^4}\delta' - \frac{dP_l}{dx}\right) - \frac{b}{\sigma}\left(\frac{dT_{lv}}{dx}\right)\delta'' \tag{6}$$

Equation (6) captures the pressure gradient and thermocapillary effects present in the thin film. Both the liquid pressure and interfacial temperature gradients are needed to solve the governing equation.

As the film is evaporating, fluid must flow into the thin-film region from the bulk meniscus, so as to maintain steady state. The flow field in the thin-film liquid is solved with the use of the lubrication approximation, given by Eq. (7) and associated boundary conditions at the wall, Eq. (8), and the liquid-vapour interface, Eq. (9).

$$\frac{dP_l}{dx} = \mu\frac{d^2u}{dy^2} \tag{7}$$

$$u = -\beta\frac{du}{dy} \tag{8}$$

$$\tau = \sigma' = \mu\frac{du}{dy} \tag{9}$$

In Eq. (8), the slip length coefficient *β*, is carried through the derivation for completeness, though later set to zero to produce a no-slip condition. The second boundary condition, Eq. (9), captures the Marangoni effects at the free liquid-vapour interface. Integrating Eq. (7), the thin-film liquid velocity is obtained as a function of the liquid pressure and interfacial temperature gradients.

$$u = \frac{1}{\mu}\frac{dP_l}{dx}\left(\frac{y^2}{2} - \delta(y-\beta)\right) + \frac{b}{\mu}\frac{dT_{lv}}{dx}(y-\beta) \tag{10}$$

With the velocity of the liquid in the thin film described by Eq. (10), the mass flow rate can be obtained through the application of continuity. After integrating over the film thickness, the mass flow rate at any point along the thin film is obtained as,

$$\dot{m} = \frac{1}{\nu}\frac{dP_l}{dx}\left(-\frac{\delta^3}{3} + \beta\delta^2\right) + \frac{b}{\nu}\frac{dT_{lv}}{dx}\left(\frac{\delta^2}{2} - \beta\delta\right) \tag{11}$$

The temperature of the channel walls is assumed be constant and uniform with heat being conducted through the film thickness normal to the planar wall. Reflecting this, the energy equation and boundary conditions take the following form,

$$\frac{\partial}{\partial y}\left(k_l\frac{\partial T}{\partial y}\right) = 0 \tag{12}$$

At $y = 0$,

$$T = T_w \tag{13}$$

At $y = \delta$,

$$-k_l\frac{\partial T}{\partial y} = \dot{m}''h_{fg} \tag{14}$$

Integrating the energy equation, Eq. (12), and applying the appropriate boundary conditions, Eqs. (13) and (14), an expression for the evaporative mass flux is obtained.

$$\dot{m}'' = \frac{k_l(T_w - T_{lv})}{\delta h_{fg}} \tag{15}$$

The evaporative mass flux can be related to the liquid mass flow by,

$$\dot{m}'' = -\frac{d\dot{m}''}{dx} \tag{16}$$

Integrating Eq. (16) with respect to film length x, the mass flow rate in the thin film is described as,

$$\dot{m} = -\int_0^x \frac{k_l(T_w - T_{lv})}{\delta h_{fg}}dx \tag{17}$$

From Eqs. (11) and (17), a relation between the liquid pressure gradient and interfacial temperature gradient is obtained.

$$\frac{dP_l}{dx} = b\frac{C_5}{C_4}\frac{dT_{lv}}{dx} + \frac{\nu k_l}{C_4 h_{fg}}\int_0^x \frac{T_w - T_{lv}}{\delta}dx \tag{18}$$

where,

$$C_4 = \frac{\delta^3}{3} - \beta\delta^2 \tag{19}$$

$$C_5 = \frac{\delta^2}{2} - \beta\delta \tag{20}$$

To form a closed solution to the governing equation, the Kelvin-Clapeyron evaporation model, introduced by Wayner Jr. (Wayner Jr., 1991), is used.

$$\dot{m}'' = \frac{1}{h_{fg}}\left(h_{lv}^{cl}(T_{lv} - T_v) - h_{lv}^{kl}(P_d + P_c)\right) \tag{21}$$

$$h_{lv}^{cl} = \eta\left(\frac{1}{T_{lv}}\right)^{3/2}\left(\frac{Mh_{fg}}{T_v}\right) \tag{22}$$

$$h_{lv}^{kl} = \eta\left(\frac{1}{T_{lv}}\right)^{3/2} V_l \tag{23}$$

$$\eta = \left(\frac{C^2M}{2\pi R}\right)^{1/2}\left(\frac{P_v h_{fg}}{R}\right) \tag{24}$$

$$C = \frac{2\gamma}{2-\gamma} \tag{25}$$

By equating the evaporative mass flux expressions, Eqs. (15) and (21), a relation is obtained where the interfacial temperature is described independently of the liquid pressure. Differentiating with respect to the thin-film length *x*, the interfacial temperature gradient is obtained as follows:

$$\frac{dT_{lv}}{dx} = \frac{2T_v}{\delta^3}\left(\frac{\chi + V_l\eta\sigma\delta^4\kappa'}{\omega}\right) \quad (26)$$

where,

$$\chi = \delta'\left[k_l\delta^2\left(T_{lv}^{5/2} - T_w T_{lv}^{3/2}\right) - 3AV_l\eta\right] \quad (27)$$

$$\omega = k_l T_v\left(5T_{lv}^{3/2} - 3T_w T_{lv}^{1/2}\right) + 2\delta\eta\left(Mh_{fg} - V_l T_v b\kappa\right) \quad (28)$$

Collecting Eqs. (6, 18, 26) and rearranging for δ''', the governing equation for the evaporating thin film is described as,

$$\delta''' = \frac{3\delta'\delta''^2}{\alpha} - \left(\frac{\omega\alpha^{\frac{3}{2}}}{\omega\alpha^{3/2} + 2T_v\psi V_l\eta\sigma\delta}\right)$$

$$\left[\frac{2T_v\psi\chi}{\omega\delta^3} - \frac{\alpha^{\frac{3}{2}}}{\sigma}\left(\frac{3A}{\delta^4}\delta' - \frac{vk_l}{C_4 h_{fg}}\int_0^x \frac{T_w - T_{lv}}{\delta}dx\right)\right] \quad (29)$$

where,

$$\psi = \frac{b}{\sigma}\left(\frac{C_5}{C_4}\alpha^{3/2} + \delta''\right) \quad (30)$$

Thus, the governing equation is obtained as a function of film thickness and associated derivatives, fluid properties and the interfacial temperature.

2.1 Initial conditions and limitations

To obtain the adsorbed film thickness, it is assumed that no evaporation is taking place in the adsorbed region. Thus, the evaporative mass flux from the Kelvin-Clapeyron model, Eq. (21), is set to zero and the interfacial temperature is assumed to be that of the wall in the adsorbed region. This yields the following equation for the adsorbed film thickness.

$$\delta_0 = \left[\frac{AV_l}{Mh_{fg}}\left(\frac{T_v}{T_w - T_v}\right)\right]^{1/3} \quad (31)$$

As it was discussed previously, to avoid a trivial solution of constant thin film thickness, small perturbations need to be applied to the initial conditions (δ, δ' and δ''). As a goal to obtain accurate solutions for thin-film region, the perturbations must be sufficiently small to ensure proximity to the adsorbed region as perturbations serve to effectively shift the coordinate system away from the adsorbed region. To avoid violating the evaporative mass flux balance, the possible limits of important parameters are searched. Thus, a maximum and minimum value range for δ'' is obtained as a function of δ and δ'.

On the lower end of the perturbation applied to δ'', the replacement of the interfacial temperature with that of the vapour temperature gives

$$\delta''_{min} = -\alpha^{3/2}\left[\frac{k_l(T_w - T_v)}{V_l\sigma\delta\eta}T_v^{3/2} + \frac{P_d}{\sigma}\right] \quad (32)$$

Upon inspection, it is seen that $\delta''_{min} < 0$ for all cases. Thus, it is imposed that the minimum value be set to $\delta''_{min} > 0$, so as to ensure a non-trivial solution of monotonically increasing film thickness.

Conversely, the upper bound for δ'' is obtained by equating the interfacial temperature to that of the wall temperature. Solving for δ''_{max},

$$\delta''_{max} = \frac{\alpha^{3/2}}{\sigma}\left(\frac{Mh_{fg}}{T_v V_l}(T_w - T_v) - P_d\right) \quad (33)$$

Inspecting Eq. (33), it is observed that the following inequality must be satisfied so as to ensure a monotonic increase in thin film thickness.

$$\frac{Mh_{fg}}{T_v V_l}(T_w - T_v) \geq \frac{A}{\delta^3} \quad (34)$$

Thus, only solutions satisfying the aforementioned perturbation limits are valid. A check must be performed when specifying the initial conditions so as to avoid any inadvertent violations of the mass flux balance or assumptions made in the analysis.

3. NUMERICAL SOLUTION PROCEDURE

The governing equation, Eq. (29), is solved with the use of a Runge-Kutta RK5 (4) 7M method, as proposed by Dormand and Prince (1980). The RK5 (4) 7M method is a modified 5th order technique designed to produce small principal truncation terms. The solution procedure is iterative. At a given solution step, the integral term contained in Eq. (29), $\int_0^x \frac{T_w - T_{lv}}{\delta}dx$, needs to be solved, which requires the knowledge of the current film thickness and interfacial temperature. These values are unknown at the current position step and need to be guessed. As a first guess, the values from the previous step are used, and the calculated values are returned from the Runge-Kutta solver and compared to the guess values. A comparison of the guess and calculated values is performed and looped until convergence criteria for both film thickness, Eq. (35) and interfacial temperature Eq. (36), are satisfied. The numerical solver is coded in MATLAB, and follows the simplified flowchart shown in Fig. 2.

$$\left|\frac{\delta_{guess} - \delta_{solved}}{\delta_{solved}}\right| \leq 0.1\% \quad (35)$$

$$\left|\frac{T_{lv_{guess}} - T_{lv_{solved}}}{T_{lv_{solved}}}\right| \leq 0.1\% \quad (36)$$

The code solves the governing equation and returns a solution when the far field curvature of the thin film has achieved the constant curvature convergence criteria (±0.01 %). The code terminates when the convergence criteria for constant curvature has been satisfied for ten consecutive steps. Though a constant curvature solution may be obtained, the desired curvature of Young-Laplace (*1/H*) is sought. Thus, an upper and lower bound are searched out, after which the search is completed by algorithm until the desired far field constant curvature condition ($\kappa = 1/H \pm 0.01\%$) is achieved. Due to the highly non-linear nature of the governing equation and the convergence criteria (±0.01 %), the perturbation value (ε_2), required a case specific precision as denoted in Table 2.

4. RESULTS AND DISCUSSION

A typical set of results obtained for pentane in a channel of 20 μm width and superheats of $(T_w - T_v)$= 0.01 K and 0.1 K at T_v= 300 K is presented in Fig. 3. This type of behaviour is representative of the thin-film development observed in other previous works, for example in Wee et al. (2005). Similarly, it is found that the numerical solution is very sensitive to the initial conditions and applied perturbations. Specifically, the results were extremely sensitive to the value of δ' and its associated perturbation ε_2. Increasing the superheats effectively steepens the thin-film profile and decreases the thin-film length. As a result of the thin-film geometry, the perturbation magnitude and sensitivity increases at larger superheats. In our solutions, we found that ε_1 had to be increased in order to seek an appropriate ε_2. However, it should be noted that applied perturbations must be sufficiently small such as to not shift the origin of the coordinate system too far into the thin-film region. A large perturbation may lead to exclusion of the near adsorbed region and thus the loss of important information for the entire solution.

Figure 4 provides a comparison to the thin-film profile generated in this study to that of Wang et al. (2007), where the thickness perturbation ε_1, was provided for one of the cases presented. Observing Fig. 4, it is seen that the profiles have the same general trend. It is of note that Wang et al. (2007) used a 4th order ODE as opposed to the 3rd order used in this study. In the 4th order model, the main sensitivity lies on the perturbation on $\delta''(\varepsilon_3)$, which needs to be iteratively determined. In a 3rd order model, the sensitivity falls on the slope perturbation, $\delta'(\varepsilon_2)$, which was determined by the numerical model to match the far field boundary condition. In the 4th order model, the slope perturbation is not sensitive so $\delta' = 1\times10^{-11}$ was used in Wang et al. (2007). In our case, the solutions were not sensitive to δ'' so it was set at 1×10^{-3}. Note that the evaporation models were also different between Wang et al. (2007) and the work presented here.

In many other works, the perturbations used are not provided. As a result, a direct comparison was not possible. Table 2 provides a list of parameters used in generating the solutions including the applied perturbations to Figs. 3 and 4 so that the future works can be directly compared against our work.

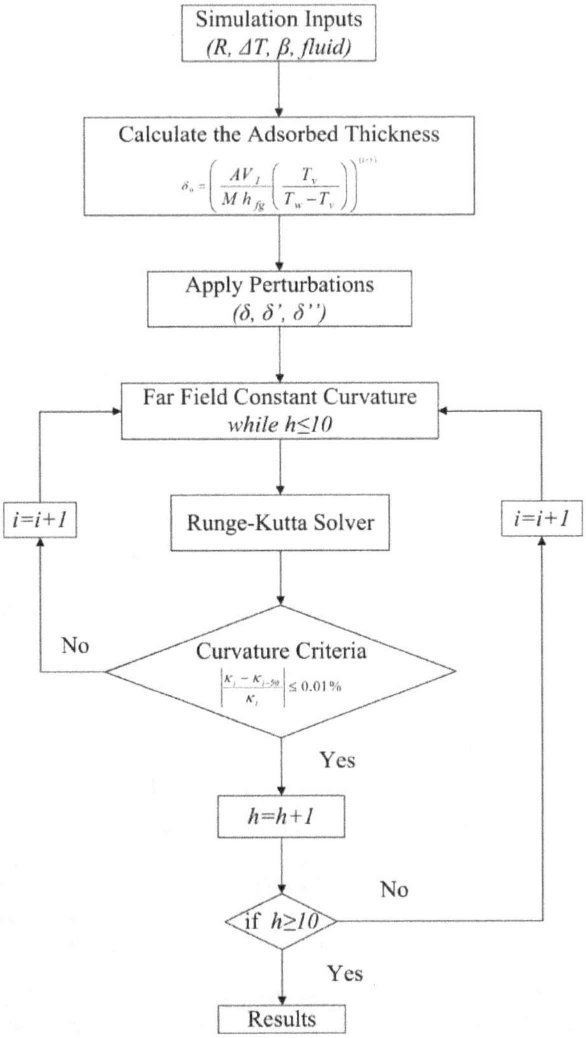

Fig. 2 Simplified flowchart outlining the solution procedure.

Table 2 Input parameters used in generating thin-film solutions.

Input	Fig. 3	Fig. 4
Fluid	Pentane	Octane
Superheat	0.01 K, 0.1 K	1 K
$2H$	20 μm	5 μm
T_v	300 K	343 K
$\delta_{0@0.01\,K}$	$4.68376868475136\times10^{-9}$ m	1.70×10^{-3} m
$\delta_{0@0.1\,K}$	$2.950758143324543\times10^{-9}$ m	
ε_1	$\delta_0 \times 0.01\%$ m	1.70×10^{-3} m
$\varepsilon_{2@0.01\,K}$	2.5205788×10^{-6}	1.477109375×10^{-2}
$\varepsilon_{2@0.1\,K}$	$2.637589900102466\times10^{-6}$	
ε_3	1×10^{-3} m^{-1}	1×10^{-3} m^{-1}
ε_4	0	0

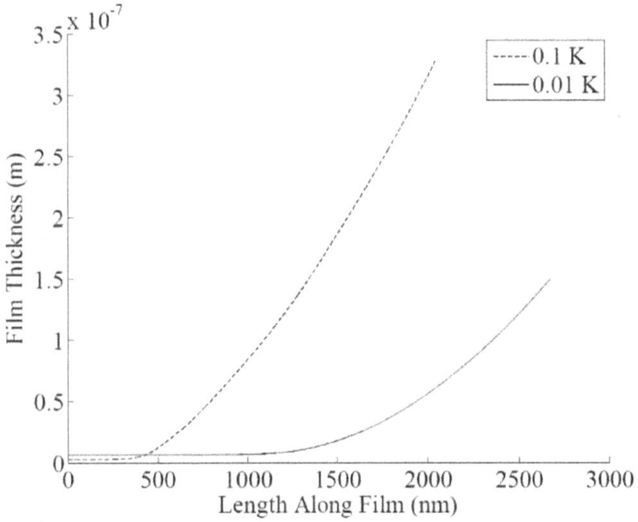

Fig. 3 Thin-film profile of pentane in a channel of 20 μm width at different superheats.

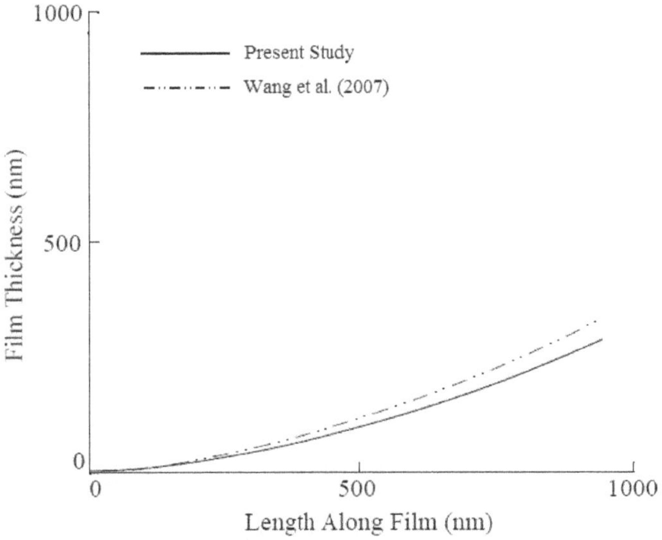

Fig. 4 Thin-film profile comparison to Wang et al. (2007).

4.1 Accommodation coefficient

The variable γ in Eq. (25) represents the accommodation coefficient. The accommodation coefficient is often used as unity in majority of the previouslsy published works: Wang et al. (2007); Wee et al. (2005); Qu et al. (2002); Schonberg et al. (1995); Stephan et al. (1992), without much discussion. An accommodation coefficient of unity implies that for every liquid molecule emitted, none are rebounded and re-absorbed, giving a perfect evaporative capacity. Mills (1965) notes that in cases of extreme fluid purity, this value should tend to unity. However, extreme purity may be unrealistic to obtain, as such the value should be lower than unity.

The effect of the accommodation coefficient is mostly omitted in the previously published thin-film solutions, yet in the case of numerically modelling a thin-film meniscus, alters the results noticeably. In some works, Wee et al. (2005); Du and Zhao (2011), $C = 2\gamma/(2-\gamma)$ was defined as accommodation coefficient and a value of 2.0 was assigned to C, resulting in the true accommodation coefficient, $\gamma = 1.0$. Panchamgam et al. (2008) referred to this C as the constant of proportionality. Thus, there is an ambiguity in the general use of this term.

Figure 5 demonstrates how the thin-film profile alters with decreasing accommodation coefficient. This analysis was performed with pentane as the working fluid in a 20 μm channel and a 0.01 K superheat. It is clear that with a decreasing value of accommodation coefficient, the film growth is delayed, resulting in a longer thin film and consequently a modified thin-film interface shape. This result was expected as a smaller accommodation coefficient would suggest less evaporative mass flux, thus extending adsorbed region of the film. The seemingly abrupt end of the thin-film profiles for $\gamma = 0.4$ and $\gamma = 0.3$ was due to the constant curvature termination condition. At these lower values of accommodation coefficient, the curvature profile loses the associated overshoot, as shown in Fig. 6.

Evaporative mass flux can be seen in Fig. 7 with decreasing accommodation coefficient. Not only does the peak magnitude of mass flux decrease, but the location is shifted along the length of the film. This corresponds with the extension of the thin-film profiles, thus the disjoining pressure decrease is retarded along the length of the film.

Observing the end of the evaporative mass flux profiles, it is clear in all cases, that evaporative mass flux at the constant curvature condition is non-zero. As mentioned earlier this deals with the use of the augmented Young-Laplace equation which entails some residual mass flux towards the intrinsic meniscus. In addition, retarding the evaporation by implementing a lower value of accommodation coefficient leaves an increased residual mass flux in order to satisfy the mass flux balance used in the derivation.

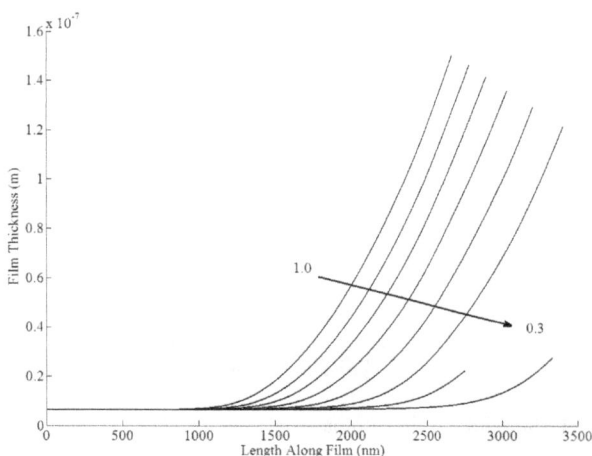

Fig. 5 Effect of accommodation coefficient on thin-film profile (Results presented in decrements of 0.1 from γ =1.0 to 0.3).

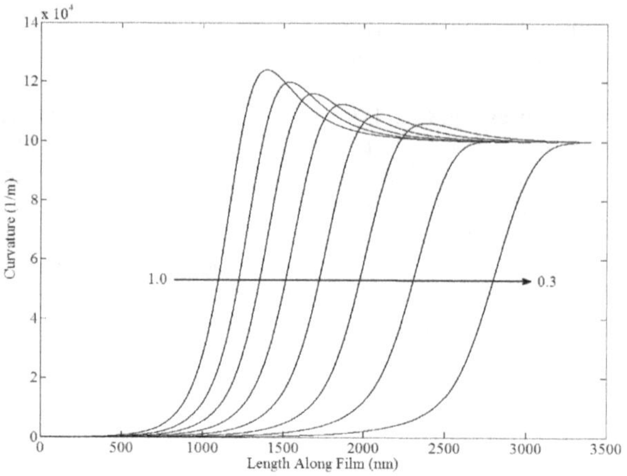

Fig. 6 Effect of accommodation coefficient on curvature (Results presented in decrements of 0.1 from γ =1.0 to 0.3).

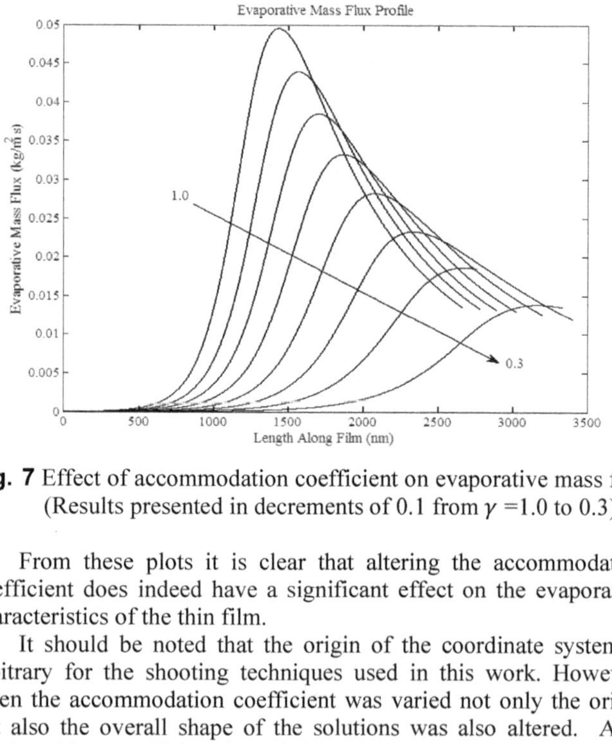

Fig. 7 Effect of accommodation coefficient on evaporative mass flux (Results presented in decrements of 0.1 from γ =1.0 to 0.3).

From these plots it is clear that altering the accommodation coefficient does indeed have a significant effect on the evaporative characteristics of the thin film.

It should be noted that the origin of the coordinate system is arbitrary for the shooting techniques used in this work. However, when the accommodation coefficient was varied not only the origin but also the overall shape of the solutions was also altered. As a result, without sufficient knowledge of accommodation coefficient, results are at best qualitative.

4.2 Hamaker constant

A similar ambiguity exists for the Hamaker constant, which had been estimated by using different methods. A survey of thin film works indicates a range of values were used for differing arrangements. This can be attributed to the many different relations proposed to estimate the Hamaker constant value for a given set of materials.

In our analysis, the Tabor-Winterton approximation is used to obtain a value for the Hamaker constant as outlined in Butt and Kapl (2010), and Israelachvili (1991). In this method, the Hamaker constant for the solid-liquid-vapour system (132) is given as a function of temperature, absorption frequency, index of refraction and dielectric constant by the following equation,

$$A_H = A_{132} = k_B T \left(\frac{\xi_1 - \xi_3}{\xi_1 + \xi_3}\right)\left(\frac{\xi_2 - \xi_3}{\xi_2 + \xi_3}\right)\frac{3}{4} + \frac{3h\nu_e}{8\sqrt{2}} \times \frac{(n_1^2 - n_3^2)(n_2^2 - n_3^2)}{(n_1^2 + n_3^2)^{1/2}(n_2^2 + n_3^2)^{1/2}\{(n_1^2 + n_3^2)^{1/2} + (n_2^2 + n_3^2)^{1/2}\}} \quad (37)$$

Of the functional terms in Eq. (37), the absorption frequency for the working fluid, ν_e, comes into question, as more than one absorption peak exists for any given material, or combination of materials. As a result of this uncertainty, a parametric study was conducted to determine the effect of variability in the Hamaker constant on the thin-film profile. By using an absorption frequency value of $\nu_e = 1.7635\times10^{15}$ Hz from Costner et al. (2009), the dispersion constant was calculated to be $A = 2.013\times10^{-21}$ J using Eq. (37), which was consistent with the values provided in Israelachvili (1991).

For the parametric study, the dispersion constant was set for a range of values around the calculated dispersion constant ($1\times10^{-21} \leq A \leq 10\times10^{-21}$) in an effort to illustrate its effect on the thin-film profile and evaporation. The geometry of the thin film, as shown in Fig. 8, illustrates that an increasing dispersion constant results in a thickening of the adsorbed layer and an elongation of the thin film length.

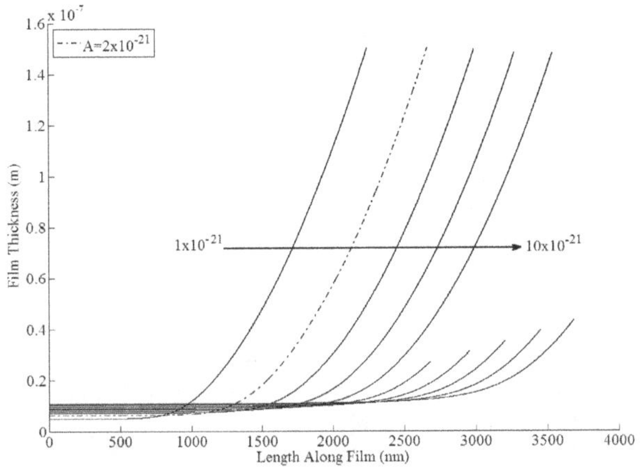

Fig. 8 Thin film profiles for increasing dispersion constants in increments of 1×10^{-21}.

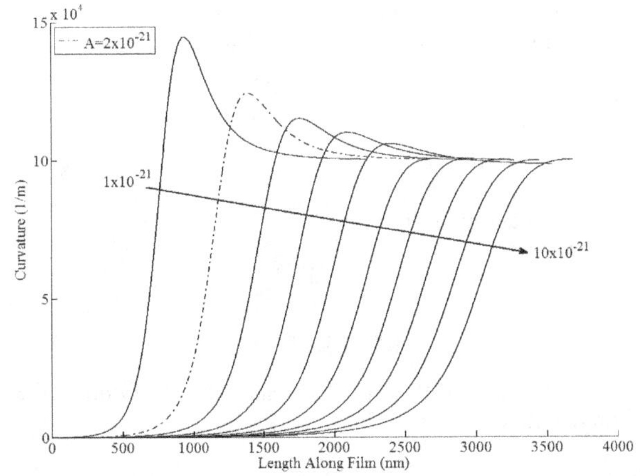

Fig. 9 Local curvature profiles for increasing dispersion constants in increments of 1×10^{-21}.

It is important to note that the thin-film termination location changes significantly for higher values of the dispersion constant. This termination can be explained by the reduction in the local curvature overshoot with increasing dispersion constant, shown in Fig. 9, similar to the trend observed previously in Fig. 6. Thus it is evident, that changes in the dispersion constant can significantly affect the thin film profile and evaporation capacity. In regards to evaporation, it can be seen in Fig. 10 that the peak evaporation rate decreases slightly with increasing dispersion constant.

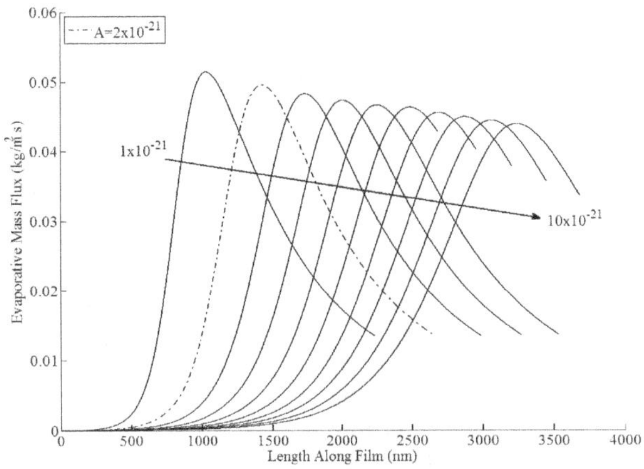

Fig. 10 Evaporative mass flux in the thin film region with increasing dispersion values in increments of 1×10^{-21}.

4.3 Absolute negative liquid pressure

At higher superheats, near the adsorbed region, disjoining effects are so pronounced that in satisfying the force-balance, Eq. (1), an absolute negative liquid pressure is generated. In the previously published thin-film solutions, the absolute negative liquid pressures were not explicitly discussed. For example, in Wang et al. (2007), a liquid pressure change (*ΔP*) was defined with respect to the pressure obtained in the assumed starting point of the thin-film solution instead of plotting the liquid pressure directly. Figure 11 shows a composite pressure profile at T_V = 300 K for a 1 K superheat, a channel width of 5 μm, and octane as the working fluid. The absolute negative liquid pressure inside the thin-film can be clearly seen in Fig. 11. As the film thickness decreases, the disjoining pressure quickly increases, leading to a decrease in the liquid pressure. For very thin films, the liquid pressure becomes negative to satisfy Eqs. (1) and (2).

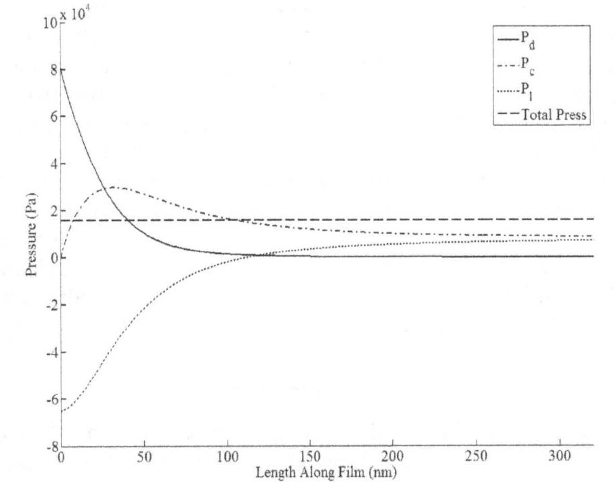

Fig. 11 Composite pressure profile for 1 K superheat, a channel width of 5 μm, and octane as the working fluid.

Maroo and Chung (2010) also reported a negative absolute liquid pressure in their analysis of evaporating meniscus using molecular dynamics. They speculated that this negative liquid pressure occurs in regions where the liquid film is being pulled by the relatively cooler liquid in the meniscus, and also towards the centre due to high disjoining pressure. The negative liquid pressures, where liquid is in a highly unstable state of tension, have been previously experimentally observed as explained in Batchelor (1979). It is possible that thin films can sustain negative pressures as the film thickness can be smaller than the critical cavitation radius. However, it is open to debate whether the negative liquid pressure originates from the simplified nature of the thin-film mathematical model or the model correctly predicts the liquid pressure in a thin film.

5. CONCLUSIONS

The non-linear ODE representing the evaporating thin film in a channel was numerically solved and the results were discusses in detail. A summary of the fundamental findings are as follows:

The numerical solution is very sensitive to the initial conditions and the applied perturbations. In particular extreme sensitivity was found with the applied value of $\delta'$and its associated perturbation ε_2. In addition, applied perturbations must be small enough such as to not shift the origin of the coordinate system to far into the thin-film region. Using too large of perturbations can result in omission of key attributes near the adsorbed region.

The accommodation coefficient and Hamaker constant had a pronounced effect on thin-film geometry and associated characteristics. Although the starting point of the solution is subject to an arbitrary shift due to the shooting techniques used, it is clear that varying these coefficients affect not only the origin but also overall shape of the resulting profiles.

Decreasing accommodation coefficient shallowed the thin-film profile. This shallowing resulted in a retardation of the decline in disjoining pressure. This corresponded in delaying the increase in liquid pressure and shifting the peak evaporative mass flux along the length of the film. The magnitude in evaporative mass flux also decreased with decreasing accommodation coefficient. Curvature overshoot was also observed to diminish with decreasing accommodation coefficient.

The variance in the Hamaker constant modified the profiles and physical characteristics of the thin film. As the Hamaker constant was increased, the thin-film length was extended, the peak evaporative mass flux was shifted towards the intrinsic meniscus, and the evaporative mass flux magnitude decreased.

Disjoining pressure effects are much more pronounced near the adsorbed layer, creating absolute negative liquid pressures. The thin films can theoretically sustain negative liquid pressures due to their small thicknesses. However, without experimental validation, the presence of negative pressures in thin films remains controversial.

It is important to highlight that from the above analyses, the current results of thin-film studies are very significant in gaining insight into the heat transfer characteristics of an evaporating thin-film. However, with the above inconsistencies identified, results can only be regarded as qualitative trends. With the ambiguity that exists within these parameters such as accommodation coefficient, Hamaker constant and applied perturbations, experimental investigation is necessary to verify the numerical results.

ACKNOWLEDGEMENTS

The support from Natural Sciences and Engineering Research Council of Canada under Discovery Grants Program is greatly appreciated.

NOMENCLATURE

a	Surface tension coefficient (N/m)
A	Dispersion constant (J)
A_H, A_{132}	Hamaker constant (J)
b	Surface tension temperature coefficient (N/m·K)
C	Constant of proportionality
C_4	Slip length relation coeff (m^3)
C_5	Slip length relation coeff (m^2)
Ca	Capillary number
h	Plank's constant (J·s)
h_{fg}	Latent heat of vapourization (J/kg·K)
h_{lv}^{cl}	Clapeyron evaporative mass flux coeff (kg/m^2·s·K)
h_{lv}^{kl}	Kelvin evaporative mass flux coeff (s/m)
H	Channel height (m)
k_l	Liquid conductivity (W/m·K)
k_B	Boltzmann's constant (J/s·m^2·K^4)
$\dot{m}$	Mass flow rate (kg/s)
$\dot{m}''$	Evaporative mass flux (kg/m^2·s)
M	Molecular mass (kg/kmol)
n	Index of refraction
P	Pressure (N/m^2)
R	Universal gas constant (J/mol·K)
T	Temperature (K)
u	Axial velocity (m/s)
V_l	Molar volume (m^3/mol)
V	Liquid velocity (m/s)
x	x direction (m)
y	y direction (m)

Greek symbols

α	Local curvature term
β	Wall slip length (m)
γ	Accommodation coefficient
δ	Thin film thickness (m)
ε_1	Thickness perturbation (m)
ε_2	Slope perturbation
ε_3	Perturbation applied to δ'' (1/m)
η	Kelvin-Clapeyron mass flux coeff (mol·K$^{3/2}$/s·m^2)
κ	Local curvature (1/m)
μ	Dynamic viscosity (Pa·s)
ν	Kinematic viscosity (m^2·s)
ν_e	Main absorption frequency (1/s)
ξ	Dielectric constant
σ	Surface tension (N/m)
τ	Liquid-vapour interfacial shear stress (N/m^2)
χ	Collection of terms in ODE (W·m·K$^{3/2}$)
ψ	Collection of terms in ODE (1/m·K)
ω	Collection of terms in ODE (W·K$^{3/2}$/m)

Subscripts

c	Capillary
d	Disjoining
l	Liquid
lv	Liquid-vapour interface
v	Vapour
w	Wall

REFERENCES

Batchelor, G.K., 1979, *An introduction to Fluid Dynamics*, Cambridge University Press, New York

Butt, Hans-Jurgen and Kappl, Michael, 2010, *Surface and Interfacial Forces*, Wiley-VCH Verlag GmbH & Co. KGaA

Costner, E.A., Long, K.B., Navar, C., Jockusch, S., Lei, X., Zimmerman, P., Campion, A., Turro, N.J., Willson, C.G., 2009,

"Fundamental Optical Properties of Linear and Cyclic Alkanes: VUV Absorbance and Index of Refraction," *Journal of Physical Chemistry*, **113**(33), 9337-9347
http://dx.doi.org/10.1021/jp903435c

DasGupta, S., Kim, I.Y., Wayner Jr.,P.C., 1994, "Use of the Kelvin-Clapeyron Equation to Model an Evaporating Curved Microfilm," *Journal of Heat Transfer*, **116**(4), 1007-1014
http://dx.doi.org/10.1115/1.2911436

DasGupta, S., Schonberg, J.A., Kim, I.Y., Wayner Jr., P.C., 1993, "Use of the Augmented Young-Laplace Equation to Model Equilibrium and Evaporating Extended Menisci," *Journal of Colloid and Interface Science*, **157**(2), 332-342
http://dx.doi.org/10.1006/jcis.1993.1194

Dormand, J.R. and Prince, P.J., 1980, "A Family of Embedded Runge-Kutta Formulae," *Journal of Computational and Applied Mathematics* **6**(1), 19-26
http://dx.doi.org/10.1016/0771-050X(80)90013-3

Du, S.-Y. and Zhao, Y.-H., 2011, "New Boundary Conditions for the Evaporating Thin-Film Model in a Rectangular Micro Channel," *International Journal of Heat and Mass Transfer*, **54**(15-16), 3694-3701
http://dx.doi.org/10.1016/j.ijheatmasstransfer.2011.02.059

Du, S.-Y. and Zhao, Y.-H., 2012, "Numerical Study of Conjugated Heat Transfer in Evaporating Thin-films Near The Contact Line," *International Journal of Heat and Mass Transfer*, **55**(1-3), 61-69
http://dx.doi.org/10.1016/j.ijheatmasstransfer.2011.08.039

Hallinan, K.P., Kim, S.J., Chang, W.S., 1994, "Evaporation from an Extended Meniscus for Nonisothermal Interfacial Conditions," *Journal of Thermophysics and Heat Transfer*, **8**(4), 709-716
http://dx.doi.org/10.2514/3.602

Israelachvili, Jacob, 1991, *Intermolecular and Surface Forces*, Academic press Harcourt Brace Jovanovich Publishers, Toronto

Kou, M.L. and Bai, Z.H., 2011, "Effects of Wall Slip and Temperature Jump on Heat and Mass Transfer Characteristics of an Evaporating Thin Film," *International Communications in Heat and Mass Transfer*, **38**(7), 874-878.
http://dx.doi.org/10.1016/j.icheatmasstransfer.2011.03.032

Maroo, S.C. and Chung, J.N., 2010, "Heat Transfer Characteristics and Pressure Variation in a Nanoscale Evaporating Meniscus," *International Journal of Heat and Mass Transfer*, **53**(15-16), 3335-3345
http://dx.doi.org/10.1016/j.ijheatmasstransfer.2010.02.030

Mills, A. F., 1965, "The Condensation of Steam at Low Pressures," *Technical Report on NSF GP-2250*, **6**(9), Space Sciences Laboratory, University of California at Berkley

Moosman, S. and Homsy, G.M., 1980, "Evaporating Menisci of Wetting Fluids," *Journal of Colloid and Interface Science*, 73(1), 212-223.
http://dx.doi.org/10.1016/0021-9797(80)90138-1

Morris, S. J. S., 2003, "The Evaporating Meniscus in a Channel," *Journal of Fluid Mechanics*, **494**, 297-317.
http://dx.doi.org/10.1017/S0022112003006153

Panchamgam, S.S., Chatterjee, A., Plawsky, J.L., Wayner Jr., P.C., 2008, "Comprehensive Experimental and Theoretical Study of Fluid Flow and Heat Transfer in a Microscopic Evaporating Meniscus in a Miniature Heat Exchanger," *International Journal of Heat and Mass Transfer*, **51**(21-22), 5368-5379
http://dx.doi.org/10.1016/j.ijheatmasstransfer.2008.03.023

Park, K. and Lee, K.-S., 2003, "Flow and Heat Transfer Characteristics of the Evaporating Extended Meniscus in a Micro-Capillary Channel," *International Journal of Heat and Mass Transfer*, **46**(24), 4587-4594
http://dx.doi.org/10.1016/S0017-9310(03)00306-5

Park, K., Noh, K.J., Lee, K.S., 2003, "Transport Phenomena in the Thin-Film Region of a Micro-Channel," *International Journal of Heat and Mass Transfer*, **46**(13), 2381-2388
http://dx.doi.org/10.1016/S0017-9310(02)00541-0

Potash Jr., M. and Wayner Jr., P.C., 1972, "Evaporation from a Two-Dimensional Extended Meniscus," *International Journal of Heat and Mass Transfer*, **15**(10), 1851-1863
http://dx.doi.org/10.1016/0017-9310(72)90058-0

Qu, W., Ma, T., Miao, J., Wang, J., 2002, "Effects of Radius and Heat Transfer on the Profile of Evaporating Thin Liquid Film and Meniscus in Capillary Tubes," *International Journal of Heat and Mass Transfer*, **45**(9), 1879-1887
http://dx.doi.org/10.1016/S0017-9310(01)00296-4

Schonberg, J. A., DasGupta, S., Wayner Jr., P.C., 1995, "An Augmented Young-Laplace Model of an Evaporating Meniscus in a Microchannel with High Heat Flux," *Experimental Thermal and Fluid Science*, **10**(2), 163-170
http://dx.doi.org/10.1016/0894-1777(94)00085-M

Stephan, P.C. and Busse, C.A., 1992, "Analysis of the Heat Transfer Coefficient of Grooved Heat Pipe Evaporator Walls," *International Journal of Heat and Mass Transfer*, **35**(2), 383-391
http://dx.doi.org/10.1016/0017-9310(92)90276-X

Wang, H., Garimella, S.V., Murthy, J.Y., 2008, "An Analytical Solution for the Total Heat Transfer in the Thin-Film Region of an Evaporating Meniscus," *International Journal of Heat and Mass Transfer*, **51**(25-26), 6317-6322
http://dx.doi.org/10.1016/j.ijheatmasstransfer.2008.06.011

Wang, H., Garimella, S.V., Murthy, J.Y., 2007, "Characteristics of an Evaporating Thin Film in a Microchannel," *International Journal of Heat and Mass Transfer*, **50**(19-20), 3933-3942
http://dx.doi.org/10.1016/j.ijheatmasstransfer.2007.01.052

Wayner Jr., P.C., Kao, Y.K., LaCroix, L.V., 1976, "The Interline Heat-Transfer Coefficient of an Evaporating Wetting Film," *International Journal of Heat and Mass Transfer*, **19**(5), 487-492
http://dx.doi.org/10.1016/0017-9310(76)90161-7

Wayner Jr., Peter C., 1991, "The Effect of Interfacial Mass Transport on Flow in Thin Liquid Films," *Colloids and Surfaces*, **52**, 71-84
http://dx.doi.org/10.1016/0166-6622(91)80006-A

Wee S.-K., Kihm, K.D., Hallinan, K.P., 2005, "Effects of the Liquid Polarity and the Wall Slip on the Heat and Mass Transport Characteristics of the Micro-Scale Evaporating Transition Film," *International Journal of Heat and Mass Transfer*, **48**(2), 265-278
http://dx.doi.org/10.1016/j.ijheatmasstransfer.2004.08.021

Zhao, J.-J., Duan, Y.Y., Wang, X.D., Wang, B.X., 2011, "Effects of Superheat and Temperature-Dependent Thermophysical Properties on Evaporating Thin Liquid Films in Microchannels," *International Journal of Heat and Mass Transfer*, **54**(5-6), 1259-1267
http://dx.doi.org/10.1016/j.ijheatmasstransfer.2010.10.026

ANALYSIS OF CHAOTIC NATURAL CONVECTION IN A TALL RECTANGULAR CAVITY WITH NON-ISOTHERMAL WALLS

Heather Dillon[a], Ashley Emery[b,†], Ann Mescher[b]

[a]*Dept of Mechanical Engineering, University of Portland, Portland, Oregon, 97203, USA*
[b]*Dept of Mechanical Engineering, University of Washington, Seattle, Washington, 98105, USA*

ABSTRACT

A computational model is presented that extends prior work on unsteady natural convection in a tall rectangular cavity with aspect ratio 10 and applies Proper Orthogonal Decomposition to the results. The solution to the weakly compressible Navier-Stokes equation is computed for a range of Rayleigh numbers between 2×10^7 and 2.2×10^8 with Prandtl number 0.71. A detailed spectral analysis shows dynamic system behavior beyond the Hopf bifurcation that was not previously observed. The wider Rayleigh range reveals new dynamic system behavior for the rectangular geometry, specifically a return to a stable oscillatory behavior that was not predicted in prior work. Proper Orthogonal Decomposition (POD) has been used to analyze the computational results. Five eigenvalue modes were required to capture correctly the basic flow structure. The POD failed to capture subtle aspects of the flow structure at high Rayleigh numbers for the model, indicating that a POD and Galerkin projection for several Rayleigh numbers will be needed to reproduce the complex behavior of the system.

Keywords: *Proper Orthogonal Decomposition, natural convection, reduced order modeling, computational fluid dynamics*

1. INTRODUCTION

Natural convection in a tall cavity with non-isothermal vertical walls has many applications including optical fiber manufacturing, optimizing computer component locations, and composite building walls. Computational work has explored a wide variety of boundary conditions, however the case of non-uniform wall temperatures has only been explored by a small group.

For the case of isothermal walls in tall cavities the flow will transition from the conduction regime to the convection regime at the critical Rayleigh number (Ra_c). As the Rayleigh number (Ra) is further increased, the system will transition to chaotic behavior. The chaotic behavior of the system has been shown to exhibit hysteresis, intermittency, a supercritical Hopf bifurcation, and period doubling. Several parameters are used to classify the systems. The geometry of the cavity is represented by the aspect ratio (A), where $A = H/W$. H is the height of the cavity and W is the width.

Convection in a tall cavity has been studied by many authors in the last 30 years as summarized in Table 1. The work of Reeve *et al.* (2004) is the only prior work to consider the same boundary conditions as in the present work. Here, the range of Rayleigh numbers has been extended. The boundary conditions for the system are specified as adiabatic on the upper and lower cavity boundaries as shown in Figure 1. The temperature on the right and left walls are constrained to vary linearly in the y-direction as shown in Figure 2. Reeve *et al.* (2004) observed that these boundary conditions, at low Rayleigh number, create a steady bi-cellular flow in contrast to uni-cellular flow for isothermal vertical walls.

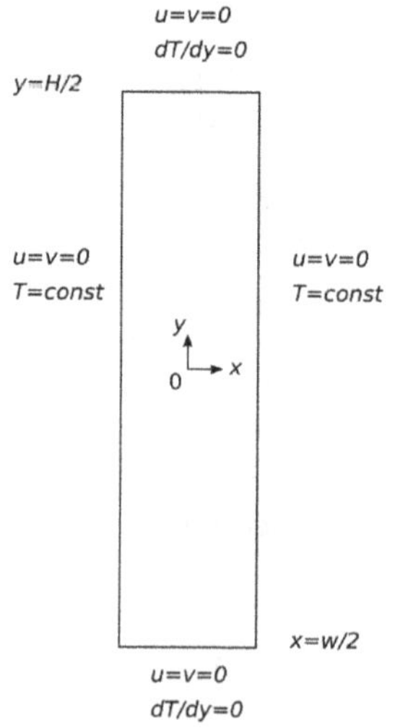

Fig. 1 Schematic of the computational domain.

All of the authors summarized in Table 1 considered isothermal vertical walls except Reeve *et al.* (2004). Some authors (Paolucci and Chenoweth (1989) and Haldenwang and Labrosse (1986)) predict an oscillatory instability for tall cavities. Chenoweth and Paolucci (1996) observed that as the temperature difference in the cavity is increased, a

†*Corresponding author. Email: emery@u.washington.edu*

Table 1 Summary of computational work for rectangular natural convection systems.

Author	Year	Ra	Pr	A	Description
Vest and Arpaci (1969)	1969	$7 \times 10^3 - 3 \times 10^5$	0.71, 1000	20-33	Galerkin method.
Korpela *et al.* (1973)	1973	$1 \times 10^2 - 1 \times 10^4$	0-50	∞	Report Ra_c as a function.
Lee and Korpela (1983)	1983	$3 \times 10^4 - 2 \times 10^5$	0-1000	15-40	Report Nusselt and streamfunctions.
Chenoweth and Paolucci (1996)	1986	$1 \times 10^5 - 1 \times 10^6$	0.71	1-10	Compare ideal gas and Boussinesq.
Chait and Korpela (1989)	1989	$5 \times 10^2 - 1.5 \times 10^4$	0.71, 1000	∞	Pseudospectral method. Reported Ra_c.
Paolucci and Chenoweth (1989)	1989		0.71	0.5-3	Frequency results, bifurcation and phase diagrams.
LeQuere (1990)	1990	$7 \times 10^3 - 4 \times 10^4$	0.71	16	Explored return to uni-cellular pattern.
Liakopoulos *et al.* (1990)	1990		0.71	10-25	Included flux wall conditions.
Suslov and Paolucci (1995)	1995	$6 \times 10^3 - 1 \times 10^4$	0.71	∞	Non-Boussinesq impact on stability and considered Ra_c with ΔT.
Xin and LeQuere (2002)	2002	$3 \times 10^5 - 5 \times 10^5$	0.71	8	Benchmark study reports Ra_c.
Reeve (2003)	2003	$2 \times 10^7 - 1 \times 10^8$	0.71	10	Commercial code FIDAP.

lower critical Rayleigh number is found. Related work in annular geometries has been explored in prior work experiementally and compuationally (Dillon *et al.*, 2011b). Early computational results for the geometry considered are presented in Dillon *et al.* (2011a) and extended with reduced order analysis in this paper.

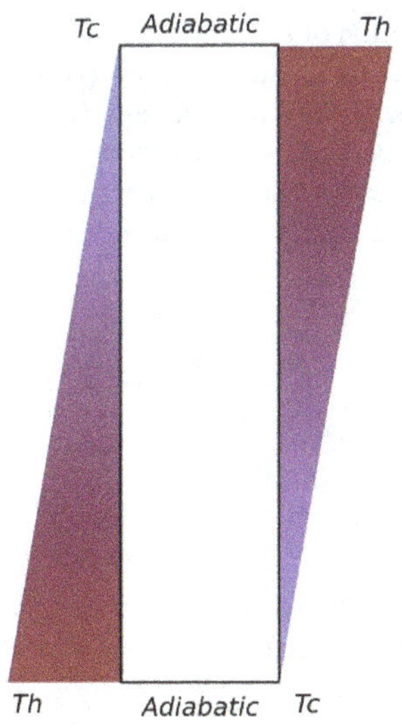

Fig. 2 Linear temperature profile boundary conditions applied to the vertical walls of the cavity.

2. METHODS

This problem was analyzed using a traditional computational tools like COMSOL and an analysis technique called proper orthogonal decomposition.

2.1. Computational Tool

The computational tool COMSOL was used to perform the simulations. Prior to this investigation the computational tool was benchmarked with experimental results from the literature, specifically Vest and Arpaci (1969), and compared to other models (Lee and Korpela (1983), Liakopoulos *et al.* (1990), Xin and LeQuere (2002), Reeve (2003)). The benchmarking work is documented in Dillon (2011).

The buoyancy driven flow is modeled as a coupled system with fluid motion (Navier-Stokes) and heat transfer. The model equations are given in Equations 1-3.

In this form u represents the velocity vector of the fluid, ρ is the fluid density, μ is the dynamic viscosity of the fluid, and f is the body force applied to the fluid. The heat transfer is governed by Equation 3. For this equation k is the thermal conductivity, c_p is the specific heat at constant pressure, and T is the temperature.

$$\rho\frac{\partial u}{\partial t} + \rho(u \cdot \bigtriangledown)u = \bigtriangledown \cdot (-pI + \mu(\bigtriangledown u + \bigtriangledown u^T) -(2\mu/3 \cdot \bigtriangledown \cdot u)I) + f \quad (1)$$

$$\frac{\partial \rho}{\partial t} + \bigtriangledown \cdot (\rho u) = 0 \quad (2)$$

$$\rho c_p \frac{\partial T}{\partial t} + \bigtriangledown \cdot (-k \bigtriangledown T) = -\rho c_p u \cdot \bigtriangledown T \quad (3)$$

The boundary conditions for the system are specified as no slip and adiabatic on the upper and lower cavity boundaries. The right and left wall temperatures have a linear profile given by $T(y) = T_h \cdot (1 - y)$ and $T(y) = T_c \cdot y$.

A grid resolution study was conducted to determine the required mesh density and to compare the results with those computed by Reeve *et al.* (2004) using FIDAP. Results for a 1120 element triangular mesh were consistent with the results of Reeve and this triangular mesh was used for computations. This is documented in more detail in Dillon (2011) and Dillon *et al.* (2009).

The density of the fluid is represented with the Boussinesq approximation to simplify the formation of the coupled Navier-Stokes equations. The height of the cavity H is the characteristic length and the temperature difference between T_h and T_c is the characteristic temperature, $\Delta T = (T_h - T_c)$. From this the dimensionless temperature becomes $\Theta = (T - T_c)/(\Delta T)$. The dimensionless time is $\tau = t\sqrt{g\beta\Delta T H^{-1}}$. The dimensionless velocity was chosen based on the work of Reeve *et al.* (2004) as $u/\sqrt{g\beta\Delta T H}$. Aspects of the dimensionless parameters are discussed in more detail in Dillon *et al.* (2010) and Dillon (2011).

2.2. Proper Orthogonal Decomposition

Proper Orthogonal Decomposition (POD) is based on the diagonalization of a matrix. The mathematical procedure linearly transforms the number of possibly correlated variables into a smaller number of uncorrelated variables. The first component contains as much of the variation in the system as possible. Assume a matrix X is an $m \times n$ matrix composed of multiple observations from a simulation.

For the POD analysis the data is centered by the mean of each row. Then the covariance matrix Cx is calculated. The covariance matrix is a square, symmetric $m \times m$ matrix whose diagonal represents the variance of particular measurements. Small diagonal terms indicate the variables are statistically independent.

$$C_x = \frac{1}{n-1} X X^T \tag{4}$$

Singular Value Decomposition (SVD) is used to diagonalize the matrix. The SVD diagonalization is shown in Equation 5, where $U \in \mathbb{C}^{m \times m}$ is unitary, $V \in \mathbb{C}^{n \times n}$ is unitary, and $\Sigma \in \mathbb{R}^{m \times n}$ is diagonal. If full SVD is used for calculations it may be applied to rank deficient matrices as described by Trefethen and Bau (1997).

$$X = U \Sigma V^* \tag{5}$$

V^* is the hermitian conjugate or adjoint of the matrix V. This means that the complex conjugate of each entry in the matrix is calculated and then the matrix is transposed.

Theorem: Every matrix $X \in \mathbb{C}^{m \times n}$ has a singular value decomposition. Furthermore, the singular values $\{\sigma_j\}$ are uniquely determined, and if X is square and the σ_j distinct, the singular vectors $\{u_j\}$ and $\{v_j\}$ are uniquely determined up to complex signs (complex scalar factors of absolute value 1) as described by Trefethen and Bau (1997).

A visual representation of the SVD process is shown in Figure 3. The following theorems Trefethen and Bau (1997) illustrate the way SVD is linked to POD, where the first confirms that the basis vectors are unique and the second theorem allows determination of N such that the Nth partial sum captures as much of matrix X as possible.

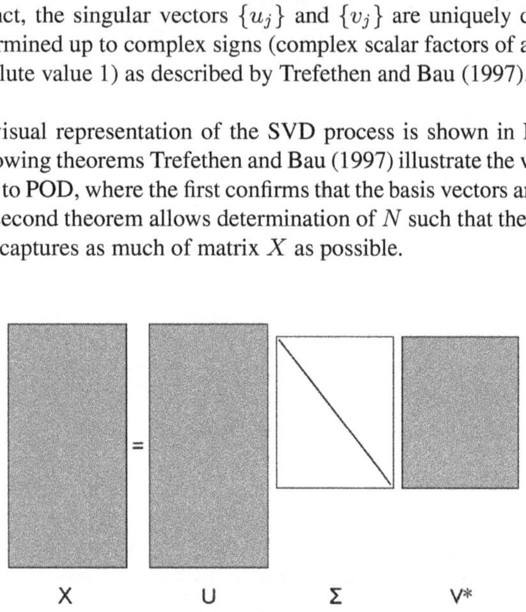

Fig. 3 Visual representation of the SVD diagonalization. Adapted from Trefethen and Bau (1997).

Theorem: X is the sum of r rank-one matrices.

$$X = \Sigma_{j=1}^{r} \sigma_r u_j v_j^* \tag{6}$$

Theorem: For any N such that $0 \leq N \leq r$, we can define the partial sum

$$X = \Sigma_{j=1}^{N} \sigma_r u_j v_j^*$$

And if $N = min\{m, n\}$, define $\sigma_{N+1} = 0$. Then

$$\|X - X_N\|_2 = \sigma_{N+1}$$

The SVD gives a type of least-square fitting algorithm, allowing us to project the matrix onto low-dimensional representations in a formal, algorithmic way as shown by Trefethen and Bau (1997). The matrix may also be projected onto other basis as considered by other authors in the literature. For this work, the matrix is projected back onto the data to investigate the contribution of each mode to the observed data to augment the frequency analysis.

2.3. Galerkin Projection

Many authors use the POD to build more elaborate models based on the calculated modes. The most common is a Galerkin projection, but some authors explored alternatives like Method of Polyargumental Systems (MPS), Blinov *et al.* (2004), Linear Stochastic Estimation (LSE), Bonnet *et al.* (1994), Equation Free (EF) modeling, Sirisup *et al.* (2005), and balanced truncation, Rowley (2005); Rowley and Marsden (2000).

For transient natural convection the Galerkin projection has been used frequently. The formulation of the Galerkin projection for natural convection problems is based on the Boussinesq equations. The Galerkin method is used to simplify the equations to a set of non-linear ordinary differential equations based on the orthogonal nature of the POD.

If $\hat{U}$ is the vector of the radial and axial velocities the Boussinesq approximation may be expressed in terms of the Grashof number (Gr) and the Prandlt number (Pr).

$$\frac{\partial \hat{U}}{\partial t} + (\hat{U} \cdot \nabla)\hat{U} + \nabla P = \Theta + \frac{1}{\sqrt{Gr}} \nabla^2 \hat{U} \tag{7}$$

$$\frac{\partial \Theta}{\partial t} + \hat{U} \cdot \nabla \Theta = \frac{1}{Pr\sqrt{Gr}} \nabla^2 \Theta \tag{8}$$

$$\nabla \cdot \hat{U} = 0 \tag{9}$$

Using the derivation of Liakopoulos *et al.* (1997), the stationary empirical eigenfunctions for the temperature (ϕ_k) and flow field (φ_k) are determined for a specific Gr and Pr, (equivalent to a specific Ra) using POD. The input temperature (Θ) and flow field $(\hat{U})$ are separated into time varying $(\Theta', \hat{U}')$ part and time averaged components prior to performing the POD, and can then be expanded in terms of the calculated eigenfunctions.

$$\hat{U}' = \sum_{k=1}^{M} a_k(t) \varphi_k \tag{10}$$

$$\Theta' = \sum_{k=1}^{M} b_k(t) \phi_k \tag{11}$$

The expansion coefficients (a_k, b_k) are calculated via integration over the spatial domain from the known eigenfunctions.

$$\begin{aligned} a_k(t) &= \int\int \hat{U}' \cdot \varphi_k d\Omega \quad k = 1, 2, ..., M \\ b_k(t) &= \int\int \Theta' \cdot \phi_k d\Omega \quad k = 1, 2, ..., M \end{aligned} \tag{12}$$

Substitution of the expansion coefficients into the Boussinesq equations provides a simplified model for the system based on a set of coefficients $(A_k, B_k, C_k, ...K_k)$ for $k = 1, 2, ..., M$ which are determined from inner products of the eigenfunctions and the flow properties.

$$\frac{da_k}{dt} = A_k + \frac{1}{\sqrt{Gr}} B_k + C_{ki} a_i + \frac{1}{\sqrt{Gr}} D_{ki} a_i + E_{kij} a_i a_j + R_{ki} b_i \tag{13}$$

$$\frac{db_k}{dt} = F_k + \frac{1}{Pr\sqrt{Gr}} G_k + H_{ki} a_i + \frac{1}{Pr\sqrt{Gr}} I_{ki} b_i + J_{kij} a_i b_j + K_{ki} b_i \tag{14}$$

This set of equations has been used by authors to represent the flow and explore the dynamic system properties. The key limitation to this approach is the restriction of the POD to a calculation at one specific Rayleigh number. The exploration of the flow near that Rayleigh number is likely to be well captured by the set of derived equations, but accurate representation of the system using the simplified expressions over a large range of Rayleigh numbers is unlikely.

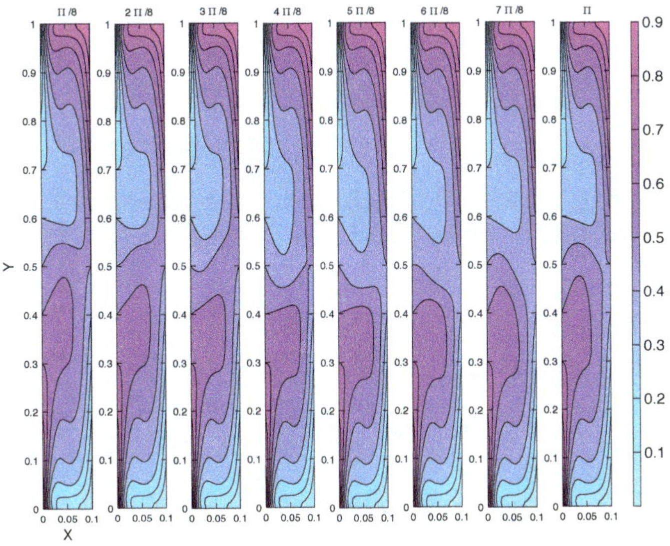

(a) Θ

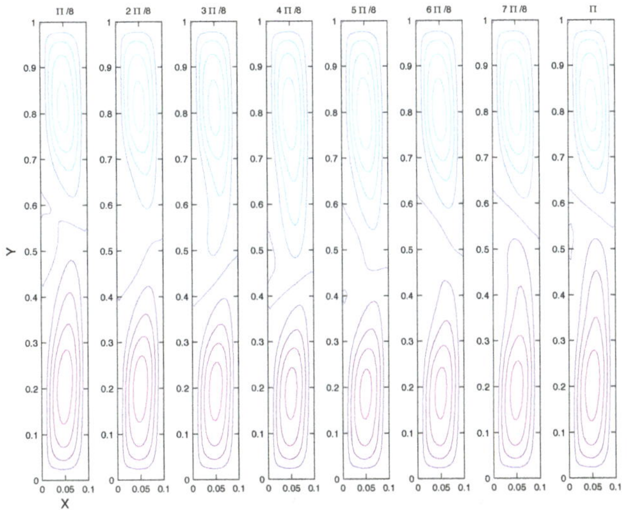

(b) ψ

Fig. 4 Sequential contour plot of the temperature Θ (upper) and stream function ψ (lower) for the rectangular cavity illustrating oscillation of the natural convection cells through one period (Π). $Ra = 2.5 \times 10^7$, $A = 10$.

3. COMPUTATIONAL RESULTS

For the rectangular cavity, the results for Rayleigh number below $Ra = 1.0 \times 10^8$ are consistent with the results from Reeve. A transition from steady to oscillatory flow is observed at $Ra = 2.2 \times 10^7$ as the system undergoes a supercritical Hopf bifurcation. Figure 4 shows one period of oscillations at $Ra = 2.5 \times 10^7$, after the bifurcation results in oscillatory flow. A second transition from oscillatory to chaotic flow is observed at $Ra = 4.5 \times 10^7$.

In the rectangular cavity as the Rayleigh number is increased the frequency of the oscillations increases. Figure 5 shows the amplitude of the temperature oscillations at the center of the cavity at several Rayleigh numbers. The local region of the oscillations near the center of the cavity also becomes smaller as the the Rayleigh number increases.

A second, unexpected, transition back to oscillatory flow is predicted at $Ra = 1.27 \times 10^8$. Figure 9 shows one period of oscillations for $Ra = 1.7 \times 10^8$ where the behavior is again oscillatory with a third cell

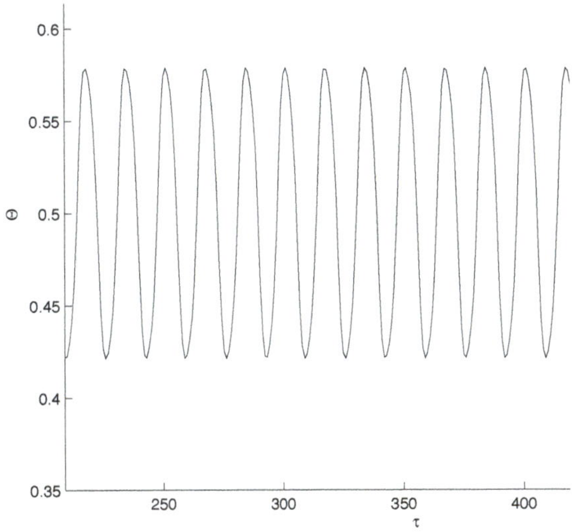

(a) $Ra = 2.5 \times 10^7$

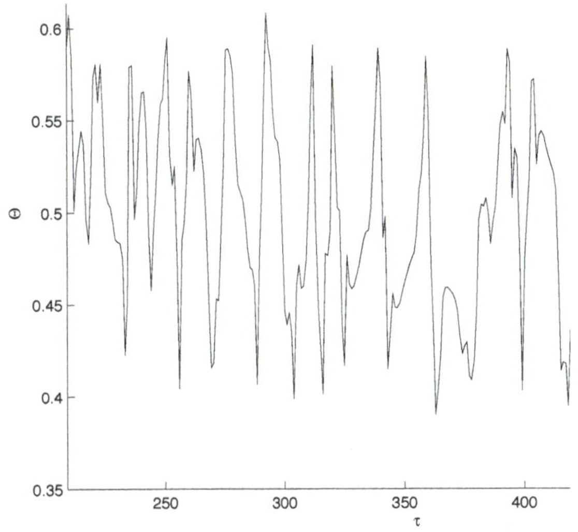

(b) $Ra = 10 \times 10^7$

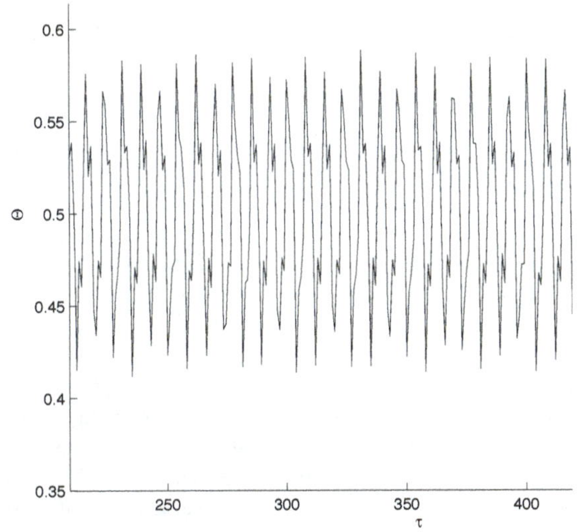

(c) $Ra = 17 \times 10^7$

Fig. 5 Temperature oscillations in the center of the cavity for $A = 10$ showing the increase in the frequency of the oscillations and the effect of the harmonics in the oscillations.

formation near the center of the cavity. In Reeve's previous research of the rectangular geometry, a tri-cellular solution path was not predicted; however, Reeve observed a similar phenomena at high Rayleigh numbers in an annular cavity.

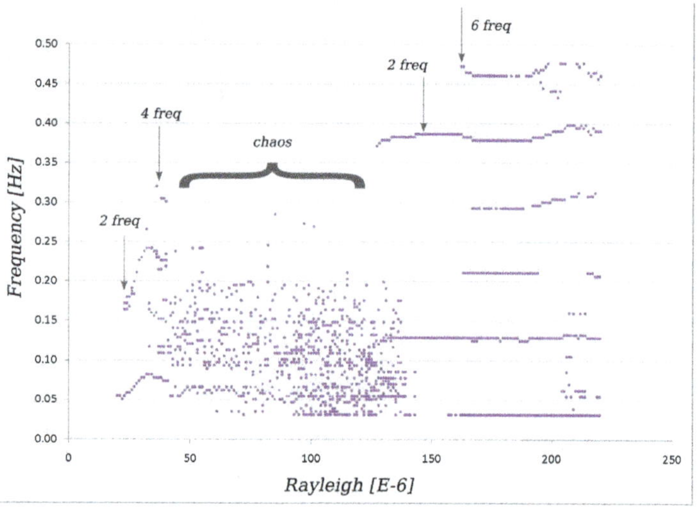

Fig. 6 Frequency summary of the temperature oscillations in the center of the rectangular cavity, $A = 10$ (listing the number of frequencies observed).

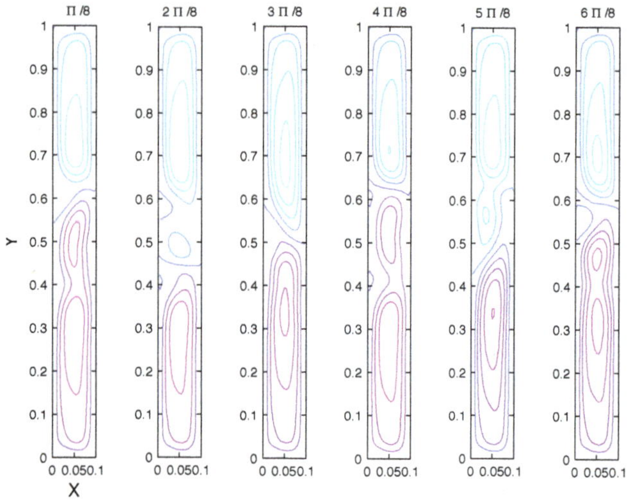

Fig. 7 Sequential contour plot of the stream function for the rectangular cavity illustrating oscillation of the natural convection cells through one period (Π) in the chaotic region. $Ra = 10 \times 10^7$, $A = 10$.

The most interesting aspects of the solution are visible when the the full frequency spectrum is shown. The peak frequencies were calculated and are shown in Figure 6. The frequency analysis reveals that the lower frequency (near 0.05 Hz) has more power than the higher frequencies in the oscillations. The initial transition to chaotic behavior is observed to occur via period halving, where two frequencies are first seen at 0.051 Hz and 0.164 Hz ($Ra = 2.2 \times 10^7$). Then at $Ra = 3.6 \times 10^7$ four frequencies are visible at 0.078, 0.160, 0.238 and 0.320 Hz. The transition to chaotic behavior at $Ra = 4.5 \times 10^7$ is shown by a collection of frequencies in the spectrum. This region ends abruptly at $Ra = 13.8 \times 10^7$ with the appearance of two frequencies at 0.129 and 0.371 Hz. The frequencies may exhibit period doubling but the Rayleigh number resolution does not show a second bifurcation, rather the system jumps to six frequencies at $Ra = 16.3 \times 10^7$. The highest frequency (0.465 Hz) bifurcates again at $Ra = 19.3 \times 10^7$ as shown in the upper right of Figure 6.

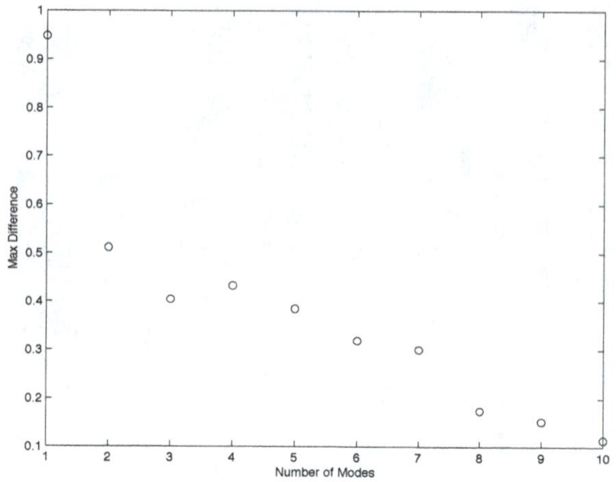

(a) POD Θ

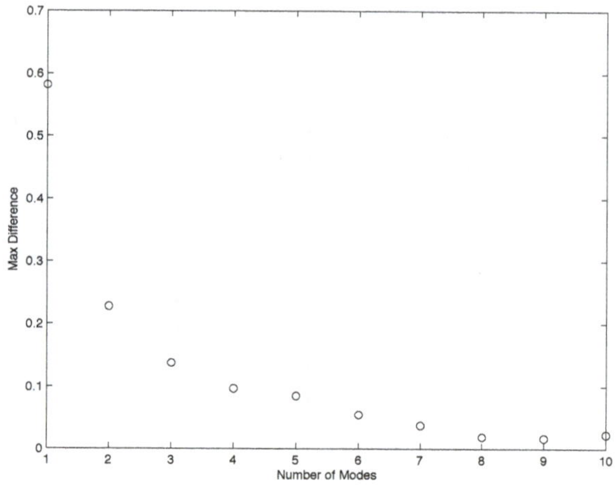

(b) POD ψ

Fig. 8 Temperature (a) and streamfunction (b) maximum difference between the POD and original solution for $Ra = 17 \times 10^7$.

The nature of the chaotic behavior is dominated by the formation of the third cell in the cavity. Figure 7 shows an example of behavior in the chaotic region ($Ra = 10 \times 10^7$). The center region of the cavity experiences chaotic temperatures and velocities but the upper and lower regions of the cavity remain relatively stable. Eventually, at higher Rayleigh number, the third cell stabilizes as shown in Figure 4 with six frequencies in the system.

4. POD RESULTS

To quantify the dynamic behavior of the system Proper Orthogonal Decomposition (POD) was applied to the temperature and velocity field of the simulation solution set. At specific Rayleigh numbers the time dependent velocity streamfunction (ψ) and temperature (Θ) data are used to calculate stationary empirical eigenfunctions (modes). The modes are determined from M snapshots representing at least four periods of oscillation based on the technique of Sirovich (1987).

Table 2 Summary of modes and energy for the POD. Eigenvalues for the five most energetic modes and the contribution to the system energy at $Ra = 2.5 \times 10^7$ and $Ra = 17 \times 10^7$ in the rectangular cavity.

$Ra = 2.5 \times 10^7$	**Temperature**	**Eigenvalues**	**Streamfunction**	**Eigenvalues**
Mode	**eigenvalue** σ	**Total Energy** *percent*	**eigenvalue** σ	**Total Energy** *percent*
1	0.59292	47.0954	0.012625	38.4593
2	0.34788	68.9592	0.0047368	76.9099
3	0.07136	93.6327	0.0018725	90.8724
4	0.056986	94.9153	0.00077265	96.2336
5	0.024877	97.7803	0.00024232	98.8188
$Ra = 17 \times 10^7$	**Temperature**	**Eigenvalues**	**Streamfunction**	**Eigenvalues**
Mode	**eigenvalue** σ	**Total Energy** *percent*	**eigenvalue** σ	**Total Energy** *percent*
1	0.71757	68.2151	0.029221	51.1865
2	0.57901	74.3525	0.015352	74.3544
3	0.22331	90.1084	0.0063467	89.3977
4	0.19776	91.2402	0.003891	93.5001
5	0.13517	94.0127	0.0015179	97.4643

$$X = \begin{bmatrix} \Theta_{1,1}(t_1) & \Theta_{1,2}(t_1) & \cdots \\ \Theta_{1,1}(t_2) & \Theta_{1,2}(t_2) & \cdots \\ \vdots & & \vdots \\ \Theta_{1,1}(t_m) & \Theta_{1,2}(t_m) & \cdots \end{bmatrix}$$

The eigenvalues were calculated for $Ra = 2.5 \times 10^7$ in the Cartesian geometry and are shown in Table 2. The 90% threshold suggested by Holmes is reached with the third eigenvalue for both parameters.

The first five modes are shown in Figure 11. These modes indicate that most of the variation in the system occurs near the center of the cavity. The structure of the modes is quantitatively similar to eigenmodes represented in prior work Liakopoulos *et al.* (1997). For $Ra = 2.5 \times 10^7$ the first five modes are used to calculate the temperature and streamfunctions of the system to visualize the error in using a reduced set of modes.

A higher Rayleigh number solution in the tri-cellular region was also analyzed. The modes are shown in Figure 12 and Table 2. The calculated temperature and streamfunction are shown in Figure 10. When compared with Figure 9 it is clear that key structures in the flow are not captured, specifically the third cell in the cavity center and the additional cell formation in the upper cavity (Π) and lower cavity ($4\Pi/8$). For complex fluid behavior in this region the POD does a relatively poor job even when 97% of the eigenvalue energy is used. The maximum differences are shown in Figure 8. In cases of very fine structure it may require retaining modes that represent as much as 99.5% of the total energy.

5. CONCLUSIONS

For natural convection with moderate driving force, COMSOL predicts the stable region of the flow and a transition to oscillatory flow as well as other computational tools do. The results for natural convection in a tall cavity with linear boundary conditions agree well with the results of Reeve *et al.* (2004) at low Rayleigh numbers. At higher Rayleigh numbers, the natural convection creates a multi-cellular flow pattern and temperature oscillations in the cavity.

These simulations over a wider Rayleigh number range show that the system transitions back to a stable oscillatory behavior as a third convection cell is formed in the cavity. The spectral analysis confirms that the system is undergoing a frequency doubling route to chaos following a supercritical Hopf bifurcation.

A POD analysis of this type has never been performed for this natural convection system. In general the 90% cumulative energy threshold for POD as suggested by other authors is not sufficient to characterize the flow or temperature structure of a system of this type. For the oscillatory region five modes were sufficient to reproduce key flow structures but at high Rayleigh numbers where the flow is more complex the POD was not able to capture the flow behavior. This indicates the need for a more robust approach to reduced order modeling for a system of this type. The limitations of POD in this problem may be reduced in part in future work by increasing the resolution of the CFD code in the time domain, but this change was not possible for consideration in the present work.

A traditional Galerkin projection was not calculated for this problem because of the dynamic system complexity. For example, if the $Ra = 2.5 \times 10^7$ results had been used as the base for a Galerkin projection it seems clear from examination of the modes that the $Ra = 18 \times 10^7$ would be poorly captured by the projection. In this case, future work based on the Galerkin projection outlined by Liakopoulos *et al.* (1997) will be modified for this problem. A projection will be derived for multiple Rayleigh numbers, representing the two cell, chaotic, and three cell regions of the flow. The authors believe that a Galerkin projection that crosses multiple Rayleigh numbers may be a robust way to represent the flow characteristics.

The POD results do provide an elegant method for understanding the complex flow behavior in the system. Future work may examine a wider Ra range for the annular geometry and additional radius ratios.

ACKNOWLEDGEMENTS

The authors wish to express sincere appreciation to the National Science Foundation for the principal funding of this work through Grant 0626533. Special thanks to Dr. Nathan Kutz, Matt Williams, Mikala Johnson and Chris Jones for technical review and mathematical suggestions.

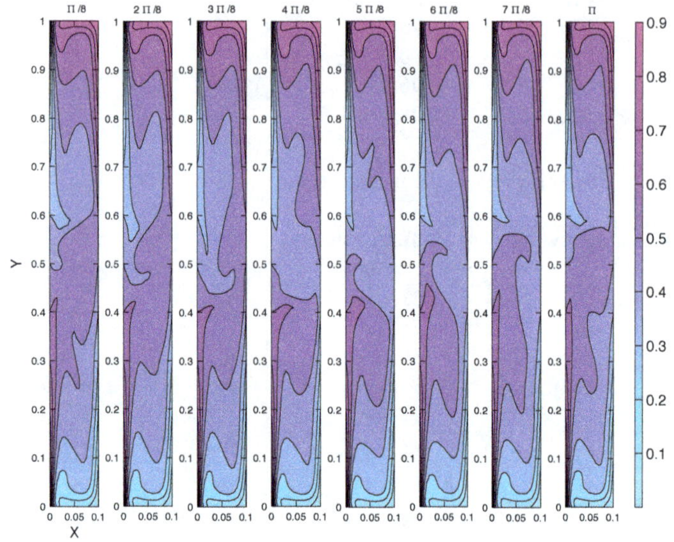

(a) Θ

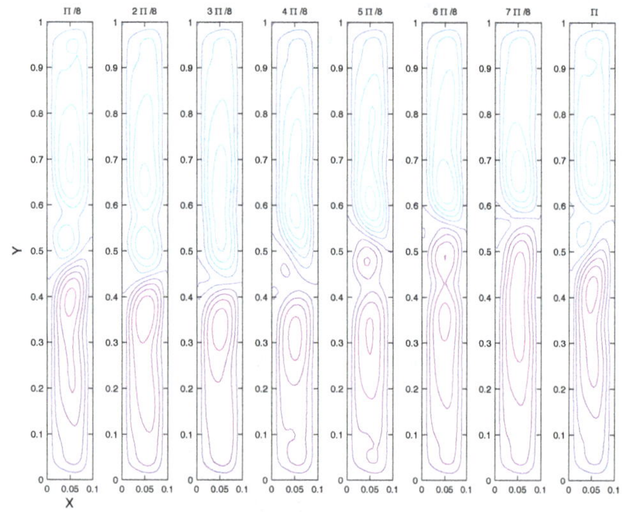

(b) ψ

Fig. 9 Sequential contour plot of the temperature Θ (upper) and stream function ψ (lower) for the rectangular cavity illustrating oscillation of the natural convection cells through one period (Π). $Ra = 17 \times 10^7$, $A = 10$.

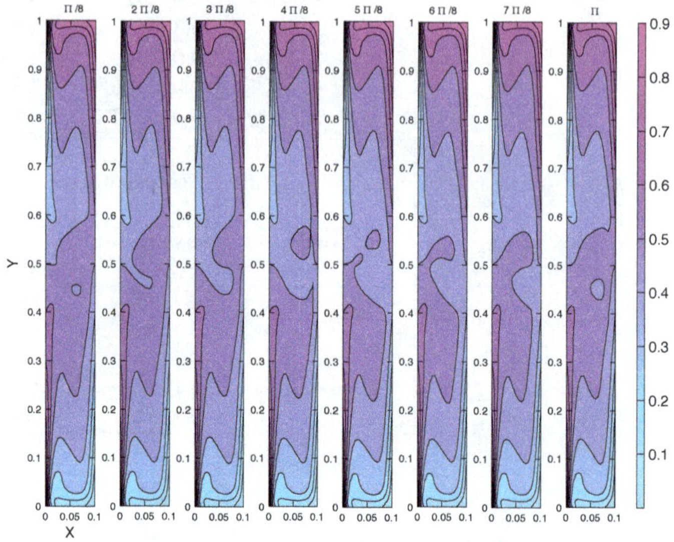

(a) POD Θ

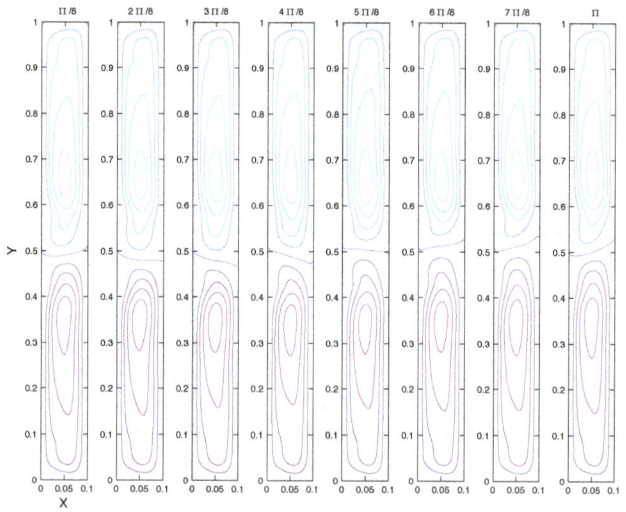

(b) POD ψ

Fig. 10 Temperature (a) and streamfunction (b) values calculated by the POD for $Ra = 17\times10^7$ with 5 modes. These should be compared with the computational results shown in Figure 9.

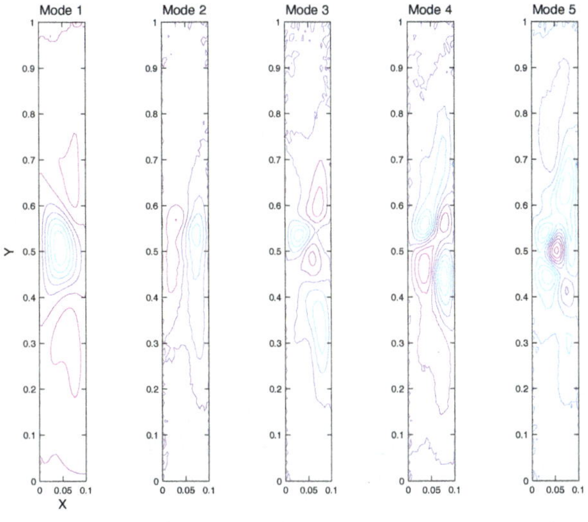

(a) Θ Modes

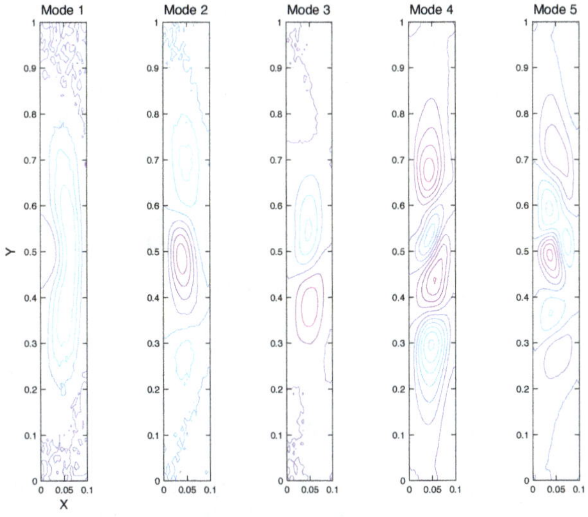

(b) ψ Modes

Fig. 11 Temperature (a) and streamfunction (b) modes calculated by the POD for $Ra = 2.5 \times 10^7$.

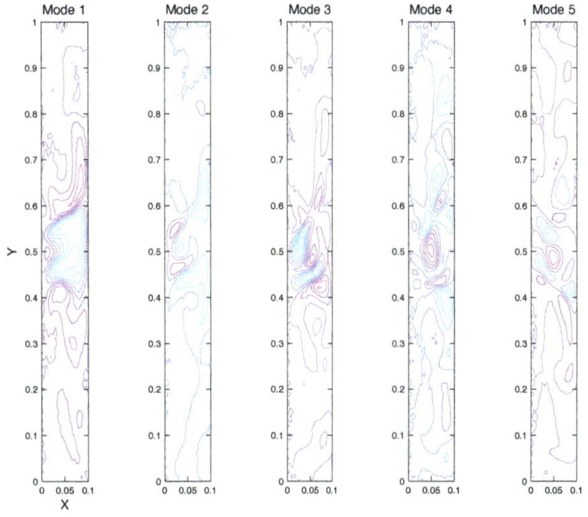

(a) Θ Modes

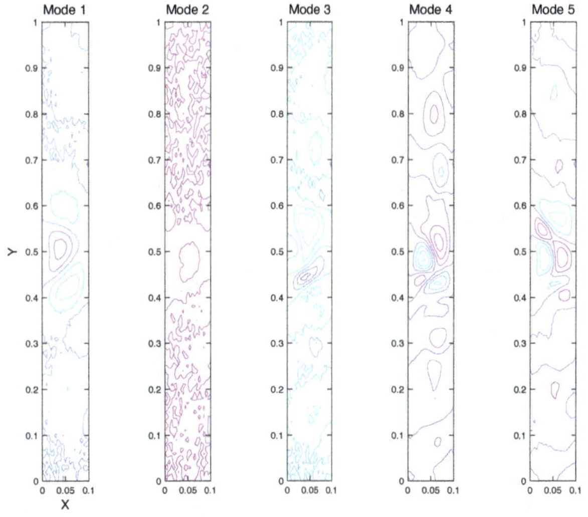

(b) ψ Modes

Fig. 12 Temperature (a) and streamfunction (b) modes calculated by the POD for $Ra = 17 \times 10^7$.

NOMENCLATURE

g	acceleration of gravity
H	height of the cavity
k	thermal conductivity
Pr	Prandlt number ($Pr = c_p\mu/k$)
Ra	Rayleigh number ($Ra = \frac{\rho^2 g c_p \beta(\Delta T) H^3}{k\mu}$)
p, P	dimensioned and dimensionless pressure
u, U	dimensioned and dimensionless velocity ($U = u/\sqrt{g\beta\Delta TH}$)
W	width of the cavity
Greek Symbols	
β	coefficient of thermal expansion
μ	dynamic viscosity
ρ	density
Θ	dimensionless temperature ($\Theta = \frac{T-T_c}{\Delta T}$)
t, τ	dimensioned and dimensionless time ($\tau = t\sqrt{g\beta\Delta TH^{-1}}$)

REFERENCES

Blinov, D., Prokopov, V., Sherenkovskii, Y., Fialko, N., and Yurchuk, V., 2004, "Effective Method for Construction of Low-dimensional Models for Heat Transfer Process," *International Journal Heat and Mass Transfer*, **47**, 5823–5828. http://dx.doi.org/10.1016/j.ijheatmasstransfer.2004.07.020.

Bonnet, J., Cole, D., Delville, J., Glauser, M., and Ukeiley, L., 1994, "Stochastic Estimation and Proper Orthogonal Decomposition: Complementary Techniques for Identifying Structure," *Experiments in Fluids*, **17**, 307–314.
http://dx.doi.org/10.1007/BF01874409.

Chait, A., and Korpela, S., 1989, "The Secondary Flow and Its Stability for Natural Convection in a Tall Vertical Enclosure," *Journal of Fluid Mechanics*, **200**, 189–216.
http://dx.doi.org/10.1017/S0022112089000625.

Chenoweth, D., and Paolucci, S., 1996, "Natural Convection in an Enclosed Vertical Air Layer with Large Horizontal Temperature Differences," *Journal of Fluid Mechanics*, **169**, 173–210. http://dx.doi.org/10.1017/S0022112086000587.

Dillon, H., Emery, A., Cochran, R., and Mescher, A., 2010, "Dimensionless versus Dimensional Analysis in CFD and Heat Transfer," *COMSOL User Conference*.

Dillon, H., Emery, A., and Mescher, A., 2009, "Benchmark Comparison of Natural Convection in a Tall Cavity," *COMSOL User Conference*.

Dillon, H., Emery, A., and Mescher, A., 2011a, "Chaotic Behavior of Natural Convection in a Tall Rectangular Cavity with Non-Isothermal Walls," *AIP Conference Proceedings*, vol. 1389, 127. http://dx.doi.org/10.1063/1.3636686.

Dillon, H., Emery, A., Mescher, A., Sprenger, O., and Edwards, S., 2011b, "Chaotic Natural Convection in an Annular cavity with Non-Isothermal Walls," *Frontiers in Heat and Mass Transfer*, **2**(2). http://dx.doi.org/10.5098/hmt.v2.2.3002.

Dillon, H., 2011, *Chaotic Natural Convection in an Annular Geometry*, Ph.D. thesis, University of Washington.

Haldenwang, P., and Labrosse, G., 1986, "2-D and 3-D Spectral Chebyshev Solutions for Free Convection at High Rayleigh Number," *Sixth International Symposium on Finite Element Methods in Flow Problems*.

Korpela, S., Gozum, D., and Baxi, C., 1973, "On the Stability of the Conduction Regime of Natural Convection in a Vertical Slot," *International Journal Heat and Mass Transfer*, **16**, 1683–1690 http://dx.doi.org/10.1016/0017-9310(73)90161-0.

Lee, Y., and Korpela, S.A., 1983, "Multicellular Natural Convection in a Vertical Slot," *Journal of Fluid Mechanics*, **126**, 91–121 http://dx.doi.org/10.1017/S0022112083000063.

LeQuere, P., 1990, "A Note on Multiple and Unsteady Solutions in Two-Dimensional Convection in a Tall Cavity," *Journal of Heat Transfer*, **112**, 965–974.
http://dx.doi.org/10.1115/1.2910508.

Liakopoulos, A., Blythe, P., and Gunes, H., 1997, "A Reduced Dynamical Model of Convective Flows in Tall Laterally Heated Cavities," *Proc R Soc London A*, **453**, 663–672.
http://dx.doi.org/10.1098/rspa.1997.0037.

Liakopoulos, A., Blythe, P., and Simpkins, P., 1990, "Convective Flows in Tall Cavities," *Simulation and Numerical Methods in Heat Transfer*, 81–87.

Paolucci, S., and Chenoweth, D., 1989, "Transition to Chaos in a Differentially Heated Vertical Cavity," *Journal of Fluid Mechanics*, **201**, 379–410.
http://dx.doi.org/10.1017/S0022112089000984.

Reeve, H., 2003, *Effect of Natural Convection Heat Transfer During Polymer Optical Fiber Drawing*, Ph.D. thesis, University of Washington.

Reeve, H.M., Mescher, A.M., and Emery, A.F., 2004, "Unsteady Natural Convection of Air in a Tall Axisymmetric, Nonisothermal Annulus," *Numerical Heat Transfer, Part A*, **45**, 625–648. http://dx.doi.org/10.1080/10407780490424262.

Rowley, C., 2005, "Model Reduction for Fluids, Using Balanced Proper Orthogonal Decomposition," *International Journal on Birfurcation and Chaos*.

Rowley, C., and Marsden, J., 2000, "Reconstruction Equations and the Karhunen-Loeve Expansion for Systems with Symmetry," *Physica D*, **142**, 1–19.
http://dx.doi.org/10.1016/S0167-2789(00)00042-7.

Sirisup, S., Karniadakis, G., Xiu, D., and Kevrekidis, I., 2005, "Equation-free/Galerkin-free POD-assisted Compuation of Incompressible Flows," *Journal of Compuational Physics*, **207**, 568–587. http://dx.doi.org/10.1016/j.jcp.2005.01.024.

Sirovich, L., 1987, "Turbulence and the Dynamics of Choerent Structures," *Quarterly of Applied Mathematics*, **45**(3), 561–590.

Suslov, S., and Paolucci, S., 1995, "Stability of Natural Convection Flow in a Tall Vertical Enclosure Under Non-Boussinesq Conditions," *International Journal of Heat and Mass Transfer*, **38**(12), 2143–2157. http://dx.doi.org/10.1016/0017-9310(94)00348-Y.

Trefethen, L.N., and Bau, D., 1997, *Numerical Linear Algebra*, Society for Industrial and Applied Mathematics.

Vest, C., and Arpaci, V., 1969, "Stability of Natural Convection in a Vertical Slot," *International Journal of Fluid Mechanics*, **36**, 1–15. http://dx.doi.org/10.1017/S0022112069001467.

Xin, S., and LeQuere, P., 2002, "An Extended Chebyshev Pseudo-Spectral Benchmark for the 8:1 Differentially Heated Cavity," *International Journal for Numerical Methods in Fluids*, **40**, 981–998. http://dx.doi.org/10.1002/fld.399.

20

THERMAL EFFICIENCY ANALYSIS OF A SINGLE-FLOW SOLAR AIR HEATER WITH DIFFERENT MASS FLOW RATES IN A SMOOTH PLATE

Foued Chabane[a,b,*,†], Noureddine Moummi[a,b], Abdelhafid Brima[a,b] , Said Benramache[c]

[a] *Mechanical Department, Faculty of Technology, University of Biskra 07000, Algeria*
[b] *Mechanical Laboratory, Faculty of Technology, University of Biskra 07000, Algeria*
[c] *Material Science Departments, Faculty of Science, University of Biskra 07000, Algeria*

ABSTRACT

This paper presents an experimental thermal efficiency analysis for a novel flat plate solar air heater with several mass flow rates. The aims are to review of designed and analyzed a thermal efficiency of flat-plate solar air heaters. The measured parameters were the inlet and outlet temperatures, the absorbing plate temperatures, the ambient temperature, and the solar radiation. Further, the measurements were performed at different values of mass flow rate of air in flow channel duct. After the analysis of the results, the optimal value of efficiency is higher level of mass flow rate equal to 0.0202 kg/s in flow channel duct for all operating conditions and the single-flow collector supplied with maximum mass flow rate appears significantly better than that another flow rate. At the end of this study, the thermal efficiency relations are delivered for different mass flow rates. Maximum efficiency obtained for the single pass air heater between the air mass flow rates from 0.0108 to 0.0184 kg/s; were 39.72% and 50.47 % respectively, with tilt angle equal 45° in location Biskra city of Algeria. The thermal efficiency correspondently the mass flow rates were 28.63, 39.69, 46.98, 55.70 and 63.61 %, respectively.

Keywords: *Solar air heater; Experimental; Exergy analysis; Single-flow; Thermal efficiency.*

1. INTRODUCTION

In this paper an attempt has been done to optimize the thermal performance of flat plate solar air heater by considering the different system and operating parameters to obtain maximum thermal performance. The report talks about thermal performance for different mass flow rates, emissivity of the plate, and tilt angle (Varun, 2010). In our design we can been found that the use of selected coatings on the absorbing plates of all the heaters considered can substantially enhance the thermal performances of the heaters, and the Plexiglas covers does not have such a significant effect on the thermal performances of the heaters (Wenfeng *et al.*, 2007). There are different factors affecting the solar collector efficiency, e.g. collector length, collector depth, type of absorber plate, glass cover plate, wind speed, etc. Increasing the absorber area or fluid flow heat-transfer area will increase the heat transfer to the flowing air (Chabane *et al.*, 2013a-e). On the other hand, it will also increase the pressure drop in the collector, thereby increasing the required power consumption to pump the air flow crossing the collector (Akpinar and Koçyigˇit, 2010; Karsli, 2007). Kalogirou (2006) estimated the performance parameters of flat plate solar collectors using ANN and results obtained are compared with actual experimental values. A number of attempts (Charters, 2006; Hollands and Shewen, 1981; Bejan *et al.*, 1981; Altfeld *et al.*, 1988; Altfeld *et al.*, 1988; Bhargava and Rizzi, 1990; Verma *et al.*, 1992; Hegazy, 1996) have been made during the last 30 years in an effort to improve the thermal performance of flat plate SAHs by optimizing air channel depth with respect to its length or width. Work reported the effect the mass flow rate in range 0.0078 to 0.0166 kg/s on the solar collector with longitudinal fins (Chabane *et al.*, 2012a).

2. EXPERIMENTAL SECTION

2.1. Collector analysis

The delivered energy output from the solar collector depends on the optical and thermal properties of the collector. This studied that contains an experimental background of the parameters used to characterize the collector thermally. The methods to characterize the collector thermally are presented. The first is to measure the interior temperature properties of the absorber plate and the bottom plate and then calculate the efficiency. The second is through outdoor measurements of temperature such as inlet, outlet and ambient temperature and the thermal characterization is also presented, indoor hot-box measurements. A typical flat-plate collector consists of an absorber in an insulated box together with transparent cover sheets (Plexiglas). The absorber is usually made of a metal sheet of high thermal conductivity, such as galvanized. Its surface is coated with a special selective material to maximize radiant energy absorption while minimizing radiant energy emission. The insulated box reduces heat losses from the back and sides of the collector (Duffie and Beckman, 1991). Plexiglas is a good material for glazing flat plate solar collectors as it transmits almost 90% of the received shortwave solar radiation. Types of plastics can also be used as covers as few of them can endure ultraviolet radiation for a long time. Polycarbonate rigid sheet, polycarbonate rigid film and corrugated sheets are plastic products available on the market. The benefit of using plastics is that they cannot be broken by hail of stones and they are flexible and light (Poulikakos, 1994).

[*] Mechanical Laboratory, Faculty of Technology, University of Biskra 07000, Algeria
[†] *Corresponding author. Email: fouedmeca@hotmail.fr and f.chabane@univ-biskra.dz*

2.2. Thermal Analysis of Solar Air Collector

$$\eta = \frac{Q_u}{Q_s} \tag{1}$$

$$Q_u = \dot{m} C_p (T_{out} - T_{in}) \tag{2}$$

Where Q_u is the accumulated useful energy extracted from the collector during the working period, which is written as

$$Q_u = \int_0^t \dot{m} C_{pa} (T_{out} - T_{in}).dt \tag{3}$$

Qs is solar input energy, determined by equation (4):

$$Q_s = \int_0^t (IA).dt \tag{4}$$

Based on energy conservation, the internal energy change of the collector is caused by thermal disturbance, expressed as:

$$Q_{ie} = Q_s - Q_u - Q_{tc,loss} - Q_{bp,loss} - Q_{re} \tag{5}$$

where $Q_{tc,loss}$ and $Q_{bp,loss}$ are the accumulated heat losses of the transparent cover and back plate respectively, which can be calculated by the following equations:

$$Q_{tc,loss} = \int_0^t (h_{1ac,out,1} + h_{1r,sky} + h_{1r,g})(T_1 - T_{a,out}) F.dt \tag{6}$$

$$Q_{tc,loss} = \int_0^t (h_{5ac,out,1} + h_{5ar,out})(T_4 - T_{a,out}) F.dt \tag{7}$$

The convective heat transfer coefficient in equations (6) and (7) can be derived from the relationship drawn by Sparrow *et al.* (1979), which is used for calculating the external heat transfer coefficient over the transparent cover and the back plate when ($Re=V_a D_h/\nu_a$)

$$Nu = 0.86\, Re^{1/2}\, Pr^{1/3} \quad , \quad Nu = \frac{h_{ac,out} D_h}{\lambda_a} \quad , \quad D_h = \frac{2WL}{W+L} \quad ,$$

$$Pr = \frac{C_{pa}\mu}{\lambda_a} \tag{8}$$

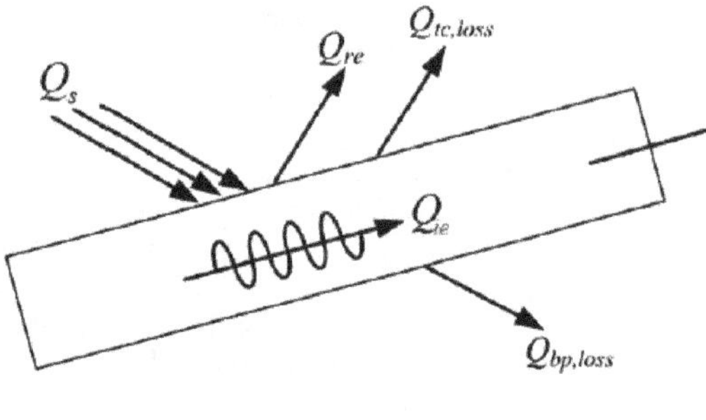

Fig. 1 Energy distributions of the solar air collector.

2.2.1. Heat transfer coefficients

The convective heat transfer coefficient $h_{1ac,\,out}$ for air flowing over the outside surface of the glass cover depends primarily on the wind velocity V_{wind}. McAdams (1954); obtained experimental result as:

$$h_{1ac,out} = 5.7 + 3.8 V_{wind} \tag{9}$$

where the units of $h_{1ac,out}$ and V_{wind} are W/m²K and m/s, respectively. An empirical equation for the loss coefficient from the top of the solar collector to the ambient was developed by Klein (1975).The heat transfer coefficient between the absorber plate and the airstream is always low, resulting in the low thermal efficiency of the solar air heater. Increasing the absorber plate shape area will increase the heat transferred to the flowing air.

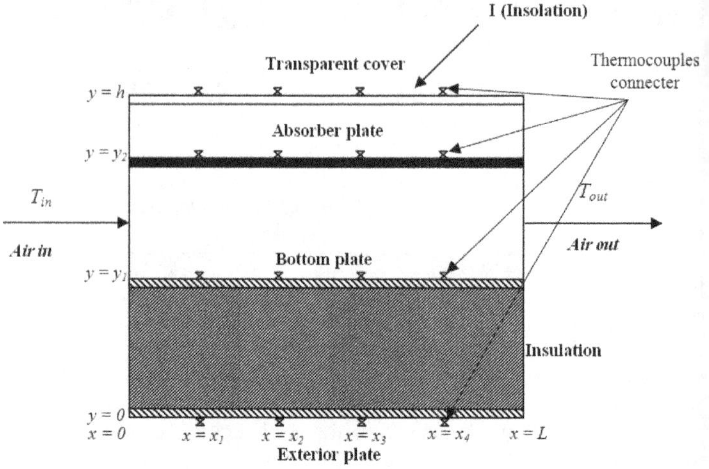

Fig. 2 Flat plate solar air heaters.

The flat-plate solar air heater (Chabane *et al.*, 2012a, b, d; Close and Dunkle, 1976; Liu and Sparrow, 1980; Seluck, 1977; Tan, 1970; Whillier, 1963) are considered to be a simple device consisting of one (transparent) covers situated above an absorbing plate with the air flowing under absorber plate (Tan, 1970; Whillier, 1963; see Fig. 2). The conventional flat-plate solar air heater has been investigated for heat-transfer efficiency improvement by introducing forced convection (Duffie and Beckman, 1980; Tonui and Tripanagnostopoulos, 2007), extended heat-transfer area (Gao *et al.*, 2007; Mohamad, 1997), and increase of air turbulence (Verma and Prasad, 2000; Yeh, 1992).

3. Description of solar air heater considered in this work

A schematic view of the constructed single flow under an absorber plate and in hollow of semi cylindrical fins which located under an absorber plate system of collector is shown in Fig. 1, the photographs of two different absorber plates of the collectors and the view of the absorber plate in the collector box are shown in Fig. 2, respectively. In this study, two modes of the absorber plates were used. The absorbers were made of galvanized iron sheet with black chrome selective coating. The plate thickness of two collectors was 0.5 mm. The cover window type the Plexiglas of 3 mm thickness was used as glazing. Single transparent cover was used of two collectors. Thermal losses through the collector backs are mainly; due to the conduction across the insulation (thickness 4 cm) (Chabane *et al.*, 2013a), those caused by the wind and the thermal radiation of the insulation are assumed negligible. After installation, the two collectors were left operating several days under normal weather conditions for weathering processes.

Thermocouples were positioned evenly, on the top surface of the absorber plates, at identical positions along the direction of flow, for both collectors. Inlet and outlet air temperature were measured by two well insulated thermocouples. The output from the thermocouples was recorded in degrees Celsius by using a digital thermocouple thermometer DM6802B: measurement range −50 to 1300 °C (−58 to 1999 °F), resolution: 1°C or 1°F, accuracy: ± 2.2 °C or ± 0.75 % of reading and Non-Contact digital infrared thermometer temperature laser gun model number: TM330: accuracy ±1.5 C/±1.5 %, measurement range −50 to 330 °C (−58 to 626 °F) resolution 0.1 °C or 0.1 °F, emissivity 0.95. A digital thermometer measured the ambient temperature with sensor in display LCD CCTV-PM0143 placed in a special container behind the collectors' body. The total solar radiation incident on the surface of the collector was measured with a Kipp and Zonen CMP 3 Pyranometer. This meter was placed adjacent to the glazing cover, at the same plane, facing due south. The measured variables were recorded at intervals of 15 min and include: insolation, inlet and outlet temperatures of the working fluid circulating through the collectors, ambient temperature, absorber plate temperatures at several selected locations and air flow rates (Lutron AM-4206M digital anemometer). All tests began at 9 AM and ended at 4 PM.

The layout of the solar air collector studied is shown in Figs. 1, 2. The collector A served as the baseline one, with the parameters as:

- The solar collecting area was 2 m (length) × 1 m (width);
- The installation angle of the collector was 45° from horizontal;
- Height of the stagnant air layer was 0.02 m;
- Thermal insulation board EPS (expanded polystyrene board), with thermal conductivity 0.037 W/(m K), was put on the exterior surfaces of the back, and side plates, with a thickness of 40 mm.
- The absorber was of a plate absorption coefficient α = 0.95, the transparent cover transmittance τ = 0.9 and absorption of the glass covers, α_g = 0.05;
- 16 positions of thermocouples connected to plates and two thermocouples to outlet and inlet flow, Fig.2.

4. RESULTS AND DISCUSSION

The single pass solar air heaters are investigated experimentally under Biskra prevailing weather conditions during the winter months, 24/01/2012, 25/01/2012, 01/02/2012, 19/02/2012, and 27/02/2012 with clear sky condition. Biskra is a city of Algeria located on 34°50'43.28"N latitude 5°44'49.11"E longitude. The performance of the solar air heater was studied and compared with the performance of a single pass solar air heater and an effect the mass flow rate of the air was varied from 0.0108 to 0.0201 kg/s.

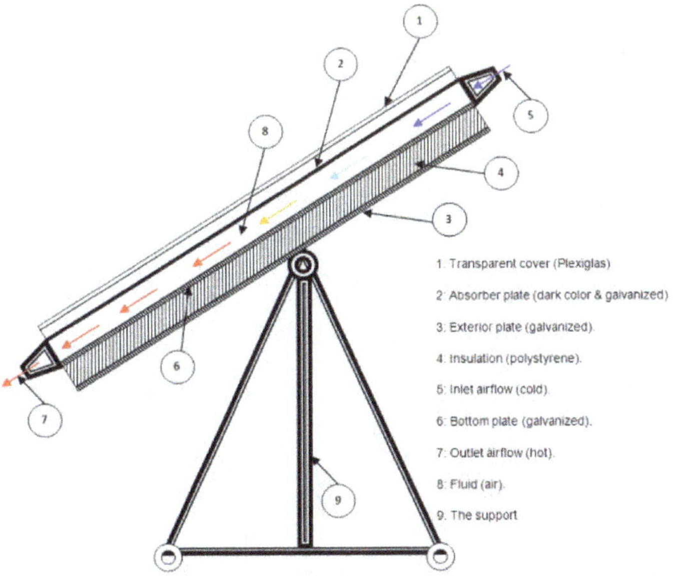

Fig. 3 Schematic view of the solar air collector

Figures 5 and 6 show the variation of the thermal efficiency and a solar intensity, respectively, with air mass flow rate. The thermal efficiency used to evaluate the performance of the solar air heater is calculated; from both figures that the thermal efficiency increases with increasing solar intensity and mass flow rate as a function of the time. The efficiencies of the rate 0.0202 kg/s are higher than inferior of 0.0202 kg/s. Figs. 5 and 6 shows the comparison of the thermal efficiency for the different mass flow rates from 0.0108 kg/s to 0.0202 kg/s. beside the results data of each value has been shown in Table 3a, 3b.

Fig. 4 The photograph of experimental set-up (Chabane *et al.*, 2013e, f)

Figures 5 and 6 show the variation of the thermal efficiency and a solar intensity, respectively, with air mass flow rate. The thermal efficiency used to evaluate the performance of the solar air heater is calculated; from both figures that the thermal efficiency increases with increasing solar intensity and mass flow rate as a function of the time. The efficiencies of the rate 0.0202 kg/s are higher than inferior of 0.0202 kg/s. Figs. 5 and 6 shows the comparison of the thermal efficiency for the different mass flow rates from 0.0108 kg/s to 0.0202 kg/s. beside the results data of each value has been shown in Table 3a, 3b.

Evidently the mean highest thermal efficiency (η = 58.02%) at solar intensity I = 898 W.m^{-2} at air flow rate 0.0161 kg.s^{-1} and 45° tilt angle (Chabane *et al.*, 2012c) at 13:10 h. The mean lowest thermal efficiency (η = 28 %) at solar intensity I = 883 W.m^{-2} at 13:00 h was obtained with air flow rate 0.0108 kg.s^{-1} and 45° tilt angle. Solar air heater were heated the air much more at the lower air rate, because the air had more time to get hot inside the collector.

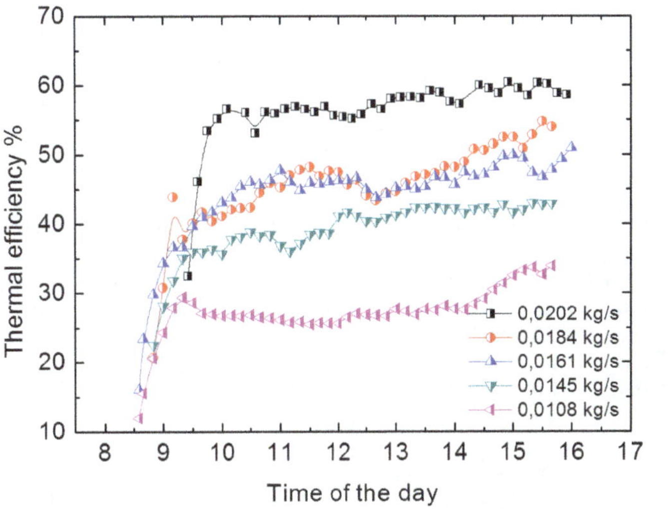

Fig. 5 Variation of collector efficiency at different mass flow rates.

Efficiency versus time at various air rates for the single pass collector in this experiment is shown in Fig. 5. The efficiencies increase to a maximum value at 12:30–16:00 h, and then start to decrease later on in the afternoon. The efficiency of a mass flow rate m = 0.0202 kg/s is higher than the others mass flow rate by 7–35 % depending on the air mass flow rate. The efficiency of a single air pass solar collector notably depends on the air mass flow rate. The maximum efficiency obtained for these single pass air collector are 63.25 % for m = 0.0202 kg/s. For solar air collectors, it is clear that the efficiency increase with an increasing air mass flow rate of air as shown in Fig. 5. The curvature of the efficiency for mass flow air at 0.0202 kg/s is wider than that of air mass flow rates 0.0108 kg/s. Fig. 8 shows the plots of thermal efficiency of the single solar air heater collectors as a function of mass flow rate m, for the period of 8:30 h to 16:00 h of the day.

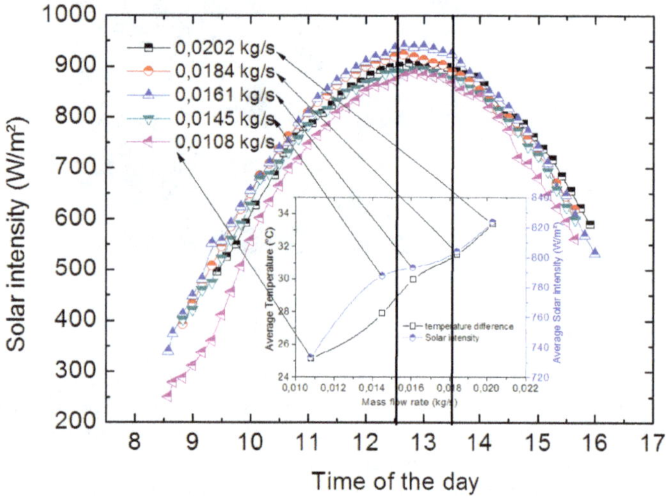

Fig. 6 Variation of solar radiation at different days.

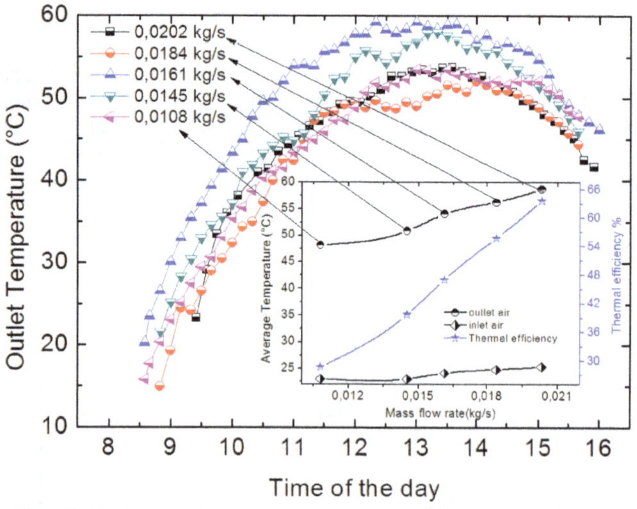

Fig. 7 Variation of outlet temperature at different mass flow rates.

The thermal efficiency of the heater improves with increasing air flow rates due to an enhanced heat transfer to the air flow while a temperature difference of fluid decreases at a constant tilt angle $\beta = 45°$ (Chabane *et al.*, 2012c). Solar intensity is at their highest values at noon (at about 13:30) as is expected. The solar intensity decreases as the time passes through the afternoon. Fig. 7 it shows overall results of experiments, including the difference of air ambient and outlet temperature and daily instantaneous solar intensity levels. The ambient temperature was between 7 and 24 °C. The inlet temperatures of solar air collectors were measurement to ambient temperature. The temperature differences between the inlet and outlet temperatures can be compared directly when determining the performance of the collectors. The highest daily solar radiation is obtained as 881.38 and 943 W/m^2 for a Flat-plate.

Figure 6 shows the solar intensity versus standard local time of the day for all the days the experiment was carried out. The solar intensity increases from the early hours of day with about 250 W/m^2 at 8:30 h to a peak value at noon and then, reduces later on during the day (Fig. 6). The highest daily solar radiation obtained with single pass solar air collector, which was for day was 940 W/m^2 at 13:40 h and the average solar intensity through that particular day was about 793 W/m^2. Calculating the mean solar intensity for each day, there was stability in the solar radiation as all mean averages are within the same and close range. The mean average solar intensity for all the days of the experiment was 733 W/m^2 and 803.50 W/m^2 for single pass solar air collector. The obtained result shows a stable amount of solar radiation measured for each day of the experiment.

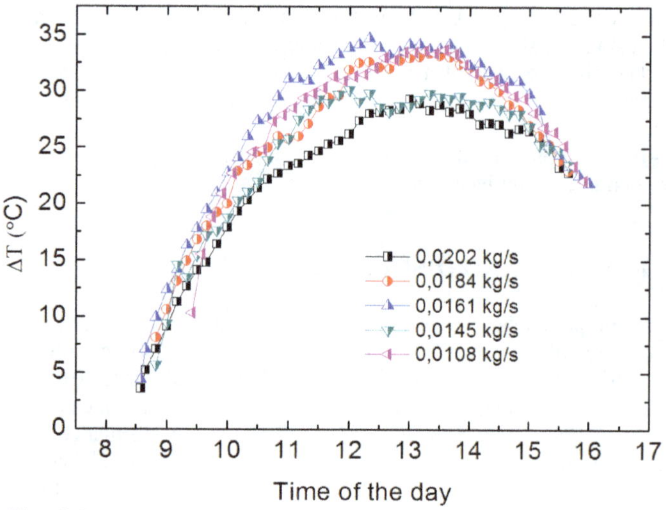

Fig. 8 Temperature difference versus standard local time of the day at different mass flow rates for double pass solar air heater.

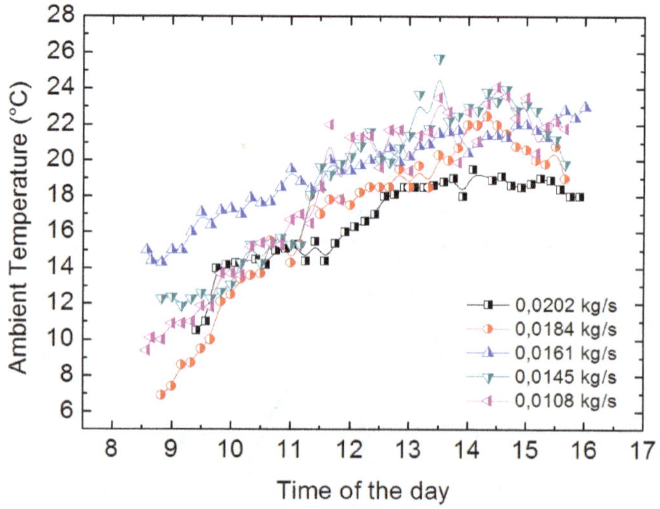

Fig. 9 Variation of ambient temperatures at different mass flow rates.

Figure 8 show the temperature differences, $\Delta T = T_{out} - T_{in}$, versus time of the day for different mass flow rates and for single pass solar air heater. The ambient temperature versus standard local time of the day for all the days the experiment was carried out is presented in Fig. 9. In general, the input temperature was found to be increasing exponential from the morning to evening with little fluctuation during some of the days. For the smooth plate of the solar air heater used in this work, ΔT was found to reduce with increasing air mass flow rate. Results show that for the same mass flow rate, the collector temperature differences increased with increasing solar radiation I, (Fig. 6) as expected. ΔT of air increases to a peak value of 34 °C for the single pass air heater of smooth plate. This peak temperature difference occurred between 12:00 h and 13:00 h for a minimum mass flow rate, m of 0.0108 kg/s. It then decreases as solar radiation drops to lower values later on during the day for the same mass flow rate of air. The changes in the peak value of the temperature difference between 12:00 h and 13:00 h is suitable to changing outdoor conditions like solar intensity (Fig. 6) and the wind speed.

5. CONCLUSION

This study shows that for a single pass solar air heater using smooth plate as absorber plate and we added the different mass flow rates there is a significant increase in the thermal efficiency of the air heater. The efficiency increases when the air mass flow increases from 0.0108 kg/s to 0.0202 kg/s. Also, the temperature difference between the outlet flow and the ambient, ΔT, reduce with an increase in the air mass flow rate. Further, results showed that for the same mass flow rate the collector temperature difference increase with increasing solar radiation, I, and decreases as solar radiation drops to lower values later on during the day. The maximum temperature difference obtained from this study is 34.70 °C for the single pass air heater about smooth plate for air mass flow rate of 0.0161 kg/s. The maximum thermal efficiency obtained is 60.40 % for single pass for air mass flow rate of 0.0202 kg/s.

ACKNOWLEDGEMENTS

The authors would like to thank, Prf. S. Youcef-Ali, Pr. H. Ben moussa, Dr.D. Bensahal, Dr. O. Belahssen and Mr. Nadjib Hamouda for helpful counseling. This study correspondently of laboratory of **LGM** (Laboratoire de Génie Mécanique de l'université Mohamed Khider de Biskra)

NOMENCLATURE

T_{ep} *temperature of exterior plat (°C)*
T_{ab} *temperature of absorber plat (°C)*
T_{pl} *temperature of transparent cover (°C)*
T_{bp} *temperature of bottom plat (°C)*
T_a *ambient temperature (°C)*
x_i *local direction longitudinal of points (m)*
y_i *local direction of thickness panel (m)*
T_{in} *temperature inlet (°C)*
T_{out} *outlet fluid temperature (°C)*
V_{wind} *wind velocity (m/s)*
$h_{ac,out}$ *convection heat transfer coefficient (W/m² K)*
C_p *specific heat of air (J/kg K)*
A_c *surface area of the collector = LW (m^2)*
i *position of thermocouple connected of 1 to 4.*
ΔT *temperature difference (°C)*
m *mass flow rate (kg/s)*
Q_u *: accumulated useful energy (W)*
Q_s *: solar input energy (W)*
$Q_{tc,loss}$*: accumulated heat losses of the transparent cover (W)*
$Q_{bp,loss}$*: accumulated heat losses of the back plate (W)*

Greek symbols
η *collector efficiency (%)*
I *global irradiance incident on solar air heater collector (W/m2)*
m *air mass flow rate (kg/s)*
ε *emissivity of absorber plate*
α_a *absorber plate absorption coefficient*
τ *transparent cover transmittance*
α_g *absorptivity of the glass covers*

REFERENCES

Agarwal, V.K., Larson, D.C., 1981, "Calculation of the Top Loss Coefficient of a Flat-Plate Collector," *Sol. Energy*, **27**, 69-71. http://dx.doi.org/10.1016/0038-092X(81)90022-0

Akpinar, E.K., Koçyigˇit, F., 2010, "Experimental Investigation of Thermal Performance of Solar Air Heater Having Different Obstacles on Absorber Plates," *Int Commun Heat Mass Transfer*, **37**, 416–21. http://dx.doi.org/10.1016/j.icheatmasstransfer.2009.11.007

Altfeld, K., Leiner, W., Fiebig, M., 1988, "Second Law Optimization of Flat-plate Solar Air Heaters Part I: The Concept of Net Exergy Flow and the Modeling of Solar Air Heaters," *Solar Energy*, **41**, 127−132. http://dx.doi.org/10.1016/0038-092X(88)90128-4

Altfeld, K., Leiner, W., Fiebig, M., 1988, "Second Law Optimization of Flat-plate Solar Air Heaters. Part 2: Results of Optimization and Analysis of Sensibility to Variations of Operating Conditions," *Solar Energy*, **41**, 309−317.
http://dx.doi.org/10.1016/0038-092X(88)90026-6

Azad, E., 2012, "Design Installation and Operation of a Solar Thermal Public Bath in Eastern Iran," *Energ Sustai Dev*, **16**, 68–73. http://dx.doi.org/10.1016/j.esd.2011.10.006

Bejan, A., Kearney, D.W., Kreith, F., 1981, "Second Law Analysis and Synthesis of Solar Collector Systems," ASME *J Solar Energy Eng*, **103**, 23−28.
http://dx.doi.org/10.1115/1.3266200

Bhargava, A.K., Rizzi, G., 1990, "A Solar Air Heater with Variable Flow Passage Width," *Energy Convers Mgmt*, **30**, 329−332. http://dx.doi.org/10.1016/0196-8904(90)90034-V

Chabane, F., Moummi, N., Benramache, S., Tolba AS, 2012a, "Experimental Study of Heat Transfer and an Effect the Tilt Angle with Variation of the Mass Flow Rates on The Solar Air Heater," *Int J Sci and Eng Inves*, **1,** 61−65.

Chabane, F., Moummi, N., Benramache, S., 2012b, "Experimental Performance of Solar Air Heater with Internal Fins Inferior an Absorber Plate: In the Region of Biskra," *Int J Energ and Tech*, **4,** 1–6.

Chabane, F., Moummi, N., Benramache, S, 2012c, "Effect of the Tilt Angle of Natural Convection in a Solar Collector with Internal Longitudinal Fins," *Int J Sci and Eng Inves*, **1**, 13−17.

Chabane, F., Moummi, N., Benramache, S., 2012d, "Performances of a Single Pass Solar Air Collector with Longitudinal Fins Inferior an Absorber Plate," *American Journal of Advanced Scientific Research*, **1**, 146-157.

Chabane, F., Moummi, N., Benramache, S, 2013a, "Experimental Study of Heat Transfer and Thermal Performance with Longitudinal Fins of Solar Air Heater," *Journal of Advanced Research*, http://dx.doi.org/10.1016/j.jare.2013.03.001

Chabane, F., Moummi, N., Benramache, S., Bensahal, D., Belahssen, O., 2013b, "Effect of Artificial Roughness on Heat Transfer in a Solar Air Heater," *Journal of Science and Engineering*, **1,** 85−93.

Chabane, F., Moummi, N., Benramache, S.,2013c, "Experimental Analysis on Thermal Performance of a Solar Air Collector with Longitudinal Fins In A Region of Biskra, Algeria," *Journal of Power Technologies*, **93**, 53−59.

Chabane, F., Moummi, N., Benramache, S., 2013d, *Experimental Study on Heat Transfer of Solar Air Heater*, LAP Lambert Academic Publishing, 112

Chabane, F., Moummi, N., Benramache, S., 2013e, "Design, Developing and Testing of a Solar Air Collector Experimental and

Review the System with Longitudinal Fins," *International Journal of Environmental Engineering Research*, **2**, 18-26.

Chabane, F., Moummi, N., Benramache, S., Bensahal, D., Belahssen, O., Lemmadi, Z.F., 2013f, "Thermal Performance Optimization of a Flat Plate Solar Air Heater," *International Journal of Energy and Technology*, **5**, 1–6.

Charters, W.W.S., 1971, "Some Aspects of Flow Duct Design for Solar-air Heater Applications," *Solar Energy*, **13**, 283–288. http://dx.doi.org/10.1016/0038-092X(71)90009-0

Close, D.J., Dunkle, R.V., 1976, "Behaviour of Adsorbent Energy Storage Beds," *Sol. Energy*, **18**, 287-292. http://dx.doi.org/10.1016/0038-092X(76)90055-4

Duffie, J.A., and Beckman, W.A., 1991, *Solar Engineering of Thermal Processes*, John Wiley and Sons, New York.

Duffie, J.A., Beckman, W.A., 1980, *Solar Engineering of Thermal Processes*, 3rd ed., Wiley, New York.

Gao, W., Lin, W., Liu, T., Xia, C., 2007, "Analytical and Experimental Studies on the Thermal Performance of Cross-Corrugated and Flat-Plate Solar Air Heaters," *Appl. Energy*, **84**, 425-41. http://dx.doi.org/10.1016/j.apenergy.2006.02.005

Hegazy, A., 1996, "Optimization of Flow-Channel Depth for Conventional Flat-Plate Solar Air Heaters," *Renewable Energy*, **7**, 15–21. http://dx.doi.org/10.1016/0960-1481(95)00117-4

Hollands, K.G.T., Shewen, E.C., 1981, "Optimization of Flow Passage Geometry for Air-Heating, Plate-Type Solar Collectors," *J Solar Energy Engng*, **103**, 323–330. http://dx.doi.org/10.1115/1.3266260

Kalogirou, S.A., 2006, "Prediction of Flat-Plate Collector Performance Parameters Using Artificial Neural Networks," *Solar Energy*, **80**, 248–59 http://dx.doi.org/10.1016/j.solener.2005.03.003

Karsli, S., 2007, "Performance Analysis of New-Design Solar Air Collectors for Drying Applications," *Renewable Energy*, **32**, 1645–60. http://dx.doi.org/10.1016/j.renene.2006.08.005

Klein, S.A., 1975, "Calculation of Flat-Plate Loss Coefficients," *Solar Energy* **17**, 79–80. http://dx.doi.org/10.1016%2f0038-092X(75)90020-1

Liu, C.H., Sparrow, E.M., 1980, "Convective-Radiative Interaction a Parallel Plate Channel-Application to Air-Operated Solar Collectors," *Int. J. Heat Mass Transf*, **23**, 1137-1146. http://dx.doi.org/10.1016/0017-9310(80)90178-7

McAdams, WH. "Heat Transmission," McGraw-Hill, New York, 1954.

Mohamad, A.A., 1997, "High Efficiency Solar Air Heater," *Sol. Energy*, **60**, 71-76. http://dx.doi.org/10.1016/S0038-092X(96)00163-6

Poulikakos D., 1994, *Conduction Heat Transfer*, Prentice Hall, Englewood Cliffs. NJ.

Seluck, M.K. 1977, "Solar Air Heaters and Their Applications," in *Solar Energy Engineering*, A.A.M. Sayigh ed; Academic Press, New York.

Sparrow, E.M.., Ramsey, J.W., Mass, E.A., 1979, "Effect of Finite Width on Heat Transfer and Fluid Flow about an Inclined Rectangular Plate," *J. Heat Transfer*, **101**,199-204. http://link.aip.org/link/doi/10.1115/1.3450946

Tan, H.M., Charters, W.W.S., 1970, "Experimental Investigation of Forced-convective Heat Transfer for Fully Developed Turbulent Flow in a Rectangular Duct with Asymmetric Heating," *Sol. Energy*, **13**, 121-125. http://dx.doi.org/10.1016/0038-092X(70)90012-5

Tonui, J.K., Tripanagnostopoulos, Y., 2007, "Improved PV/T Solar Collectors with Heat Extraction by Forced or Natural Air Circulation," *Renew. Energy*, **32**, 623-637. http://dx.doi.org/10.1016/j.renene.2006.03.006

Varun, S., 2010, "Thermal Performance Optimization of a Flat Plate Solar Air Heater Using Genetic Algorithm," *Applied Energy*, **87**, 1793–1799 http://dx.doi.org/10.1016/j.apenergy.2009.10.015

Verma, R., Chandra, R., Garg, H.P., 1992, "Optimization of Solar Air Heaters of Different Designs," *Renewable Energy*, **2**, 521–531. http://dx.doi.org/10.1016/0960-1481(92)90091-G

Verma, S.K., Prasad, B.N., 2000, "Investigation for the Optimal Thermo Hydraulic Performance of Artificially Roughened Solar Air Heaters," *Renew. Energy*, **20**, 19-36. http://dx.doi.org/10.1016/S0960-1481(99)00081-6

Wenfeng, G., Wenxian, L., Tao, L., Chaofeng, X., 2007, "Analytical and Experimental Studies on the Thermal Performance of Cross-Corrugated and Flat-Plate Solar Air Heaters," *Applied Energy*, **84**, 425–441. http://dx.doi.org/10.1016/j.apenergy.2006.02.005

Whillier, A., 1963, "Plastic Covers for Solar Collectors," *Sol. Energy*, **7** (3), 148-154. http://dx.doi.org/10.1016/0038-092X(63)90060-4

Yeh, H.M., 1992, "Theory of Baffled Solar Air Heaters," *Energy*. **17**, 697-702. http://dx.doi.org/10.1016/0360-5442(92)90077-D

OPTIMIZATION OF COMPRESSION RATIO OF JATROPHA OIL BLEND WITH DIESEL FUELLED ON VARIABLE COMPRESSION RATIO ENGINE

Biswajit De*, Rajsekhar Panua

Mechanical Engineering Department, National Institute of Technology, Agartala, Tripura, 799055, India

ABSTRACT

As the world is facing crisis due to the dwindling resources of fossil fuels, rapid depletion of conventional energy is a matter of serious concern for the mankind. So there is a necessity to find alternate fuels. Vegetable oils, because of their agricultural origin, due to less carbon content compared to mineral diesel are producing less CO_2 emissions to the atmosphere is used as an alternate fuel in substitute to diesel fuel. In the present study optimum compression ratio for VCR diesel engine fuelled with Jatropha oil blends with diesel (30%) has been determined at 203 bars injector opening pressure, 23^0 CA BTDC injection timing and at 1500 rev/min rated speed. The test results revealed that compression ratio 19 exhibited better performance and lower emissions and hence, is considered as optimum compression ratio.

Keywords: Variable compression ratio engine; optimum; blend; emissions;

1. INTRODUCTION

Energy is the most important component of human life and is an essential input for every activity. With ever growing population, improvement in the living standard of humanity, industrialization of developing countries, the global demand for energy is expected to rise rather significantly in the near future. The rapid depletion of petroleum reserves and rising oil prices has led to the search for alternative fuels. Non edible oils are promising fuels for power requirement of agricultural applications. Thomas and Maurin (1993) explained that vegetable oil is renewable, environmental- friendly and produced easily in rural areas, where there is an acute need for modern forms of energy. Therefore in recent years systematic efforts have been made by several research workers to use vegetable oils as fuel in engines. From previous studies, it is evident that there are various problems associated with vegetable oils being used as fuel in compression ignition engines, mainly caused by their high viscosity. Agarwal reported that due to the high viscosity, in long term operation, vegetable oils normally (2007) introduce the development of gumming, the formation of injector deposits, ring sticking, as well as incompatibility with conventional lubricating oils. Therefore, a reduction in viscosity is of prime importance to make vegetable oils a suitable alternative fuel for diesel engines. Rakopoulos et al. (2006) performed tests to evaluate and compare the use of a variety of vegetable oil of various origin as a substitute to conventional diesel fuel at blend ratios of 10/90 and 20/80 in a DI (direct injection) diesel engine. The obtained data was analyzed for various parameters such as thermal efficiency, BSFC (brake specific fuel consumption), smoke opacity, and CO2, CO and HC emissions. It was found that NOx emissions were reduced with use of vegetable oil in the diesel engine. Altuna and Bulut (2008) reported that their experimental results show that the engine power and torque of the mixture of sesame oil–diesel fuel are close to the values obtained from diesel fuel and the amounts of exhaust emissions are lower than those of diesel fuel. Hence, it is seen that blend of sesame oil and diesel fuel can be used as an alternative fuel successfully in a diesel engine without any modification and also it is an environmental friendly fuel in terms of emission parameters. Chen and Wang (2008) stated that hydrocarbon is apparently decreased when the engine was fueled with ethanol–ester–diesel blends. Hydrocarbon Fuelling the engine with oxygenated diesel fuels showed increased carbon monoxide (CO) emissions at low and medium loads, but reduced CO emissions at high and full loads, when compared to pure diesel fuel. Jindal and Nandwana (2010) found that the combined increase of compression ratio and injection pressure increases the BTHE and reduces BSFC while having lower emissions. For small sized direct injection constant speed engines used for agricultural applications (3.7 kW), the optimum combination was found as CR of 19 with IP of 250 bar. Pradhan and Raheman (2014) reduced the viscosity and density of CJO (crude Jatropha oil) by heating it using the heat from exhaust gas of a diesel engine with an appropriately designed helical coil heat exchanger. Experiments were conducted to evaluate the combustion characteristics of a DI (direct injection) diesel engine using PJO (preheated Jatropha oil). The results indicated that BSFC (brake specific fuel consumption) and EGT (exhaust gas temperature) increased while BTE (brake thermal efficiency) decreased with PJO as compared to HSD (high speed diesel) for all engine loadings. The reductions in CO_2 (carbon dioxide), HC (hydrocarbon) and NOx (nitrous oxide) emissions were observed for PJO along with increased CO (carbon monoxide) emission as compared to those of HSD. Pilusa and Mollaggee (2012) used Vehicle Emissions Analyser to measure the emissions at each stage, and a similar procedure was followed to measure the emissions after installation of the Whale filter. The results showed a significant average reduction in carbon monoxide CO (35.3%), nitrogen oxides NOx (26.1%) and hydrocarbons HC (34.3%) emissions after the Whale filter was installed in the four vehicles.

* Corresponding author. Email: biswajitde62@gmail.com

2 EXPERIMENTAL SETUP

A single cylinder, direct injection, four-stroke, vertical, water-cooled, naturally aspirated variable compression ratio diesel engine with a bore of 80 mm and a stroke of 110 mm is selected for the present study. Labview based Engine Performance Analysis software package "EnginesoftLV" is provided for on line performance evaluation. The engine is connected to eddy current type dynamometer for loading. The compression ratio can be changed without stopping the engine and without altering the combustion chamber geometry by specially designed tilting cylinder block arrangement. The technical specification of the test engine is given in Table 1 and the different properties of Diesel and Jatropha oil are given in Table 2. The schematic diagram of the experiment set up is shown in Fig. 1.The set up has stand-alone panel box consisting of air box, two fuel tanks for duel fuel test, manometer, fuel measuring unit, transmitters for air and fuel flow measurements, process indicator and engine indicator. 30% Jatropha oil blend with diesel is prepared. One litre of 30% Jatropha oil blend contains 700 ml of diesel and 300 ml of jatropha oil. For warm-up the engine, it is started with diesel and then switched over to blended oil.

Table 1 Specification of the test engine.

S. No	*Parameters*	*Specification*
1.	General Details	Single cylinder, four stroke compression ignition engine, constant speed, vertical, water cooled, direct injection
2.	Stroke	110 mm
3.	Bore	87.5 mm
4.	Displacement	661 cc
5.	Compression ratio	17.5
6.	Rated output	3.7 KW
7.	Rated speed	1500 rpm

Table 2 Properties of mineral diesel and Jatropha oil.

Property	*Mineral diesel*	*Jatropha oil*
Density(kg/m^3)	841	917.5
API gravity	37.123	22.8925
Kinematic viscosity at 40^0 C (cSt)	2.575	36.63
Pour point (^{0}C)	-6	4.5
Fire point (^{0} C)	104	275
Cloud point (^{0} C)	3.5	9.5
Flash point (^{0} C)	72	230
Calorific value (MJ/kg)	44.864	38.6355
Carbon (%, w/w)	80.32	76.113
Hydrogen (%, w/w)	12.358	10.517
Nitrogen (%, w/w)	1.758	0

Fuel from valve enters into the engine through fuel measuring unit, which enables the volumetric flow of the fuel to be measured easily. Engine start at no load conditions at CR of 15 and varying the load from idle to rated load of 3.7 kW in a number of steps and a set of readings are obtained for fuel consumption, rpm, exhaust temperature, NO_X, CO_2, smoke opacity and power output. Similar set of readings are recorded for CR of 16, 17, 18 and 19. The emissions (NO_X, CO_2, concentrations) are recorded by using Gas Analyzer (AVL Di Gas 444) and the opacity is recorded by smoke meter (AVL 437). The accuracy of the measured values is ensured by calibrating the gas analyzer using reference gases before each measurement and allowing the smoke meter to adjust its zero point before each measurement.For each setting, the emission values and the other values are recorded thrice and a mean of these is taken for comparison. The performance of the engine at different loads and settings are evaluated in terms of BSFC, brake thermal efficiency and emissions of carbon dioxide, oxides of nitrogen, smoke opacity and exhaust gas temperature.

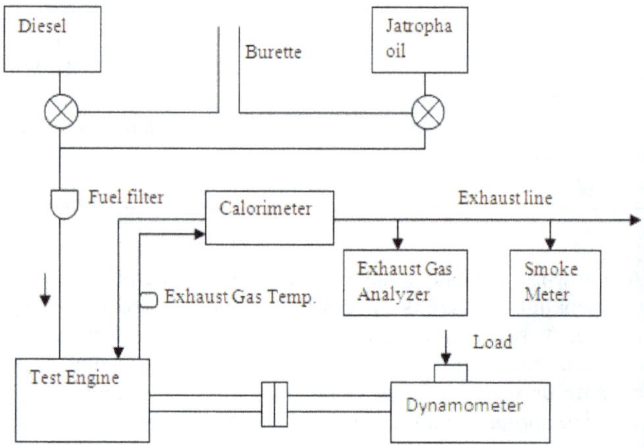

Fig. 1 Schematic diagram of experimental setup.

3 RESULTS AND DISCUSSIONS

In order to determine the optimum compression ratio for variable compression ratio diesel engine fuelled with Jatropha oil blend with diesel (30%) and the engine is started at no load conditions at CR of 15 and varying the load from idle to rated load of 3.7 kW in a number of steps and a set of reading is obtained. Fuel consumption, rpm, exhaust temperature, NO_X, CO and power output are measured. Similar set of readings are recorded for CR of 15, 16, 17, 18 and 19 by changing it using the tilting cylinder head arrangement. For all settings, the emission values and the other values are recorded thrice and a mean of these is taken for comparison for entire load conditions and compression ratios of 15, 16, 17, 18 and 19 while maintaining the injection pressure of 203 bars and 23^0 CA BTDC injection timing at rated speed of 1500 rev/min. The optimization is done on the basis of maximum brake thermal efficiency.

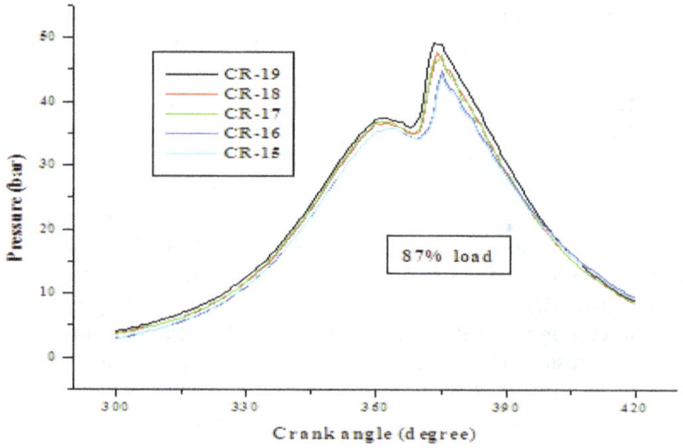

Fig. 2 Pressure rise diagram at different CR.

Figure 2 shows that in-cylinder pressure increase with increase of engine load and compression ratio. At CR of 19, the pressure rise inside the engine cylinder is higher and lower at CR of 15.It happened due to the reason that at CR of 19 better combustion of air-fuel mixture takes place. Fig. 3 shows the variations of brake thermal efficiency with respect to load at different compression ratios for diesel fuel engine operation. The value of brake thermal efficiency of dual fuel engine is low at low loads but significantly high at higher engine loads because at low loads the fuel air ratio of the air-fuel mixture is very less, resulting in incomplete flame propagation and most of the fresh air-gas mixture remains unburnt.

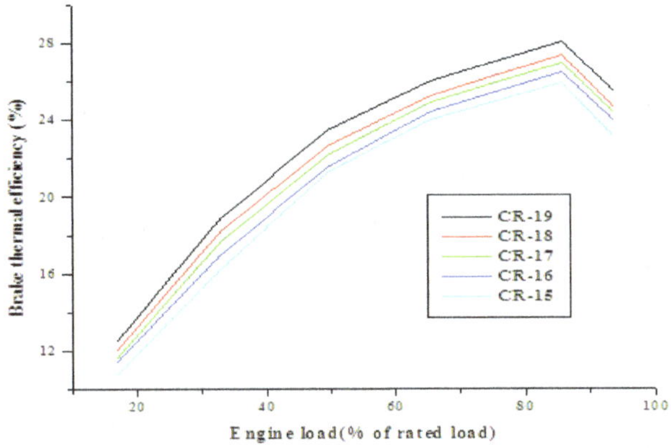

Fig. 3 Variation of brake thermal efficiency with respect to load at different CR.

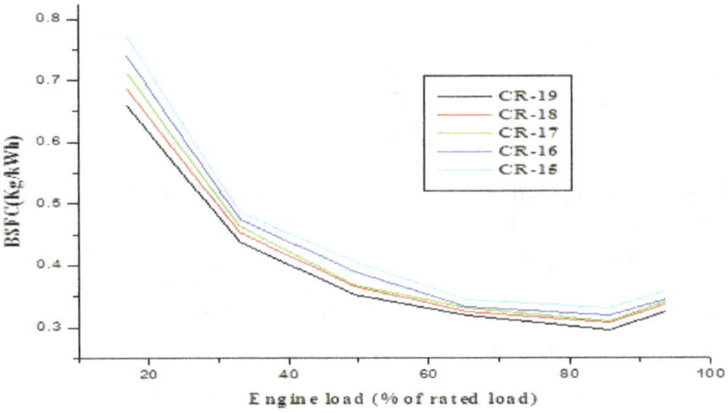

Fig. 4 BSFC with load at different CR.

At higher engine loads, fuel-air ratio increases, resulting in complete combustion and increase in brake thermal efficiency. The maximum brake thermal efficiency is obtained at a CR of 19, due to the superior combustion and better intermixing of the fuel. The brake thermal efficiency at CR of 18 is also very close to that of maximum brake thermal efficiency, particularly at higher loads. The least brake thermal efficiency is obtained at a CR of 15. The CR of 19, is found to be the best for all load conditions. The change of compression ratio from 15 to 19 resulted in, 2.81%, 3.8%, 4.01%, and 5.14% increase in brake thermal efficiency at 65% load and 2.61%, 3.12%, 4.87%, and 5.48% at 87% load respectively. This improved performance of the engine at higher compression ratio may be due to higher temperatures resulted in the combustion chamber which reduces the volatility and ignition delay of the blended fuel and the maximum thermal efficiency can be attributed to the superior combustion and better intermixing of air and fuel at optimum compression ratio. Rakopoulos et al obtains the same results. The comparison of brake specific fuel consumption with varying load is presented in Fig. 4 at 1500 rev/min. The least fuel consumption is obtained at compression ratio of 19. The fuel consumption at CR=18 is found very close to optimum value. At the lower sides of the compression ratios, the fuel consumption is higher due to higher viscosity and poor volatility of the blended fuel.

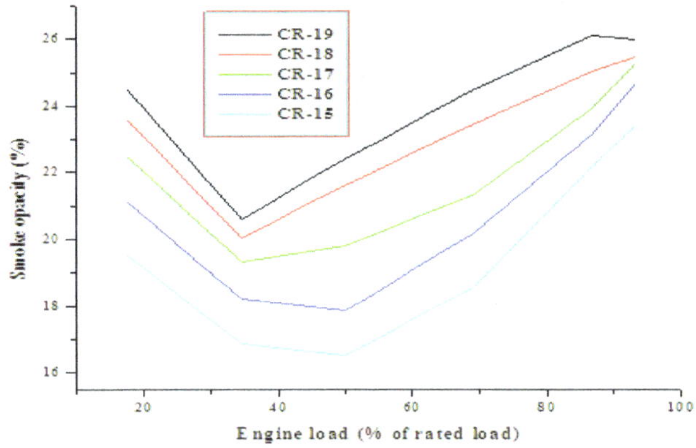

Fig. 5 Variation of Smoke opacity with respect to load at different CR.

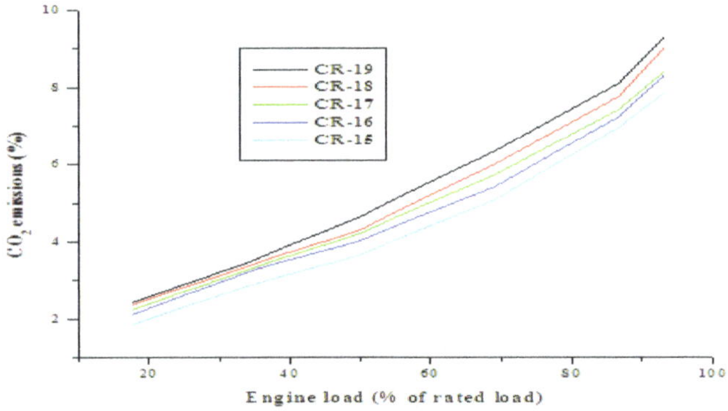

Fig. 6 Comparison of CO_2 emissions with load at different CR.

The variation of smoke opacity with varying engine load is presented in Fig. 5. Smoke opacity increases with the increase in load. This may be due proper mixing at fuel rich region and rising of combustion temperature. Smoke opacity at CR of 19 is found highest and this is because of better combustion at optimum compression ratio. Altuna and Bulut obtained the same results.

Figure 6 presents the variation of carbon dioxide emission with load at different compression ratios. It is observed that the carbon dioxide emission at compression ratio of 19 is higher. This is due to superior combustion which takes place at optimum compression ratio. More amount of CO_2 is an indication of complete combustion of carbon molecule present in fuel. However, more amount of CO_2 is not much harmful to human beings but is leading to higher ozone depletion potential and global warming.

NOx emission with varying engine load is shown in Fig.7. NOx emission increases with increase in load at all the compression ratios. At low load, rich air-fuel mixture causes incomplete combustion, but at higher load conditions proper mixing at fuel rich region results in complete combustion and rising of combustion temperature. NOx emission at compression ratio of 19 is found highest and this is because of better combustion at optimum compression ratio. NO_X emissions at compression ratio of 19 are found about 445.3 ppm which is equivalent to 2.83 (g/kWh) according to the Eq. (1) (Pilusa et al., 2012) and this value is within the permissible range of NO_X emissions with respect to **Bharat stage emission standards** are emission standards instituted by the Government of India to regulate the output of air pollutants from internal combustion engine equipment, including motor vehicles. The standards and the timeline for implementation are set by the Central Pollution Control Board under the Ministry of Environment & Forests (http://cpcb.nic.in/Functions.php).

$$EP_{i,w}\left(g/kWh\right)=\frac{EV_{i,w}\left(ppm\right)}{1\times10^{6}}\times\left(\frac{M_i}{28.84g/mol}\times4160g/kWh\right) \quad (1)$$

where EPi,w is pollutant mass, i, referenced to Peff (g/kWh), EVi,w is exhaust emission value of components on wet basis, i is volume share (ppm), and Mi is molecular mass of the components, i, (g/mol).

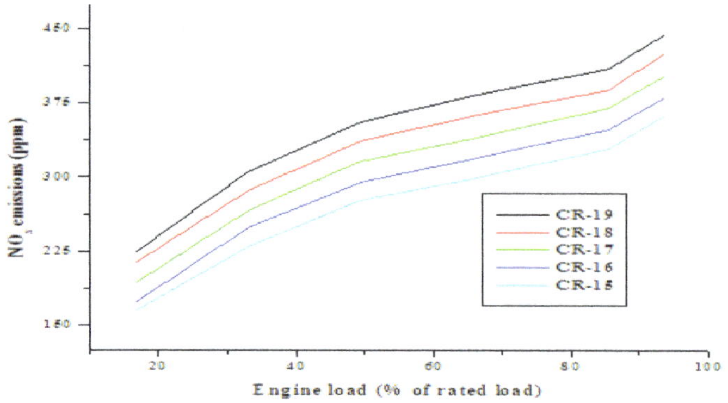

Fig. 7 Variation of NO_X emissions with load at CR.

4 CONCLUSIONS

The following are the main conclusions drawn after extensive tests results on Jatropha-diesel blend (30%) operation on variable compression ratio diesel engine.

- At rated speed of 1500 rev/min, the compression ratio 19 shows highest brake thermal efficiency, and hence, may be considered as optimum compression ratio for variable compression ratio diesel engine. CR of 18 exhibited marginally lower brake thermal efficiency compared to optimum compression ratio.
- Better fuel economy is observed at CR of 19 compared to other compression ratios.
- Highest smoke opacity, carbon dioxide and NOx emissions are observed at compression ratio of 19 for blending fuel operation. Hence, the test results show that the compression ratio 19 seems to be over all optimum compression ratio for blending fuel operation.

NOMENCLATURE

VCR	Variable compression ratio
CA	Crank angle
BTDC	Before Top Dead Centre
BSFC	Brake specific fuel consumption
CR	Compression ratio
CO_2	Carbon dioxide
NO_x	Oxides of Nitrogen
J30	30% Jatropha oil and 70% Diesel oil in blends

REFERENCES

Agarwal, A. K., 2007, "Biofuels (Alcohols and Biodiesel): Applications as Fuels for Internal Combustion Engines," *Progress in Energy and Combustion Science,* 33, 233–271. http://dx.doi:10.1016/j.pecs.2006.08.003

Altuna, S., Bulut, H. and Oner, C., 2008 "The Comparison of Engine Performance and Exhaust Emission Characteristics of Sesame Oil–Diesel Fuel Mixture with Diesel Fuel in a Direct Injection Diesel Engine," *Renewable Energy*, 33, 1791-1799. http://dx.doi:10.1016/j.renene.2007.11.008

Chen, H., Wang, J. and Shuai, S., 2008 "Study of Oxygenated Biomass Fuel Blends on a Diesel Engine," *Fuel*, 87, 3462-3468. http://dx.doi:10.1016/j.fuel.2008.04.034

Jindal, S., Nandwana, B. P. and Rathore, N. S., 2010 "Experimental Investigation of the Effect of Compression Ratio and Injection Pressure in a Direct Injection Diesel Engine Running on Jatropha Methyl Ester," *Applied Thermal Engineering*, 30, 442-448. http://dx.doi:10.1016/j.applthermaleng.2009.10.004

Pilusa, T. J. and Mollaggee, M. M., 2012, "Reduction of Vehicle Exhaust Emissions from Diesel Engines Using the Whale Concept Filter," *Aerosol and Air Quality Research,* 12, 994–1006. http://dx.doi:10.4209/aaqr.2012.04.0100

Pradhan, P. and Raheman, H., 2014, "Combustion and Performance of a Diesel Engine with Preheated Jatropha Curcas Oil Using Waste Heat from Exhaust Gas," *Fuel*,115, 527-533. http://dx.doi.org/10.1016/j.fuel.2013.07.067

Rakopoulos, C. D, Antonopoulos, K. A., Rakopoulos, D. C., Hountalas, D. T., Giakoumis, E. G., 2006 "Comparative Performance and Emissions Study of a Direct Injection Diesel Engine Using Blends of Diesel Fuel with Vegetable Oils or Biodiesels of Various Origins," *Energy Conversion and Management*, 47, 3272–3287. http://dx.doi:10.1016/j.enconman.2006.01.006

Rakopoulos, C. D, Antonopoulos, K. A., Rakopoulos, D. C., Hountalas, D. T., Giakoumis, E. G., 2006 "Development and Application of Multi-Zone Model for Combustion and Pollutants Formation in Direct Injection Diesel Engine Running with Vegetable Oil or Its Bio-Diesel," *Energy Conversion and Management*, 48, 1881–1901. http://dx.doi:10.1016/j.enconman.2007.01.026

Thomas, R. W. and Maurin, B. O., 1993, "Identification of Chemical Change Occurring during the Transient Injection of Selected Vegetable Oils," *SAE Technical Paper Series, 930933.*

22

COMBUSTION EFFICIENCY INSIDE CATALYTIC HONEYCOMB MONOLITH CHANNEL OF NATURAL GAS BURNER START-UP AND LOW CARBON ENERGY OF CATALYTIC COMBUSTION

Shihong Zhang*, Zhihua Wang

School of Environment and Energy Engineering, Beijing University of Civil Engineering and Arch., Beijing, China

ABSTRACT

This article discussed exhaust gas temperature and pollutant emissions characteristics of the combustion of rich natural gas-air mixtures in Pd metal based honeycomb monoliths burner during the period of start-up process. The burner needs to be ignited by gas phase combustion with the excessive air coefficient (a) at 1.3. The chemistry at work in the monoliths was then investigated using the stagnation point flow reactor or SPFR. The experimental results in catalytic monolith can be explained from SPFR. The exhaust gas temperature and pollutant emissions were measured by thermocouple K of diameter 0.5 and the analyser every 1 minute, respectively. Meanwhile combustion efficiency were calculated. Catalytic combustion of natural gas plays an important role for low carbon energy in industrial applications.

Keywords: *catalytic combustion; temperature; exhaust gas pollutant emissions; start-up; combustion efficiency.*

1. INTRODUCTION

Rising concentrations of greenhouse gases(GHG) in the atmosphere have been associated with global climate change. This phenomenon, which has resulted in an increase of global mean surface temperature, has serious negative effects on climatic systems, on the natural environment, and upon human society (Solomon *et al.*, 2007). Along with the development of society, vast quantities of these gases have been discharged into the atmosphere-namely carbon dioxide (CO_2), methane and other non-CO_2 gases (Nakata *et al.*, 2011).These emissions are generated primarily from the combustion of fossil fuels, such as petroleum, coal, and natural gas. With climate change threats, the levels of GHG need to be stabilized and eventually reduced. Clearly, our consumption of fossil fuels must decrease, partly due to a limited and uncertain future supply and partly because of undesirable effects on the environment (International Energy Agency, 2009).

Essentially, a sustainable supply of energy for societal needs must be secured in long-term for our future generations. With well-founded scientific supports and international agreement, renewable energy sources must be urgently developed and widely adopted to meet environmental and climate related targets and to reduce our dependence on oil and secure future energy supplies (Sawangphol and Pharino (2011). To reduce CO_2 emissions into the atmosphere, alternatives for "CO_2-free" utilisation of natural gas should be developed. Natural gas is recognized as the fossil fuel causing least damage to the environment. This is because it is clean, has a low carbon ratio (Lynum, 1997).

Catalytic combustion has so far found limited applications. However, the need for distributed and portable power generation that relies on modularity and small scales may render catalytic combustion an appealing technology (Deshmukh and Vlachos, 2007). Catalytically assisted combustion can greatly improve the performance of combustion devices and aid the development of new energy generation technologies (Wiswall *et al.*, 2009). Catalytic combustion in small devices is preferred because the reaction occurring at much lower temperatures and being sustained at much leaner fuel/air ratios compared to homogeneous combustion, thereby easing the design constraints of the system (Bijjula and Vlachos, 2011).

Catalytic combustion can be applied in commercial class gas engines to reduce NOx emission and micro gas turbines to improve flame stability (Yuan *et al.*, 2008). When used in burners, these oxidation catalysts increase the stability of ultra-lean combustion while generating near-zero pollutant emissions, thus offering a sound alternative to less stable low-NOx combustion technologies such as lean burn (Dupont *et al.*, 2000). Introduction of the hybrid catalytic combustor concept has also been able to overcome the material constraints associated with fully catalytic combustors to meet the higher turbine inlet temperatures (Andrae *et al.*, 2005).

In recent years, research efforts have focused on portable, hydrocarbon-fueled power generating devices. Catalytic microreactors, in particular, have received a lot of attention due to their operational benefits at small scales. While all such studies, both numerical and experimental, have provided valuable insight on the steady-state behavior of catalytic micro-combustors, studies on their transient behavior and in particular on the crucial issue of micro-reactor start-up remain limited (Karagiannidis and Mantzaras, 2009).

An extensive experimental investigation was carried out to investigate transient behavior of catalytic combustion burner during start-up process. Exhaust gas temperature and pollutant emissions could be measured by thermocouple K of diameter 0.5 and the analyser every 1 minute, respectively. Gas temperature and emissions mechanism with the increasing of time during start-up process were studied. At the same time, combustion efficiency inside catalytic honeycomb monolith channel of natural gas burner VI was calculated.

*Corresponding author. Email: shihongzhang@bucea.edu.cn.

2. EXPERIMENTAL SET-UP

Figure 1 illustrates the exhaust gas analysis system of catalytic combustion burner, The square honeycomb monoliths were 150mm wide in sides of the square and 20mm long, with square-shaped cells which sectional area was 1mm×1mm. The support for all the monoliths tested here was cordierite. The four square catalytic honeycomb monoliths were installed in the burner each time. The lengths of catalytic honeycomb monoliths were 20mm for the catalytic combustion burner.

In order to decrease the temperature of mixtures in chamber connected with the monolith's entrance, the 20mm long blank monoliths were inserted between the chamber and the Pd based catalytic monolith's entrance as assembly of monolith. At the chamber outlet we recorded length of catalytic honeycomb monolith as zero as shown in Fig. 2.

In the experiment, the reactant gas feeds of natural gas and air were regulated via GMS005 0BSRN200000 natural gas meter and CMG400A080100000 air meter with 0~50 L/min and 0~80m3/h of full-scale ranges, respectively. The two meters were provided electric current through manostat.

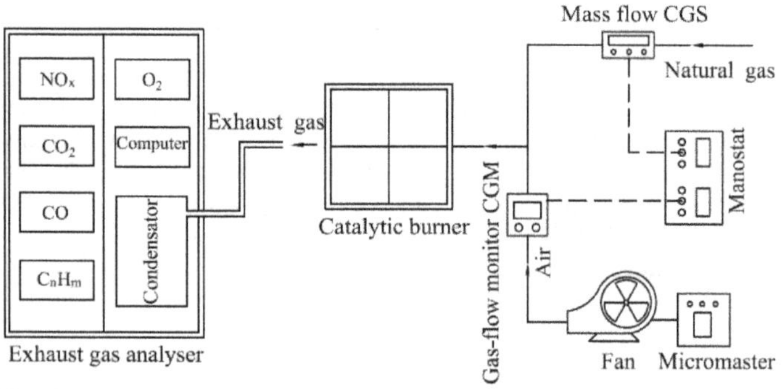

Fig. 1 Exhaust gas analysis system of catalytic combustion burner VI

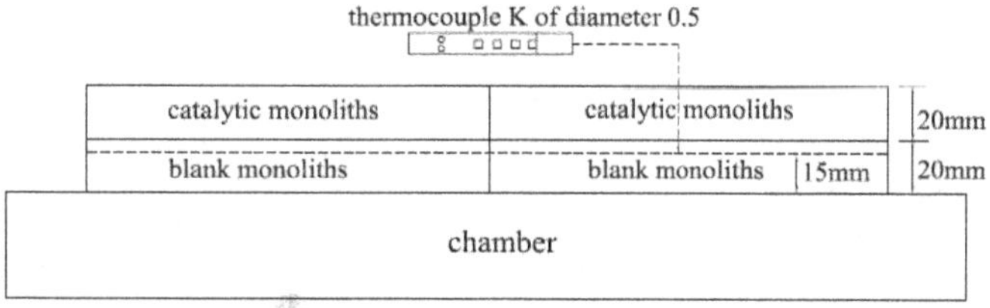

Fig. 2 Schematic of exhaust gas temperature measurement of catalytic honeycomb channel.

In the process of ignition, we need to swept the inside of burner by air for five minutes to ensure that there was no residual natural gas. The burner must be ignited by gas phase combustion with the excessive air coefficient (a) at 1.3 during the period of start-up process under the condition of invariable natural gas flow rate (9.5 L/min). The blue flame was achieved above the monolith and gradually disappeared when the catalyst started glowing red internally as the temperatures inside the monolith increased Then the excessive air coefficient(a) should be adjusted to 2.0 under fuel lean condition while the catalyst reached steady state.

Exhaust gas temperature inside catalytic honeycomb channel within the first 15mm of monolith could be measured by thermocouple K of diameter 0.5 which sensor was supposed to measure temperature for transient changes, and pollutant emissions could be measured by the NO-NO_2-NOx thermo electron analyser, CO/CO_2 thermo electron analyser every 1 minute. At the same time we observed and recorded the data.

The emissions characteristics of the combustion of fuel-rich mixtures of methane (main composition of natural gas) and air were studied in steady-state conditions in a catalytic honeycomb monolith burner. In order to investigate the parameters controlling the kinetics and products selectivities of the heterogeneous (solid-gas) oxidation of methane on a 'model' noble metal, and the homogeneous ignition inhibition phenomenon, fundamental work on a small-scale reactor was carried out (the stagnation point flow reactor or SPFR). The combustion of rich $CH_4/O_2/N_2$ mixtures on a polycrystalline platinum foil in a stagnation point flow reactor at atmospheric pressure and in steady-state was investigated. For the SPFR, Fig. 3 showed the diagram of the reactor.

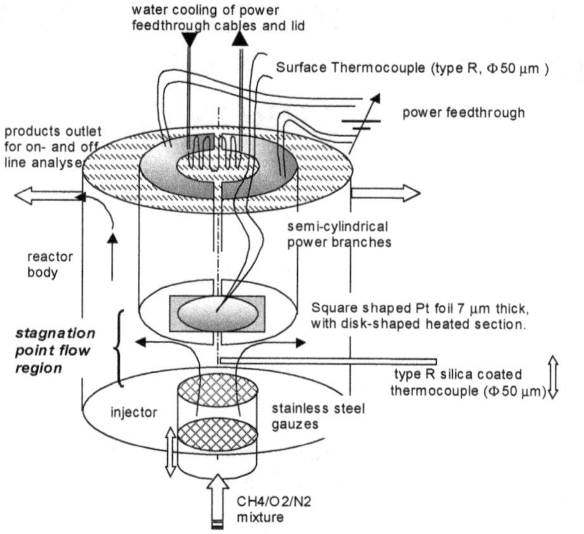

Fig. 3 Stagnation point flow reactor

3. EXPERIMENTAL RESULT AND DISCUSSION

3.1 Mechanism of catalytic combustion of methane

The following equations (1) and (2) for the percent fuel conversion and the selectivity of products SELk for the species 'k' were derived (on a mol basis) (Dupont *et al.*, 2001):

$$CV_{CH_4} = -100\frac{F_{CH_4}C_S + \int_0^L W_{CH_4}\dot{\omega}_{CH_4}Cdx}{\rho_0 Y_{CH_4,0}U_0} \quad (1)$$

$$SEL_k = -100\frac{\frac{F_k}{W_k}C_S + \int_0^L \dot{\omega}_k Cdx}{[\frac{F_{CH_4}}{W_{CH_4}}C_S] + \int_0^L \dot{\omega}_{CH_4}Cdx} \quad (2)$$

where W_k is the molar mass of the species k, $\dot{\omega}_k$ (in mol/ cm^3 • s), is the molar production rate of k, F_k (in kg/ cm^2 • s), is the mass flux of k at the foil surface, the subscript 0 means 'at the injector outlet' (x = 0 cm). ρ_0, Y_{CH_4} 0, and U_0 are the gas density, fuel mass fraction and axial velocity at the injector outlet, their product being the mass flux of fuel in the reactor. The coefficient C is a correction factor which accounts for a slight radial expansion of the control volume used to perform the species balances in the calculation of the fuel conversion and products selectivities. This control volume, originally cylindrical with a radius r at its basis (location of injector) increasing to a radius $r^*_s=r(1+KLT_s/T_i)$ at the catalytic surface. Cs corresponds to x = L = 1cm, i.e, the value of C at the foil surface. For our SPFR, we show the conversion curves obtained this time without correction factor, therefore it is one. The proportionality constant K was chosen to match the predicted fuel conversion of a sample point with its corresponding experimental value.

In order to describe the foil condition for each experiment, two parameters were varied in order to investigate their effects on the fuel conversions and CO selectivities; these are:

1. The fuel mixture strength by varying the parameter $\alpha = \dot{V}_{CH_4}/(\dot{V}_{CH_4}+\dot{V}_{O_2})$, where $\dot{V}_k$ is the input volume flow rate of the relevant species k,α was varied from 0.35 to 0.58.
2. The N_2 the mol fraction of nitrogen gas in the inlet mixture, $XN_2 = (\dot{V}_{N_2}+0.79\dot{V}_{air})/\dot{V}_{tot}$.

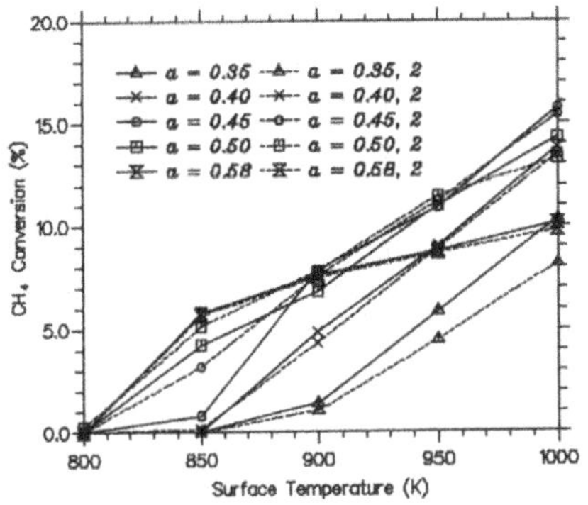

(a)

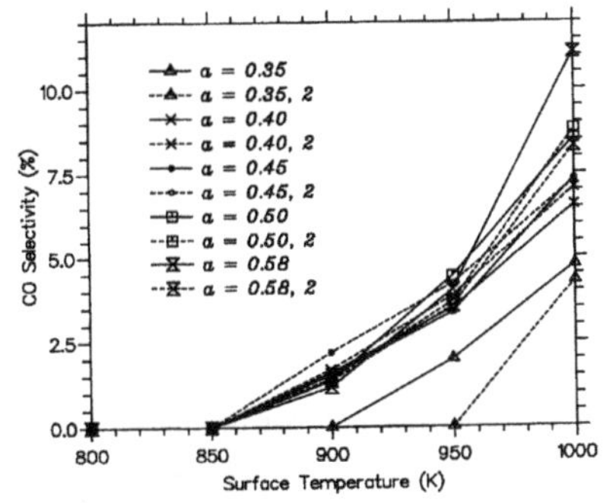

(b)

Fig. 4 CH_4 conversions inhibition of gas phase ignition phenomenon and CO selectivities.

Figure 4 show the CH_4 conversions inhibition of gas phase ignition phenomenon, as will be seen later, and CO selectivities for the fuel-rich concentrations of α (0.35, 0.4, 0.45, 0.5 and 0.58) and XN_2 (0.91), i.e. conditions of partial oxidation on Pt foil for the SPFR. The CH_4 conversion increases with an increase in temperature for the same value of the fuel mixture strength (α).

The main results from these runs are the large CO selectivities (more than 10%) obtained at low catalyst temperatures (800-1000 K). No amounts of C_2H_4 and C_3H_8 were detected and negligible concentrations of C_2H_6 were detected (lower than 8 ppm).

3.2 Experimental results

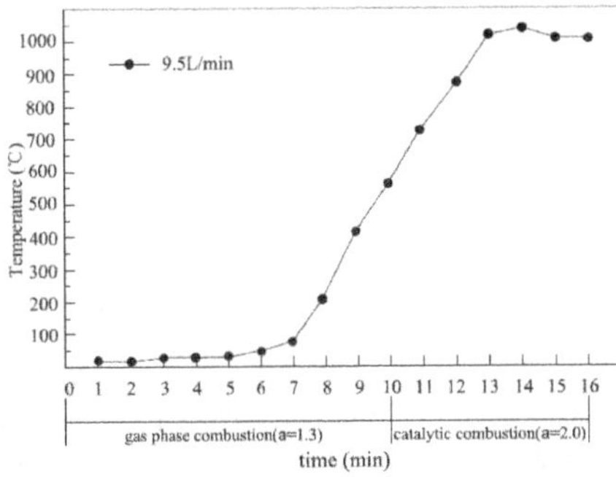

Fig. 5 Exhaust gas temperature in catalytic combustion burner VI during the start-up process with the passage of time (a is excessive air coefficient).

Under the condition of invariable natural gas flow rate (9.5 L/min), the experiment was run for 16 minutes. At the time of the ignition we recorded it as zero at the open end of monolith which was connected with pipes of analysers.

Figure 5 plots exhaust gas temperature inside catalytic honeycomb monolith channel of natural gas burner between 0 and 16 minutes. As we can see from the profiles, exhaust gas temperature ascended gradually with the passage of time firstly, then gas temperature was stable when the time reached 13 minutes. The highest temperature among all the time was measured at 14th minute which was 1080°C.

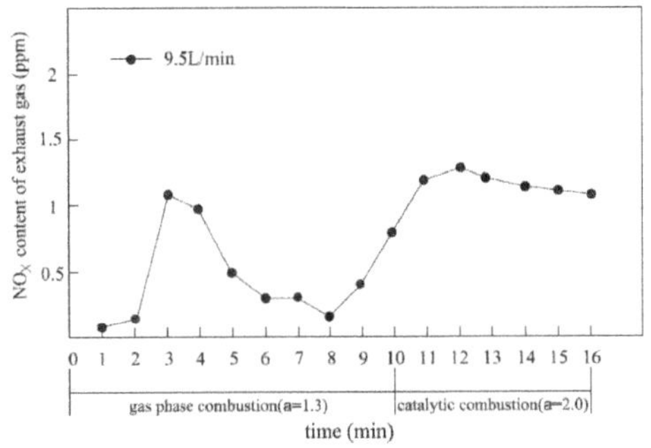

(a) NO_X content of exhaust gas

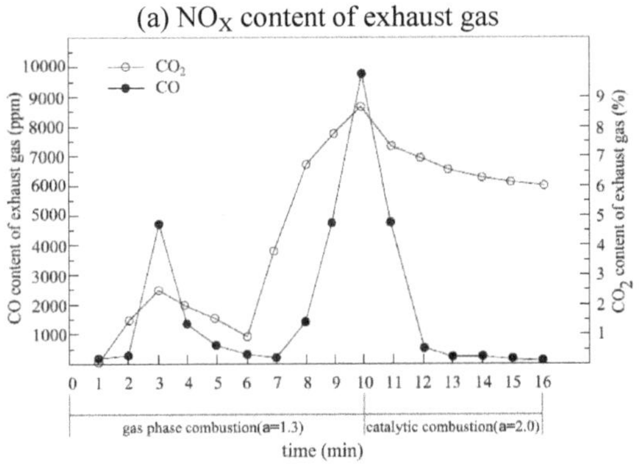

(b) CO and CO_2 content of exhaust gas

Fig. 6 Exhaust gas pollutant emissions in catalytic combustion burner VI during the start-up process with the passage of time (a is excessive air coefficient).

Figure 6 plots exhaust gas pollutant emissions in catalytic combustion burner VI during the start-up process as time went on. It was seen from Fig. 6(a) that the emission of NOx increasing during the first portion of gas combustion between 0 and 3 minutes at the open end of burner was detected. Then NOx decreased between 3 and 8 minutes. Because the flame appeared during the early period of start up process inside honeycomb monolith channels. Then the combustion flame was disappeared gradually with the decreasing NOx after 3 minutes. After 8 minutes, the emission of NOx was rising again, at the same time the temperature was increasing quickly inside honeycomb monolith channels. However the channel temperature was too low to product a larger number of NOx (below 2 ppm).

In this gas phase combustion of rich natural gas /air mixtures, about 7 percent of natural gas/air mixtures was used in the process of start-up. The total volume flow rate of reactants was the same as that of products via the reaction $CH_4 + 2O_2 = CO_2 + 2H_2O$. So the percentage of CO_2 should remain unchanged according to the excessive air coefficient. As shown in Fig. 5(b) the percentage of CO_2 was about 8% without vapor by the CO_2 analyzer inside honeycomb monolith channels during the gas phase combustion. When the burner reached the state of catalytic combustion, the percentage of CO_2 was decreased to about 6% approximately. It was clear that CO_2 measurement was sure to the above analysis.

It was also shown in Fig. 6(b) that theCO content of exhaust gas ascended dramatically as time went on between 0 and 10 minutes. During the start-up process, there were CO selectivities for the fuel-rich concentrations of a (1.3) (conditions of partial oxidation). The main results from these runs were that the large CO selectivities obtained at low catalyst temperatures inside honeycomb monolith channels. When the burner reached the state of catalytic combustion, the CO content of exhaust gas dropped to near zero.

CO and CO_2 were increasing during the first portion of gas combustion due to the appeared flame influence and then it decreased with disappeared flame between 3 and 7 minutes. Also CO and CO_2 concentrations had been significantly diluted in 7 minutes due to existing the air in the pipes. After 7 minutes exhaust gas concentrations were distributed all of the pipes.

From the chemical kinetics it can be seen the large CO selectivities for the fuel-rich concentrations. After local catalytic ignition, a steady state was sure to be reached under fuel lean condition.

The exhaust gas content was shown in table 1 between 11 minutes and 16 minutes when the catalyst reached the steady state.

It was shown that these oxidation catalysts increased the stability of ultra-lean combustion while generating near-zero pollutant emissions. The temperature inside the monoliths channels indicated that the catalyst's role was to enable the ignition of nature gas mixtures below flammability limits, to ensure the complete oxidation of the nature gas to CO_2 via surface reactions in the steady state.

Table 1 Exhaust gas content (9.5L/min)

Time(min)	C_nH_m (ppm)	CO (ppm)	CO_2 (%)
11	5.22	5088	7.4
12	5.03	830	7
13	4.25	145	6.7
14	4.22	162	6.3
15	4.14	6.06	6.2
16	4.27	5.41	6

3.3 Calculation of combustion efficiency during the start-up process.

It was evidenced that the catalytic combustion efficiency was almost close to 100%. But there were a lot of CnHm and CO from the exhaust gas in catalytic combustion burner during the start-up process.

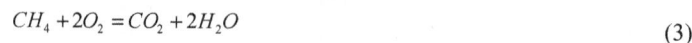

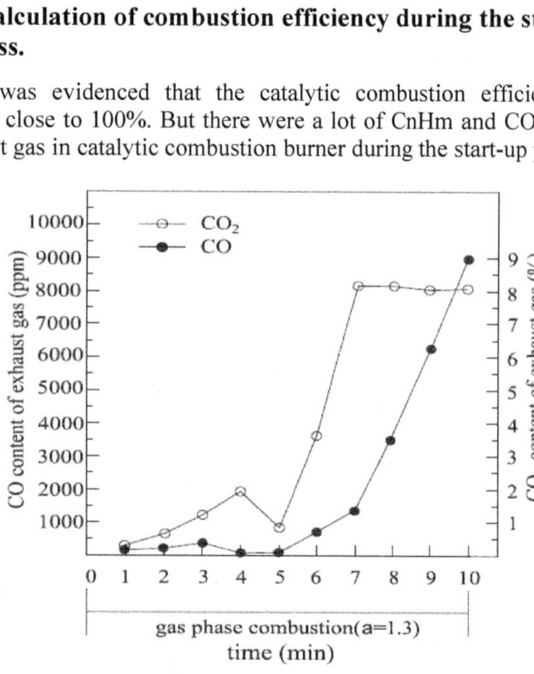

Fig. 7 Exhaust gas pollutant emissions in catalytic combustion burner VI during the start-up process with the passage of time (a is excessive air coefficient).

It was proved that gas-phase combustion had not been oxidized completely and its combustion efficiency should be calculated.

The experiment of gas-phase combustion is run for 10 minutes at an invariable natural gas flow rate of 10 L/min. At the time of ignition we recorded it as zero.

Figure 7 plots the content of CO and CO_2 in exhaust gas inside catalytic honeycomb monolith channel of natural gas burner VI between 1 and 10 minutes.

The un-burnt CnHm, CO and CO_2 content of exhaust gas inside honeycomb monolith channels was listed in Table 2.

Table 2 Exhaust gas content (10L/min)

Time(min)	CnHm (ppm)	CO (ppm)	CO_2 (%)
1	3110	86.6	0.66
2	6450	334	0.83
3	13550	505	1.75
4	13900	40.1	2.44
5	10500	63.5	1.27
6	3100	552	3.8
7	470	1430	8.3
8	96	3580	8.3
9	12.8	6330	8.1
10	6.32	9060	8.2

The content of CO in gas-phase combustion was higher than that of catalytic combustion. The maximum of CO emission reached about 9060 ppm, which indicated that natural gas of gas-phase combustion hadn't been oxidized completely.

It was revealed that the content of CnHm in gas-phase combustion was measured at 13900ppm~6.32ppm. After an interval of 4 minutes C_nH_m was decreased rapidly with the passage of time. It proved that the conversion of gas-phase combustion was lower than that of catalytic combustion.

As the main composition of natural gas was methane, the chemical reaction equation is (3) in the following:

$$CH_4 + 2O_2 = CO_2 + 2H_2O \tag{3}$$

CO was a kind of intermediate which was generated during combustion of hydrocarbon. The number of C atom remained unchanged in the reaction process. The total volume of CO and CO_2 was the same as that of CH_4 via the reaction equation (3) ($V_{CO}+V_{CO2}=V_{CH4}$).The volume of methane was given 1 Nm3 ($V_{CH4}=1$). According to equation (4), the volume of CO and un-burnt CH_4 was alculated as:

$$\begin{cases} V_{CO} + V_{CO_2} = 1Nm^3 \\ \dfrac{V_{CO}}{V_f^d} = \gamma_{CO} \\ \dfrac{V_{CO_2}}{V_f^d} = \gamma_{CO_2} \\ \dfrac{V_{CH_4}^1}{V_f^d} = \gamma_{CH_4}^1 \end{cases} \tag{4}$$

where V_{CO} and V_{CO2} are the volume of CO and CO_2 in exhaust gas(m^3), respectively. V_f^d is the total volume of exhaust gas. γ_{CO} and$\gamma_{CO\ 2}$ are ratio of CO volume and CO_2 volume to that of exhaust gas, respectively. V_{CH4}^1 is the volume of un-burnt CH_4 in exhaust gas(m^3). γ_{CH4}^1 is ratio of CH_4 volume to that of exhaust gas. γ_{CO}, γ_{CO2}, γ_{CH4}^1 are measured by the analyser.

According to equation (5): 1 m^3 CO oxidized completely to CO_2 could generate 12644 kJ heat. It proved that the content of CO had an important influence for utilization of thermal energy of the fuel.

$$CO + 0.5O_2 = CO_2 + \Delta T \quad (5)$$

Therefore, the heat released of un-burnt CH_4 and CO were calculated as:

$$Q_{CH_4} = V^1_{CH_4} H_1 \quad (6)$$

$$Q_{CO} = V_{CO} H_2 \quad (7)$$

H_l is net calorific value of methane under standard conditions which is 35.88MJ/m3. H_2 is calorific value of CO which is 12.64MJ/m3.

The following equations (8) for heat released percent of unburnt CH_4 and CO to reactant (natural gas) was derived:

$$K = \frac{Q_{CH_4} + Q_{CO}}{V_{CH_4} H_1} \quad (8)$$

The combustion efficiency of gas-phase was calculated by equation (9):

$$\eta = 1 - \frac{V^1_{CH_4} + V^2_{CH_4}}{V_{CH_4}} \quad (9)$$

V_{CH4}^2 is the volume of CH_4 which has been used in generating CO. Here $V_{CH4}^2 = V_{CO}$ was known.

According to the aforementioned equations and experimental data(table 1 and table 2), the ratio of heat released of un-burnt CH_4 and CO and combustion efficiency of burner VI were calculated during the start-up process with the passage of time in table 3.

It was shown from table 3 that part of the energy was wasted in gas combustion which hadn't been oxidized completely. It was proved that the combustion efficiency of gas-phase combustion was lower than that of catalytic combustion by calculated data.

Table 3 Ratio of heat released (K) and combustion efficiency (η) during start-up process with the increasing of time.

(a) Exhaust gas content (10L/min)

Time(min)	K(%)	η (%)
1	0.47	51.7
2	0.75	21.4
3	0.77	22.2
4	0.57	42.8
5	0.82	17.5
6	0.085	90.6
7	0.011	97.8
8	0.016	95.9
9	0.027	92.8
10	0.037	90

(b) Exhaust gas content (9.5L/min)

Time(min)	K(%)	η (%)
11	0.024	93.6
12	0.004	98.8
13	0.001	99.7
14	0.001	99.7
15	0	100
16	0	100

For all tests of the catalytic combustion, only extremely small amount of CO, unburned fuel and NOx was detected inside the monolith channels and over the open end of the burner VI. The pollutant emissions can be reduced to a minimum 'near-zero' by the catalytic combustion process.

4. CONCLUSIONS

This article discussed exhaust gas temperature and pollutant emissions characteristics of the combustion of rich natural gas-air mixtures in Pd metal based honeycomb monoliths by means of experiments on a practical burner during the start-up process. It was shown that exhaust gas temperature ascended gradually firstly with the passage of time, then gas temperature was stable when the time reached 13 minutes. There were large CO selectivities for the fuel-rich concentrations of a (1.3), i.e. conditions of partial oxidation.

Simultaneously, the emission of NOx was near zero inside honeycomb monolith channels as time went on. This article played a key role for further start-up study of catalytic combustion burner.

It proved that the concentration of pollutant emissions in gas-phase combustion was higher than that of catalytic combustion. It was shown that the conversion of gas-phase combustion was lower than that of catalytic combustion by calculation of combustion efficiency.

The depth concern of catalytic combustion of natural gas can restructure energy policy and reduce CO_2 emissions.

ACKNOWLEDGMENT

The project sponsored by the Beijing Municipality Key Lab of Heating, Gas Supply, Ventilating and Air Conditioning Engineering; Funding Project (Building environment and facilities engineering).

NOMENCLATURE

C	correction factor
C_m	specific heat of water J/(kg•°C)
CV	fuel conversion (%)
F_k	mass flux of k at the foil surface, (kg/ cm^2•s)
G	mass flow rate (kg/s)
H_h	gross calorific value (MJ/Nm^3)
L_g	gas volum (m^3)
M_{rH2O}	molar mass of water (g/mol)
Q_i	thermal output (kW)
SEL_K	selectivity of products for the species 'k'
U_0	axial velocity 'at the injector outlet' (x = 0 cm), (m/s)
V_g	volume of nature gas (Nm^3)
$\dot{V}_k$	input volume flow rate of the relevant species k
$Y_{CH4,0}$	fuel mass fraction 'at the injector outlet' (x = 0 cm)
W_k	molar mass of the species k

Greek Symbols

ρ_0	gas density at the injector outlet' (x = 0 cm), (kg/m^3)
η	thermal efficiency (%)
$\dot{\omega}_k$	molar production rate of k (mol/ cm^3•s)

REFERENCES

Andrae, J.C.G., Johansson, D., Bursell, M., and Fakhrai, R., 2005, "High-Pressure Catalytic Combustion of Gasified Biomass in a Hybrid Combustor," *Applied Catalysis A: General*, **293**, pp. 129 - 136. http://dx.doi.org/10.1016/j.apcata.2005.07.003

Bijjula, K., and Vlachos, D.G., 2011, "Catalytic Ignition and Autothermal Combustion of JP-8 and Its Surrogates over a Pt/c-Al_2O_3 Catalyst," *Proceedings of the Combustion Institute*, **33**(2), pp.1801–1807. http://dx.doi.org/10.1016/j.proci.2010.05.008

Deshmukh, S.R., and Vlachos, D.G., 2007, "A Reduced Mechanism for Methane and One-step Rate Expressions for Fuel-Lean Catalytic Combustion of Small Alkanes on Noble Metals," *Combustion and Flame*, **149**(4), pp.366–383.

http://dx.doi.org/10.1016/j.combustflame.2007.02.006

Dupont, V., Zhang, S.H., and Williams, A., 2000, "Catalytic and Inhibitory Effects of Pt Surfaces on the Oxidation of $CH_4/O_2/N_2$ Mixtures," *Int. J. Energy Res.*, 2000, pp.1291-1309.
http://dx.doi.org/10.1002/1099-114X(200011)24:14<1291::AID-ER716>3.0.CO;2-F

Dupont, V., Zhang, S. H., Williams, A., 2001, "Experiments and Simulations of Methane Oxidation on a Platinum Surface," *Chemical Engineering Science*, **56** (8), 2659-2670.
http://dx.doi.org/10.1016/S0009-2509(00)00536-4

International Energy Agency. 2009, World Energy Outlook. World Energy Outlook. Paris: The International Energy Agency; pp. 698.

Karagiannidis, S., and Mantzaras, J., "Numerical Investigation of Methane-Fueled, Catalytic Microreactor Start-Up," Paul Scherrer Institute, Combustion Research Laboratory, CH-5232, 2009, Villigen PSI, Switzerland.
http://www.combustion.org.uk/ECM_2009/P810317.pdf

Lynum, S., 1997, "Natural Gas Utilization without CO_2 Emissions," *Energy Convers. Mgmt.*, **38**, Suppl., pp. S165-S172.

Nakata, T., Silva, D., Rodionov, M., 2011, "Application of Energy System Models for Designing a Low-Carbon Society," *Progress in Energy and Combustion Science*, **37**(4), pp. 462-502.
http://dx.doi.org/10.1016/j.pecs.2010.08.001

Sawangphol, N., and Pharino, C., 2011, "Status and Outlook for Thailand's Low Carbon Electricity Development," *Renewable and Sustainable Energy Reviews*, **15**(1), pp. 564–573.
http://dx.doi.org/10.1016/j.rser.2010.07.073

Solomon, S., Qin, D., Manning, M., Chen, Z., Marquis, M., Averyt, K.B., Tignor M., and Miller, H.L., 2007 Climate Change: The Physical Science Basis, Contribution of Working Group 1 to the Fourth Assessment Report of the Intergovernmental Panel on Climate Change. Cambridge University Press, Cambridge.

Wiswall, J.T., Wooldridge, M.S., and Im, H.G., 2009, "An Experimental Study of the Effects of Platinum on Methane/Air and Propane/Air Mixtures in a Stagnation Point Flow Reactor," *Journal of Heat Transfer*, **131**, 111201
http://dx.doi.org/10.1115/1.3156788

Yuan, T., Lai., Y.H., Chang, C.K., 2008, "Numerical Studies of Heterogeneous Reaction in Stagnation Flows Using One-dimensional and Two-dimensional Cartesian Models," *Combustion and Flame*, **154**, pp. 557–568.
http://dx.doi.org/10.1016/j.combustflame.2008.06.005

23

COMPUTATIONAL STUDIES OF SWIRL RATIO AND INJECTION TIMING ON ATOMIZATION IN A DIRECT INJECTION DIESEL ENGINE

Renganathan Manimaran[a], Rajagopal Thundil Karuppa Raj[b,*]

[a]*Thermal and Automotive Research Group, School of Mechanical and Building Sciences, VIT Chennai, Tamilnadu, India*

[b]*Energy Division, School of Mechanical and Building Sciences, VIT Vellore, Tamilnadu, India*

ABSTRACT

Diesel engine combustion modeling presents a challenging task with the formation and breakup of spray into droplets. In this work, 3D-CFD computations are performed to understand the behaviour of spray droplet diameter and temperature during the combustion by varying the swirl ratio and injection timing. After the validation and grid and time independency tests, it is found that increase in swirl ratio from 1.4 to 4.1 results in peak pressure rise of 8 bar and an advancement of injection timing from 6 deg bTDC to 20 deg bTDC results in increase of peak pressure by 15 %.

Keywords: *Spray droplet parameters, Computational Fluid Dynamics, Combustion, Emissions, 4S-Direct injection diesel engine.*

1. INTRODUCTION

During this decade, the demanding stringent exhaust emission regulations have prompted for innovative spray technologies and better combustion control strategies especially in diesel engines due to NO_x and particulate emissions. Generally the NO_x and particulate matter are controlled after the combustion in the exhaust pipeline using catalytic converters. The primary cause for these emissions lies behind the distribution of fuel droplets inside the combustion chamber in ensuring complete combustion. Trade-off between the power output and the NO_x emissions is better achieved using controlled feedback injection timing (Heywood 1988) mainly in compression ignition engines. The time period between the spray of diesel fuel and actual start of combustion is generally referred as ignition delay period. This ignition delay period is a crucial task during experimental investigation of diesel engines. The study of these processes by experimental approach involves expensive instruments with high level of skill and moreover, consumes a lot of time. Nowadays computational techniques evolved such that modeling these processes can contribute to better understanding of spray penetration, combustion and pollutant formation.

Reitz and Diwakar (1986) implemented an Eulerian-Lagrangian spray and atomization model for diesel sprays. Their numerical study on internal flow characteristics for a multi-hole fuel injector gives better agreement with the available experimental data. This indicates the capability of numerical model for studying diesel spray characteristics. Magnussen and Hjertager (1976) developed a model based on the eddy break-up concept. This model relates the combustion rate to the eddy dissipation rate. This model expresses the rate of reaction by the mean mass fraction of the reacting species, the turbulence kinetic energy and the rate of dissipation.

Hossainpour and Binesh (2009) highlighted the prediction of droplet spray models in a CFD code. The spray calculations are based on statistical method referred as discrete droplet method. The results are validated with the experimental data. They reported that spray penetration which plays a dominant role in combustion and emission characteristics are predicted better with modeling methodologies.

Prasad et al. (2011) carried out simulation on different bowl configuration to analyze the effect of swirl on combustion. They found that re-entrant piston bowl could create highest turbulent kinetic energy and swirl in the cylinder. They also studied the effects of injector sac volume on the combustion and emission. The studies indicate that sac-less injector could result in lower emissions. Many literature (Arcoumanis et al. 1997; Rakopoulos et al. 2010; Kar et al. 2012; Torregrosa et al. 2012; Thurnheer et al. 2011) insist that spray dynamics plays a strong role on evaporation rate, flow field, combustion process and emissions. As a result, the atomization of fuel affects the combustion efficiency and pollutant formation. Modeling the atomization process during diesel combustion requires careful validation with the experimental results.

The in-cylinder turbulent motion of air is characterized by swirl, squish and tumble phenomena. Swirl is varied by designing the intake port and shaping the piston bowl for re-entrant combustion. For combustion chamber of re-entrant effects, the turbulent kinetic energy is intensified at TDC of compression stroke due to the conservation of angular momentum. Combustion is efficient and leads to low soot and high NO_x emissions (Costa et al. 2012; Chmela et al. 2007; Catania et al. 2011; Kondoh et al. 1985). The effect of variation of injection timing in diesel engine was studied by Sayin and Canakci (2009). They found that NO_x and CO_2 emissions increased while the unburned HC and CO emissions decreased when injection timing is advanced. Han et al. (1996) investigated numerically the multiple injections and split injection cases. They found that split injection reduces the soot significantly without the change in NO_x emissions whereas multiple injections reduce NO_x significantly. The numerical study on diesel engine simulation with respect to injection timing and the air boost pressure was carried out by Jayashankara et al (2010) using commercial CFD code. They validated the results of flow-field from CFD simulation with the experimental work of Payri et al (2004). From the CFD simulation, they observed that increase in cylinder pressure, cylinder temperature and NO_x emissions results from advancing the injection timing. They also found from simulation that the supercharged and inter-cooled engine results in higher NO_x emissions as compared to naturally aspirated engines.

Corresponding Author E-mail: tkraj75@gmail.com

Several experimental and computational studies are performed to understand the spray, combustion and pollutant formation processes. However, the study of droplet parameters towards combustion and emissions are rarely reported in literature (R. Manimaran, R. Thundil Karuppa Raj, 2013). Further, the study related to the behaviour of droplet parameters during combustion on the effect of swirl ratio and injection timing is not explored further in the literature. This paved the way for carrying out a task to study the droplet variables, combustion and emission characteristics by varying the engine parameters. Droplet variables like droplet mass, droplet sauter mean diameter, droplet temperature, droplet velocity and spray penetration can be measured experimentally but leads to a tedious task. To avoid the laborious task by experiments, CFD modeling gives better understanding on the processes to study the droplet variables on the variation of swirl and injection timing. However, the accuracy of the models and schemes employed should be ascertained and the result has to be validated with experimental results. Hence the aim of the present work is to understand the behavior of droplet variables towards the combustion and pollutant formation. The commercial CFD code, STAR-CD is used to simulate the in-cylinder processes such as spray, auto-ignition, combustion and pollutant formation. The results of the simulation are validated with the experiments data available from the literature after the suitable grid and time scales are observed. The in-cylinder averaged quantities and droplet parameters are analyzed from the simulation. Similar studies are continued to understand the behavior of droplet variables and predict the performance and emissions by varying the swirl ratio and advancing the injection timing.

2. METHODOLOGY

A commercial CFD code *STAR-CD* is used to model and simulate the combustion process and emissions in a direct injection Diesel engine. The CFD simulation involves the three steps as outlined in the following sections.

2.1 GRID GENERATION, GRID INDEPENDENCE AND TIME INDEPENDENCE STUDIES

The geometry of the piston bowl is obtained from Colin et al (2003). The piston bowl shape is prepared from a standard computer-aided-design package. After the piston bowl is generated, a spline is created from the bowl profile and used for the creation of in-cylinder mesh. The meshing of the in-cylinder fluid domain is performed using *es-ICE (Expert System – Internal Combustion Engine)* grid generation tool. In this study, a 45° sector mesh is considered due to symmetry nature of the in-cylinder domain and thereby the computational time can be reduced considerably. The in-cylinder grid thus obtained is checked for negative volumes at all locations between BDC and TDC. The meshed geometry of the moving fluid domain at 40° after TDC i.e. 760 deg CA is shown in Fig. 1.

The boundary of the domain consists of moving wall at the bottom, periodic zones at the sides, cylinder wall at the end side, cylinder head wall at the top, axis and the injector. Hexahedral cells are created in the in-cylinder fluid domain and a few tetrahedral cells near the fuel injector.

The hexahedral meshes are placed very fine to the wall and thereby both the hydrodynamic boundary layer and thermal boundary layers are captured more precisely. The total number of cells in the moving domain amounts to 45,000 at TDC. This cell count is verified after carrying out a series of grid independent tests as shown in Fig. 2. It can be observed from Fig. 2 that increasing the cells beyond 45000 cells does not alter the in-cylinder peak pressure and other process variables. Thus the numerical simulations are grid independent beyond 45000 cells at TDC location. Time independent study is carried out by varying the time step from 0.5 deg CA to 0.02 deg CA as shown by logarithmic scale in Fig. 2. It can be observed that in-cylinder averaged peak pressure does not get varied even the crank angle step interval is reduced below 0.025 deg. Hence the optimum crank angle step interval is maintained at 0.025 deg for all simulations in this study.

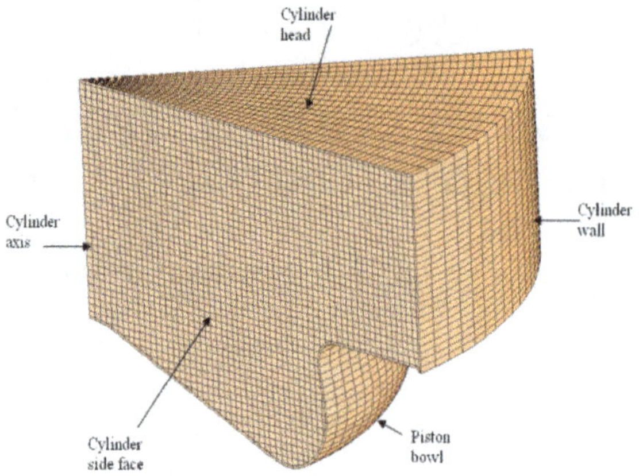

Fig. 1 Computational grid with boundary surfaces at 760 deg CA (40 deg after TDC)

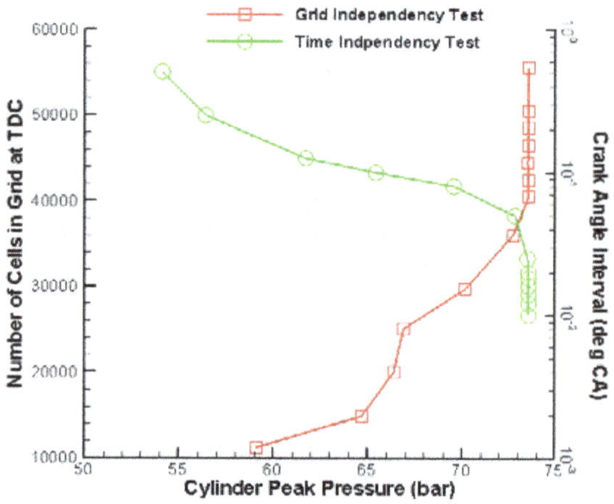

Fig. 2 Variation of peak pressure with grid density and crank angle interval

2.2 SOLVER DETAILS

Once the in-cylinder fluid domain mesh is available for simulation, the meshed geometry file is considered for combustion analysis in STAR-CD code. Lagrangian multiphase treatment is activated in the simulation of droplet break-up and spray penetration phenomena. The turbulent dispersion model is included for the droplet to experience randomly varying velocity field in the cylinder. Collision model (P.J. O'Rourke 1981) is also considered to detect the collision of parcels for every time step. Gravitational force is also accounted on the droplet parcels. The number of droplet parcels considered in this work is limited to 50 million which is more sufficient to capture the trajectory, spray penetration and collision physics (Munnannur 2007). RNG k-ε turbulence model (Tahry 1983) is used for modeling the turbulent Eulerian flow-field in the cylinder. The flame surface density equation is solved by adopting extended coherent flame model for 3 zones namely, the unmixed fuel zone, the mixed gases zone and the unmixed zone of air together with EGR (Colin and Benkenida 2004) .

The injections of fuel start at 714° which is equal to 6° before TDC. The injection of droplets at three crank angles viz. 718° CA to 720 deg CA are as shown in Fig. 3 (a), (b) and (c) respectively.

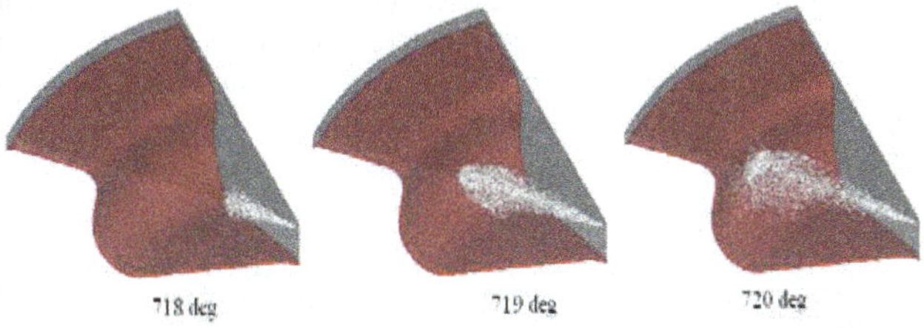

a) (b) (c)

Fig. 3 Spray visualization : (a) 2 deg CA before TDC, (b) 1 deg CA before TDC and (c) at TDC

The breakup of spray as observed in Fig. 3 is common in diesel engines due to surface tension and aerodynamic shear between the fuel and surrounding air in turbulent motion inside the cylinder at high pressures.

Reitz and Diwakar (1986) spray droplet model is considered for spray formation. According to this model, break-up of droplets occur in two modes.

(a) Bag Break Up: The non-uniform pressure field in the neighborhood of droplet causes the droplet to expand in the low-pressure or wake region and eventually disintegrate when surface tension forces are overcome. This happens typically when the Weber number, $W_e > 12$. The lifetime of the droplet in this mode is given as

$$\tau_b = (\pi\, \rho_d^{1/2} D_d^{3/2})/(4\sigma_d^{1/2}) \quad (1)$$

(b) Stripping Break Up: The liquid droplet is sheared or stripped from the droplet surface due to the large amplitude waves of small or large wavelengths. At high amplitudes, this is called as catastrophic break up. This mode happens typically when Weber number satisfies the condition,

$$We/Re^{0.5} = 0.5 \quad (2)$$

The lifetime of the droplet in this mode

$$\tau_b = (20\rho_d^{1/2} D_d)/(2u_{rel}\rho_g^{1/2}) \quad (3)$$

Spray impingement model is formulated within the framework of the Lagrangian approach to reflect the stochastic nature of the impingement process. A random procedure is adopted to determine the droplet post-impingement quantities. This allows secondary droplets resulting from a primary droplet splash to have a droplet size and droplet velocity distributions.

Table 1 lists the boundary conditions applied to the in-cylinder fluid domain. The STAR-CD (CD Adapco, 2010) code computes by discretizing the fluid domain using finite volume approach under implicit formulation mode. The PISO algorithm (Pressure Implicit by Splitting of operators) is used to provide pressure velocity coupling to compute the flow-field and other transport equations. Second order upwind discretization schemes are chosen for computing the conservation equations of mass, momentum and energy.

Table 1 Boundary Conditions

Boundary	Momentum boundary condition	Thermal boundary Condition
Cylinder head	Wall	450 K
Cylinder wall	Wall	400 K
Piston bowl	Moving wall	450 K
Cylinder side face	Periodic	450 K

The ECFM-3Z combstion model (Colin and Benkenida 2004) is chosen for the simulation of complex mechanisms like turbulent mixing, flame propagation, diffusion combustion and pollutant formations. A small amount of exhaust gas is mixed with fresh air and then introduced into the combustion chamber. This modifies the fuel/air ratio and lowers the peak temperature so that the chemical reaction rate between nitrogen and any unused oxygen is strongly reduced. Species concentrations involved in combustion reactions can be written as a function of mixture fraction within the presumed probability density function model of combustion.

Table 2 lists the models accompanied in the code for simulation. The liquid film model (Bai and Gosman 1996) accounts for convective transport of conserved quantities within the film and from/to the gas phase. The standard pool boiling (Rohsenow 1952) is used to model liquid film boiling, when the wall temperature exceeds the saturation temperature of the liquid as the film starts to boil when the heat flux passes from the wall to the film.

2.3 POST-PROCESSING

Time accurate computations are allowed till the residual values of the conservation equations of continuity, momentum and energy reach 10^{-5}. Auxiliary equations involving the turbulence, spray models and models for combustion and soot emissions are also computed at every time step. Once a time step is completed, the code outputs the in-cylinder averaged data such as pressure, temperature, heat release rate, NO_x and soot emissions to an ASCII file output for further analysis. The contours of the same quantities are also obtained by storing the information at preset crank-angles.

3. VALIDATION

Table 3 lists the specification of engine dimensions, injection timing and combustion parameters. The 45° sector CFD model of Colin and Benkenida (2004) experimental engine cylinder is modeled and a series of grid and time independency tests are carried out as shown in Fig. 2. Crank angle step interval of 0.025° CA (i.e. 4.167 x 10^{-6} seconds) and mesh with 45000 cells at TDC position are obtained as key information for further simulation from these tests. Validation of the current simulation work is carried out with the experimental pressure data of Colin and Benkenida (2004) from the literature.

Figure 4 shows the comparison of the simulation results with the experimental in-cylinder pressure under firing conditions. The computed in-cylinder pressure data from numerical simulation are in good agreement with the experimental data. The in-cylinder averaged pressure during the non-firing mode of simulation is also shown in Fig. 4. The numerically simulated pressure values are in close agreement with the experimental data and the maximum deviation in peak pressure is less than 0.2%.

Table 2 Models accompanied in code

Phenomena	Model
Droplet breakup	Reitz-Diwakar (1986)
Turbulence	RNG k-ε model (Tahry 1983)
Combustion	ECFM-3Z (Colin and Benkenida 2004)
Liquid Film	Angelberger et al (1997)
Droplet wall interaction	Bai and Gosman (1996)
Atomization	Huh (1991)
Boiling	Rohsenow (1952)
NO_x mechanism	Hand (1989), De Soete (1975)
Soot	Mauss (2006)

Table 3 Engine specification

Bore	0.085 m
Stroke	0.088 m
Compression ratio	18
Connecting Rod Length	0.145 m
Valves/Cylinder	4
Engine Speed (N)	1640 RPM
Fuel	n-Dodecane
Start of injection (deg bTDC)	6.0
Injection duration (deg.)	8.03
Injected mass (g)	0.0144
F/A equivalence ratio	0.67
EGR rate (%)	31
Swirl ratio (SR)	2.8
Injector hole diameter	148 x 10^{-6} m
Spray angle	152 deg
Intake valve opening (lift at 0.5 mm)	360 deg (TDC)
Intake valve closing (lift at 0.5 mm)	54 deg
Exhaust valve opening (lift at 0.5 mm)	860 deg

4. RESULTS AND DISCUSSION

The parameters such as in-cylinder temperature, heat release rate, NO_x and soot emissions are predicted numerically for the same geometry of Beard and Colin (2003). The in-cylinder temperature increases till 736 deg CA due to diffusion combustion and thereafter decreases as expected. It is found that the peak temperature during the simulation reaches nearly 1700 K at nearly 740° CA.

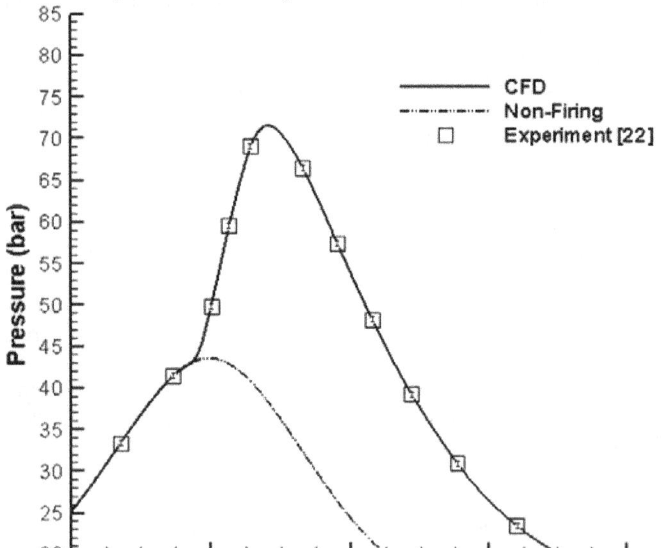

Fig. 4 Comparison of computed and experimental pressures with crank angle

The in-cylinder heat release rate curve rises after 716 deg CA steeply due to the rapid rise in pressure. During this period, the mixture may be homogeneous such that premixed combustion can happen. Due to the sudden rise in pressure, the firing inside the cylinder leads to uncontrolled combustion. After the peak heat release, the combustion is controlled due to diffusion between air and fuel particles.

The soot level rises up earlier than 720 deg CA while NO_x emissions rise little later than 720 deg CA. The NO_x emissions are found to be higher than soot emissions. Nitrogen oxides are strongly dependent on temperature (primary dependence), oxygen concentration and duration of combustion. NO_x is mainly formed during the diffusion rather than the premixed phase of combustion. Soot is formed from unburned fuel that nucleates from the vapor phase to a solid phase in fuel-rich regions at elevated temperatures. Hydrocarbons or other available molecules may condense on, or be absorbed by soot depending on the surrounding conditions.

The numerical tool thus employed here, is able to predict the various engine parameters like engine temperature, heat release rate and emissions for every degree of crank angle. This study mainly concentrates on the effect of fuel droplet mass distribution, droplet diameter and spray penetration which includes physical processes like atomization, mixing, evaporation and boiling phenomena, which are very cumbersome to measure and record experimentally. The fuel droplet traces a nearly linear path from the time of formation, often breaking and coalescing with other drops in the neighbourhood. The coalescence is however not applicable to the drops on the outer envelope of spray because the droplets are formed first and hence do not interact with other droplets on the outside. The trajectory and breakup of droplet depends on ambient pressure, neighbourhood velocity too. Break-up of these drops is negligible if the drops are small as in high-pressure sprays. Thus, the droplets on the spray surface can be said to reduce in size only by vaporisation.

Evaporation of fuel depends on the temperature and relative velocity between droplet and continuous phase medium. The aerodynamic forces on a droplet depend on droplet mass. As a result, smaller droplets undergo more rapid acceleration than larger droplets. Heating times and vaporization times will be shorter for smaller droplets. The liquid sheet disintegration or atomization typically results in liquid ligaments or droplets with a characteristic dimension that is smaller than the original length scale associated with the stream. Disintegration will continue in a cascade fashion until the decreased length scale brings the Weber number for the resulting droplets below the critical value for the droplets. From the start of injection to the combustion period considered in this study, it can be observed that the droplet mass and diameter increase initially due to coalescence and later the break-up involves the mass of the droplet to decrease later 720 deg CA. The Sauter mean diameter of the drops decreases as a consequence of increasing aerodynamic interactions (increasing the relative velocity) between liquid fuel ligaments or bigger drops and the surrounding fluid medium. The increase in SMD at short peaks may be due to rise in ambient pressure in cylinder.

The peak droplet temperature is obtained nearly a few degrees of CA after TDC due to heat transfer from the surrounding fluid medium. The droplet temperature lowers thereafter due to evaporation and heat transfer from the droplet to surrounding medium. Droplet velocity is maximum at 3 deg before TDC. Higher droplet velocities assist the droplet to reach the end of bowl and also help in shearing or breakup of droplets. This results in greater penetration of fuel in the bowl. Droplet velocity increases initially due to higher momentum and later decreases because of the rise in in-cylinder pressure. The fluctuations in velocity and spray penetration are attributed to the turbulent nature of flow-field in the in-cylinder volume. The numerical simulation is able to predict the fuel spray characteristics, droplet diameter which is very cumbersome to measure by experimental techniques for every degree of crank angle rotation. Thus numerical study of in-cylinder engine characteristics provides a better understanding of actual physical process involved in spray distribution, mixing and combustion processes.

5. PARAMETRIC STUDIES

As the studies on droplet parameters gave fruitful information on the combustion and emission characteristics, the study is continued further to understand the flow physics and combustion phenomena in the cylinder by varying the swirl ratio and injection timing . For both of these cases, the engine dimensions and boundary conditions are considered to be same as in Table 3. The swirl ratio is varied from 1.4 to 4.1 and injection timing is varied between 6 deg bTDC to 20 deg bTDC.

5.1. EFFECT OF THE SWIRL RATIO (SR)

The swirl inside the cylinder is varied by changing the piston bowl profile as listed in the literature (Prasad et al 2011, Colin and Benkenida 2004). The bowl shape is carefully chosen (Beard and Colin 2003) to obtain the desired swirl ratio. Five cases of piston bowl are created and swirl ratio is varied as 1.4, 2.3, 3.2, 4.1 and 4.5. The swirl ratio reported here, is computed to be the highest at the end of compression stroke. Swirl enhances the mixing of air and fuel in the cylinder and therefore the combustion efficiency can be increased further. As swirl ratio is increased in the engine cylinder, the in-cylinder pressure and temperature are increased due to better fuel mixing with surrounding air and better combustion with higher heat release rates. Fig. 5 shows the peak in-cylinder average pressure rises from 73 bar to 81 bar as swirl ratio is increased from 1.4 to 4.1. However, the peak pressure falls to 76 bar when the swirl ratio is further increased to 4.5. The in-cylinder turbulence can be increased at higher swirl ratio of 4.5 and this leads to decrease in the cylinder temperature. As the peak pressure is highest at a swirl ratio of 4.1 amongst the cases considered, the in-cylinder and droplet variables are compared between swirl ratios of 1.4 to 4.1. The timing of maximum pressure or peak pressure inside the cylinder occurs nearly at 727 deg CA. The peak in-cylinder averaged temperature increases

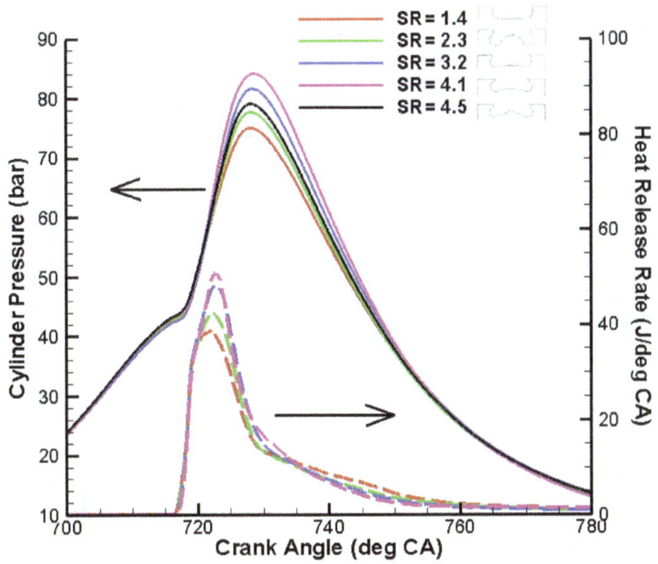

Fig. 5 Variation of pressure and heat release rate with crank angle for different swirl ratio

The heat release rates for increasing swirl ratio are also plotted in Fig. 5. It can be understood that there is 37 % increase in heat release rate when swirl ratio is increased from 1.4 to 4.1. Table 4 shows ignition delay is higher at lower swirl ratio.

Table 4 Ignition delay for various swirl ratios

Swirl Ratio	Ignition Delay (deg)
1.4	4.650
2.3	4.225
3.2	3.750
4.1	3.275
4.5	3.125

from 1667 K to 1808 K when swirl ratio is increased from 1.4 to 4.1. The ratio of change in temperature and pressure between swirl ratio matches nearly with the literature (Prasad et al 2011). The in-cylinder temperature increases with the swirl ratio.

It is also to be considered that higher swirl ratio leads to lower ignition delay in both the models due to reduced physical delay period. Ignition delay period is calculated as the difference between the start of injection timing and the start of auto-ignition for every simulation case. Although the ignition delay is lowered, the presence of better re-entrant piston bowl geometry (to account for higher swirl ratio) leads to better mixing of fuel with air followed by combustion and thereby heat release is maximum for swirl ratio of 4.1. The cumulative heat release is computed and increases with swirl ratio as 589.91 J, 603.64 J, 621.79 J, and 625.84 J for swirl ratios of 1.4, 2.3, 3.2 and 4.1 respectively.

The NO_x emissions are compared for various swirl ratio as shown in Fig. 6. Since the temperature in the cylinder increases with the swirl ratio, the NO_x emission levels are also observed to be higher. As swirl ratio increases from 1.4 to 4.1, the NO_x levels increases from 4.4 g/kg of fuel to 8.6 g/kg of fuel respectively. The soot emissions exhibit reverse trend with the NO_x emissions as in Fig. 6. The soot levels decrease with the increase in swirl ratio from 1.4 to 4.1 due to better mixing of fuel and air, leading to lower fuel accumulation and deposition. Although the soot levels increases with time, the overall soot level reduction is 21% from swirl ratio of 1.4 to 4.1.

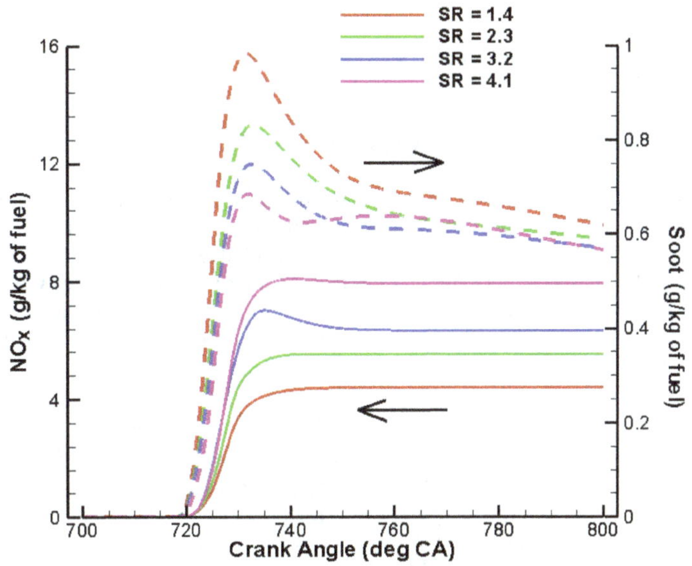

Fig. 6 Variation of NO_x, soot for different swirl ratio

The droplet parameters are studied by considering the increase in swirl ratio from 1.4 to 4.1. The increase in swirl ratio leads to the additional break-up of droplets and interaction between the surrounding air and droplet is increased at higher swirl ratio. This leads to the break-up of droplets due to shear. As the droplet break up continues till the start of combustion, the resulting droplets that are not involved in primary combustion exhibit a relative change in diameter of the droplet after 725 deg CA. Evaporation is followed by final stage of droplet break up, leading to the reduction of droplet mass. Sauter mean diameter increases for swirl ratios of 3.2 and 4.1 twice as compared to swirl ratios of 1.4 and 2.3 due to the chance of coalescence at higher swirl ratio as shown in Fig. 7.

The reason for lower SMD for swirl ratios of 3.2 and 4.1 is due to the higher relative velocity caused by the swirl ratio for 3.2 and 4.1, the droplet SMD gets affected in the same time period. The SMD falls once the diffusion combustion initiates.

The droplet temperature is observed to be highest for swirl ratio of 4.1 as shown in Fig. 7. This is due to the higher in-cylinder temperature at highest swirl ratio of 4.1. Heat transfer from surrounding air to the droplet is significant at higher in-cylinder temperatures for the swirl ratio of 4.1. From the computations, it is observed that the spray penetration becomes insignificant as the ambient gas pressure increases. As the ambient gas pressure increases the pressure drop across the nozzle decreases and so the spray penetration also decreases.

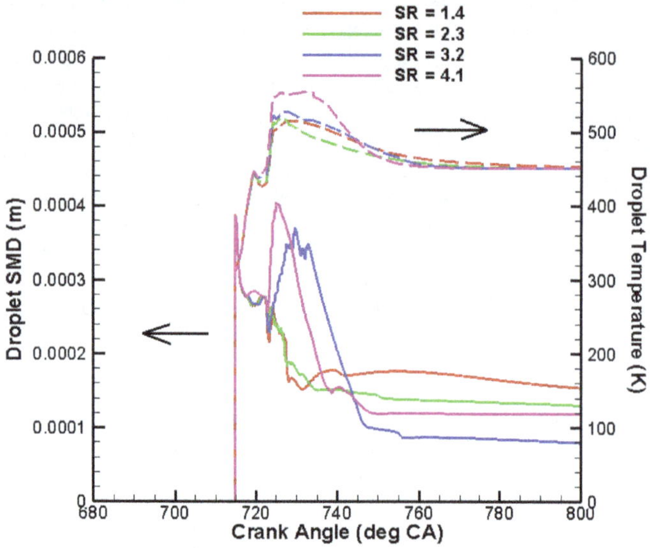

Fig. 7 Variation of droplet SMD and droplet temperature for different swirl ratio

5.2. EFFECT OF THE INJECTION TIMING (IT)

The effect of variation in injection timing is carried out numerically with the optimized swirl ratio of 4.1. The injection timings considered are 6 deg before TDC (or) 714 deg CA, 13 deg bTDC (or) 707 deg CA,20 deg bTDC (or) 700 deg CA and 27 deg bTDC (or) 693 deg CA.

The in-cylinder averaged pressure is shown in Fig. 8 for different injection timing. The delay period for every case of injection timing are given in Table 5.

Table 5 Delay period for different injection timings

Injection Timing (deg CA)	Delay Period (deg CA)
714	3.275
707	6.150
700	7.275
693	8.125

Advancing the injection timing with respect to TDC results in increase in the cylinder pressure due to increased delay period. The in-cylinder pressure and temperature is higher for the case when the time of start of injection is 20 deg bTDC as compared to the case of 6 deg bTDC. This leads to a 15 % increase in cylinder pressure from injection timing of 6 deg bTDC to 20 deg bTDC. These numerically simulated values are in good agreement with Jayashankara and Ganesan (2010). The delay period is almost doubled when the injection timing is varied from 6 deg bTDC to 13 deg bTDC, resulting in increase of peak cylinder pressure from 81 bar to 89 bar. However, the peak pressure falls to 86 bar as injection timing is ther advanced to 693 deg CA i.e. 27 deg bTDC. This is because the in-cylinder temperature might not be sufficient at 693 deg CA for the fuel-air mixture to attain auto-ignition and thereby leads to a slightly lower peak pressure. Hence, the in-cylinder and droplet variables are compared between 6 to 20 deg bTDC injection timings. As in-cylinder peak pressure is higher with injection timing of 20 deg bTDC, the in-cylinder peak temperature is also higher at the same injection timing. It is observed that there is an increase in cylinder temperature with the advancement of injection timing till 700 deg CA and thereafter the temperature decreases.

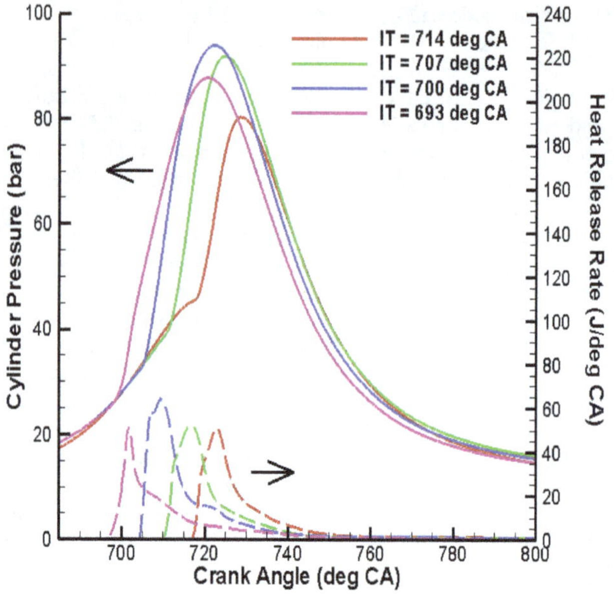

Fig. 8 Computed pressure, heat release rate with crank angle for different injection timings

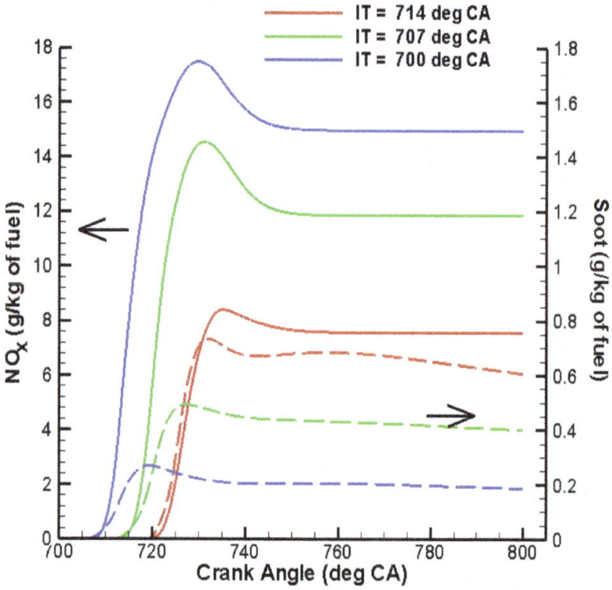

Fig. 9 Computed NO_x , soot for different injection timings

The ignition delay period is longer as the injection timing is advanced since the required in-cylinder pressure and temperature are not sufficient to start the auto-ignition process. The delay period is listed against the injection timing in Table. 5. The heat release rates at three injection timings are also shown in Fig. 8. The peak heat release rate is observed to be highest with injection timing of 700 deg CA than the remaining cases. The slope of the rising curve is highest at injection timing of 700 deg CA, and thereby the heat release rate is rapid during this uncontrolled combustion period. The cumulative heat release increases with advancing the injection timing. These values are found to be 625.84 J, 667.61J and 687.83 J at injection timings of 714 deg CA, 707 deg CA and 700 deg CA respectively.

The NO_x and soot emissions are shown in Fig. 9. As the in-cylinder temperature is higher at injection timing of 700 deg CA, the NO_x emissions are higher for the same case. It is observed that the NO_x levels increases nearly twice between the injection timing of 714 deg

CA and 700 deg CA respectively. The soot levels are observed to decrease with the advance in injection timing due to reduction in droplet diameter and longer ignition delay period. The soot levels decreases nearly by one-third when the injection timing is advanced to 700 deg CA from 714 deg CA. The variation of soot emissions as observed in Fig. 9 are in correspondence with the droplet diameter of Fig. 10.

The droplet mass and droplet diameter increases with advancement of injection timing. Due to the increase in delay period with the advance in injection timing, the droplet undergoes break-up to a greater extent and hence overall mass of the droplet can be lowered as compared to the other injection timings considered. The same trend is observed with the droplet SMD, whereas another rise in peak occurs for later injection timing. This is because the relative velocity between the fuel droplet and surrounding air is lower. This is verified by observing the droplet velocity variation after the computations. Higher droplet velocities are obtained for an injection timing of 20 deg CA bTDC (or) 700 deg CA. This is due to lower in-cylinder pressure at the time of injection and thereby higher drag force is experienced by the droplet as compared to other injection timings.

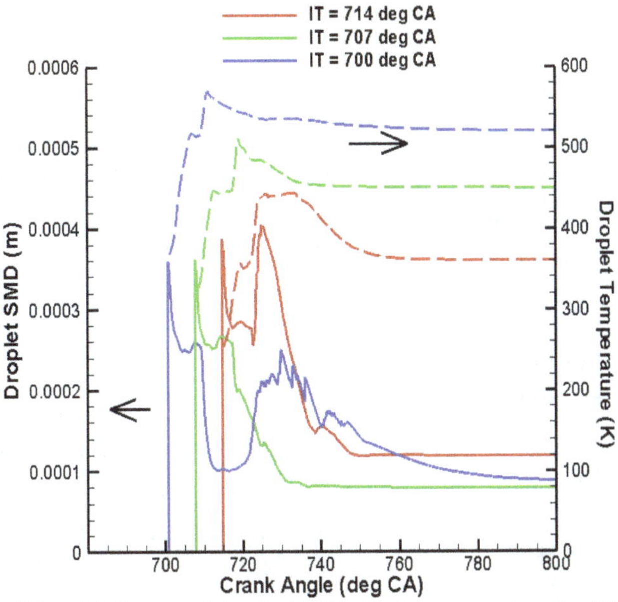

Fig. 10 Computed droplet SMD, droplet temperature for different injection timing

The droplet temperature is observed to be maximum when the injection timing is advanced as shown in Fig. 10. This is due to higher in-cylinder temperature during combustion, caused by longer delay period. Heat transfer due to combustion increases the droplet temperature further. The spray penetration is affected by the in-cylinder pressure as discussed earlier and the spray penetration is maximum with higher injection angle of 700 deg CA or 20 deg CA bTDC. This is because the resistance offered by air pressure inside the cylinder is reduced when the injection timing is advanced.

6. CONCLUSIONS

In the present work, different models governing the direct injection diesel engine combustion and pollutant formation are studied. Grid and time independent tests are carried out and the results are validated with the literature experimental data. In-cylinder flow-field, temperature and heat release rate are investigated. The variation of droplet parameters such as droplet mass, droplet diameter, droplet velocity, droplet temperature and spray penetration are also studied, however few are reported here. The analyses are extended towards understanding the droplet behavior, combustion and pollutant formation by varying the in-cylinder swirl ratio and injection timing. From the results, the following conclusions are obtained.

1. When the swirl ratio is increased from 1.4 to 4.1, the peak in-cylinder pressure increases by 8 bar thereby resulting in better combustion. The peak pressure falls by 5 bar as swirl ratio is increased beyond 4.1.
2. Heat release rate occurs nearly at 722 deg CA and increases by 37 % when swirl ratio is increased from 1.4 to 4.1.
3. Due to higher temperature the NO_x emissions are doubled, while soot emissions are halved when the swirl ratio is increased to 4.1 from 1.4. Decrease in soot levels occur at lower Sauter Mean Diameter.
4. Advancing the injection timing leads to increase in in-cylinder averaged quantities like pressure and temperature considerably. The pressure rise is 15 % over the injection timing advancement of 14 deg CA. This is due to longer ignition delay period as the in-cylinder pressure and temperature is not sufficient at the end of compression stroke for the fuel-air mixture to attain auto-ignition. There is a decrease in peak pressure and heat release rate when the injection timing is advanced to 27 deg bTDC.
5. Heat release increases by 40% by advancing the injection timing from 714 deg CA to 700 deg CA. This results in better combustion due to prolonged combustion period and ignition delay period.
6. Nitrogen oxides and soot emissions show inversing trend with the advancement of injection timing. NO_x levels are doubled and soot emissions are decreased by one-third from 714 deg CA to 700 deg CA.
7. Droplet parameters are studied and found to affect the combustion process and emission formation significantly by varying the swirl ratio and injection timings.

NOMENCLATURE

bTDC	before Top Dead Centre
BDC	Bottom Dead Centre
CA	Crank Angle
CFD	Computational Fluid Dynamics
D	Diameter
HC	Hydrocarbons
IT	Injection timing
J	Joule
k	Turbulent kinetic energy
NO_x	Oxides of nitrogen
SR	Swirl ratio
SMD	Sauter Mean Diameter
T	Break-up time
TDC	Top Dead Centre
u	Velocity

Greek symbols

ε	Turbulent eddy dissipation rate
μ	Dynamic viscosity
ρ	Density
σ	Surface tension
τ	Life time

Suffixes

b	breakup
d	droplet
g	gas phase (Eulerian)
rel	relative

REFERENCES

Heywood J. B., 1988, *Internal Combustion Engine Fundamentals*, McGraw-Hill, New York, 491-566.

Reitz R.D., Diwakar R., 1986, "Effect of Drop Breakup on Fuel Sprays," *SAE Technical Paper Series* 860469. http://dx.doi.org/10.4271/860469

Magnussen B.F., Hjertager B.H., 1976, "On Mathematical Modeling of Turbulent Combustion with Special Emphasis on Soot Formation and Combustion," *16th Symp. on Combustion, The Combustion Institute*, 719-729. http://dx.doi.org/10.1016/S0082-0784(77)80366-4

Hossainpour S., Binesh A.R., 2009, "Investigation of fuel Spray Atomization in a DI Heavy-Duty Diesel Engine and Comparison of Various Spray Breakup Models," *Fuel* **88**, 799-805. http://dx.doi.org/10.1016/j.fuel.2008.10.036

Prasad B.V.V.S.U., Sharma C.S., Anand T.N.C., Ravikrishna R.V., 2011, "High Swirl-inducing Piston Bowls in Small Diesel Engines for Emission Reduction," *Applied Energy* **88,** 2355-2367. http://dx.doi.org/10.1016/j.apenergy.2010.12.068

Arcoumanis C., Gavaises M., French B., 1997, "Effect of Fuel Injection On The Structure Of Diesel Sprays," *SAE Paper* 970799. http://dx.doi.org/10.4271/970799

Rakopoulos C.D., Kosmadakis G.M., Pariotis E.G., 2010, "Investigation of Piston Bowl Geometry and Speed Effects in a Motored HSDI Diesel Engine Using A CFD Against a Quasi-Dimensional Model," *Energy Conversion and Management* **51-3**, 470-484. http://dx.doi.org/10.1016/j.enconman.2009.10.010

Kar M.P., Hoon K.N., Gan S., 2012, "Simulation of Temporal and Spatial Soot Evolution in an Automotive Diesel Engine Using the Moss–Brookes Soot Model," *Energy Conversion and Management* **58**, 171-184. http://dx.doi.org/10.1016/j.enconman.2012.01.015

Torregrosa A.J., Bermúdez V., Olmeda P., Fygueroa O., 2012, "Experimental Assessment for Instantaneous Temperature and Heat Flux Measurements under Diesel Motored Engine Conditions," *Energy Conversion and Management* **54-1**, 57-66. http://dx.doi.org/10.1016/j.enconman.2011.10.009

Thurnheer T., Edenhauser D., Soltic P., Schreiber D., Kirchen P., Sankowski A.,2011, "Experimental Investigation On Different Injection Strategies in A Heavy-Duty Diesel Engine: Emissions and Loss Analysis," *Energy Conversion and Management* **52-1**, 457-467. http://dx.doi.org/10.1016/j.enconman.2010.06.074

Costa M., Sorge U., Allocca L., 2012, "Increasing Energy Efficiency of a Gasoline Direct Injection Engine Through Optimal Synchronization of Single or Double Injection Strategies," *Energy Conversion and Management* **60** , 77-86. http://dx.doi.org/10.1016/j.enconman.2011.12.025

Chmela F.G., Pirker G.H., Wimmer A., 2007, "Zero-Dimensional ROHR Simulation for DI Diesel Engines – A Generic Approach," *Energy Conversion and Management* **48(11)**, 2942-2950. http://dx.doi.org/10.1016/j.enconman.2007.07.004

Catania A.E., Finesso R., Spessa E., 2011, "Predictive Zero-Dimensional Combustion Model for DI Diesel Engine Feed-Forward Control," *Energy Conversion and Management* **52(10)**, 3159-3175. http://dx.doi.org/10.1016/j.enconman.2011.05.003

Kondoh T., Fukumoto A., Ohsawa K., Ohkubo Y., 1985, "An Assessment of a Multidimensional Numerical Method to Predict the Flow in Internal Combustion Engines," *SAE Paper* 850500. http://dx.doi.org/10.4271/850500

Cenk Sayin, Mustafa Canakci, 2009, "Effects of Injection Timing on the Engine Performance and Exhaust Emissions of A Dual-Fuel Diesel Engine," *Energy Conversion and Management*, **50**, 203–213. http://dx.doi.org/10.1016/j.enconman.2008.06.007

Han Z., Uludogan A., Hampson G.J., Reitz R.D., 1996, "Mechanism of Soot and NOx Emission Reduction Using Multiple-Injection in a Diesel Engine," *SAE* paper no. 960633. http://dx.doi.org/10.4271/960633

Jayashankara B., Ganesan V., 2010, "Effect of Fuel Injection Timing and Intake Pressure on the Performance of A DI Diesel Engine – A Parametric Study Using CFD," *Energy Conversion and Management* **51**, 1835–1848. http://dx.doi.org/10.1016/j.enconman.2009.11.006

Payri F., Benajes J., Margot X., Gil A., 2004, "CFD Modeling of the In-Cylinder Flow in Direct-Injection Diesel Engines," *Computers and Fluids* **33**, 995–1021. http://dx.doi.org/10.1016/j.compfluid.2003.09.003

Béard P., Colin O., Miche M., 2003, "Improved Modeling of DI Diesel Engines Using Sub-Grid Descriptions of Spray and Combustion," *SAE Paper* 2003-01-0008. http://dx.doi.org/10.4271/2003-01-0008

O"Rourke P.J., 1981, "Collective Drop Effects on Vaporising Liquid Sprays," *PhD Thesis*, University of Princeton.

Munnannur A, 2007, "Droplet Collision Modeling in Multi-Dimensional Engine Spray Computations," Ph.D. Dissertation, University of Wisconsin-Madison.

El Tahry S.H., 1983, "k-ε Equation for Compressible Reciprocating Engine Flows," *AIAA J. Energy* **7**, 345–353. http://dx.doi.org/10.2514/3.48086

Colin O., Benkenida A., 2004, "3-Zone Extended Coherent Flame Model (ECFM3Z) for Computing Premixed/Diffusion Combustion," *Oil and Gas Science and Technology – Rev. IFP*, **59**,593-609.

Angelberger C., Poinsot T., Delhay B., 1997, "Improving Near-Wall Combustion and Wall Heat Transfer Modeling in SI Engine Computations," *SAE Technical Paper Series* 972881, 113-130. http://dx.doi.org/10.4271/972881

Bai C., Gosman A.D., 1996, "Mathematical Modeling of Wall Films Formed by Impinging Sprays," *SAE Technical Paper Series* 960626. http://dx.doi.org/10.4271/960626

Huh K.Y., Gosman A.D., 24-27 September 1991, "A Phenomenological Model of Diesel Spray Atomisation," *Proc. Int. Conf. on Multiphase Flows (ICMF "91)*, Tsukuba.

Rohsenow, W.M., 1952, "A Method of Correlating Heat Transfer Data for Surface Boiling Liquids," *Transactions of the ASME*, **74**, 969.

STAR methodology for internal combustion engine applications, 2010, CD-adapco version 4.16.

Manimaran R., Thundil Karuppa Raj R., 2013, "Numerical Investigations on Combustion and Emission Characteristics in a Direct Injection Diesel Engine at Elevated Fuel Temperatures," *Frontiers in Heat and Mass Transfer*, **4**, 013008.
http://dx.doi.org/10.5098/hmt.v4.1.30081871

Hand G., Missaghi M., Pourkashanian M., Williams A., 1989, "Experimental Studies and Computer Modelling of Nitrogen Oxides in a Cylindrical Natural Gas Fired Furnace," *9th Members Conf., International Flame Research Foundation*, Noordwijkerhout, The Netherlands .

De Soete G.G., 1975, "Overall Reaction Rates of NO and N2 Formation from Fuel Nitrogen," 15th Symp. (Int.) on Combustion, *The Combustion Institute*, 1093-1102.

Mauss F., Netzell K., Lehtiniemi H., 2006 "Aspects of Modeling Soot Formation in Turbulent Diffusion Flames," *Combust. Sci. and Tech.* **178,** 1871-1885.
http://dx.doi.org/10.1080/00102200600790888

LARGE EDDY SIMULATION OF THE DIFFUSION PROCESS OF NUTRIENT-RICH UP-WELLED SEAWATER

Shigenao Maruyama[a], Masud Behnia[b], Masasazumi Chisaki[c], Takuma Kogawa[c,*], Junnosuke Okajima[a], and Atsuki Komiya[a]

[a] *Institute of Fluid Science, Tohoku University, Katahira, Aoba-ku, Sendai 980-8577, Japan*
[b] *School of Aerospace, Mechanical and Mechatronic Engineering, The University of Sydney, NSW 2006,Australia*
[c] *School of Engineering, Tohoku University, Aoba 6-6, Aramaki-aza, Aoba-ku, Sendai 980-8579, Japan*

ABSTRACT

The diffusion process of deep seawater drawn up by a vertical pipe deployed in the ocean is investigated. This vertical pipe is based on the principal of perpetual salt fountain. Numerical simulations of seawater upwelling from the pipe are performed based on experiments conducted in the Mariana trench region. Two turbulence modeling approaches were examined: *k-ε* model and Large Eddy Simulations (LES). The results in both models show that diffusion of the deep seawater diffusion after ejection from the pipe. The LES results show a 50% lower vertical penetration compared to the *k-ε* model as well as well as predicting that the horizontal diffusion is stronger than the vertical one.

Keywords: *Perpetual salt fountain, Deep seawater, Nutrient transport, Eddy diffusion*

1. INTRODUCTION

A rapid growth of world's population has led to a need for more food, however an expansion of farm production has been somewhat limited. In contrast, there is a large ocean area which area that has not been used for food production. Increasing of food production in this area needs to be explored. The upwelling of deep seawater provides the euphotic surface layer with nutrients of deeper ocean, resulting in an increase in the oceanic productivity. A large area of the ocean, except some upwelling regions and high latitude domains, is characterized by a low productivity where low nutrient levels of the surface water prevent phytoplankton growth. Therefore, this large area does not play a role in supplying biological and fisheries resources.

Maruyama *et al.* (2004) have proposed a concept for increasing of food production by artificial upwelling of the deep seawater using a perpetual salt fountain (Stommel *et al.*, 1956). Perpetual salt water fountain is a principle which allows bringing the deep seawater to the surface,(note that which "deep" seawater is defined as the seawater deeper than 200 m containing rich nutrients such as NO_3 and PO_4 Sunlight does not reach this depth, preventing phytoplankton from its photosynthesis and thus its growth. When a vertical pipe is deployed in the ocean where the hydrographic structure is characterized by a salinity minimum and filled with the deep seawater less saline than upper layers, the heating by outside warm water results in a buoyancy force leading to an upwelling flow in the pipe (see Fig. 1). Since this upward flow is accompanied with an uptake of less saline deep seawater into the pipe, the artificial upwelling is perpetually maintained. One of the advantages of this upwelling method is that it does not require any energy input except for the initial filling of the pipe with the deep seawater. This mechanism can be adopted in a large tropical and subtropical region where the minimum salinity layer is observed at depths of 300 to 600 m (Reid, 1965; Talley, 1993). This salinity minimum is usually referred to as "intermediate water" in oceanography (Reid, 1965; Talley, 1993, 1999, 2003; Schmitz, 1995, 1996a, 1996b; Yasuda *et al.*, 1996).

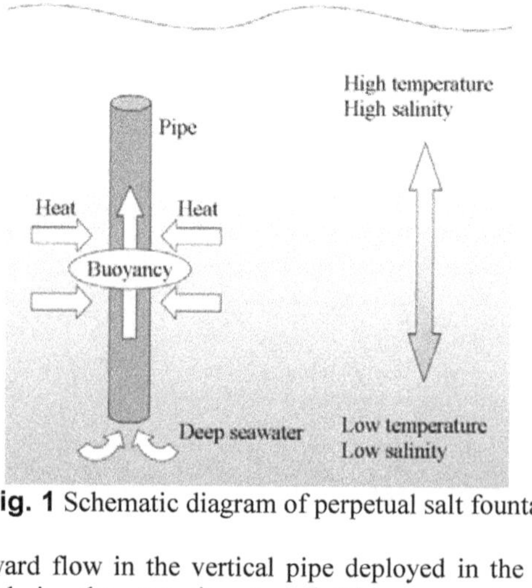

Fig. 1 Schematic diagram of perpetual salt fountain

The upward flow in the vertical pipe deployed in the ocean was first observed in the experiments of Maruyama *et al.* (2004), demonstrating the theoretical study of Stommel *et al.* (1956). The experiments were conducted in the Mariana Trench region in the Pacific Ocean (location coordinates: 11.43°N, 142.42°E). Figure 1 shows the composite image of surface chlorophyll concentration around the upwelling pipe at the experiment in 2005. As shown in Fig. 1, it was observed that the chlorophyll concentration at the pipe outlet was about 100 times larger than that in the surrounding surface seawater (Maruyama *et al.*, 2011, POPULAR SCIENCE, 2011). In addition, Maruyama *et al.* (2011), Zhang *et al.* (2004 and 2006) conducted the

* *Corresponding author. Email: takuma@pixy.ifs.tohoku.ac.jp*

numerical simulation inside the upwelling pipe and Maruyama *et al.* (2011) estimated upward flow velocity in the pipe was 2.45 mm/s (212 m/day). Similar values were also earlier obtained by Tsubaki *et al.* (2007). Figure 2(a),(b) show the simulated trajectories and simulated upwelling velocities in each depth (Maruyama et al. 2011). As shown in this Fig. 2, numerical simulation predicted the reverse flow at the central section of pipe. Zhang *et al.* (2004) predicted that the deep seawater descends 10 m after the ejection from the pipe outlet. In consideration of the practical use of the upwelling flow, one should focus on the nutrients, and its diffusion process from the pipe outlet to the ocean surface layers. It is known that in the ocean, turbulent diffusion is much larger than the molecular diffusion (Ledwell *et al.*, 1998), therefore diffusion should be treated as turbulent diffusion in a numerical simulation.

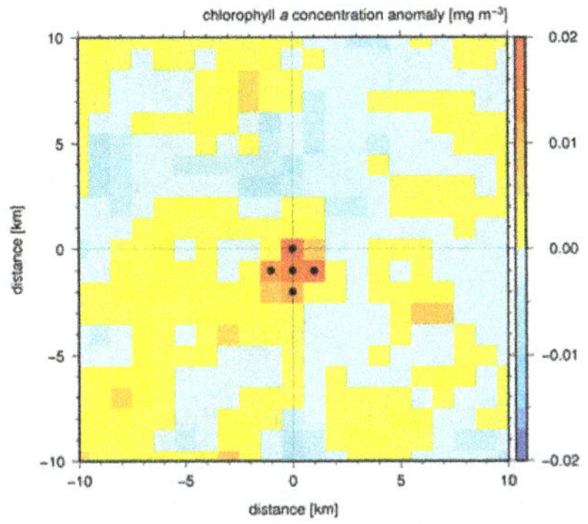

Fig. 1 Composite image of surface chlorophyll concentration anomaly around floating pipe system (Dots represent the grid at which signal is statistically significant level of 1%.) Maruyama *et al.* (2011)

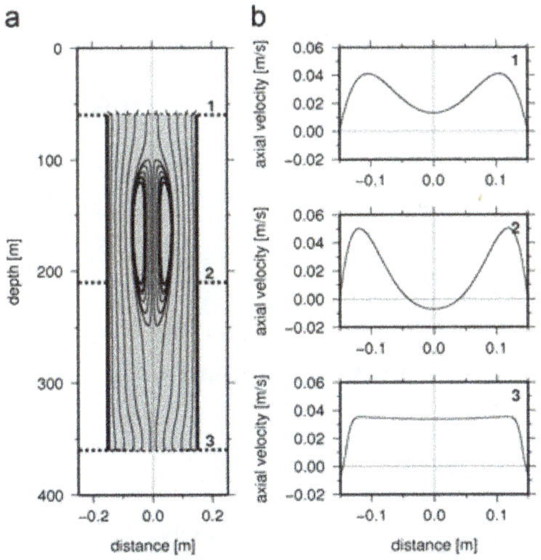

Fig. 2 (a) Simulated trajectories and (b) simulated upwelling velocities inside the pipe. The inside areas of the pipe are illustrated as the shaded area of (a). Upwelling velocities at 3 sections (1: outlet, 2: central section, 3: inlet) are shown. The location of 3 sections are labeled in (a). Maruyama *et al.* (2011)

In order to estimate the impact of the deep seawater fountain on increasing the oceanic productivity, it is necessary to predict the diffusion process of the deep seawater ejecting from the pipe outlet. Williamson *et al.* (2009) simulated the process of outflow from the pipe deployed in the Mariana trench by adopting a k-ε turbulence model (Launder and Spalding, 1973), which assumes isotropic turbulence viscosity. Their k-ε simulations were carried out with various turbulent Schmidt and Prandtl numbers. Their results showed that the horizontal advection is dominant in the diffusion process of the nutrients. However, turbulent Schmidt and Prandtl numbers had little effect on the results. Further, they noted that the assumption of isotropic turbulence viscosity caused an overestimation of the vertical diffusion. The oceanic diffusivity is still a topic of contemporary oceanography, however, it is well-known that the horizontal diffusion is much larger than the vertical diffusion (Ledwell *et al.*, 1998; Toole and McDougall, 2001). Therefore, this requires that in numerical simulations the turbulence anisotropy is taken into account. Hence, one could argue that the Direct Numerical Simulation (DNS) will be the best way to overcome this problem. Of course, DNS is computationally very expensive, in particular when both the spatial and temporal scales of the calculation is large, such as the present case.

Previously k-ε results have been shown not to reproduce the physical phenomena. Therefore in order to check and see if the turbulence model has been the cause of this discrepancy, a more sophisticated LES would have been tested. LES model (Smagorinsky, 1963; Liu *et al.*, 1997; Yuan *et al.*, 1999; Lewis, 2005) is a less computationally intensive approach than DNS. Here, we adopt an LES model for the simulation of the diffusion process of deep seawater after leaving the pipe outlet. In LES, it is assumed that, only the turbulence viscosity smaller than the grid scale (i.e. sub grid scale - SGS) is isotropic and the turbulence larger than the grid scale (GS) is directly simulated. Thus, LES may provide more realistic results and overcome some of the issues encountered when using a k-ε type model.

Recently, due to the advent of more powerful computers, LES has become a common tool to simulate turbulence in a variety of flow types. Of relevance is the study of Lewis (2005) who coupled LES with a model for plankton population dynamics to simulate the dispersion nutrients in the North Pacific. Liu *et al.* (1997) simulated a temperature stratified channel flow using LES, with taking into account the fluctuations due to the buoyancy force. Flöhlich *et al.* (2004) also used LES to simulate the turbulent flow around a cylinder (note: a similar geometry to our present study). However, their related cross-flow Reynolds number was much higher than that of present study. Yuan *et al.* (1999) simulated a round jet in cross-flow also by using LES. The performed their calculations at two Reynolds numbers, 1050 and 2100, which is close to the present study. Sun et al (2007) simulated buoyancy-driven convection in a rotating disk cavity using LES and unsteady Reynolds-Averaged Navier-Stokes (RANS) modeling, resulting in LES solution in better agreement with velocity and heat transfer measurements.

Very little LES work is published relevant to the oceanic environment which takes into account the temperature stratification and concentration fields in depths of 10–100 m. In this study, the diffusion process of outflow from the outlet of the vertical pipe deployed in the Mariana region is simulated by using LES. The objectives of this study are to investigate whether the behavior seen in oceans is observed using an LES model and to compare the LES results with the more commonly used k-ε turbulent model simulations.

2. GENERAL GUIDELINES

2.1 Governing Equations

In Large Eddy Simulation, the grid scale (GS) and sub-grid scale (SGS) are separated by filtering the Navier-Stokes equations as shown in Eqs. (1)–(4) below. Here, only the SGS turbulence is modeled, while GS turbulence is directly solved.

$$\frac{\partial \overline{u}_i}{\partial x_i} = 0, \tag{1}$$

$$\frac{\partial \overline{u}_i}{\partial t} + \overline{u}_j \frac{\partial \overline{u}_i}{\partial x_j} = -\frac{1}{\rho_0}\frac{\partial \overline{p}}{\partial x_i} + \frac{\partial}{\partial x_j}\left(\nu_{sgs}\frac{\partial \overline{u}_i}{\partial x_j} - \tau_{ij}\right) + \left(\frac{\rho - \rho_{ref}}{\rho_0}\right) g_i, \tag{2}$$

$$\frac{\partial \overline{T}}{\partial t} + \overline{u}_j \frac{\partial \overline{T}}{\partial x_j} = \frac{\partial}{\partial x_j}\left(\alpha + \frac{\nu_{sgs}}{Pr_t}\right)\frac{\partial \overline{T}}{\partial x_j}, \tag{3}$$

$$\frac{\partial \bar{C}}{\partial t} + \bar{u}_j \frac{\partial \bar{C}}{\partial x_j} = \frac{\partial}{\partial x_j}\left(D + \frac{\nu_{sgs}}{Sc_t} \right)\frac{\partial \bar{C}}{\partial x_j}, \qquad (4)$$

Where the bar represents the mean component for each variable. It should be emphasized that LES averaging is spatial based, whereas in k-ε type models a temporal or ensemble average is considered. The SGS stress component τ_{ij} in Eq. (2) is expressed as follows,

$$\tau_{ij} - \frac{1}{3}\tau_{kk}\delta_{ij} = -2\nu_{sgs}\bar{S}_{ij}, \qquad (5)$$

Where the strain rate tensor $\bar{S}_{ij}$ is

$$\bar{S}_{ij} = \frac{\partial \bar{u}_i}{\partial x_j} + \frac{\partial \bar{u}_j}{\partial x_i}. \qquad (6)$$

The SGS turbulent kinematic viscosity ν_{sgs} is obtained by using the standard Smagorinsky-Lilly model,

$$\nu_{sgs} = L_{sgs}{}^2 \sqrt{2\bar{S}_{ij}\bar{S}_{ij}}, \qquad (7)$$

Where the SGS mixing length L_{sgs} is obtained by using the following equation,

$$L_{sgs} = \min\left(\kappa d, C_s V^{1/3}\right). \qquad (8)$$

The model constants used in our simulations were chosen as noted below.

$$C_s = 0.1, \kappa = 0.4187, Pr_t = 0.85, Sc_t = 0.7. \qquad (9)$$

These constant values have been tested and are known to apply to a wide range of flows (Launder *et al.*, 1974; Yoshizawa *et al.*, 1995, FLUENT, 2006) and were therefore chosen for the present simulations.

It should be noted that the SGS model in LES still assumes an isotropic turbulence. However, this is reasonable because the scale of the SGS turbulence is small enough to be treated as steady and isotropic in a statistical sense. Further, it is known that the large eddies of the GS affect most part of the solution and they primarily depend on the shape of the system and the boundary conditions, however the small eddies do not depend on the shape of the system and they can be treated as isotropic. Therefore, a general model can be obtained by modeling only the small eddies of the SGS. Further, since the spatial scale in this study is large, small eddies of the SGS are not expected to have much impact on obtaining an accurate numerical result.

2.2 Solution Method

The numerical simulation was conducted using a commercial CFD package (FLUENT 6.3). The governing equations were discretized using the second-order upwind scheme. An explicit time-advancement scheme with the fractional step method was employed.

The calculation model considered in this study is shown in Fig. 3. The domains dimensions are 17 m × 15 m × 6 m (see Fig. 2) and pipe diameter was 0.3 m, which is the actual pipe diameter in the Mariana trench region experiments of Maruyama *et al.* (2004). The pipe length used in their experiment was 300 m, though in order to focus on the pipe outlet, only the upper 10 m of the pipe is included in this domain. The grid was made to be finer near the pipe outlet with the finest scale being 0.015 m. The computational grid consisted of 324,420 elements.

For the initial and boundary conditions of temperature and salinity (note: Practical Salinity Scale is used here; see UNESCO [1981]), the experimentally measured profiles shown in Fig. 4 were used. The velocity boundary condition for the domain inlet was assumed to be a uniform and constant (0.015 m/s), which is the measured relative velocity between the pipe and the ocean flow. In addition we neglected the turbulent in inlet because the inlet velocity is very small. The outflow boundary condition was set as fully-developed. At the bottom of the pipe domain, the velocity profile shown in Fig. 5 was used. This velocity profile was based on the upwelling simulation results of Sato *et al.* (2007). For the nutrient, NO_3 concentration at 300 m depth was applied (Garcia *et al.*, 2006). Other boundaries were set as a free-slip condition.

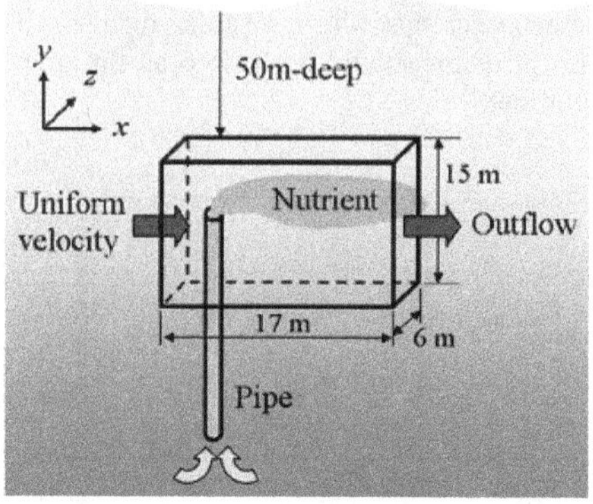

Fig. 3 The computational model

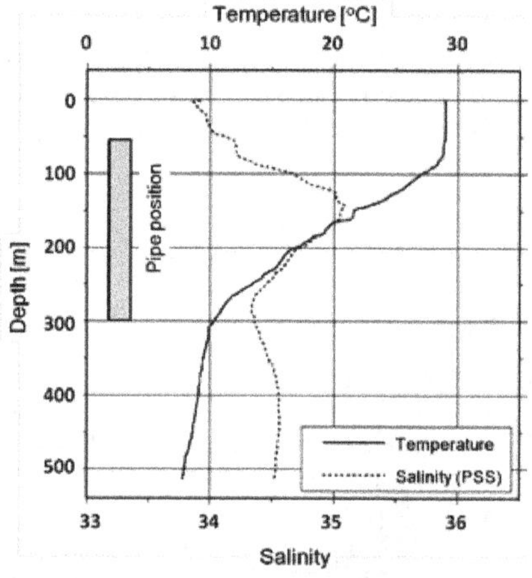

Fig. 4 Vertical profiles of temperature and salinity in Mariana trench region

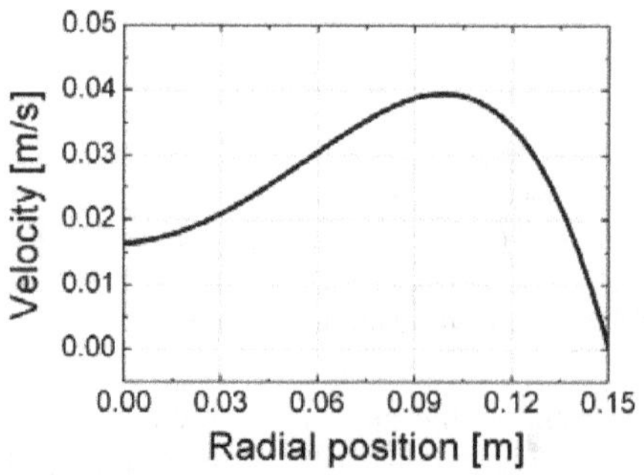

Fig. 5 Velocity profile employed for the bottom of the pipe domain

3. RESLUTS AND DISCUSSION

The simulations results obtained by using both the *k-ε* and LES models are presented and discussed here. In the simulations, the concentration profile was considered to have reached steady state when the volume of integral mass fraction of nutrients over the whole domain was not changing any more. The simulations time step was set to $\Delta t = 0.5$ s for the k-ε model and 0.1 s for the LES. The concentration profile reached its steady state value at t = 10,000 s for the k-ε model and 2,500 s for the LES.

Figures 6 (a) and (b) show the cross-sectional view (along the x-y plane) of nutrient contours for the LES and k-ε, respectively. Here, for comparison purposes and in order to clarify the differences, the time-averaged contours are shown for LES. The colored contours show the nutrient concentration percentages versus the distance.

Firstly, in order to show the effect of vertical diffusion, attention is drawn to the spread of the 2% contour level (see Fig. 6). It is noted that, downstream of the pipe outlet the predicted spreading width is 1 and 3 m for the LES and k-ε, respectively. Secondly, the 2% nutrient concentration contour level in the other direction (horizontal – top view) is examined. Here, the horizontal diffusion as shown in Figs. 7 (a) and (b) leads to a spread width of 3 and 5 m for the LES and k-ε, respectively. It should be noted that these figures are not the cross-sectional view at a certain depth, but the 2% nutrient concentration contour viewed from the top. Figures 7 (a) and (b) indicate that the ratio of the vertical diffusion to that of the horizontal one is about 33% for the LES, whereas in the case of the k-ε model it is about 50%.

As noted previously, the diffusion behavior seen in the ocean indicates that the horizontal diffusion is larger than the vertical one. The results here confirm that, at least as far as the ratio of the diffusion components is concerned, the LES results are more in line with the observations in the ocean.

The results here for both cases show that the nutrients ejected from the pipe descend downwards, and downstream of the pipe reach a neutral buoyancy. Furthermore, it is seen that the horizontal advection caused by the ocean flow is dominant, especially downstream of the pipe. The reason for the nutrient descent is that the salinity at the location (depth) of the pipe outlet is lower than that of the deep seawater ejected from the pipe outlet (see Fig. 4). However, as seen in Fig. 6 the vertical penetration of the two simulations is considerably different (i.e. 8 m for the k-ε model and 4 m for the LES).

In this calculation condition, there were some strong stratifications for temperature, salinity, nutrient. As stated in previous research (Filippo M. Denaro *et al.* 2007), the strong stratification with vertical direction makes the diffusion process with vertical direction anisotropic From the results presented here, it can be concluded that the k-ε model predicts a stronger diffusion (in both horizontal and vertical directions) than the LES. In the k-ε models, the flow was assumed isotropic with both vertical direction and horizontal direction and this model could not take into account the stratification effect. Because of this ignorance of stratification effect, the k-ε models predicted stronger diffusion with the vertical direction. Whereas, the LES model can directly simulate the flow and take into account the stratification effect at the resolved scale. Because of this reason, the calculation result of LES model showed the weaker diffusion with vertical direction compared to the k-ε model as shown in Fig. 6 and it can be concluded that LES model qualitatively result can better represent the diffusion process than the k-ε model.

Of course, in order to ascertain this behavior quantitatively, future experiments to measure the nutrient concentrations are needed. Moreover, once the horizontal and vertical turbulent diffusion coefficients in the ocean are experimentally measured, then the LES model can be fine-tuned for better predictions. A strong diffusion of nutrients in the upstream direction is observed in Fig. 6(b). The k-ε model may over-predict the turbulent diffusion in the present flow situation. It is expected that the temporal average model may not be appropriate for simulation of the turbulent flow for the oceanic flow of the present scale.

In order to further examine the feasibility of a real perpetual salt fountain, future simulations need to focus on a larger scale of the order of 100 m. Of course, as the accuracy of the LES model depends on the grid size, the computational cost of such a large scale model may prove to be prohibitive. Further, the real ocean flow strongly affects the simulation results, however in the model used here a constant and uniform ocean flow was assumed. In fact, the ocean characteristics are such that the flow direction and magnitude vary at all times. Introducing a more realistic ocean flow characteristics in the model will lead to more accurate results.

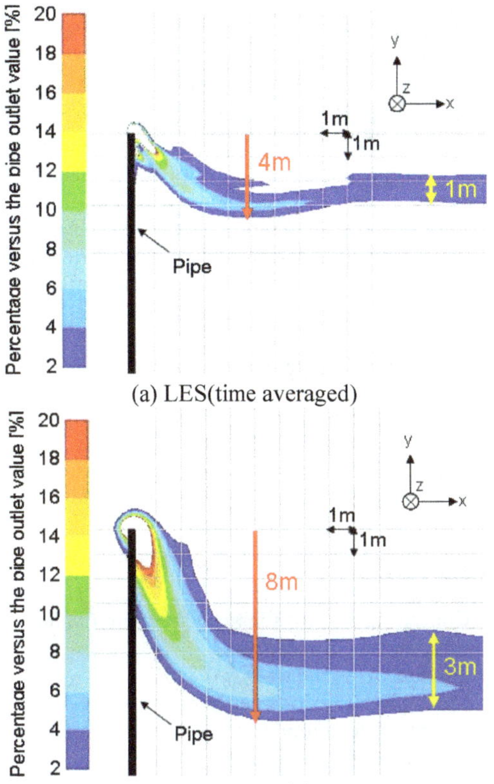

(a) LES(time averaged)

(b) k-ε model

Fig. 6 Nutrient concentration contours (cross-section view along x-y plane)

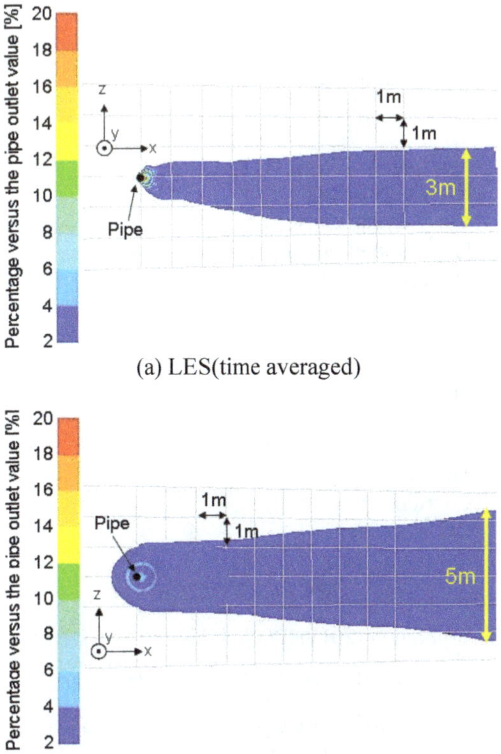

(a) LES(time averaged)

(b) k-ε model

Fig. 7 The 2% dilution line of nutrient concentration (top view)

4. CONCLUDING REMARKS

In this study, we have performed turbulent flow simulations using the k-ε and LES models for a perpetual salt fountain. The diffusion process of the nutrient-rich deep seawater ejecting from a vertical pipe deployed in the ocean was numerically investigated. Based on the results obtained the following conclusions were reached.

(1) The LES model can qualitatively represent the behavior seen in real ocean compared to the k-ε model.

(2) A comparison of the LES and k-ε results show that the k-ε model predicts a stronger diffusion in both the horizontal and vertical directions.

(3) In order to quantitatively represent the real behavior seen in the ocean, future experiments to measure the nutrient concentrations as well as, horizontal and vertical turbulent diffusion coefficients are needed.

ACKNOWLEDGEMENTS

The calculations were conducted using the supercomputer SGI Altix at the Institute of Fluid Science, Tohoku University. This work was partially supported by a Grant-in-Aid for Scientific Research (B). The project was conducted collaboratively with Tokyo Metropolis.

NOMENCLATURE

C	Concentration for salinity and nutrient (wt%)
Cs	Smagorinsky constant (-)
D	Distance from the wall (m)
d	Molecular diffusion coefficient (m^{2}s)
G	Gravity acceleration (m/s^2)
L	Mixing length (m)
p	Pressure (N/m^2
Pr	Prandtl number (-)
Sc	Schmidt number (-)
$\overline{S_{ij}}$	Strain rate tensor (1/s)
t	Time (s)
T	Temperature (°C, K)
u	Velocity (m/s)
V	Cell volume (m^3)

Greek symbols

α	Thermal diffusivity (m^2/s)
δ_{ij}	Kronecker delta (-)
κ	Von Karman constant (-)
ν	Kinematic viscosity (m^2/s)
ρ	Density (kg/m^3)
τ	Shear stress (m^2/s^2)

Subscripts

i, j, k	Tensor index
ref	Reference
sgs	Sub grid scale
t	Turbulent
x, y, z	Coordinates
0	Ocean surface

REFERENCES

BONNIER corporation (2011): POPULAR SCIENCE May 2011.

FLUENT Inc. (2006): FLUENT6.3 User's Guide.

Filippo M. Denaro, Giuliano De Stefano, Daniele Iudicone, Vinecenzo Botte (2007): Afinite volume dynamic large-eddy simulation method for buoyancy driven turbulent geophysical flows. J Ocean Modeling, **17**, 199-218. http://dx.doi.org/10.1016/j.ocemod.2007.02.002

Fröhlich, J., W. Rodi, P. Kessler, S. Parpais, J. P. Bertoglio and D. Laurence (2004): Large eddy simulation of flow around circular cylinders on structured and unstructured grids. *Notes on Numerical Fluid Mechanics*, **66**, 319–338.

Garcia, H.E., R. A. Locarnini, T. P. Boyer and J. I. Antonov (2006):World Ocean Atlas 2005, Volume 4: Nutrients (phosphate, nitrate, silicate), U.S. Government Printing Office.

Launder, B. E. and D. B. Spalding (1974): The numerical computation of turbulent flows. *Computer Methods in Applied Mechanics and Engineering*, **3**, 269–289. http://dx.doi.org/10.1016/0045-7825(74)90029-2

Ledwell, J. R., A. J. Watson and C. S. Law (1998): Mixing of a tracer in the pycnocline. *J. Geophys. Res.*, **103**, C10, 21499–21529. http://dx.doi.org/10.1029/98JC01738

Lewis, D. M. (2005): A simple model of plankton population dynamics coupled with a LES of the surface mixed layer. *Journal of Theoretical biology*, **234**, 565–591. http://dx.doi.org/10.1016/j.jtbi.2004.12.013

Liu, N. Y., X. Y. Lu, S. W. Wang and L. X. Zhuang (1997): Large-eddy simulation of stratified channel flow. *ActaMechanicaSinica*, **13**, 331–338. http://dx.doi.org/10.1007/BF02487192

Maruyama, S., K. Tsubaki, K. Taira and S. Sakai (2004): Artificial upwelling of deep seawater using the perpetual salt fountain for cultivation of ocean desert. *J. Oceanogr.*, **60**, 563–568. http://dx.doi.org/10.1023/B:JOCE.0000038349.56399.09

Maruyama, S., T Yabuki, T. Sato, K. Tsubaki, A. Komiya, M. Watanabe, H. Kawamura and K. Tsukamoto (2011): Evidence of increasing primary production in the ocean by Stommel's perpetual salt fountain, *Deep-Sea Res. Part I-Oceanogr. Res. Pap.*, **58**, 567–574. http://dx.doi.org/10.1016/j.dsr.2011.02.012

Reid, J. L. (1965): Intermediate waters of Pacific Ocean, *Johns Hopkins Oceanogr. Stud.*, **5**, 96 pp.

Sato, T., S. Maruyama, A. Komiya and K. Tsubaki (2007):Numerical simulation of upwelling flow in pipe generated by perpetual salt fountain, *Proceedings of16th Australian Fluid Mechanics Conference*, 394–397.

Schmitz, W. J. (1995): On the interbasin-scale thermohaline circulation. *Rev. Geophys.*, **33**, 151–173.

Schmitz, W. J. (1996a): On the world ocean circulation. Volume I, some global features/North Atlantic circulation, Woods Hole Oceanographic Institution, Technical Report, WHOI-96-03.

Schmitz, W. J. (1996b): On the world ocean circulation. Volume II, the Pacific and Indian Oceans/a global update, Woods Hole Oceanographic Institution, Technical Report, WHOI-96-08.

Smagorinsky, J. (1963): General Circulation Experiments with the Primitive Equations I, The basic experiment. *Monthly Weather Review*, **91**, 99–164. http://dx.doi.org/10.1175/1520-0493(1963)091%3C0099:GCEWTP%3E2.3.CO;2

Stommel, H., A. B. Arons and D. Blanchard (1956): An oceanographical curiosity: the perpetual salt fountain. *Deep-Sea Res.*, **3**, 152–153. http://dx.doi.org/10.1016/0146-6313(56)90095-8

Sun, Z., K.Lindblad, J. W. Chew and C. Young (2007): LES and RANS Investigations Into Buoyancy-Affected Convection in a Rotating Cavity

With a Central Axial Throughflow. *J. Eng. Gas Turbines Power*, **129**, 318–325,
http://dx.doi.org/10.1115/1.2364192

Talley, L. D. (1993): Distribution and formation of North Pacific Intermediate Water. *J. Phys. Oceanogr.*, **23**, 517–537.
http://dx.doi.org/10.1175/1520-0485(1993)023<0517:DAFONP>2.0.CO:2

Talley, L. D. (1999): Some aspects of ocean heat transport by the shallow, intermediate and deep overturning circulations. p. 1–22. *In Mechanisms of Global Climate Change at Millenial Time Scales,* Geophys.Mono. Ser. 112, ed. by P. U. Clark, R. S. Webb and L. D. Keigwin, American Geophysical Union
http://dx.doi.org/10.1029/GM112p0001.

Talley, L. D. (2003): Shallow, intermediate, and deep overturning components of the global heat budget. *J. Phys. Oceanogr.*, **33**, 530–560.
http://dx.doi.org/10.1175/1520-0485(2003)033<0530:SIADOC>2.0.CO;2

Toole, J. M. and T. J. McDougall (2001): Mixing and Stirring in the Ocean Interior. p. 337–355. In *Ocean Circulation and Climate: Observing and Modelling the Global Ocean,* ed. by G. Siedler, J. Church and J. Gould, Academic Press, San Diego.

Tsubaki, K., S. Maruyama, A. Komiya and H. Mitsugashira (2007): Continuous measurement of an artificial upwelling of deep sea water induced by the perpetual salt fountain. *Deep-Sea Res. I*, **54**, 75–84.
http://dx.doi.org/10.1016/j.dsr.2006.10.002

UNESCO (1981): The Practical Salinity Scale 1978 and the International Equation of State of Seawater 1980, Unesco Technical Papers in Marine Science, **36**, 25 pp.

Yasuda, I., K. Okuda and Y. Shimizu (1996): Distribution and modification of the North Pacific Intermediate Water in the Kuroshio-Oyashiointerfrontal zone. *J. Phys. Oceanogr.*, **26**, 448–465.
http://dx.doi.org/10.1175/1520-0485(1996)026<0448%3ADAMONP>2.0.CO%3B2

Yoshizawa, A., S. Murakami, T. Kobayashi, N. Taniguchi, Y. Dai, A. Kuroda, K. Kamemoto, S. Kato, Y. Nagano and T. Tsuji (1995):*Analysis of Turbulent Flows*, University of Tokyo Press., (in Japanese).

Yuan, L. L., R. L. Street and J. H. Ferziger (1999): Large-eddy simulations of a round jet in crossflow.*Journal of Fluid Mechanics*, **379**, 71–104.
http://dx.doi.org/10.1017/S0022112098003346

Williamson, N., A. Komiya, S. Maruyama and M. Behnia (2009): Nutrient transport from an artificial upwelling of deep seawater. *J. Oceanogr.*, **65**, 349–359.
http://dx.doi.org/10.1007/s10872-009-0032-x

Zhang, X., S. Maruyama, S. Sakai, K. Tsubaki and M. Behnia (2004): Flow prediction in upwelling deep seawater — the perpetual salt fountain. *Deep-Sea Res. I*, **51**, 1145–1157.
http://dx.doi.org/10.1016/j.dsr.2004.03.010

Zhang, X., S. Maruyama, K. Tsubaki, S. Sakai and M. Behnia (2006): Mechanism for enhanced diffusivity in the deep-sea perpetual salt fountain. *J. Oceanogr.*, **62**, 133–142.
http://dx.doi.org/10.1007/s10872-006-0039-5

STUDY OF INTERNAL FLOW CHARACTERISTICS OF INJECTOR FUELLED WITH VARIOUS BLENDS OF DIETHYL ETHER AND DIESEL USING CFD

Vijayakumar Thulasi*, Thundil Karuppa Raj Rajagopal

School of Mechanical and Building Sciences, VIT University, Vellore 632014, India

ABSTRACT

Researchers across the world are exploring the potential of using diethyl ether as an alternate fuel to meet the stringent emission norms due to the high oxygen content in the fuel. The spray characteristics of any injected fuel are highly influenced by its physical properties. Due to high injection pressure in CI engines the fuel tends to cavitate inside the nozzle greatly. The change in fuel properties will affect the cavitating behavior of the fuel. In this paper computational technique is used to study and compare the internal flow characteristics of a fuel injector for different blends of diethyl ether and diesel fuel. Multi phase flow model considering the fuel as a mixture of vapor and liquid is adopted for the simulation study. The percentage volume of diethyl ether is varied from 0 to 100% and the flow characteristics are studied. Results indicate that as the percentage volume of diethyl ether increases in the fuel, the cavitating phenomenon also increases resulting in decrease in mass of fuel injected into the cylinder.

Keywords: *Cavitation, Diethyl ether, Diesel, Injector, fuel, CI engine*

1. INTRODUCTION

The major challenges faced by the developed nations today are: the economical use of the existing fossil fuels and the development of suitable alternative and renewable fuels considering their environmental impacts. The diesel engine, though provide high power output with better fuel economy, produce high NOx and smoke emissions. Reduction of emissions, to any great extent, without sacrificing fuel economy will be an enormous challenge. The emissions from the diesel engines can be reduced by treating the emissions before letting of to the atmosphere, by improving the combustion characteristics or by improving the fuel properties before inducting in to the cylinder. The diesel fuel properties can be improved by adding additives or blending with another fuel. Guru et al (2002, 2011) studied the effect of organic based metallic additives on the performance and emission characteristics of diesel engine. The authors reported that use of additives improved the specific fuel consumption by 2.5%. The authors also reported that CO and smoke emissions are reduced by around 20% with a marginal increase in the NOx emissions. Labeckas et al (2005) studied the effect of additives on the diesel engine fuelled with shale oil and reported that the NOx emissions are reduced by 23%. They also reported that CO and smoke emissions are increased by 15% and around 35% respectively. Cataluna et al (2006) studied the usage of ethers as an additive to the diesel fuel and reported that ter-amyl ethyl ether improved the diesel performance effectively.

Among all the ethers, diethyl ether has a greater potential to be used as diesel fuel additive. The effects of using diethyl ether and ethanol as additive to biodiesel and diesel blends were studied by Qi et al (2011). The authors reported that addition of 5% diethyl ether by volume reduces the smoke emission, due to its higher volatility. The authors also reported that the NOx emissions are slightly increased when 5% of ethanol was used as the additive. The use of diethyl ether along with diesel-water emulsion in direct injection diesel engine was studied by Ramesh et al (2002). The authors reported that NOx level at full load operation is reduced substantially.

The performance and emission characteristics of any engine are highly affected by the spray characteristics of the fuel. In diesel engine the fuel is injected at high pressure to overcome the air resistance (back pressure) to get penetrated into the chamber. High pressure is also needed to enhance the atomization and spray penetration of the injected fuel and also to improve the combustion efficiency. The high fuel pressure available at the nozzle seat (200 bar) is converted into kinetic energy at the loss of pressure energy as it passes through the nozzle orifice. The drop in pressure at the entry of the nozzle is very high, leading to cavitation, and it reduces as moving towards the nozzle exit. The fuel pressure available at the nozzle exit is little higher than the in-cylinder air pressure. Cavitation is the formation of vapor bubbles in the liquid fuel when the pressure rapidly drops below the saturation pressure of the liquid fuel. The nucleation process of cavitation and the formation of vapor bubbles are experimentally studied by Takenaka et al (2004) using neutron radiography. The effect of cavitation on the fuel spray characteristics is studied by Lee et al (2008) experimentally. He reported that the primary breakup regime of the fuel is highly influenced by the turbulence created inside the nozzle. J.M. Desantes et al (2010) also reported the cone angle of the fuel spray is increased due to the formation of vapor inside the nozzle.

The injection flow characteristics of the fuel are greatly affected by the fuel density, vapor pressure and surface tension. Hosny et al (1996) studied that the cavitation phenomenon are more sensitive to the changes in fuel properties and developed correlation between cavitation and fuel properties. The thermophysical and transport properties of diethyl ether are different from diesel; hence different injection flow characteristics can be expected. Vijayakumar et al (2011) studied the cavitating phenomenon of diethyl ether, dimethyl ether and diesel fuel

*Corresponding author : vijayakumar.t@vit.ac.in

numerically and reported that the fuel velocity at the nozzle exit for ether fuels is greater than for the diesel fuel. They also reported that the mass flow rates for the ether fuels are substantially reduced when compared with diesel fuel at same injection pressure. It can be said that the rate of injection of the fuel, cavitation and the turbulence at the nozzle exit are affected by the injector flow characteristics, which in turn affects the spray atomization and penetration and hence the performance.

In the present study, the injector flow characteristics for different blends of diethyl ether and diesel fuel are studied numerically using Computational Fluid Dynamics. The effects of physical properties on the cavitation, injection velocity, coefficient of discharge and mass flow rate at the nozzle exit are simulated for different blends (0 – 100 % by volume in steps of 10) of diethyl ether and diesel fuel. The fuel injection pressure is taken as 200 bar and the pressure inside the combustion chamber (back pressure) is taken as 30 bar and a comparative study of flow characteristics is done for all the blends.

2. INJECTOR FLOW MODEL

The nozzle flow simulations were performed using ANSYS Fluent. The fluid is assumed to be a mixture comprising of liquid fuel (diesel and diethyl ether) and its vapor. Multi phase flow analysis using Schnerr and Sauer cavitation model is performed with no-slip condition between the liquid and vapor. The Schnerr and Sauer model can be used for both mixture and Eulerian multiphase models. Schnerr and Sauer model can also be used well along with all turbulence models available in Fluent, but this model doesn't takes into account the effect of non condensable gases present in the mixture.

RNG k-ε model derived using a rigorous statistical technique called renormalization group theory, with non-equilibrium wall conditions is used in order to account for the large pressure differential across the nozzle. This model takes into account the effect of swirl well also it is capable of predicting the effective viscosity of the mixture even at low Reynolds number flow.

The vapor formation, the bubble growth and condensation are solved by considering Rayleigh-Plesset equation (2006). A three-hole injector with an orifice diameter of 184 μm and an included angle of 90° is considered for the analysis. The flow is considered to be symmetrical across all the nozzles and hence only one nozzle is considered for analysis as shown in Fig 1. The fluid domain is characterized by 441787 tetrahedral cells with 84878 nodes. The inlet and outlet conditions are provided with pressure values and symmetry conditions are employed to demarcate the 120° sector mesh. Wall boundary conditions, with no slip between the fuel-vapor mixture and the wall surface, are adopted for all the other surfaces. The flow simulation is performed at the full needle lift condition of 0.2 mm

2.1 Validation of computational model

The experimental data from Som et.al (2010) for a nozzle orifice of diameter 169μm with an included angle of 120° fuelled with diesel was used to validate the computational flow model. Fuel is injected at pressures of 1100 bar and 1300 bar for 3 ms. The back pressure is taken as 30 bar. The comparison of the mass of the fuel injected per nozzle and discharge coefficient between the experimental values and simulation values are shown in Table 1.

Table 1 Comparison of experimental and simulation values

Property	Injection Pressure, (bar)	Exp.	Sim.	% error
Mass flow rate, kg/s	1100	0.00621	0.0067	-7.89
	1300	0.00693	0.0073	-5.34
Coefficient of Discharge	1100	0.64	0.69	-7.81
	1300	0.64	0.693	-8.28

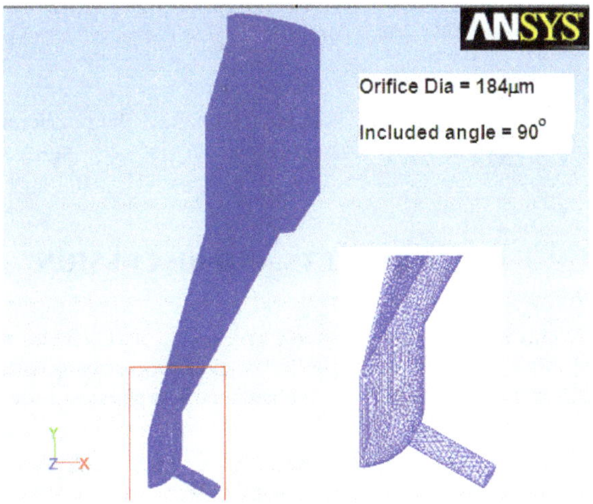

Fig. 1 Injector model

It is observed that the deviations of simulation data are well within the acceptable limits from the experimental values. It can be said that the computational model is able to capture the inner flow characteristics of the injector well and hence can be used for comprehensive parametric study of the injector flow.

3. FLOW CHARACTERIZATION

The injector flow characteristics are studied by the cavitation number (K), discharge coefficient (C_d), velocity coefficient (C_v), area coefficient (C_a), and Reynolds number (Re) as described below: Singhal et.al (2002). The cavitation number, K is calculated from

Error! Bookmark not defined.
$$K = \frac{2(P_i - P_v)}{\rho_f V^2} \quad (1)$$

where P_i is the injection pressure (Pa), P_v is the saturation vapor pressure of the fuel (Pa), ρ_f is fuel density (kg/m^3) and V is the characteristic velocity of the fuel (m/s). The discharge coefficient, C_d is calculated using the following equation

$$C_d = \frac{M_{act}}{A_{th} \cdot \sqrt{2 . \rho_f . \Delta P}} \quad (2)$$

where M_{act} is the actual mass flow rate (kg/s) which is obtained from the simulation, A_{th} is the nozzle exit area (m^2), ρ_f is the fuel density (kg/m^3) and ΔP is the pressure differential across the nozzle orifice.

The velocity coefficient, C_v is calculated from the following equation

$$C_v = \frac{V_{act}}{\sqrt{2 . \Delta P / \rho_f}} \quad (3)$$

where V_{act} is the actual velocity at the nozzle exit (m/s).

The area coefficient is calculated as

$$C_a = \frac{C_d}{C_v} \quad (4)$$

The Reynolds number, Re is calculated from the following equation

$$R_e = \frac{\rho_f . V . D_{ex}}{\mu_f} \quad (5)$$

where V is the average flow velocity (m/s) along the nozzle orifice, D_{ex} is the nozzle exit diameter (m), μ_f is the fuel viscosity (kg-m/s).

4. RESULTS AND DISCUSSION

The thermo physical and transport properties of the fuels: diesel and diethyl ether are listed in Table 2. The fuel properties of diethyl ether are taken from CRC handbook of chemistry and physics: Lide (2003).

Table 2 Fuel properties

Fuel property	DEE	Diesel
Carbon weight %	64.7	83
Hydrogen weight %	13.5	17
Oxygen weight %	21.6	0
Density @ 25°C (kg/m3)	713.4	822
Viscosity @ 25°C (kg-m/s)	0.0002448	0.00224
Surface tension @ 25°C (N/m)	0.017	0.0020
Vapor pressure @ 25°C (Pa)	58660	1280

The injector flow simulation is performed for 120° sector mesh for an injection pressure of 200 bar with a back pressure of 30 bar. The fuel temperature is taken as 298 K for all the fuel blends.

4.1 Cavitation

Figures 2 and 3 show the vapor fractions of diesel and diethyl ether fuel. The cavitation inception is found in all the blends. It is observed that the vapor volume fraction of diethyl ether is more than the diesel due to the higher saturation pressure.

The vapor that is formed at the nozzle entry for diesel fuel is found to be collapsed immediately inside the orifice itself before reaching the nozzle exit, whereas for diethyl ether it is found that some small amount of vapor is carried along the orifice to the nozzle exit at 100% blend. Higher viscosity and lower saturation pressure of the diesel fuel helps in collapsing the vapor formed immediately. The similar kind of observation was made by Shi and Arafin (2010). The authors reported the cavitation phenomenon is highly influenced by the fluid viscosity.

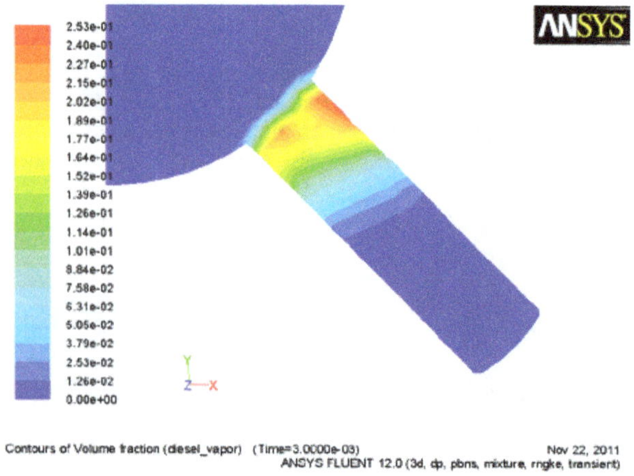

Fig. 2 Diesel vapor volume fraction at 0% blend

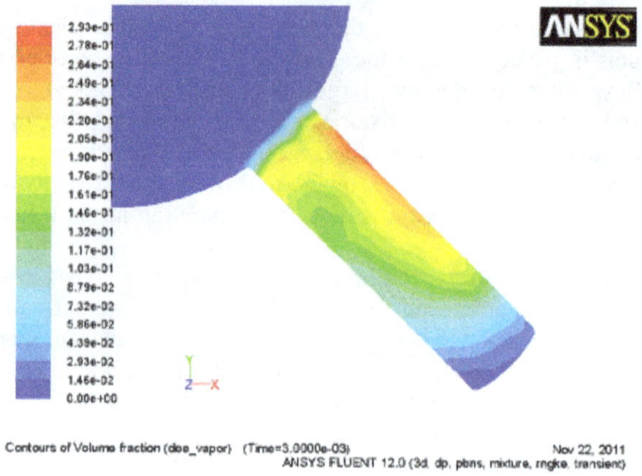

Fig. 3 Diethyl ether vapor volume fraction at 100% blend

Figure 4 shows the variation of vapor volume fraction for diesel and diethyl ether inside the injector at different blends of the fuel. It is observed that at higher blends of diethyl ether the vapor volume fraction of diethyl ether increases at little higher rate than the rate at which the diesel vapor decreases. As the percentage of the diethyl ether increases in the fuel the saturation pressure and viscosity of the fuel is also changes.

The variation of cavitations number for different blends of the fuel is shown in Fig 5. It is observed that as the percentage volume of diethyl ether increases the cavitating behavior of the fuel increases. It is also observed that the cavitation number decreases gradually up to 90% of blend. The presence of diesel fuel up to 90% blend controls the cavitating behavior to some extent. At 100% diethyl ether the cavitation number decreases sharply.

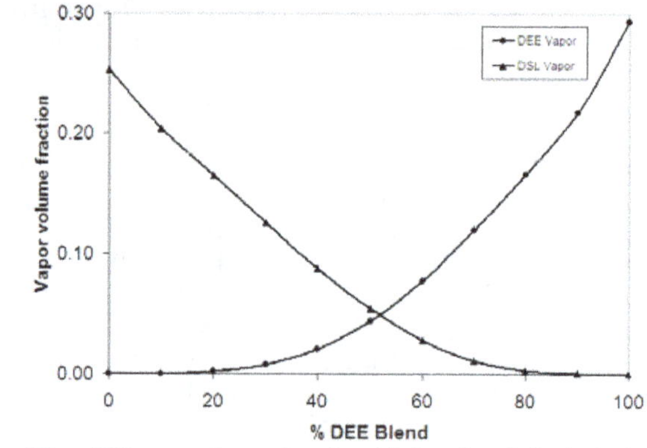

Fig. 4 Vapor volume fraction of diesel and diethyl ether

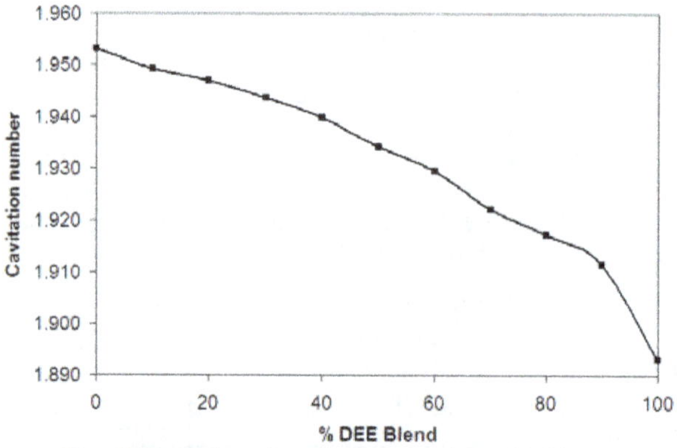

Fig. 5 Variation of cavitation number with blends

Figure 6 shows the variation of Reynolds number with cavitation number. Higher the cavitation number lesser is the Reynolds number.

At lower cavitation number more amount of vapor is formed which increases the turbulent behavior of the fuel inside the orifice. It can be said that diethyl ether creates more turbulence inside the orifice than the diesel fuel due to lower viscosity

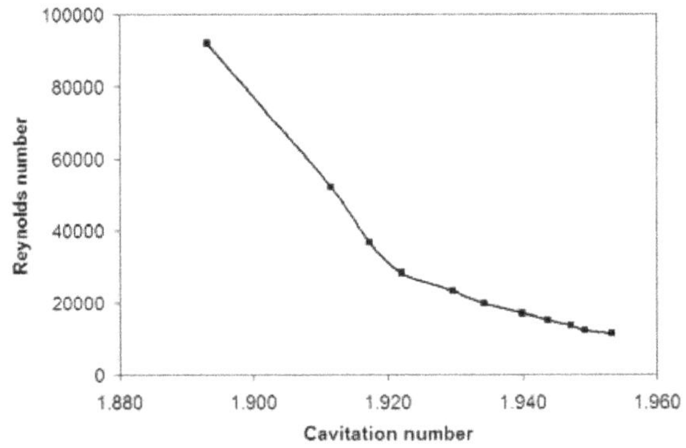

Fig. 6 Variation of Reynolds number with cavitation number

4.2 Nozzle exit parameters

The variation of coefficient of discharge with different blends is shown in Fig. 7. It is observed that as the volume percentage of diethyl ether increases the fuel viscosity and fuel density gradually decreases. Due to this the total vapor (diesel + diethyl ether) formed at the nozzle entry increases, which reduces the area (Fig. 8) available for liquid fuel to pass through: Park et.al (2010). This increases the fuel velocity at the nozzle exit. This is evident from Fig 9 which shows the variation of exit velocity with blends. Fig 10 shows the variation of mass flow rate with blends.

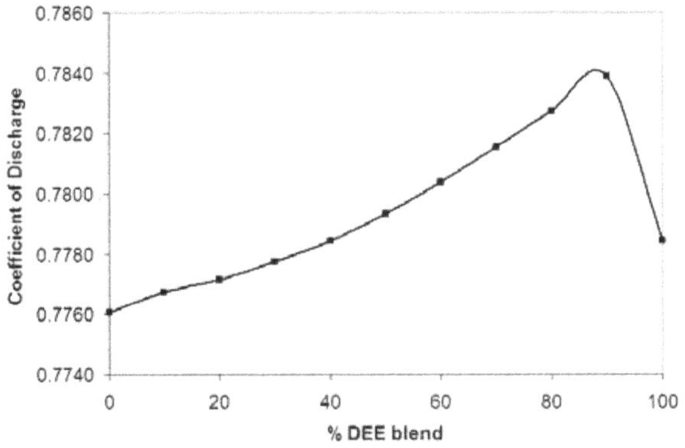

Fig. 7 Variation of coefficient of discharge

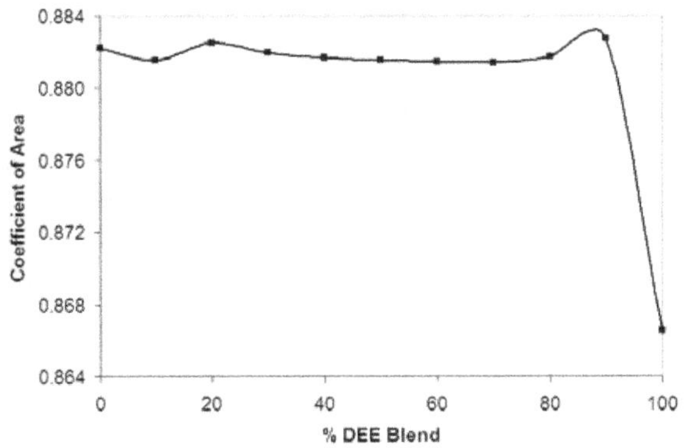

Fig. 8 Variation of coefficient of area

The mass flow rate is observed to be reducing due to the decrease in fuel density. Though the mass flow rate decreases with increasing blends, the coefficient of discharge is found to increase up to 90% of blend. At 100 % diethyl ether due to some vapors are convected along the flow up to the nozzle exit thereby reducing the mass flow rate and also the coefficient of discharge and area decreases appreciably.

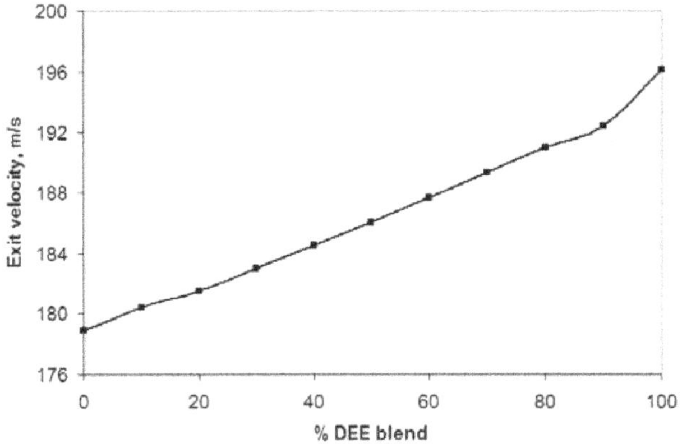

Fig. 9 Variation of exit velocity with blends

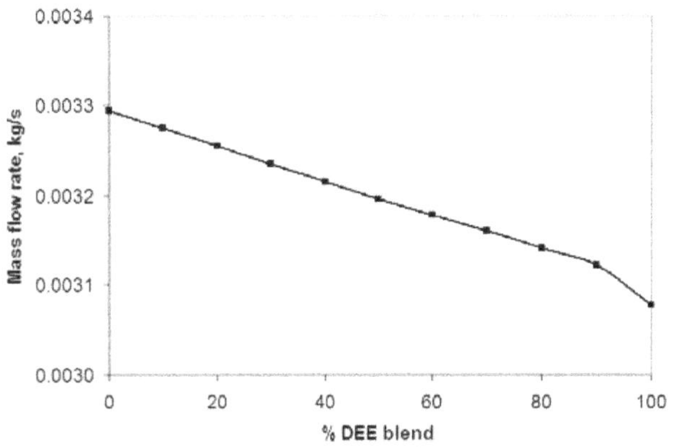

Fig. 10 Variation of mass flow rate with blends

5. CONCLUSION

The cavitation behavior and the nozzle exit parameters for different blends of diethyl ether and diesel are studied using the computational technique. The major conclusions are as follows.

- The cavitating behavior of the fuel increases with increasing percentage of diethyl ether
- The exit velocity of the fuel increases with increase in blends
- The mass flow rate of the fuel is found to be decreasing with increasing diethyl ether volume in fuel

REFERENCES

Desantes, J.M, Payri, R., Salvador F.J., and De la Morena, J, 2010, "Influence of Cavitation Phenomenon on Primary Break-up and Spray Behavior at Stationary Conditions," *Fuel*, **89**, 3033–3041 http://dx.doi.org/10.1016/j.fuel.2010.06.004

FLUENT v6.3 documentation

Guru, M, Karakaya, U, Altıparmak, D, Alıcılar, A, 2002, "Improvement of Diesel Fuel Properties by using Additives," *Energy Conversion and Management*, **43**, 1021–1025. http://dx.doi.org/10.1016/S0196-8904(01)00094-2

Hosny, D.M, Hudgens, D., and Cox, T, 1996, "Cavitation Correlation to Fluid Media Properties," SAE Paper no: 960882, 1996-02-01.

Keskin, A, Gürü, M, Altıparmak, D, 2011, "Influence of Metallic Based Fuel Additives on Performance and Exhaust Emissions of Diesel Engine," *Energy Conversion and Management,* **52**, 60–65. http://dx.doi.org/10.1016/j.enconman.2010.06.039

Labeckas, G., Slavinskas, S., 2005, "Influence of Fuel Additives on Performance of Direct-Injection Diesel Engine and Exhaust Emissions when Operating on Shale Oil," *Energy Conversion and Management* **46**, 1731–1744
http://dx.doi.org/10.1016/j.enconman.2004.09.001

Lee, C.S, Suh, H. K., Park, S. H, 2008, "Experimental Investigation of Nozzle Cavitating Flow Characteristics for Diesel and Biodiesel Fuels," *International Journal of Automotive Technology*, **9**(2), 217-224
http://dx.doi.org/10.1007/s12239-008-0028-3

Lide, D.R, 2003, CRC Handbook of Chemistry and Physics, 84th edition, CRC Press.

Menezes, E.W., da Silva, R., Cataluna, R., Ortega, R.J.C., 2006, "Effect of Ethers and Ether/Ethanol Additives on the Physicochemical Properties of Diesel Fuel and on Engine Tests," *Fuel,* **85**, 815–822, http://dx.doi.org/10.1016/j.fuel.2005.08.027

Park, S.H, Suh, H.K., and Lee, C.S., 2010, "Nozzle Flow and Atomization Characteristics of Ethanol Blended Biodiesel Fuel," Renewable Energy **35**(2010) 144–150
http://dx.doi.org/10.1016/j.renene.2009.06.012

Qi, D.H., Chen, H., Geng, L.M., Bian, Y.Z., 2011, "Effect of Diethyl Ether and Ethanol Additives on the Combustion and Emission Characteristics of Biodiesel-Diesel Blended Fuel Engine," *Renewable Energy,* **36**, 1252-1258
http://dx.doi.org/10.1016/j.renene.2010.09.021

Shi, J.M. and Arafin, M.S., 2010, "CFD Investigation of Fuel Property Effect on Cavitating Flow in Generic Nozzle Geometries," ILASS – Europe 2010, 23rd *Annual Conference on Liquid Atomization and Spray Systems*, Brno, Czech Republic, September 2010

Singhal, A.K., Athavale, A.K., Li H, Jiang, Y., 2002 "Mathematical basis and validation of the full cavitation model". *Journal of Fluid Engineering*; **124**, 617–24.
http://dx.doi.org/10.1115/1.1486223

Som, S., Longman, D.E., Ramírez, A.I., Aggarwal, S.K., 2010, "A Comparison of Injector Flow and Spray Characteristics of Biodiesel with Petrodiesel," Fuel, **89**, 4014–4024
http://dx.doi.org/10.1016/j.fuel.2010.05.004

Subramanian, K.A. and Ramesh, A., 2002, "Use of Diethyl Ether Along with Water-Diesel Emulsion in a Di Diesel Engine," *SAE paper*, 2002-01-2720

Takenaka, N., Kadowaki, T., Kawabata, Y., Lim, I.C., Sim, C.M., 2004, "Visualization of Cavitation Phenomena in a Diesel Engine Fuel Injection Nozzle by Neutron Radiography," *Nuclear Instruments and Methods in Physics Research Section A: Accelerators, Spectrometers, Detectors and Associated Equipment,* **52**(1-3), 129-133.
http://dx.doi.org/10.1016/j.nima.2005.01.089

Vijayakumar, T, Thundil Karuppa Raj R, and Nanthagopal, K., 2011, "Effect of Injection Pressure on the Internal Flow Characteristics for Diethyl and Dimethyl Ether and Diesel Fuel Injectors," *Thermal Science*, **15**(4), 1123-1130
http://dx.doi.org/10.2298/tsci100717091v

26

CFD MODELLING AND VALIDATION OF COMBUSTION IN DIRECT INJECTION COMPRESSION IGNITION ENGINE FUELLED WITH JATROPHA OIL BLENDS WITH DIESEL

Biswajit De*, Rajsekhar Panua

Mechanical Engineering Department, National Institute of Technology, Agartala, Tripura, 799055, India

ABSTRACT

This paper presents a pre-mixed combustion model for diesel and Jatropha oil blends combustion studies. Jatropha oil blends are considered as a mixture of diesel and Jatropha oil. CFD package, FLUENT 6.3 is used for modeling the complex combustion phenomenon in compression ignition engine. The experiments are carried out on a single cylinder, four strokes, water cooled direct injection compression ignition engine at compression ratio of 17.5 at full load condition at constant speed of 1500 rpm fuelled with diesel and jatropha oil blends with diesel. The numerical model is solved by considering pressure based, implicit and unsteady solver and the effect of turbulence has taken into account. For turbulence modeling, RNG k-ε model is used. Sub models such as droplet collision model and TAB model are used for spray modeling. Species transport and pre-mixed combustion model are used for in-cylinder combustion. Computational results for a four stroke single cylinder diesel engine are comparing favorably against experiments. The results are in good agreement with experimental data. The present development yields a basis for detailed CFD studies of diesel and Jatropha oil blends combustion in a four stroke single cylinder diesel engine.

Keywords: *k-ε model; pre-mixed combustion model; simulation;*

1. INTRODUCTION

CFD is an efficient tool for studying fluid flow, mixture formation and combustion in internal combustion engines, where size makes experimentation very expensive. As a rapid and cost effective tool, CFD is being increasingly used in different stages of engine design, optimization and performance analysis. Difficulties arise when simulating a DI engine where the combustion can take place under partially mixed conditions. The combustion model is required to be able to handle with both the premixed and non-premixed burning and their transition. Goldsworthy (2006) investigated a simplified Heavy fuel oil (HFO) evaporation and combustion model, in which the fuel was considered as a mixture of a heavy (residual) and a light (cutter) component. The model was implemented in the CFD code Star-CD, and tested against experimental data for constant volume combustion chambers, for two representative heavy fuels, one of poor combustion quality and one of good combustion quality. The results were compared to experiments and a good agreement was found between measured and computed data for ignition delay, burning rate, and spray and flame structure, including flame liftoff length. A more detailed approach of HFO modeling is presented in the studies of Struckmeier *et al.* (2009, 2010). In their work, evaporation, ignition and combustion models were further developed, still accounting for a two-component fuel. Their CFD results, obtained with a KIVA-based code, report a good agreement with experimental data in constant volume combustion chambers, in terms of spray and combustion development. Shundoh *et al.* (1992) investigated that multiple injections divide the total quantity of fuel into two or more injections per combustion event. A pilot injection is also usually defined as an injection where 15% or less of the total mass of fuel is injected in the first injection. Many researchers are now investigating pilot and split injection as an effective means to simultaneously reduce NOx and soot emissions. He also reported that NOx could be reduced by 35%, and smoke by 60 to 80%, without a penalty in fuel economy if pilot injection was uses in con junction with high pressure injection. Nehmer *et al.* (1994) studied the effect of split injection in a heavy-duty diesel engine by varying the amount of fuel in the first injection from 10% to 75% of the total amount of fuel. They found that split injection better utilized the air charge and allowed combustion to continue later into the power stroke than for a single injection case, without increased levels of soot production. Tow *et al.* (1994) found that using a double injection with a relatively long dwell on a heavy duty engine resulted in a reduction of particulate emissions by a factor of three with no increase in NOx and only a slight increase in BSFC compared to a single injection. Zhang (1999) used a single cylinder HSDI diesel engine to investigate the effect of pilot injection with EGR on soot, NOx and combustion noise, and found that pilot injection increased soot emissions. The author also showed that reducing the amount of fuel in the pilot injection and increasing the interval between pilot and main injections could reduce the pilot flame area when the main injection starts, resulting in lower soot emissions.

Montgomery(1996) simulated a single cylinder version of a Caterpillar 3400 series heavy duty DI diesel engine using Fluent 6.2 and compared to published experimental data. The objective of the study was to validate an ignition model in conjunction with the existing dynamic mesh and spray models against an established data set consisting of six different load and speed conditions (modes) from a federal transient test procedure. An additional focus of the work was the evaluation of the applicability of the current models for predicting production of nitrogen

* Currently in New Business or Academic Affiliation, City, State/Province, Zip Code, Country

oxides at high temperatures (thermal NO_x). Patterson and Reitz (1998) has been developed a new spray model to improve the prediction of model in conjunction with the existing dynamic mesh and spray diesel engine combustion and emissions using the KIVA-II CFD code. The accuracy of modeling the spray breakup process has been improved by the inclusion of Rayleigh-Taylor accelerative instabilities, which are calculated simultaneously with a Kelvin-Helmholtz wave model. This model also improves the prediction of the droplet sizes within a diesel spray and provides a more accurate initial condition for the evaporation, combustion, and emissions models. The objective of the present study is to validate a pre-mixed combustion models against an experimental data set consisting of diesel and jatropha oil blends at full load conditions and at constant speed of 1500 rpm.

2. CFD MODELS

The modeling of flow field of continuous and dispersed phases of combustion are carried out in detail using a commercial CFD package, FLUENT. The two dimensional in-cylinder, transient and reacting flow system in a direct injection Diesel engine is modeled by solving a set of governing equations from the law of conservation of mass, momentum, energy and species.

2.1 Turbulence Model

Turbulence is distinguished by fluctuation of velocity field. The RNG k-ε model (EI Tahry *et al.*, 1983) is which the turbulent Reynolds number forms of the k and ϵ equation are used in conjunction with the algebraic 'law of the wall' representation of flow, heat and mass transfer for the near wall region. The RNG k-ε model having an advantage to include the effect of swirl compared to the standard k-ε model, which is important for internal combustion engine combustion analysis. Transport equations for the RNG k-ε model is defined as,

$$\frac{\partial}{\partial t}(\rho k)+\frac{\partial}{\partial x_i}(\rho k u_i)=\frac{\partial}{\partial x_j}\left(\alpha_k \mu_{eff}\frac{\partial k}{\partial x_j}\right)+G_k+G_b-\rho\in -Y_M+S_k \quad (1)$$

$$\frac{\partial}{\partial t}(\rho\in)+\frac{\partial}{\partial x_i}(\rho\in u_i)=\frac{\partial}{\partial x_j}\left(\alpha_\in \mu_{eff}\frac{\partial \in}{\partial x_j}\right)+C_{1\in}\frac{\in}{k}(G_k+C_{2\in}G_b)$$

$$-C_{2\in}\rho\frac{\in^2}{k}-R_\in+S_\in \quad (2)$$

In these equations, G_k characterizes the generation of turbulence kinetic energy; G_b is the generation of turbulence kinetic energy due to buoyancy. Y_M represents the contribution of the fluctuating dilatation in compressible turbulence. The quantities α_k and α_ϵ are the inverse effective Prandtl numbers for k and ε respectively. S_k and S_ϵ are user defined source terms. The model constants $C_{1\epsilon}$ and $C_{2\epsilon}$ in equation have values derived analytically by the RNG theory.

2.2 Spray breakup Model

There are mainly two Spray breakup models in fluent, the TAB and the wave model. The distorting droplet effect of TAB model (Huh *et al.*, 1991) is considered in the present study. The equation governing a damped force oscillator is,

$$F-k^x-d\frac{dx}{dt}=m\frac{d^2x}{dt^2} \quad (3)$$

where x is the displacement of the droplet equator from its spherical position and the coefficients of this equation are taken from Taylor's analogy;

$$\frac{F}{m}=C_F\frac{\rho_g u^2}{\rho_l r},\quad \frac{k}{m}=C_k\frac{\sigma}{\rho_l r^2}\ \text{and}\ \frac{d}{m}=C_d\frac{\mu_l}{\rho_l r^2} \quad (4)$$

where ρ_g and ρ_l are the continuous phase and discrete phase densities, u is the relative velocity of the droplet, r is the undisturbed droplet radius, σ is the droplet surface tension and μ_l is the droplet viscosity, C_F, C_k and C_d are dimensionless constants.

2.3 Droplet collision Model

There Droplet collision model (O'Rourke *et al.*, 1981) is which includes tracking of droplets; for estimating the number of droplet collisions and their outcomes. When two parcels of droplets collide then algorithm further establish the type of collision. Only coalescence and bouncing outcomes are measured. The probability of each outcome are calculated from the collision Weber number (We_l) and fit to experimental observation. The Weber number is given as,

$$We_l=\frac{\rho U_{rel}^2\overline{D}}{\sigma} \quad (5)$$

where U_{rel} the relative velocity between two is parcels and $\check{D}$ is the arithmetic mean diameter of the two parcels. The state of the two colliding parcels is modified based on the outcome of the collision.

2.4 Wall Film Model

There Spray-wall interaction is an important element of the mixture creation process in diesel engines. In a DI diesel engine, fuel is injected directly into the combustion chamber, where the spray can impinge upon the piston. The modelling of the wall-film inside a DI engine is compounded by the occurrence of carbon deposits on the surfaces of the combustion chamber. This carbon deposit soak up the liquid layer. It is understood that the carbon deposits absorb the fuel later in the cycle. The wall-film model (Bai *et al.*, 1996) in Fluent allows a single constituent liquid drop to impinge upon a boundary surface and form a thin film. Interaction during impact with a boundary and the criteria by which the regimes are detached are based on the impact energy and the boiling temperature of the liquid. The impact energy is defined by,

$$E^2=\frac{\rho V_r^2 D}{\sigma}\left(\frac{1}{\min\left(\frac{h_o}{D_1}\right)+\frac{\delta_{bl}}{D}}\right) \quad (6)$$

where ρ is the liquid density, V_r is the relative velocity of the particle in the frame of the wall; D is the diameter of the droplet and σ is the surface tension of the liquid. Here δ_{bl} is a boundary layer thickness.

2.5 Combustion Model

The combustion model is combined with species transport and pre-mixed combustion to simulate the overall combustion process in a diesel engine. This approach is based on the solution of transport equation for species mass fraction. The reaction rates that emerge as source terms in the species transport equation are computed from well known Arrhenius rate expressions. The turbulent premixed combustion model, involves the solution of a transport equation for the reaction progress variable. The premixed combustion model (Colin *et al.*, 2004) thus considers the reacting flow field to be divided into regions of burnt and unburnt species, separated by the flame sheet. The progression of the reaction is therefore the same as the progression of the flame front. The flame front propagation is modeled by solving a transport equation for the scalar quantity *c*, the (Favre averaged) reaction progress variable:

$$\frac{\partial}{\partial t}(\rho c)+\nabla.(\rho\bar{v}c)=\nabla.\left(\frac{\mu_t}{S_{ct}}\nabla_c\right)+\rho S_c \quad (7)$$

where c = reaction progress variable, Sc_t = turbulent Schmidt number for the gradient turbulent flux, Sc = reaction progress source term (s−1). The progress variable is defined as

$$c=\frac{\sum_{i=1}^{n}Y_i}{\sum_{i=1}^{n}Y_{i,ad}} \quad (8)$$

where n = number of products, Y_i = mass fraction of species i, $Y_{i,ad}$ = mass fraction of species i after complete adiabatic combustion. Based on this definition, $c = 0$ where the mixture is unburnt and $c = 1$ where the mixture is burnt: $c = 0$: unburnt mixture, $c = 1$: burnt mixture. The value of c is defined as a boundary condition at all flow inlets. It is usually specified as either 0 (unburnt) or 1 (burnt).

3 MODEL DEVELOPMENT

3.1 Geometry Development and Meshing of Computational Domain

In present work, the geometry has been modeled and meshed in pre-processor Gambit 2.4.6 and then it is exported to fluent 6.3 for simulation and prediction. Figure 1 shows the computational domain of two dimensional combustion chamber geometry of the diesel engine with inlet and exhaust ports. Both intake and exhaust ports have been meshed with quadrilateral structured mesh in the zone upstream of the valves and the combustion chamber with triangular structured mesh. The combustion chamber is bowl-in-piston type, which having a hemispherical groove on piston top.

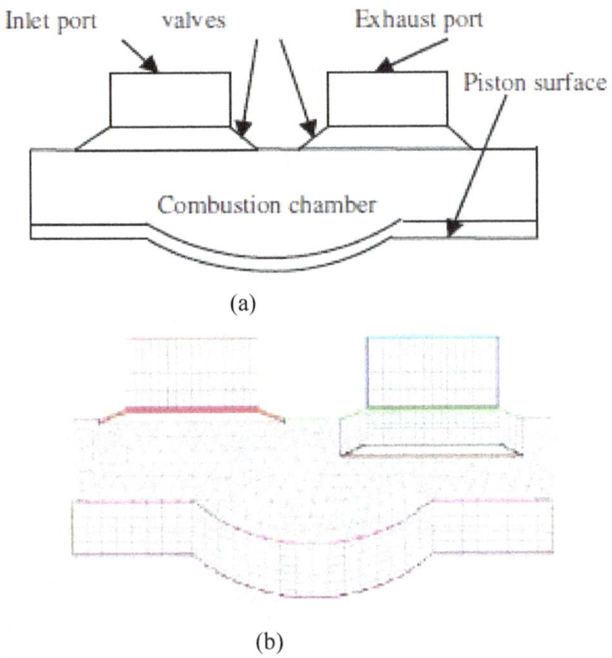

Fig. 1 Physical model and mesh: (a) Geometry of combustion chamber with valves and **(b)** Mesh structure of computational domain for model geometry at 600 degree Crank angle

The geometry has been modeled at its zero degree crank angle position at TDC as shown in figure 1(a). Figure 1(b) shows the mesh structure of computational domain for model geometry at 600 degree Crank angle.

4 MODELING METHODOLOGY

In the present study, the numerical model is solved as unsteady, first order implicit considering turbulence effects to simulate the internal combustion for the engine. The numerical methodology is segregated pressure-based solution algorithm. For solving species, the discrete phase injection with species transport equation and pre-mixed combustion equations are used. The upwind scheme is used for the discretization of the model equations and a finite-volume-based technique to convert the governing equations to algebraic equations that can solve numerically. The governing equations for mass, momentum and energy equations used and appropriate initial boundary conditions are chosen for combustion analysis.

4.1 Grid independency tests

The grid independency test of the model is carried out. It can be observed that increasing the cells beyond 30000 cells does not alter the in-cylinder peak pressure and other variables. Thus the grids that are solved by final volume method are independent beyond 30000 cells at TDC condition.

5 VALIDATION

Experiments are performed on a fully instrumented, single cylinder, four strokes, water cooled direct injection compression ignition engine at compression ratio of 17.5 at full load condition at constant speed of 1500 rpm fuelled with diesel and jatropha oil blends with diesel. The specification of test engine is given in the Table 1 and the analyzed results for various physical, chemical and thermal properties of Diesel and Jatropha oil are given in Table 2. Pressure is recorded with crank angle sensor revolution /degree. For digital load measurement strain gauge sensor, range 0-50 Kg with eddy current dynamometer is used. Labview based Engine Performance Analysis software package "EnginesoftLV" is provided for on line performance evaluation. For all settings, the cylinder pressure values are recorded thrice and a mean of these is taken for comparison.

To verify the results from simulation, the pressure data computed is compared against experimental pressure data from experiments. Fig. 2 shows that the computed in-cylinder pressure data from numerical simulation are in good agreement with the experimental data.

Table 1 Specification of the test engine.

S. No	*Parameters*	*Specification*
1.	General Details	Single cylinder, four stroke compression ignition engine, constant speed, vertical, water cooled, direct injection
2.	Stroke	110 mm
3.	Bore	87.5 mm
4.	Displacement	661 cc
5.	Compression ratio	17.5
6.	Rated output	3.7 KW
7.	Rated speed	1500 rpm

Table 2 Properties of mineral diesel and Jatropha oil.

Property	*Mineral diesel*	*Jatropha oil*
Density(kg/m^3)	841	917.5
API gravity	37.123	22.8925
Kinematic viscosity at 40° C (cSt)	2.575	36.63
Pour point (°C)	-6	4.5
Fire point (°C)	104	275
Cloud point (° C)	3.5	9.5
Flash point (°C)	72	230
Calorific value (MJ/kg)	44.864	38.6355
Carbon (%, w/w)	80.32	76.113
Hydrogen (%, w/w)	12.358	10.517
Nitrogen (%, w/w)	1.758	0

6 RESULTS AND DISCUSSIONS

6.1 Cylinder pressure rise results

Figure 2 shows the modeling and experimental in-cylinder pressure at CR of 17.5 operating at full load condition. The modeled cylinder pressure data shows good agreement with experimental results. The maximum pressure rise depends upon the quantity of fuel vaporized during the delay period and occurs in the state of combustion, some degrees after the beginning of combustion.

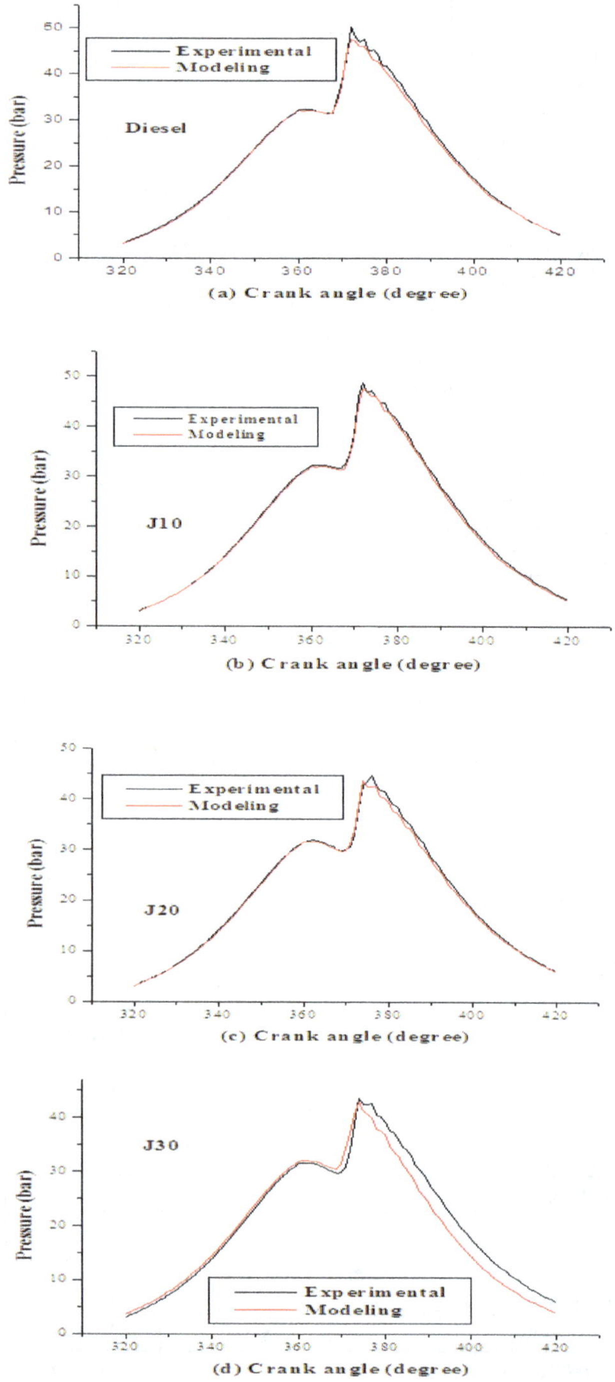

Fig. 2 Comparison between modeling and experimental pressure diagram: (a) diesel fuel, (b) 10% jatropha oil blend, (c) 20% jatropha oil blend, (d) 30% jatropha oil blend.

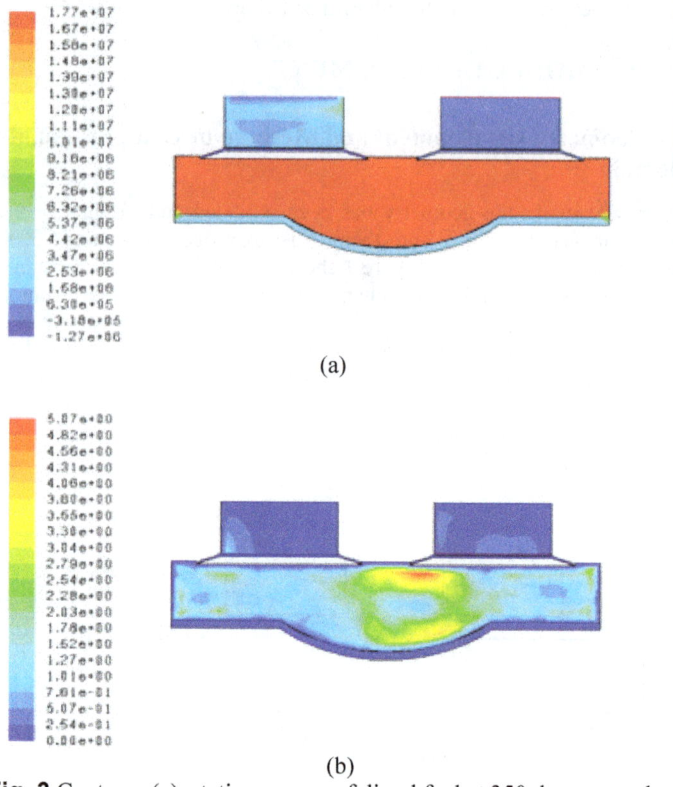

Fig. 3 Contours (a): static pressure of diesel fuel at 350 degree crank angle, (b): velocity magnitude of 10% Jatropha oil blend at 370 degree crank angle.

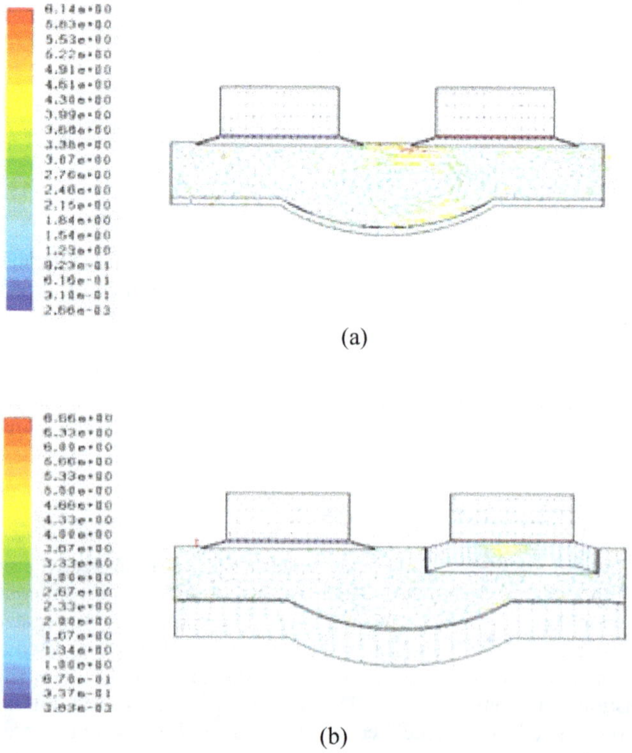

Fig. 4 Velocity vectors by velocity Magnitude (a): 20% Jatropha oil blend at 370 degree crank angle, (b): 30% Jatropha oil blend at 370 degree crank angle.

Therefore both scale and timing of occurrence of peak pressure are precisely predicted by the model. It can be noted that experimental peak pressure is about 50.24 bars at 372 degree CA and modeling peak pressure is 49.18 bars at 373 degree CA for base line diesel fuel.

The values of the experimental peak pressure for different Jatropha oil blends is such as that for 10%, 20% and 30% is found to be less than that of diesel fuel by 3.23%, 9.08%, and 12.27%. The corresponding modeling pressure for the Jatropha blends is found to be less by 3.17%, 10.12%, and 12.58% with respect to diesel fuel. Figure 3(a) shows the contours of static pressure of diesel fuel at 350 degree crank angle, (b): Contours of velocity magnitude of 10% Jatropha oil blend at 370 degree crank angle and 4(a): Velocity vectors by velocity Magnitude of 20% Jatropha oil blend at 370 degree crank angle, (b): Velocity vectors by velocity Magnitude of 30% Jatropha oil blend at 370 degree crank angle respectively.

7 CONCLUSIONS

The numerical model integrated with sub models includes spray, droplet collision, wall film and combustion model with species transport and pre-mixed combustion theory. The bowl-in piston combustion geometry is used for model construction. In this study RNG k-ε model is implemented to confine in-cylinder turbulence. Simulated results of in-cylinder pressure for pure diesel, 10%, 20% and 30% jatropha oil blends with diesel have been analyzed. A good agreement between the modeling and experimental data ensures the accuracy of the numerical prediction of this work. The comparison shows that the present model manages to predict the combustion characteristics quite well. The result reported in this paper illustrate that the numerical simulation can be one of the most powerful and beneficial tool to compute the essential features of combustion parameters for ICE development, optimization and performance analysis.

NOMENCLATURE

RNG	Renormalization Group Theory
TAB	Taylor Analogy Breakup
CFD	Computational Fluid dynamics
G_k	the generation of turbulence kinetic energy
G_b	the generation of turbulence kinetic energy due to buoyancy
Y_M	the contribution of the fluctuating dilatation in compressible turbulence
α_k and α_ε	the inverse effective Prandtl numbers for k and ε respectively
S_k and S_ε	the user defined source terms
$C_{1\varepsilon}$ and $C_{2\varepsilon}$	the model constants derived analytically by the RNG theory
x	the displacement of the droplet equator from its spherical position
ρ_g and ρ_l	the continuous phase and discrete phase densities
u	the relative velocity of the droplet
r	the undisturbed droplet radius
ρ	the droplet surface tension
μ_l	the droplet viscosity
C_f, C_k and C_d	dimensionless constants
U_{rel}	the relative velocity between two is parcels
$\check{D}$	the arithmetic mean diameter of the two parcels
ρ	the liquid density
V_r	the relative velocity of the particle in the frame of the wall
D	the diameter of the droplet and ρ is the surface tension of the liquid
ρ_{bl}	a boundary layer thickness
c	reaction progress variable
Sct	turbulent Schmidt number for the gradient turbulent flux
Sc	reaction progress source term (s−1).
n	number of products
Yi	mass fraction of species *i*
Yi,ad	mass fraction of species *i* after complete adiabatic combustion
ICE	Internal combustion engine
DI	Direct injection
CA	Crank angle
J10	10% Jatropha oil and 90% Diesel oil in blends
J20	20% Jatropha oil and 80% Diesel oil in blends
J20	20% Jatropha oil and 80% Diesel oil in blends

REFERENCES

Andreadis, P., Zompanakis, A., Chryssakis, C. and Kaiktsis, L., 2011, "Effects of Fuel Injection Parameters on the Performance and Emissions in a Large-bore Marine Diesel Engine," *International Journal of Engine Research*, 2(1),14–29. http://dx.doi:10.1243/14680874JER511

Bai,C., and Cosman, A.D., 1996, "Mathematical Modeling of Wall Films Formed by Impinges Sprays," *SAE Technical Paper Series*, 960626.

Colin, O. and Benkenida, A., 2004, "The 3-Zone Extended Coherent Flame Model (ECFM3Z) for Computing Premixed/Diffusion Combustion," *Oil & Gas Science and Technology-Rev, IFP*, 59(6), 593-609. http://dx.doi.org/10.2516/ogst:2004043

EI Tahry, S.H., 1983, "K-ε Equation for Compressible Reciprocating Engine Flows," *AIAA, Journal of Energy*, 7(4), 345-353. http://dx.doi:10.2514/3.48086

Goldsworthy, L., 2006, "Computational Fluid Dynamics Modelling of Residual Fuel Oil Combustion in the Context of Marine Diesel Engines," *International Journal of Engine Research*, 7(2), 181–199. http://dx.doi:10.1243/146808705X30620

Huh, K.Y. and Gosman, A.D., 1991, "A Phenomenological Model of Diesel Spray Atomization," ICMF1991, *Proceedings of International Conference on Multiphase Flow*, Tsukuba.

Ng, H. K. and Mahamed Ismail,H., 2012, "Evaluation of Non-Premixed Combustion and Fuel Spray Models for in-cylinder Diesel Engine Simulation," *Applied Energy*, 90, 271-279. http://dx.doi:10.1016/j.apenergy.2010.12.075

Montgomery, D. and Reitz, R. D.,1996, "Six Mode Cycle Evaluation of the Effect of EGR and Multiple Injections on Particulate and NO Emission from a DI Diesel Engine," *SAE paper*, 960316.

Mobasheri, R. and Peng, Z., 2012, "Analysis the Effect Of Advanced Injection Strategies on Engine Performance and Pollution Emissions in A Heavy Duty Di-diesel Engine by CFD Modeling," *International Journal of Heat and Fluid Flow*, 33, 59-69. http://dx.doi:10.1016/j.ijheatfluidflow.2011.10.004.

Nehmer, D.A., and Reitz, R.D., 1994, "Measurement of the Effect of injection Rate and Split Injections on Diesel Engine Soot and NOx Emissions'" *SAE Paper*, 940668.

O'Rourke, P.J., 1981, "Collective Prop Effects on Vaporizing Liquid Sprays," *PhD Thesis*, University of Princeton.

Patterson, M.A. and Reitz, R.D., 1998, "Model the Effects of Fuel Spary Characteristics on Diesel Engine Combustion and Emission," SAE *paper*, 980131.

Stamoudis, N. and Chryssakis, C., 2014, "A Two Component Heavy Fuel Oil Evaporation Model for CFD Studies in Marine Diesel Engines," *Fuel*, 115,145-153. http://dx.doi.org/10.1016/j.fuel.2013.06.035

Struckmeier, D., Tsuru, D., Kawauchi, S., and Tajima, H., 2009, "Multi-Component Modeling of Evaporation, Ignition and Combustion Processes of Heavy Residual Fuel Oil," SAE Technical Paper Series 2009-01-2677, *SAE International Powertrains, Fuels and Lubricants Meeting*. San Antonio, Texas.

Struckmeier, D., Tajima, H., and Tsuru, D., 2010, "New Application and Modeling of Low Ignitability Fuel for Marine Engines," *CIMAC 2010*, Paper No. 117, Bergen, Norway.

Shundoh, S., Komori, M., Tsujimura, K., and Kobayashi, S., 1992, "Nox Reduction from Diesel Combustion Using Pilot Injection with High Pressure Fuel Injection," *SAE paper*, 920461.

Tow, T.C., Pierpont, A., and Reitz, R.D., 1994, "Reducing Particulate and NOx Emissions by Using Multiple Injections in a Heavy Duty D.I. Diesel Engine," *SAE Paper* 940897.

Zhang, L., 1999 "A Study of Pilot Injection in a DI Diesel Engine," *SAE paper*, 01-3493.

NUMERICAL ANALYSIS OF NATURAL CONVECTION IN A RIGHT-ANGLED TRIANGULAR ENCLOSURE

Manoj Kr. Triveni*, Dipak Sen, RajSekhar Panua

National Institute of Technology, Agartala, Tripura, 799055, India

ABSTRACT

A numerical investigation has been performed for heat transfer analysis in a right-angled triangular enclosure filled with water. The side wall of the enclosure is maintained at high temperature compare to the base wall while hypotenuse is kept thermally insulated. Two - dimensional steady-state continuity, momentum and energy equations along with the boussinesq approximation are solved by finite volume method using commercial available software, FLUENT 6.3. The computational results are shown in terms of isotherms, streamlines and velocity contour for Rayleigh number ($10^5 \leq Ra \leq 10^7$). The heat transfer is presented in terms of local and average Nusselt number. The result encapsulates that both flow field and temperature distributions are affected with Rayleigh number. The simulated results are validated with the experimental and numerical results and it shows a good agreement with the published results. Finally, a correlation for Nusselt number (Nu) with Rayleigh number (Ra) has been developed for vertical hot wall.

Keywords: *Natural convection, Triangular enclosure, Rayleigh number, Numerical simulation.*

1. INTRODUCTION

Many researchers have been working on natural convection using a triangular cavity due to its wide applications in building insulation, solar collector and electronic equipments. Akinsete and Coleman (1982) were used a numerical technique to investigate the laminar natural convection in air contained in a long horizontal right-triangular enclosure. Steady-state solutions had been obtained for height-base ratios of $0.0625 </ H/B \leqslant 1.0$ for Grashof number of $800 \leqslant Gr_{(B)} \leqslant 64\,000$. Results were reported that the heat transfer across the base wall increases towards the hypotenuse/base intersection such that the third of the base length nearest the intersection accounts for about 60% of the heat transferred across the base. Asan and Namli (2000) have carried out a numerical study for laminar natural convection in a pitched roof cross section under summer day boundary conditions. Problem was solved for different height – base ratio and Ra = $10^3 – 10^6$. Aydin and Yesiloz (2011) have studied the buoyancy induced flow and heat transfer mechanisms in a water-filled quadrantal cavity experimentally and numerically. It has been concluded from the investigation that the influence of Rayleigh number is insignificant at Ra< 10^3 and it becomes significant beyond 10^4.

Basak et al. (2012) have worked on entropy generation due to natural convection in right-angled triangular enclosures filled with porous media. It was observed that the total entropy generation is increasing function of Darcy number. Ghasemi and Aminossadati (2010) have done a numerical study on natural convection in right triangular enclosure with heat source on its vertical wall and water-CuO nanofluid as a source medium. Parameters such as Rayleigh number, solid volume fraction, heat source location and enclosure aspect ratio which affect the heat transfer rate were examined. Enhancement in heat transfer was observed for upper location of heat source and high aspect ratio for high Rayleigh number. Kaluri et al. (2010) have analyzed natural convection in right-angled triangular enclosures with various top angles. It was observed that for lower angle (15^0), the average nusselt number remains invariant with increase in Ra, but the enhanced convection at higher Ra significantly affects the heat flow distribution. Kent (2009) has done a numerical analysis in an isosceles triangular cavity by varying the aspect ratio and base angle from 15° to 17°. The lower aspect ratio has high heat transfer rate from the bottom surface of the triangular enclosure. Ridouane et al. (2005) have numerically computed laminar natural convection in a right-angled triangular cavity filled with air. The vertical side wall is considered as hot wall while cooling is done from hypotenuse of the triangular enclosure. It was examined that the heat transfer enhancement get decreased when both apex angle and Rayleigh number diminishes.

Saha (2011) has inspected the heat transfer and fluid flow in a triangular enclosure for instantaneous heating on the inclined walls. Investigation shows that the nusselt number was very high initially due to conduction effect and after that it decreases gradually and become steady state. Sun and Pop (2011) were numerically investigated the heat transfer behavior of nanofluids in a triangular enclosure filled with porous medium. Heat transfer rate was solved for different parameters such as Rayleigh number (Ra), size of heater (Ht), position of heater (Yp) and enclosure aspect ratio. The maximum value of average Nusselt number is obtained for decreasing aspect ratio, lowering the heater position with the highest value of Rayleigh number and the largest size of heater. Also, heat transfer is enhancing with the increasing of solid volume fraction at low Rayleigh number. Tzeng et al. (2005) proposed the Numerical Simulation Aided Parametric Analysis method to solve natural convection equations in streamline-vorticity form.

Varol et al. (2006) used a flush mounted heater on side wall of the triangular cavity. Aspect ratio of the triangle, location of heater, length of heater and Rayleigh number are some parameters which considered for heat transfer analysis. It was reported that the heat transfer is

* Corresponding author. Email: triveni_mikky@yahoo.com

affected due to both position and location of heater and increases with the increase in Rayleigh number. Varol et al. (2007) attached a thin solid adiabatic fin in porous right-angled triangular enclosure and solved numerically using finite volume method. They found that the Nusselt number is an increasing function of Rayleigh number and with the increasing of dimensionless solid adiabatic fin height, the heat transfer rate decreases. Varol et al. (2008) examined heat transfer and fluid flow in a triangular enclosure filled with fluid-saturated porous medium with a conducting thin fin on the hot vertical wall. The investigation explored that the increasing value of Y_p and W_p leads to decrease in heat transfer. Natural convective flow and heat transfer in an inclined quadrantal cavity is studied experimentally and numerically by Yesiloz and Aydin (2011). The effects of the inclination angle, φ and the Rayleigh number, Ra on fluid flow and heat transfer are investigated for the range of angle of inclination between $0^0 \le \varphi \le 360^0$, and Ra from 10^5 to 10^7. It was revealed that heat transfer changes dramatically according to the inclination angle which affects convection currents inside, i.e. flow physics inside. Yesiloz and Aydin (2013) have investigated natural convection heat transfer experimentally and numerically in a right-angled triangular cavity. The cavity is heated from below and cooled from the side wall. A new approach is used to overcome the singularity effect. From the result, it is observed that the heat transfer rate increases with Rayleigh number and a correlation has been developed for Nu.

From the above literature assessment, it is noticed that the plenty of work has been done on a right-angle triangular cavity considering vertical wall as hot wall. Different heat transport medium such as air, water and nanofluid used as working medium for heat transport but no correlation has been developed for vertical wall to understand the phenomena of heat transfer. Hence a correlation is developed between average Nusselt number and Rayleigh number for vertical hot wall of triangular enclosure filled with water.

2. PROBLEM DESCRIPTION

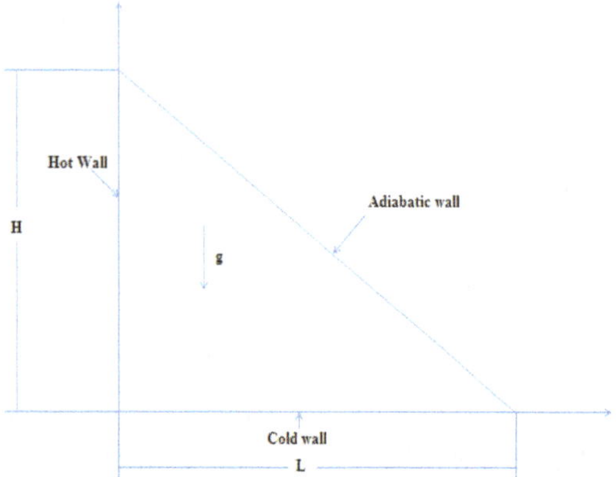

Fig. 1 Graphical presentation of triangular cavity.

The schematic diagram of the triangular enclosure is shown in figure 1. Copper is used for hot and cold walls while Plexiglas as an insulating material for inclined wall. Side and base walls of the triangular cavity are thermally active and the temperature of the side wall is higher than the base wall ($T_h > T_c$) while the hypotenuse is considered as an adiabatic wall.

2.1 Mathematical Formulation

The governing equations are obtained on certain assumptions such as:

- All walls are assumed to be impermeable.
- Radiation effect is negligible.
- Fluid is considered as steady and laminar.
- All properties of the fluid are presumed to be constant except density which changes with temperature (Boussinesq approximation).

With these assumptions the dimensional governing equations can be written as:

$$\frac{\partial u}{\partial x}+\frac{\partial v}{\partial y}=0 \tag{1}$$

$$u\frac{\partial u}{\partial x}+v\frac{\partial u}{\partial y}=-\frac{1}{\rho}\frac{\partial p}{\partial x}+\nu\left(\frac{\partial^2 u}{\partial x^2}+\frac{\partial^2 u}{\partial y^2}\right) \tag{2}$$

$$u\frac{\partial v}{\partial x}+v\frac{\partial v}{\partial y}=-\frac{1}{\rho}\frac{\partial p}{\partial y}+\nu\left(\frac{\partial^2 v}{\partial x^2}+\frac{\partial^2 v}{\partial y^2}\right)+g\beta(T-T_0) \tag{3}$$

$$u\frac{\partial T}{\partial x}+v\frac{\partial T}{\partial y}=\alpha\left(\frac{\partial^2 T}{\partial x^2}+\frac{\partial^2 T}{\partial y^2}\right) \tag{4}$$

By using dimensionless parameters such as

$$X=\frac{x}{H}, Y=\frac{y}{H}, U=\frac{uH}{\alpha}, V=\frac{vH}{\alpha}, \theta=\frac{T-T_c}{T_h-T_c},$$
$$P=\frac{pH^2}{\rho\alpha^2}, \Pr=\frac{\upsilon}{\alpha}, Ra=\frac{g\beta(T_h-T_c)H^3}{\alpha\upsilon} \tag{5}$$

The dimensionless governing equation can be written as

$$\frac{\partial U}{\partial X}+\frac{\partial V}{\partial Y}=0 \tag{6}$$

$$U\frac{\partial U}{\partial X}+V\frac{\partial U}{\partial Y}=-\frac{\partial P}{\partial X}+\Pr\left(\frac{\partial^2 U}{\partial X^2}+\frac{\partial^2 U}{\partial Y^2}\right) \tag{7}$$

$$U\frac{\partial V}{\partial X}+V\frac{\partial V}{\partial Y}=-\frac{\partial P}{\partial Y}+\Pr\left(\frac{\partial^2 V}{\partial X^2}+\frac{\partial^2 V}{\partial Y^2}\right)+Ra\Pr\theta \tag{8}$$

$$U\frac{\partial \theta}{\partial X}+V\frac{\partial \theta}{\partial Y}=\left(\frac{\partial^2 \theta}{\partial X^2}+\frac{\partial^2 \theta}{\partial Y^2}\right) \tag{9}$$

And dimensionless boundary conditions are:

at the hot wall: θ = 1, U = V = 0 at X = 0 and 0 < Y < 1

at the cold wall: θ = 0, U = V = 0 at Y = 0 and 0 < X < 1

at the adiabatic wall: ∂θ/∂n = 0, U = V = 0 at 0 < X < 1 and 0 < Y < 1 (10)

2.2 Numerical Analysis

The problem has solved by commercial available software; FLUENT 6.3 which hire finite volume method. For fluid based problem, FVM is providing better flexibility and preciseness. SIMPLE algorithm coupled with pressure and velocity is used to solve the dimensionless governing equations (6-9) with the corresponding boundary conditions (10). Momentum and energy equations are discretized by second order upwind technique while the pressure interpolation is done by PRESTO scheme. The local nusselt number is calculated with standard nusselt number definition (Nu_x = h.L/*k*) and the average Nusselt number defined by Yesiloz and Aydin (2013) for triangular cavity is:

$$Nu=\left.\frac{dT}{d\phi}\right|_{\phi=0}\frac{1}{T_h-T_c}\frac{\Pi}{2} \tag{11}$$

The stream function can defined as

$$u=\frac{\partial\psi}{\partial x}, v=-\frac{\partial\psi}{\partial x} \tag{12}$$

And the non-dimensional stream function calculate by

$$\Psi = \psi/\alpha \quad (13)$$

3. PROBLEM VALIDATION

3.1 Grid Independency Test

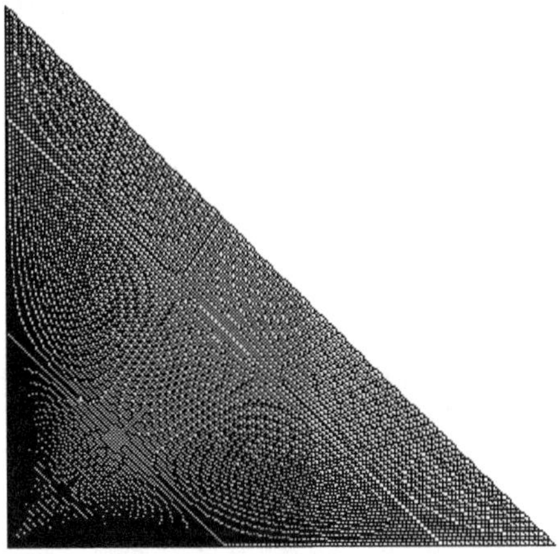

Fig. 2 Grid structure for physical domain.

The local Nusselt number and stream functions have calculated for the triangular cavity by creating meshes but the size and structure of the grid or mesh should be appropriate according the cavity for highly accurate results. Hence, grid independency test is conducted for different sizes such as 1600, 3600, 6400, 10000, 14400 and 19600 to examine the suitable grid density and recommended grid is shown in figure 2.

Table1 Relative error analysis with different grid sizes.

Cells	Ψ_{max}	Deviation (%)	Nu	Deviation (%)
3600	3.3623		5.63	
6400	3.4482	2.49	5.99	6.075
10000	3.5261	2.21	6.25	4.089
14400	3.4163	3.12	6.40	2.400
19600	3.3547	1.8	6.38	0.417
25600	3.5613	5.8	6.68	4.500

The independency on grid is considered on the basis of less deviation in non-dimensional stream function and Nusselt number which calculated by relative error analysis. Table 1 explicates the less deviation at grid 14400 and 19600 compare to others. Hence, 14400 grids have recommended to the present work for less conservation time.

3.2 Validation of Results

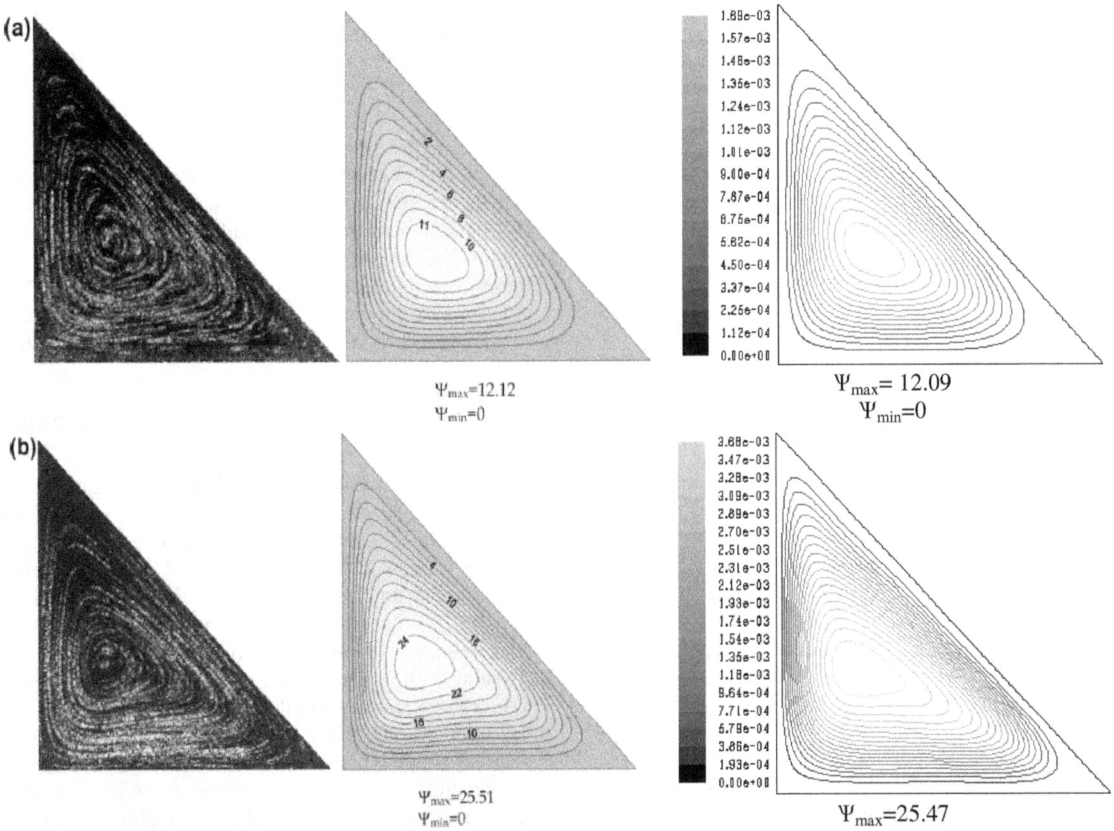

Fig. 3 Experimental and Numerical streamlines (left and middle) from Yesiloz and Aydin (2013) and numerical simulation (right) from the present study for a) Ra = 10^5 and b) Ra = 5×10^5.

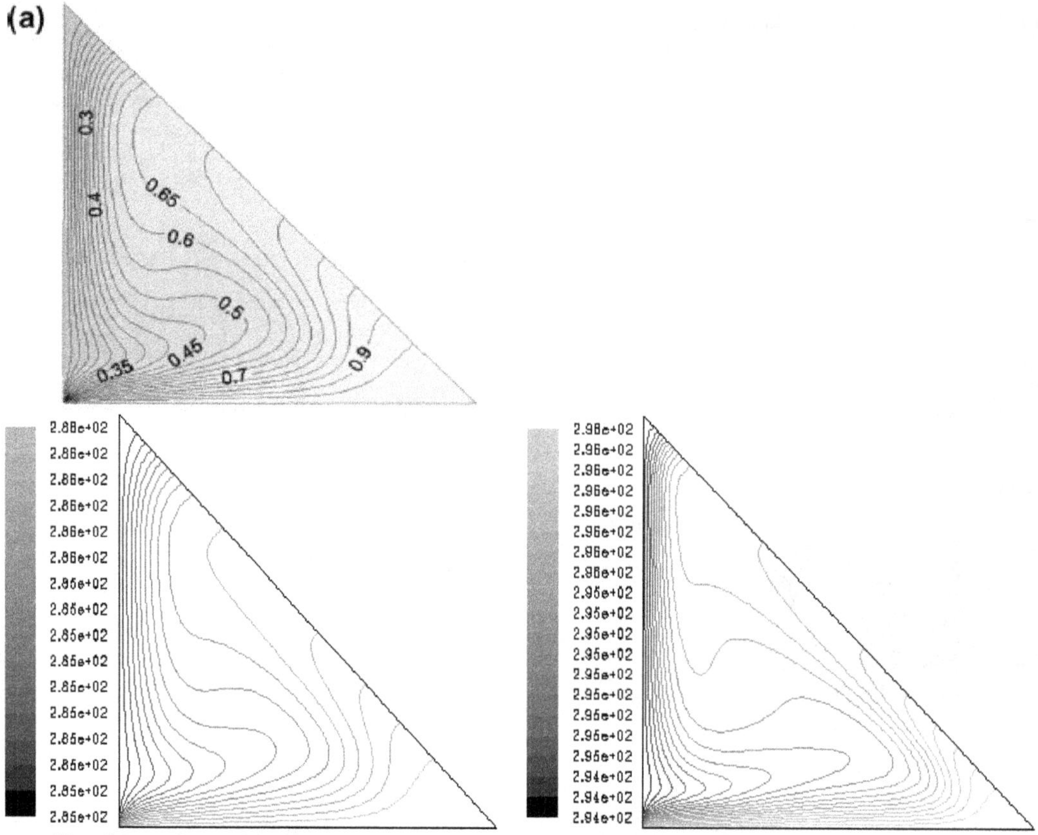

Fig. 4 Experimental and Numerical isotherms (above) from Yesiloz and Aydin (2013) and numerical isotherms (below) from the present study for a) Ra = 10^5 and b) Ra = 5×10^5.

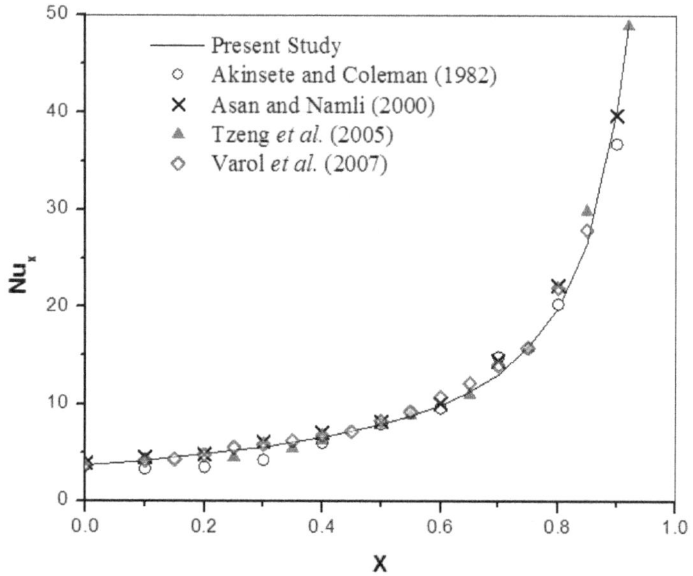

Fig. 5 Validation of present work with other published data for a triangular cavity filled with air at Ra = 2772, Pr = 0.73.

The numerical investigation of laminar natural convection in a triangular cavity has done for different Rayleigh number. The obtained results have been validated with the existing results for right-angle triangular enclosure filled with water when the bottom wall is heated and cooling is done from the side wall. The results obtained at Ra = 10^5 and Ra = 5×10^5 are validated in terms of streamlines and isotherms with the existing experimental and numerical outcomes of Yesiloz and Aydin (2013) keeping same boundary conditions which is shown in figure 3 and 4 respectively. Also, variation in local Nusselt number with position for triangular cavity is validated with the published data which displayed in Fig. 5. Both results are showing quite good agreement with the published results.

4. RESULTS AND DISCUSSIONS

The present problem intended to investigate the heat transfer for different Rayleigh number (10^5 to 10^7) in a triangular cavity. The results have shown by streamlines, isotherms and velocity contour.

Fig. 6 and 7 shows the effect of Rayleigh number on flow field, temperature and velocity distribution. Results unfold that the heat transfer is mainly depending on the fluid flow and temperature distributions. At low Rayleigh number, figure 6a, streamline is nearly circular at the center and are far from the hot wall. As the Rayleigh number increases, streamlines are getting denser at the lower portion of the hot wall. It is no longer circular at the center and turns into a peculiar shape which can be shown in fig. 6 (b) - (e). The outer rings of the streamlines become curvilinear, grow into jet like shape and getting closure to the right corner of the triangular cavity. It is noticeable that the vortex flow of the fluid is cumulative with higher Rayleigh number which indicates the rate of heat transfer is enhancing between hot wall and fluid with increasing Ra.

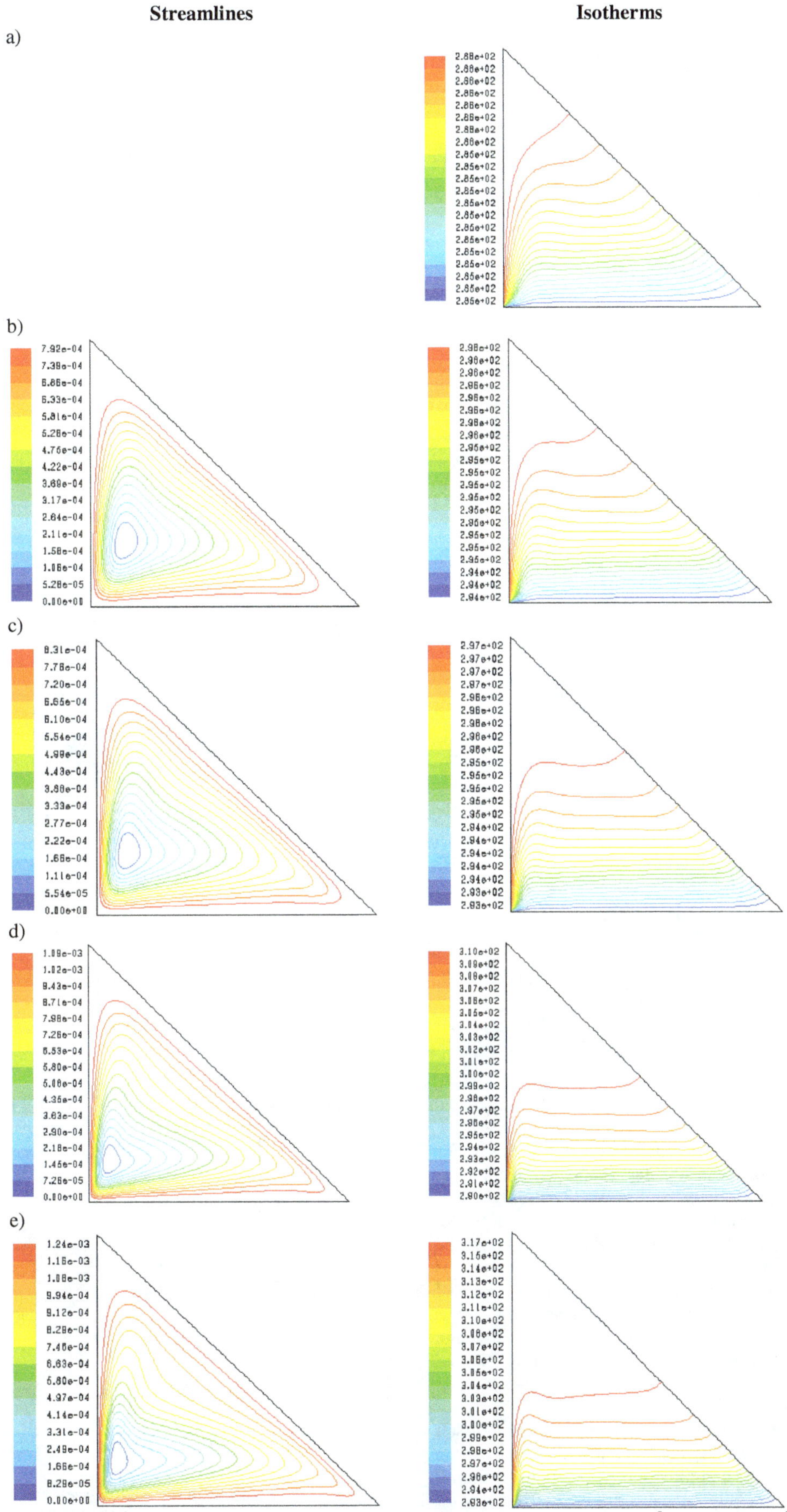

Fig. 6 Streamline (left column) and Isotherms (right column) for; a) Ra = 10^5, b) Ra = 5×10^5, c) Ra = 10^6, d) Ra = 5×10^6, e) Ra = 10^7.

The right column of the figure 6 presented the distribution of the isotherms inside the cavity. At Ra = 10^5, isotherms are distributed over the entire cavity. At higher Rayleigh number (Ra = 5×10^6 and Ra = 10^7) isotherms are separated from the upper corner and getting crowded towards the cold wall and boundary layer formation occurs near the cold wall. It is happening because density of the fluid come to be lighter with growing temperature difference which transfer heat to the cold wall carried by the fluid from the hot wall and thus enhancing the heat transfer rate.

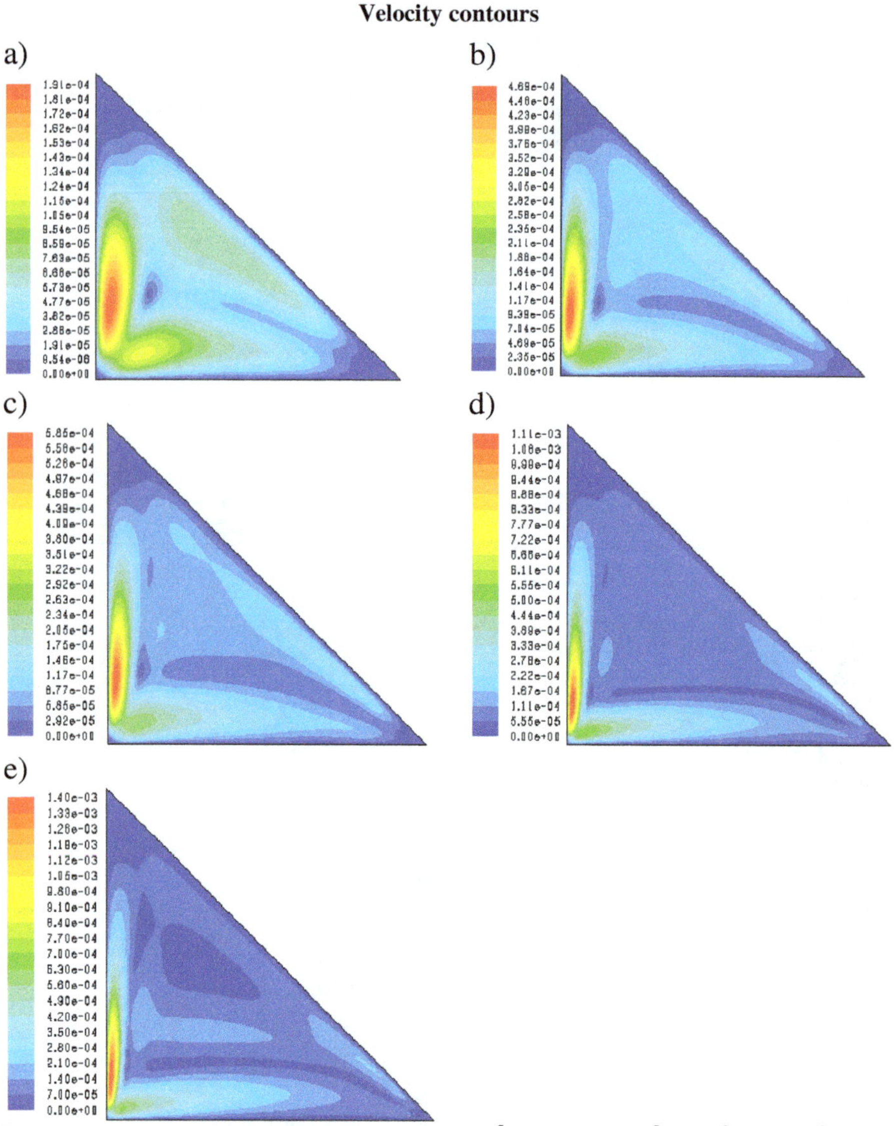

Fig. 7 Velocity contour for Rayleigh number; a) Ra = 10^5, b) Ra = 5×10^5, c) 10^6, d) 5×10^6, e) Ra = 10^7.

Figure 7 presents the velocity contour in triangular cavity for variable Rayleigh number. The heat transfer at the corner of the triangular caity is due to conduction because of negligilbe fluid movment. There are two rings present in velocity contour digram for Ra = 10^5 which shown in figure 7a. However, at high Rayleigh number the vertical ring become thinner and thinner as well as the lower ring is vanishing. From Fig. 6 and 7, it is apparent that density and viscocity of the fluid are decreasing with the increment of fluid velocity and is resonsible for enrichment of convective heat transfer at high rayleigh number. The enhancement in convection attenuates the thermal boundary layer of the fluid will be responsible for heat transfer augmentation.

Figure 8 describes the heat transfer in terms of local Nusselt number (Nu_x) for the hot wall. The Nu_x is almost equal to 0.5 near the hot wall for all Ra. It is higher for non-dimensional position between 0 to 0.2 and after that Nu_x is gradully decreasing with the increase of dimensionless postion in y-direction.

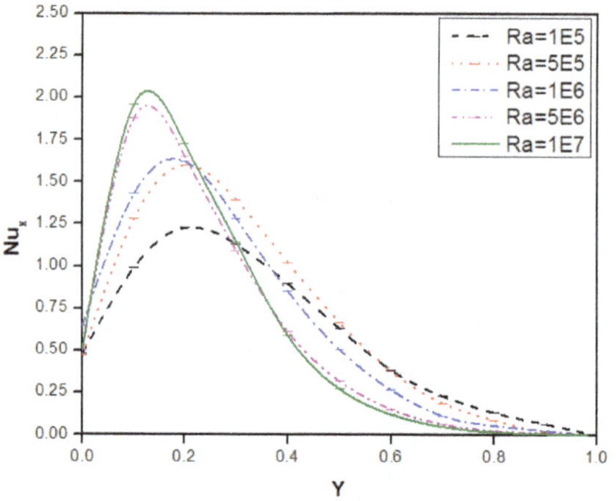

Fig. 8 Variation of local Nusselt number along hot wall for different Ra.

Based on the numerical results obtained, a correlation equation can be established the dependence of Nu on Ra as follows:

$$Nu = 0.3219\ Ra^{1/4}$$

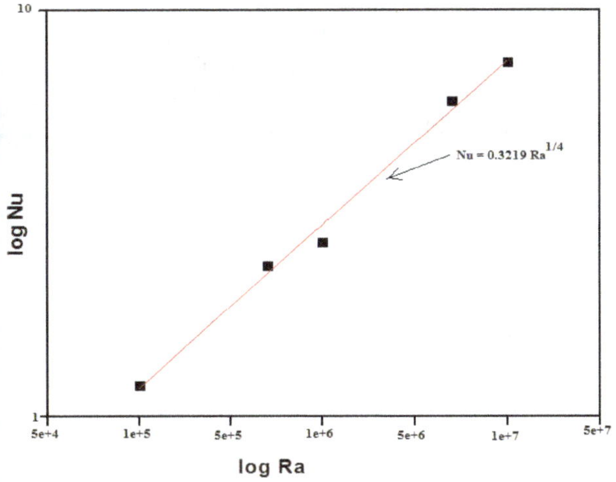

Fig. 9 Correlation curve of average Nu with Ra.

5. CONCLUSIONS

A numerical analysis is done in a water filled right-angle triangular enclosure for different Ra. The results are shown by streamline, isotherms and velocity contour. The heat transfer rate has shown in terms of local and average nusselt number. From the analysis it is observed that the average Nusselt number increases with the increase of Rayleigh number. The local Nusselt number is gradually decreasing with increasing non-dimensional parameter. Also, a correlation has been developed between Nusselt number and Rayleigh number for right-angled triangular cavity with vertical hot wall.

NOMENCLATURE

g	acceleration due to gravity (m/s^2)
k	thermal conductivity (W/m·K)
H	height of the triangular cavity (m)
L	width of the triangular cavity (m)
p	pressure (Pa)
P	dimensionless pressure
u, v	velocity component in x and y direction (m/s)
U, V	dimensionless velocity component in x and y dirction
T	temperature (K)
T_h	hot wall temperature
T_c	cold wall temperature
Pr	prandtl number
Ra	rayleigh number
Nu_x	local Nusselt number
Nu	average Nusselt number

Greek Symbols

α	thermal diffusivity (m^2/s)
β	thermal expansion coefficient (1/K)
θ	dimensionless temperature
ρ	density of water (kg/m3)
μ	dynamic viscosity of water ($kg.m/s^2$)
υ	kinematic viscosity (m^2/s)
ψ	stream function (1/s)
Ψ	dimensionless stream function

Subscripts

h	hot wall
c	cold wall

ACKNOWLADGEMENT

The authors wish to express their very sincerely thanks to the reviewers for their valuable comments and suggestions.

REFERENCES

Akinsete, V.A. and Coleman, T.A., 1982, "Heat Transfer by Steady Laminar Free Convection in Triangular Enclosures," *International Journal of Heat and Mass Transfer*, **25** (7), 991 – 998. http://dx.doi.org/10.1016/0017-9310(82)90074-6

Asan, H. and Namli, L., 2000, "Laminar Natural Convection in a Pitched Roof of Triangular Cross-Section: Summer Day Boundary Conditions," *Energy and Buildings*, **33** (1), 69 – 73. http://dx.doi.org/10.1016/S0378-7788(00)00066-9

Aydin, O. and Yesiloz, G., 2011, "Natural Convection in a Quadrantal Cavity Heated and Cooled on Adjacent Walls," *ASME Journal of Heat Transfer*, **133**, 0525011–7. http://dx.doi.org/10.1115/1.4003044

Basak, T., Gunda, P. and Anandalakshmi, R., 2012, "Analysis of Entropy Generation during Natural Convection in Porous Right-angled Triangular Cavities with Various Thermal Boundary Conditions," *Int. J. of Heat and Mass Transfer*, **55**, 4521-4535. http://dx.doi.org/10.1016/j.ijheatmasstransfer.2012.03.061

Ghasemi, B. and Aminossadati, S.M., 2010, "Brownian Motion of Nanoparticles in a Triangular Enclosure with Natural Convection," *Int. Journal of Thermal Science*, **49**, 931-940. http://dx.doi.org/10.1016/j.ijthermalsci.2009.12.017

Kaluri, R.S., Anandalakshmi, R. and Basak T., 2010, "Benjan's Heatline Analysis of Natural Convection in Right-angled Triangular Enclosure: Effects of Aspect-Ratio and Thermal Boundary Conditions," *Int. J. of Thermal Sciences*, **49**, 1576-1592. http://dx.doi.org/10.1016/j.ijthermalsci.2010.04.022

Kent, E.F., 2009, "Numerical Analysis of Laminar Natural Convection in Isosceles Triangular Enclosures for Cold Base and Hot Inclined Walls," *Mechanics Research Communication,* **36**, 497-508. http://dx.doi.org/10.1016/j.mechrescom.2008.11.002

Ridouane, E.I., Campo, A. and Chang, J. Y., 2005, "Natural Convection Patterns in Right-angled Triangular Cavities with Heated Vertical Sides and Cooled Hypotenuses," *Journal of Heat Transfer,* **127**, 1181- 1186. http://dx.doi.org/10.1115/1.2033903

Saha, S. C., 2011, "Unsteady Natural Convection in a Triangular Enclosure under Isothermal Heating," *Energy and Buildings*, **43**, 695-703. http://dx.doi.org/10.1016/j.enbuild.2010.11.014

Sun, Q. and Pop, I., 2011, "Free Convection In A Triangle Cavity Filled With A Porous Medium Saturated With Nanofluids With Flush Mounted Heater On The Wall," *Int. Journal of Thermal Science*, **50**, 2141-2153. http://dx.doi.org/10.1016/j.ijthermalsci.2011.06.005

Tzeng, S.C., Liou, J.H. and Jou, R.Y., 2005, "Numerical Simulation-aided Parametric Analysis of Natural Convection in a Roof of Triangular Enclosures," *Heat Transfer Engineering*, **26**, 69 – 79. http://dx.doi.org/10.1080/01457630591003899

Varol, Y., 2006, "Natural Convection in A Triangle Enclosure with Flush Mounted Heater on the Wall," *Int. Communication in Heat and Mass Transfer*, **33**, 951-958. http://dx.doi.org/10.1016/j.icheatmasstransfer.2006.05.003

Varol, Y., Oztop, H.F. and Yilmaz, T., 2007, "Natural Convection in Triangular Enclosures with Protruding Isothermal Heater," *Int. J. Heat Mass Transfer*, **50**, 2451-2462. http://dx.doi.org/10.1016/j.ijheatmasstransfer.2006.12.027

Varol, Y., 2008, "Natural Convection in Porous Media-filled Triangular Enclosure with a Conducting Thin Fin on the Hot Vertical Wall," *Proceedings of the Institution of Mechanical Engineers, Part C: Journal of Mechanical Engineering Science,* **222**(9), 1735-1743. http://dx.doi.org/10.1243/09544062JMES1031

Yesiloz. G. and Aydin, O., 2011, "Natural Convection in an Inclined Quadrantal Cavity Heated and Cooled on Adjacent Walls", *Experimental Thermal and Fluid Science*, **35**, 1169–1176. http://dx.doi.org/10.1016/j.expthermflusci.2011.04.002

Yesiloz, G. and Aydin, O., 2013, "Laminar Natural Convection in Right-angled Triangular Enclosures Heated and Cooled on Adjacent Walls," *Int. J. of Heat and Mass Transfer*, **60**, 365-374. http://dx.doi.org/10.1016/j.ijheatmasstransfer.2013.01.009

THERMAL CONDUCTIVITY OF BINARY MIXTURES OF GASES

Etim S. Udoetok*

Mechanical Engineering Department, University of Uyo, Uyo, Nigeria

ABSTRACT

A model for the coefficient of thermal conductivity of binary mixtures of gases has been derived. The theory presented is based on the assumption of random fluctuations between two possible extreme arrangements of a binary gas mixture. The results obtained from the new model compared favorably with published experimental results. The proposed new model provides a simple approach without sacrificing much accuracy compared to previous models. It is applicable to any binary mixture of gases which includes monoatomic gas mixture, polyatomic gas mixtures and mixtures involving rare gases. The new model can be very useful in analysis like combustion where the main equations are already sophisticated.

Keywords: *heat, mixture, gas, conduction and resistance.*

1. INTRODUCTION

Thermal analysis involving gas mixtures usually involves the evaluation of specific heats and thermal conductivities of the gas mixture. While the method for evaluating the specific heat of gas mixture is simple and straight forward, the evaluation of thermal conductivities of gas mixtures have been more sophisticated and several models have been developed (Shapiro, 2004; Li et al., 2011; Lindsay and Bromley, 1950; Mason and Saxena, 1958; Cheung et al., 1962; Simon et al., 1998; Tipton et al., 2009; Papari, 2009). The best approaches are rigorous and poses certain difficulties, so simple predictive models are desirable (Shapiro, 2004). Many thermal systems involves binary mixtures of gases like fuel-oxidizer and carbon dioxide-nitrogen, and the binary mixture gas is the simplest mixture level so some works have been focused on it (Simon et al., 1998; Tipton et al., 2009; Imaishi et al., 1984; Song et al., 2010; Barua et al., 1962; Hirschfelder, 1957; Gambhir and Saxena, 1966). There is also a wealth of published experimental data on the conductivity of binary gas mixture systems (Simon et al., 1998; Tipton et al., 2009; Imaishi et al., 1984; Song et al., 2010; Gambhir and Saxena, 1966; Papari et al., 2005; Imaishi and Kestin, 1984; Saxena et al., 1965).

Works on simplified models for thermal conductivities of binary mixtures of gases includes the Wassijewa model which was based on kinetic theory and is given as (Wasiljewa, 1904)

$$k_m = \frac{k_1}{1+\Lambda_{12}\dfrac{x_2}{x_1}} + \frac{k_2}{1+\Lambda_{21}\dfrac{x_1}{x_2}} \tag{1}$$

where the Λs are based on viscosity and conductivity. An equation of the same form was developed by Sutherland (1895) for the viscosity of gaseous mixtures. The simplified equation by Wassijewa later became complex as the best method for evaluating Λs were investigated (Lindsay and Bromley, 1950; Hirschfelder, 1957). Hirschfelder et al. also proposed a simplified thermal conductivity model which was for gas mixtures as well as for pure gases (Hirschfelder et al., 1948). The Hirschfelder et al equation is a modified form of the Euken equation (Euken, 1913) and is given as

$$k \cong \mu\left(c_P + \frac{5}{4}\frac{R}{M}\right) \tag{2}$$

The modified Euken equation failed to accurately predict thermal conductivities for gas mixtures. However, it accurately predicted the thermal conductivity of pure gases, since it was derived for polyatomic gas based on contributions due to degrees of freedom so having than one component of gas will amount to a difference in the degrees of freedom (Poling et al., 2000).

Another form of simplified equation was recommended by Kennard (1938) and is given as

$$k_m = k_1(x_1)^2 + K(x_1x_2) + k_2(x_2)^2 \tag{3}$$

The Kennard equation requires the evaluation of a constant K from experimental work. Lindsay and Bromley (1950) showed that even when K is determined by least square method, the equation has error of up to 12%. The Kennard equation is more of a curve fit that requires sufficient data from experiment.

Mason and Saxena (1958) later proposed a simplified model given as

$$k_{mix} = \sum_{i=1}^{n} k_i\left[1 + \sum_{k=1,k\neq i}^{n} G_{ik}\frac{x_k}{x_i}\right]^{-1} \tag{4}$$

which for the case of a binary gas mixture reduces to the Wassijewa equation form. The Gs are found as a function of molar masses and viscosities.

The purpose of this work is to develop the simplest possible model for the thermal conductivity of a binary mixture of gases that will have an acceptable error or no error at all.

2. METHOD

Consider 1-dimensional heat conduction of a mixture of two gases A and B contained between two wall of temperatures T_1 and T_2 (see Fig. 1). The insulation is used to ensure illustration of heat conduction in x-direction only and allow for the control of mixture component fractions. The following assumptions are used in this analysis:

i. Heat transfer is mainly by conduction and heat transfers by other modes are negligible.
ii. The two gases are perfectly mixed.

* Email: etim.udoetok@yahoo.com

iii. The gas mixture is a continuum and at the walls, the mixture makes perfect contact with the wall so intermolecular and wall collisions are neglected and expected to reflect in the values of conductivities of the mixture components.

iv. Heat conduction and variation of temperature is in the x-direction only.

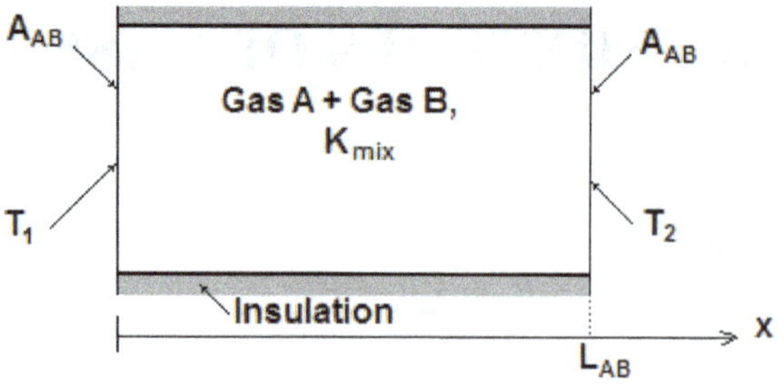

Fig. 1 Mixture of two conducting gases A and B

The molecules of both gases are moving in random motion, so it can be assumed that the gas mixture is fluctuating between two extreme heat conduction arrangements of the gases. These two separate arrangements of the gases are series and parallel arrangement. Therefore, the heat conductivity of the gas mixture can be taken as the average of the heat conductivities obtained for the two heat conduction extremes.

i. Heat conductivity of series arrangement: In this arrangement, the gases are separated and arranged in series as shown in Fig. 2.

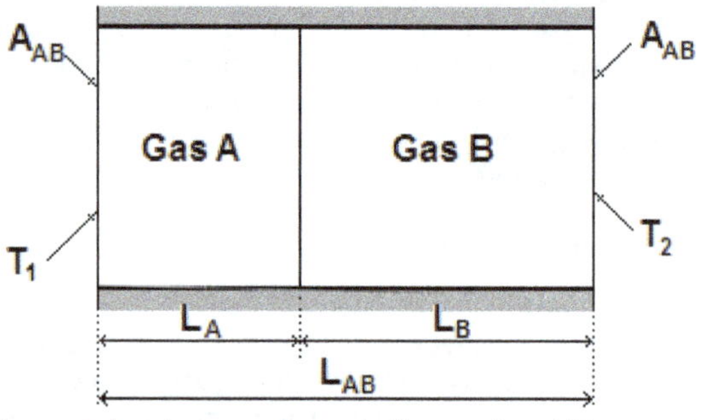

Fig. 2 Separated series arrangement of gases A and B

Thermal resistance method can be used to estimate the overall thermal conductivity of the resulting composite conductor. The overall thermal resistance is the sum of the thermal resistances of the two gases and is give as

$$\Re_{A-B} = \Re_A + \Re_B \tag{5}$$

This implies that

$$\frac{L_{AB}}{k_{A-B}A_{AB}} = \frac{L_A}{k_A A_{AB}} + \frac{L_B}{k_B A_{AB}} \tag{6}$$

Eq. (6) simplifies to

$$\frac{1}{k_{A-B}} = \frac{L_A A_{AB}}{k_A L_{AB} A_{AB}} + \frac{L_B A_{AB}}{k_B L_{AB} A_{AB}} \tag{7}$$

It can be seen in Eq. (7) that $L_A A_{AB}/L_{AB}A_{AB}$ and $L_B A_{AB}/L_{AB}A_{AB}$ represents the mole fractions of A and B respectively. Therefore, Eq. (7) becomes

$$\frac{1}{k_{A-B}} = \frac{X_A}{k_A} + \frac{X_B}{k_B} \tag{8}$$

Thus, the thermal conductivity of the series arrangement is

$$k_{A-B} = \frac{k_A k_B}{X_A k_B + X_B k_A} \tag{9}$$

ii. Heat conductivity of parallel arrangement: In this arrangement, the gases are arranged in parallel with the x-coordinate (See Fig. 3). By thermal resistance method, the effective resistance of the parallel composite arrangement is

$$\frac{1}{\Re_{A\|B}} = \frac{1}{\Re_A} + \frac{1}{\Re_B} \tag{10}$$

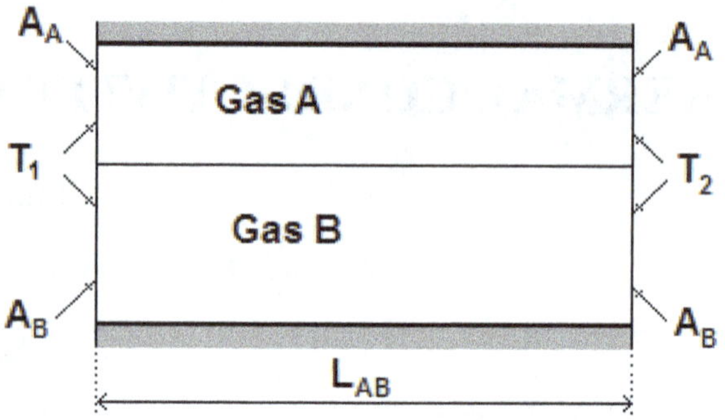

Fig. 3 Separated parallel arrangement of gases A and B

This implies that

$$\frac{1}{\left(\dfrac{L_{AB}}{k_{A\|B}(A_A + A_B)}\right)} = \frac{1}{\left(\dfrac{L_{AB}}{k_A A_A}\right)} + \frac{1}{\left(\dfrac{L_{AB}}{k_B A_B}\right)} \tag{11}$$

Eq. (11) can also be written as

$$\frac{k_{A\|B}(A_A + A_B)}{L_{AB}} = \frac{k_A A_A}{L_{AB}} + \frac{k_B A_B}{L_{AB}} \tag{12}$$

Eq. (12) can be simplified as

$$k_{A\|B} = \frac{k_A A_A L_{AB}}{L_{AB}(A_A + A_B)} + \frac{k_B A_B L_{AB}}{L_{AB}(A_A + A_B)} \tag{13}$$

It can be seen in Eq. (13) that $A_A L_{AB}/(L_{AB}(A_A + A_B))$ and $A_B L_{AB}/(L_{AB}(A_A + A_B))$ are the mole fractions of gases A and B respectively, thus

$$k_{A\|B} = k_A X_A + k_B X_B \tag{14}$$

Therefore, the average of the two heat conductivities from Eqs. (9) and (14) gives the mixture heat conductivity.

$$k_{mix} = 0.5k_{A-B} + 0.5k_{A\|B} \tag{15}$$

which implies that

$$k_{mix} = 0.5\frac{k_A k_B}{X_A k_B + X_B k_A} + 0.5(k_A X_A + k_B X_B) \tag{16}$$

This new equation, Eq. (16), is a function of mole fractions and thermal conductivities of the pure component gases at the mixture temperature. No special constants or coefficients are involved, except the 0.5 coefficients which were obtained from taking the average between the two heat conductivity extremes.

3. COMPARISON OF MODEL RESULT WITH PUBLISHED EXPERIMENTAL DATA

Tables 1-8 shows results of the proposed new model compared with experimental results that have been previously published. The individual gas conductivities were directly taken from the experimental data i.e. $k_A = k_{exp}$ when $X_A = 1$, $X_B = 0$ and $k_B = k_{exp}$ when $X_B = 1$, $X_A = 0$. The results showed that the proposed new model for estimating thermal conductivities of binary gas mixtures offers results with acceptable degree of accuracy for all cases of binary gas mixtures. High errors of more than 10% occurred at some combinations of N_2-He (see Table 2 and 5) and O_2-He (see Table 4). N_2-He experimental results from two different authors were checked and similar error by the new model was observed. Therefore, the high error is not directly due to experimental error. However, under the influence of gravity, He is very light and tends to go up or go in the direction opposite the action of gravity when released in a heavier gas. Therefore, the N_2-He and O_2-He mixtures have higher tendencies to violate the number two assumption used in the derivation of Eq. (16). In fact, the N_2-He and O_2-He mixtures will be perfect mixtures in zero g or when a mixing device is used to ensure perfect mixing. The errors observed in the N_2-He and O_2-He mixture results were due to indirect experimental errors caused by the imperfection of the mixing of the two gases. In order to check the validity of this explanation of the high errors observed in the N_2-He and O_2-He results, a mixture of He and another light gas was checked (see Table 8) and a mixture of N_2 and O_2 was checked (see Table 6). In both cases it was observed that the new model, Eq. (16), results have errors of less than 5%.

Additionally, more published experimental data were checked and it was observed that the proposed model, Eq. (16) has errors averaging less than 2% in most cases (Dael and Cauwenbergh, 1968; Ghambir and Saxena, 1966; Imaishi et al., 1984; Imaishi and Kestin, 1984; Lindsay and Bromley, 1950; Mason and Saxena, 1958; Papari et al., 2009; Song et al., 2010; Tipton et al., 2009).

Table 1 Thermal conductivity of NO-CO mixture at various temperatures. (k is in cal/°C·cm·s)

Temperature	X_{NO}	$k_{exp} \times 10^5$ (Barua et al., 1969)	$k_{mix} \times 10^5$ Eq. (16)	% Error
0 °C	0.000	5.5	5.5000	0.00
	0.067	5.38	5.5073	2.37
	0.246	5.44	5.5269	1.60
	0.496	5.5	5.5543	0.99
	0.741	5.56	5.5813	0.38
	0.959	5.6	5.6054	0.10
	1.000	5.61	5.6100	0.00
80 °C	0.000	6.82	6.8200	0.00
	0.052	6.79	6.8363	0.68
	0.246	6.88	6.8974	0.25
	0.501	6.93	6.9785	0.70
	0.752	7.03	7.0593	0.42
	0.949	7.19	7.1233	0.93
	1.000	7.14	7.1400	0.00
120 °C	0.000	7.6	7.6000	0.00
	0.058	7.65	7.6046	0.59
	0.273	7.7	7.6218	1.02
	0.506	7.8	7.6404	2.05
	0.739	7.82	7.6590	2.06
	0.948	7.83	7.6758	1.97
	1.000	7.68	7.6800	0.00
160 °C	0.000	8.22	8.2200	0.00
	0.050	8.13	8.2274	1.20
	0.499	8.44	8.2945	1.72
	0.744	8.54	8.3313	2.44
	0.942	8.40	8.3612	0.46
	1.000	8.37	8.3700	0.00
200 °C	0.000	8.61	8.6100	0.00
	0.057	8.51	8.6399	1.53
	0.255	8.74	8.7446	0.05
	0.504	9.12	8.8781	2.65
	0.734	9.20	9.0031	2.14
	0.947	9.35	9.1205	2.45
	1.000	9.15	9.1500	0.00

Table 2 Thermal conductivity of N_2-He mixture at 30 °C and 45 °C. (k is in cal/°C·cm·s)

Temperature	X_{He}	$k_{exp} \times 10^5$ (Barua, 1959)	$k_{mix} \times 10^5$ Eq. (16)	% Error
30 °C	1.0000	36.35	36.3500	0.00
	0.8864	28.43	28.0966	1.17
	0.7432	21.52	22.3014	3.63
	0.6041	17	18.3364	7.86
	0.4681	13.02	15.1420	16.30
	0.2893	9.904	11.4583	15.69
	0.1528	8.044	8.8711	10.28
	0.0000	6.116	6.1160	0.00
45 °C	1.0000	37.64	37.6400	0.00
	0.8651	28.23	28.0441	0.66
	0.7362	22.06	22.9081	3.84
	0.6241	18.08	19.5693	8.24
	0.4981	14.38	16.4179	14.17
	0.2962	10.14	12.0523	18.86
	0.1562	8.337	9.2989	11.54
	0.0000	6.382	6.3820	0.00

Table 3 Thermal conductivity of N_2-Ne mixture at 30 °C and 45 °C. (k is in cal/°C·cm·s)

Temperature	X_{Ne}	$k_{exp} \times 10^5$ (Barua, 1959)	$k_{mix} \times 10^5$ Eq. (16)	% Error
30 °C	1.0000	11.62	11.6200	0.00
	0.9026	10.66	10.8806	2.07
	0.7754	9.716	10.0187	3.12
	0.6954	9.914	9.5239	3.93
	0.4496	7.938	8.1692	2.91
	0.3277	7.291	7.5668	3.78
	0.1286	6.524	6.6534	1.98
	0.0000	6.1	6.1000	0.00
45 °C	1.0000	11.98	11.9800	0.00
	0.9000	10.88	11.2175	3.10
	0.7612	9.878	10.2750	4.02
	0.6504	9.365	9.5968	2.47
	0.4937	8.381	8.7221	4.07
	0.3038	7.543	7.7619	2.90
	0.1480	7.032	7.0352	0.04
	0.0000	6.384	6.3840	0.00

Table 4 Thermal conductivity of O_2-He mixture at 30 °C. (k is in cal/°C·cm·s)

Temperature	X_{O2}	$k_{exp} \times 10^5$ (Saxena et al., 1965)	$k_{mix} \times 10^5$ Eq. (16)	% Error
30 °C	0.0000	36.37	36.3700	0.00
	0.1500	27	26.6563	1.27
	0.3000	20.14	21.2918	5.72
	0.4500	15.64	17.3342	10.83
	0.6000	12.3	14.0070	13.88
	0.7500	9.642	11.0163	14.25
	0.9000	7.621	8.2262	7.94
	1.0000	6.441	6.4410	0.00

Table 5 Thermal conductivity of N_2-He mixture at 30 °C. (k is in cal/° C·cm·s)

Temperature	X_{N2}	$k_{exp} \times 10^5$ (Saxena et al., 1965)	$k_{mix} \times 10^5$ Eq. (16)	% Error
30 °C	0.0000	36.5	36.5000	0.00
	0.1500	26.6	26.4152	0.69
	0.3000	19.83	21.0064	5.93
	0.4500	15.34	17.0401	11.08
	0.6000	11.52	13.7057	18.97
	0.7500	9.31	10.7040	14.97
	0.9000	7.401	7.8991	6.73
	1.0000	6.102	6.1020	0.00

Table 6 Thermal conductivity of N_2-O_2 mixture at 319 °C. (k is in cal/° C·cm·s)

Temperature	X_{O2}	$k_{exp} \times 10^5$ (Saxena et al., 1965)	$k_{mix} \times 10^5$ Eq. (16)	% Error
319 °C	0.0000	10.7	10.7000	0.00
	0.6098	11.19	11.2519	0.55
	1.0000	11.62	11.6200	0.00

Table 7 Thermal conductivity of SO_2-Ar mixture at various temperatures. (k is in cal/° C·cm·s)

Temperature	X_{SO2}	$k_{exp} \times 10^5$ (Gupta, 1967)	$k_{mix} \times 10^5$ Eq. (16)	% Error
39 °C	0.000	4.389	4.3890	0.00
	0.146	3.995	3.9999	0.12
	0.362	3.533	3.5085	0.69
	0.572	3.228	3.0955	4.11
	0.764	2.81	2.7560	1.92
	1.000	2.374	2.3740	0.00
80.1 °C	0.000	4.854	4.8540	0.00
	0.151	4.419	4.4806	1.39
	0.28	4.233	4.1923	0.96
	0.504	3.798	3.7418	1.48
	0.79	3.214	3.2330	0.59
	1.000	2.893	2.8930	0.00
121.3 °C	0.000	5.22	5.2200	0.00
	0.166	4.964	4.8830	1.63
	0.344	4.609	4.5517	1.24
	0.608	4.171	4.1044	1.60
	0.798	3.838	3.8079	0.78
	1.000	3.511	3.5110	0.00
161.1 °C	0.000	5.699	5.6990	0.00
	0.164	5.3	5.3716	1.35
	0.381	4.982	4.9732	0.18
	0.556	4.684	4.6752	0.19
	0.750	4.34	4.3645	0.63
	1.000	3.989	3.9890	0.00
200.6 °C	0.000	6.047	6.0470	0.00
	0.215	5.698	5.6809	0.30
	0.373	5.479	5.4284	0.92
	0.558	5.220	5.1477	1.38
	0.782	4.848	4.8264	0.45
	1.000	4.530	4.5300	0.00

Table 8 Thermal conductivity of H_2-He mixture at 30 °C. (k is in cal/°C·cm·s)

Temperature	X_{He}	$k_{exp} \times 10^5$ (Cauwenberg and Dael, 1971)	$k_{mix} \times 10^5$ Eq. (16)	% Error
30 °C	0.0000	44.380	44.3800	0.00
	0.1006	42.967	43.5306	1.31
	0.2516	41.169	42.2905	2.72
	0.5066	38.502	40.2821	4.62
	0.7001	37.138	38.8221	4.53
	0.8001	36.689	38.0870	3.81
	0.8523	36.590	37.7082	3.06
	0.8799	36.483	37.5093	2.81
	0.8823	36.528	37.4921	2.64
	0.9133	36.535	37.2698	2.01
	0.9213	36.510	37.2126	1.92
	0.9400	36.523	37.0792	1.52
	0.9600	36.563	36.9370	1.02
	0.9800	36.578	36.7953	0.59
	1.0000	36.654	36.6540	0.00

4. CONCLUSIONS

The proposed new equation for finding the thermal conductivities of binary mixtures of gases gives results with good degree of accuracy. It is applicable to any binary mixture of gases which includes mixtures of monoatomic, mixtures of polyatomic and mixtures involving rare gases. Unlike other models which are mostly direct functions of mole fractions, component thermal conductivities, viscosities, molar masses, densities and/or temperatures, the proposed new model is just a direct function of mole fractions and component thermal conductivities. The proposed new model still retains high degree of accuracy since thermal conductivities of the mixture components are already functions of viscosity, molar masses and/or temperature as can be seen in Eq. (2) that mainly works for pure gases. It can also be shown from kinctic theory that for pure gases (Turns, 2000)

$$k = \left(\frac{K_B^3}{\pi^3 m \sigma^4} \right)^{\frac{1}{2}} T^{\frac{1}{2}} \tag{17}$$

The new method eliminates some of the complexities involved in models that were previously developed by other authors. However, the new method is not recommended for mixtures involving a high density gas mixed with a very low density gas say, He or H_2, in non-zero gravity conditions.

NOMENCLATURE

A	area (m^2)
c_p	specific heat (J/kg·K)
G_{ik}	coefficients base on molar mass and viscosities
K	constant (W/m·K)
K_B	Boltzmann constant (J/K)
k_i	thermal conductivity of specie i (W/m·K)
L	length (m)
M	molar mass (kg/kmol)
m	molecular mass (kg)
R	gas constant (kJ/kg·K)
T	temperature (K)
u	interfacial velocity (m/s)
X_i	mole fraction of specie i
x	coordinate (m)

Greek Symbols

Λ_{ij}	coefficients based on viscosity and conductivity

$\Re$	thermal resistance (K/W)
μ	viscosity (kg/m·s)
σ	molecular diameter (m)

Subscripts

A	gas A
B	gas B
AB	gas A and B mixture
A-B	A in series with B
A||*B*	A in parallel with B
exp	experiment
mix	mixture
1	wall 1
2	wall 2

REFERENCES

Barua, A. K., 1959, "Thermal Conductivity and Eucken Type Correction for Binary Mixtures of N2 with some Rare Gases," *Physica,* **25**, 1275-1286. http://dx.doi.org/10.1016/0031-8914(59)90049-7

Barua, A. K., Gupta, A. D. and Mukhopadhyay, P., 1969, "Thermal Conductivity of Nitric Oxide, Carbon Monoxide and their Binary Mixtures," *International Journal of Heat and Mass Transfer,* **12**, 587-593. http://dx.doi.org/10.1016/0017-9310(69)90040-4

Cauwenbergh, H. and Dael, W. V., 1971, "Measurements of the Thermal Conductivity of Gases III: Data for Binary Mixture Helium and Argon with Hydrogen Isotopes," *Physica,* **54**, 347-360. http://dx.doi.org/10.1016/0031-8914(71)90182-0

Cheung, H., Bromley, L. A. and Wilke, C. R., 1962, "The Thermal Conductivity of Gas Mixtures," *AIChE Journal,* **8**(2), 221-228. http://dx.doi.org/10.1002/aic.690080219

Dael, V. W. and Cauwenbergh, H., 1968, "Measurements of the Thermal Conductivities of Gases II: Data for Binary Mixtures of He, Ne and Ar," *Physica,* **40**, 173-181. http://dx.doi.org/10.1016/0031-8914(68)90015-3

Euken, A., 1913, "On the Thermal Conductivity, the Specific Heat and the Viscosity of Gases," *Physikalische Zeitschrift,* **14**, 324-332.

Gambhir, R. S. and Saxena, S. C., 1966, "Thermal Conductivity of Gas Mixtures: Ar-D2, Kr-D2 and Ar-Kr-D2," *Physica,* **32**, 2037-2043. http://dx.doi.org/10.1016/0031-8914(66)90166-2

Gupta, A. D., 1967, "Thermal Conductivity of Binary Mixtures of Sulphur Dioxide and Inert Gases," *International Journal of Heat and Mass Transfer*, **10**, 921-929. http://dx.doi.org/10.1016/0017-9310(67)90069-5

Hirschfelder, J. O., Bird, R. B. and Spotz, E. L., 1948, "Transport Properties for Non-Polar Gases," *Journal of Chemical Physics,* **16**(10), 968-980. http://dx.doi.org/10.1063/1.1746696

Hirschfelder, J. O., 1957, "Thermal Conductivity in Polyatomic Electronically Excited or Chemically Reacting Mixtures III," *Sixth International Symposium on Combustion,* 351-365.

Imaishi, N., Kestin, J., and Wakeham, W. A., 1984, "Thermal Conductivity of two Binary Mixtures of Gases of Equal Molecular Weight," Physica, **123A**, 50-71.

Imaishi, N., and Kestin, J., 1984, "Thermal Conductivity of Methane with Carbon Monoxide," *Physica,* **126A**, 301-307.

Kennard, E. H., 1938, *Kinetic Theory of Gases*, McGraw-Hill, New York.

Li, H., Wilhelmsen, O., Lv, Y., Wang, W., and Yan, J., 2011, "Viscosities, Thermal Conductivities and Diffusion Coefficients of CO2 Mixtures: Review of Experimental Data and Theoretical Models," *International Journal of Greenhouse Gas Control,* **5**, 1119-1139. http://dx.doi.org/10.1016/j.ijggc.2011.07.009

Linsay, A. L., and Bromley, L. A., 1950, "Thermal Conductivity of Gas Mixtures," *Industrial and Engineering Chemistry*, **42**(8), 1508-1511. http://dx.doi.org/10.1021/ie50488a017

Mason, E. A., and Saxena, S. C., 1958, "Aproximate Formula for the Thermal Conductivity of Gas Mixtures," *The Physics of Fluids,* **1**(5), 361-369. http://dx.doi.org/10.1063/1.1724352

Papari, M. M., Mohammad-aghaiee, D., Haghighi, B. and Bousheri, A., 2005, "Transport Properties of Argon-Hydrogen Gaseous Mixture from an Effective Unlike Interaction," *Fluid Phase Equilibria*, **232**, 122-135. http://dx.doi.org/10.1016/j.fluid.2005.03.022

Papari, M. M., Khordad, R., and Akbari, Z., 2009, "Further Property of Lennard-Jones Fluid: Thermal Conductivity," *Physica A*, **388**, 585-592. http://dx.doi.org/10.1016/j.physa.2008.11.003

Poling, B. E., Prausnitz, J. M., and O'Connell, J. P., 2000, *The Properties of Gases and Liquids,* 5th ed., McGraw-Hill, New York, pp. 10.1-10.70.

Saxena, S. C., Saksena, M. P., Gambhir, R. S. and Gandhi, J. M., 1965 "The Thermal Conductivity of Nonpolar Polyatomic Gas Mixtures," *Physica*, **31**, 333-341. http://dx.doi.org/10.1016/0031-8914(65)90038-8

Shapiro, A. A., 2004, "Fluctuation Theory for Transport Properties in Multicomponent Mixtures: Thermodiffusion and Heat Conductivity," *Physica A*, **332**, 151-175. http://dx.doi.org/10.1016/j.physa.2003.10.014

Simon, J. M., Dysthe, D. K., Fuchs, A. H. and Rousseau, B., 1998, "Thermal Diffusion in Alkane Binary Mixtures: A Molecular Dynamics Approach," *Fluid Phase Equilibra*, **150-151,** 151-159. http://dx.doi.org/10.1016/S0378-3812(98)00286-6

Song, B., Wamg, X., Wu, J. and Liu, Z., 2010, "Prediction of Transport Properties of Pure Noble Gases and some of their Binary Mixtures by ab initio Calculations," *Fluid Phase Equilibria,* **290**, 55-62. http://dx.doi.org/10.1016/j.fluid.2009.09.010

Sutherland, W., 1895, "The Viscosity of Mixed Gases," *Philosophical Magazine Series 5*, **40**(246), 421-431. http://dx.doi.org/10.1080/14786449508620789

Tipton, E. L., Tompson, R. V. and Loyalka, S. K., 2009, "Chapman-Enskog Solutions to Arbitrary Order in Sonine Polynomials III: Diffusion, Thermal Diffusion, and Thermal Conductivity in Binary, Rigid-Sphere, Gas Mixture," *European Journal of mechanics B/Fluids*, **28**, 353-386. http://dx.doi.org/10.1016/j.euromechflu.2008.12.002

Turns, S. R., 2000, *An Introduction to Combustion: Concepts and Applications,* Second ed., McGraw-Hill, New York, pp. 86-90.

Wassiljewa, A., 1904, "Heat Conduction in Gas Mixtures," *Physikalische Zeitschrift*, **5** (22), 737-742.

EFFECTS OF THERMAL AND SOLUTAL STRATIFICATION ON MIXED CONVECTION FLOW ALONG A VERTICAL PLATE SATURATED WITH COUPLE STRESS FLUID

K. Kaladhar[a], D. Srinivasacharya[b]

[a]*Department of Mathematics, National Institute of Technology Puducherry, Karaikal-609605, India*
[b] *Department of Mathematics, National Institute of Technology, Warangal-506004, India*

ABSTRACT

The effect of heat and mass stratification on mixed convection along a vertical plate embedded in a couple stress fluid has been presented. The nonlinear system of equations with appropriate boundary conditions is primarily reduced to non-dimensional form by pseudo-similarity transformations. Keller-box implicit finite difference scheme is employed to solve the resultant system of dimensionless equations. The validation of this scheme is shown through the comparison between the present and available literature under special case of the present problem. The couple stress parameter, mixed convection parameter and the double stratification parameter effects on the rates of heat and mass transfer for diverse values of the emerging flow parameters are illustrated in tabular form. The present results show that the couple stress parameter, mixed convection parameter, thermal and the solutal stratification parameters influences the flow significantly.

Keywords: *Mixed convection, couple stress fluid, double stratification.*

1. INTRODUCTION

A flow with free and forced convection is of considerable interest due to its wide range in manufacturing and practical applications that consist of, metallurgical processes, nuclear reactors, heat exchangers, geothermal systems, geothermal systems, crystal growth, nuclear waste materials, and many. In literature, convection in heat and mass transfer flow along a non-isothermal vertical plane for Newtonian fluids with boundary layer estimates have been studied by many authors. The theoretical results for heat and solutal transfer from a vertical flat plate have been presented by Somers (1956). Szewczyk (1964) analyzed the effects of mixed convection laminar flow. Mixed convection flow along a flat plate with local similarity method was studied by Lloyd and Sparrow (1970). It has been produced the solutions through pure to mixed convection. Detailed explanation of many works can be found in Bejan (1994). Most recently, The effect on homogeneous chemical reaction on mixed convection in a porous medium saturated with polar fluid have been studied numerically by Patil and Chamkha (2013).

Analysis of non-Newtonian fluids with heat and mass transfer is momentous in practical situations. For instance, slurries, foodstuffs, polymeric liquids, thermal design of industrial equipment dealing with molten plastics, etc. The examples of such fluids are blood at low shear rate, lubricants containing small amount of polymer additives, electro-rheological fluids, paints, fiber solutions and synthetic fluids, etc. The nonlinear relationship between the rate of strain and stress can be found in non-Newtonian fluid models. Stokes (1966) introduced this model, in which body couples, non-symmetric tensors and couple stresses exist. The classical viscous theories fails to describe the size dependant effect but which can be found with the effect of couple stresses. The free and mixed convection flow of couple stress fluid in a vertical channel have been presented by Srinivasacharya and Kaladhar (2012a,b). The magnetohydrodynamics, viscous dissipation and heat mass transfer effects on horizontal wavy channel in a porous channel saturated with couple stress fluid have been discussed by Muthuraj et al. (2013). The couple stress fluid with melting heat transfer under stagnation point flow was offered by Hayat et al. (2013). Srinivasacharya and Kaladhar (2013) described the analytical solution for mixed convection flow of couple stress fluid with MHD, Hall and ion-slip effects between circular rotating parallel disks. Makinde and Eegunjobi (2013) studied the nature of couple stress fluid in a vertical channel with entropy generation and porous medium. Najeeb et al. (2013) reported the flow of couple stress fluid in an contracting and expanding porous channel with an approximate solutions. Most recently, the size-dependent and consisting creeping flow of couple stress fluid was presented by Hadjesfandiari et al. (2013). The presence of different fluids or variations in temperature/ concentration leads to the stratification of a fluid. Although there is minute literature and in view of applications, Prandtl (1952), Jaluria and Himasekhar (1983), Murthy et al. (2004), Srinivasacharya et al. (2011) and few more are presented the stratification effects in different cases.

It is the objective of the present work to consider the mixed convection flow of couple stress fluid along a vertical plate with solutal and thermal stratification effects. The medium is linearly stratified and the wall concentration and wall temperature are constants. The present boundary conditions are close realistic to the practical interest, like the heat mass transfer characteristics around a cooling magmatic intrusion or around a hot radioactive subsurface storage site where the theory of convection is involved. The governing nonlinear system of equations are solved by using the Keller box method (Cebeci and Bradshaw (1984)). The effects of the couple stress parameter, mixed convection, and the stratification parameters are examined and are presented graphically.

2. ANALYSIS

The geometry of the problem and the coordinate system are shown in Fig. 1, in which x-axis and y-axis are along and normal to the vertical plate respectively. Steady and incompressible couple stress fluid along a plat plate is considered. Stratification and mixed convection are also taken into consideration. Density change in the fluid is neglected

* Corresponding author. Email: kkr.nitpy@gmail.com

everywhere except in the buoyancy, and all the other physical properties of the fluid are assumed constant. The uniform temperature and concentrations of the plate is T_w and C_w respectively. The free stream velocity, temperature and concentration parallel to the vertical plate are maintained at u_∞, $T_{\infty,0}$ and $C_{\infty,0}$ respectively. At any arbitrary reference point in the medium (inside the boundary layer), the values of $T_{\infty,0}$ and $C_{\infty,0}$ are smaller than T_w and C_w respectively. $T_\infty(x)=T_{\infty,0}+Ax$ and $C_\infty(x)=C_{\infty,0}+Bx$ are the vertical linear temperature and concentration stratification parameters in an ambient medium. Where A and B are constants, these are varied to alter the intensity of stratification in the medium.

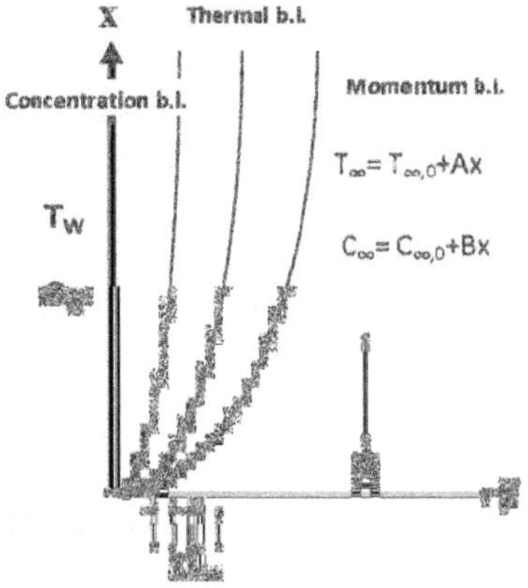

Fig 1 Physical model and coordinate system

The governing equations for the couple stress fluid, boundary layer approximations and Boussinesq approximations are given by

$$\frac{\partial u}{\partial x}+\frac{\partial v}{\partial y}=0 \tag{1}$$

$$\rho\left(u\frac{\partial u}{\partial x}+v\frac{\partial u}{\partial y}\right)=\mu\frac{\partial^2 u}{\partial y^2}+\rho g\left(\beta_T(T-T_\infty)+\beta_C(C-C_\infty)\right)-\eta_1\frac{\partial^4 u}{\partial y^4} \tag{2}$$

$$u\frac{\partial T}{\partial x}+v\frac{\partial T}{\partial y}=\alpha\frac{\partial^2 T}{\partial y^2} \tag{3}$$

$$u\frac{\partial C}{\partial x}+v\frac{\partial C}{\partial y}=D\frac{\partial^2 C}{\partial y^2} \tag{4}$$

where the velocities in x and y directions are u and v respectively, T is the temperature, C is the concentration, g is the acceleration due to gravity, ρ is the density, μ is the dynamic coefficient of viscosity, β_T is the coefficient of thermal expansion, β_C is the coefficient of solutal expansions, α is the thermal diffusivity and D is the solutal diffusivity of the medium and η_1 is the couple stress fluid parameter. The subscript ∞ indicates the condition at the outer edge of the boundary layer.

The physical boundary conditions for this problem are

$$u=0,\ v=0,\ v_x-u_y=0, T=T_w,\ C=C_w \text{ at } y=0 \tag{5a}$$

$$u=0,\ v_x-u_y=0,\ T=T_\infty(x),\ C=C_\infty(x) \text{ as } y\to\infty \tag{5b}$$

where the boundary condition $v_x - u_y = 0$ imply that the couple stress fluid is irrotational at the boundaries (Type A condition for Couple stress fluid; Stokes (1966)) and k is the thermal conductivity of the fluid.

The stream function ψ is introduced in view of continuity equation (1), which is

$$u=\frac{\partial\psi}{\partial y},\qquad v=-\frac{\partial\psi}{\partial x} \tag{6}$$

Substituting Eq. (6) in Eqs. (2)-(4) and then using the following transformations

$$\xi=\frac{x}{L},\ \eta=\left(\frac{\mathrm{Re}}{\xi}\right)^{\frac{1}{2}}\frac{y}{L},\ f(\xi,\eta)=\left(\frac{\mathrm{Re}}{\xi}\right)^{\frac{1}{2}}\frac{\psi}{Lu_\infty}$$

$$\theta(\xi,\eta)=\frac{T-T_{\infty,0}}{T_w-T_{\infty,0}}-\frac{Ax}{T_w-T_{\infty,0}} \tag{7}$$

$$\phi(\xi,\eta)=\frac{C-C_{\infty,0}}{C_w-C_{\infty,0}}-\frac{Bx}{C_w-C_{\infty,0}}$$

The final non-linear system of differential equations is

$$\mathrm{Re}\,S^2 f^{v}-\xi f'''-\frac{1}{2}\xi f f''-\xi^2 Ri(\theta+\mathrm{N}\varphi)=\xi^2\left(f'\frac{\partial f}{\partial\xi}-f'\frac{\partial f'}{\partial\xi}\right) \tag{8}$$

$$\frac{1}{\mathrm{Pr}}\theta''+\frac{1}{2}f\theta'-\varepsilon_1\xi f'=\xi\left(f'\frac{\partial\theta}{\partial\xi}-\theta'\frac{\partial f}{\partial\xi}\right) \tag{9}$$

$$\frac{1}{Sc}\phi''+\frac{1}{2}f\phi'-\varepsilon_2\xi f'=\xi\left(f'\frac{\partial\phi}{\partial\xi}-\phi'\frac{\partial f}{\partial\xi}\right) \tag{10}$$

where the primes indicate partial differentiation with respect to η alone, $\mathrm{Re}=\frac{u_\infty L}{\nu}$ is the Reynolds number, $Gr=\frac{g\beta_T\left(T_w-T_{\infty,0}\right)L^3}{\nu^2}$ is the thermal Grashof number, $\mathrm{Pr}=\frac{\nu}{\alpha}$ is the Prandtl number, $Sc=\frac{\nu}{D}$ is the Schmidt number, η_1 is the couple stress fluid parameter and $N=\frac{\beta_C\left(C_w-C_{\infty,0}\right)}{\beta_T\left(T_w-T_{\infty,0}\right)}$ is the buoyancy ratio. It should be noted that $N > 0$ indicates aiding buoyancy where both the thermal and solutal buoyancies are in the same direction and $N < 0$ indicates opposing buoyancy where the solutal buoyancy is in the opposite direction to the thermal buoyancy. When $N = 0$, the flow is driven by thermal buoyancy alone. $Ri=\frac{Gr}{\mathrm{Re}^2}$ is the mixed convection parameter, which represents the ratio of buoyancy forces to the inertia forces and is used to delineate the free, forced and mixed convection regimes. $Ri \ll 1$ corresponds to pure forced convection, whereas $Ri \gg 1$ corresponds to pure free convection. $\varepsilon_1=\frac{AL}{T_w-T_{\infty,0}}$ and $\varepsilon_2=\frac{BL}{C_w-C_{\infty,0}}$ are the thermal and solutal stratification parameters and are constants. $S=\frac{1}{L}\sqrt{\frac{\eta_1}{\mu}}$ is the couple stress parameter, the effects of couple-stress are significant for large values of S (= l/L), where $l=\sqrt{\frac{\eta_1}{\mu}}$ is the material constant. If

Boundary conditions (5) in terms of f, θ and ϕ become

$$\eta=0:\ f'(\xi,0)=0,\ f(\xi,0)=f''(\xi,0)=0, \theta(\xi,0)=1-\varepsilon_1\xi,$$
$$\varphi(\xi,0)=1-\varepsilon_2\xi$$

(11a)

$$\eta\to\infty:\ f'(\xi,\infty)=1,\quad f''(\xi,\infty)=0,\quad \theta(\xi,\infty)=0, \phi(\xi,\infty)=0 \tag{11b}$$

The heat and mass transfers from the plate are given respectively by

$$q_w = -k\left[\frac{\partial T}{\partial y}\right]_{y=0}, \quad q_m = -D\left[\frac{\partial C}{\partial y}\right]_{y=0} \tag{12}$$

The local Nusselt number $Nu_x = \dfrac{q_w x}{k(T_w - T_{\infty,0})}$ and local Sherwood number $Sh_x = \dfrac{q_m x}{D(C_w - C_{\infty,0})}$ are given by

$$Nu_x = -\mathrm{Re}_x^{1/2}\,\theta'(\xi,0), Sh_x = -\mathrm{Re}_x^{1/2}\,\varphi'(\xi,0) \tag{13}$$

where $\mathrm{Re}_x = \dfrac{u_\infty x}{\nu}$ is the local Reynolds number.

3. RESULTS AND DISCUSSIONS

The energy equation (9) and concentration (10) are coupled with the flow equation (8). Since the system of differential equations is nonlinear and non-homogenous, the closed form solutions are not obtained and hence the system is solved using the implicit Keller-box method (Cebeci and Bradshaw (1984)). This method has been proven to be adequate and give accurate results for boundary layer equations. In boundary conditions, the value of η at ∞ is replaced by a sufficiently large value of η where the temperature, concentration profiles approach zero and velocity approaches to one. For that the value is considered as η = 8 with a grid size of η of 0.01. The dimensionless velocity, temperature and concentration function are evaluated and are presented through plots, which are shown in Figs. 2-13. The effects of emerging parameters of the flow on dimensionless velocity, temperature and concentration are discussed. In throughout the computations the constant values are assumed as Re = 100, Pr = 0.71, Sc = 0.22 and ξ =0.1.

By taking S = 0, ε_1 = 0, ε_2 = 0, and N = 0, the present system of equations reduces to Newtonian fluid with mixed convection flow of vertical plate of Lloyd and Sparrow (1970).

Table 1: Comparison of $Nu_x Re_x^{-1/2}$ for mixed convection between a vertical flat plate and Newtonian fluids (Lloyd and Sparrow (1970))

	Pr = 0.72		Pr = 10		Pr = 100	
Ri	Lloyd and Sparrow	Present	Lloyd and Sparrow	Present	Lloyd and Sparrow	Present
0.00	0.2956	0.2956	0.7281	0.7281	1.5720	1.5743
0.01	0.2979	0.2979	0.7313	0.7313	1.5750	1.5766
0.04	0.3044	0.3044	0.7404	0.7410	1.5850	1.5891
0.10	0.3158	0.3158	0.7574	0.7581	1.6050	1.6065
0.40	0.3561	0.3561	0.8259	0.8272	1.6910	1.6962
1.00	0.4058	0.4058	0.9212	0.9212	1.8260	1.8293
2.00	0.4584	0.4584	1.0290	1.0302	1.9940	1.9985
4.00	0.5258	0.5258	1.1730	1.1738	2.2320	2.2385

Thus, using the principle of local similarity and in absence of couple stress parameter S, buoyancy ratio, stratification parameters ε_1 and ε_2 with ξ =1, the comparison has been made by the special case of Lloyd and Sparrow (1970) with the outcome of the governing equations (2)-(4) and it is noticed that they are in good agreement, as shown in Table 1. Therefore, the present code has been used with great confidence to this study.

The dimensionless velocity profile for different values of mixed convection parameter at fixed S, ε_1 and ε_2 is shown in Fig. 2. It can be observed that the dimensionless velocity rises as the mixed convection parameter increases. Compared with the pure forced convection case (Ri=0.0, i.e. the limiting case), higher values of Ri implies the higher velocity. This is because of increase in mixed convection parameter leads to the higher buoyancy effects in mixed convection and on temperature profile is examined in Fig. 3. It can be seen from this figure that as Ri increases, the dimensionless temperature decreases. By comparing with pure forced convection the temperature is less in case of mixed convection. The temperature reduces when the convection cooling effect increases, this happens when Ri (buoyancy effects) increases. Figure 4 depicts that the concentration profile at different values of Ri. It is noticed that as Ri increases, the concentration decreases.

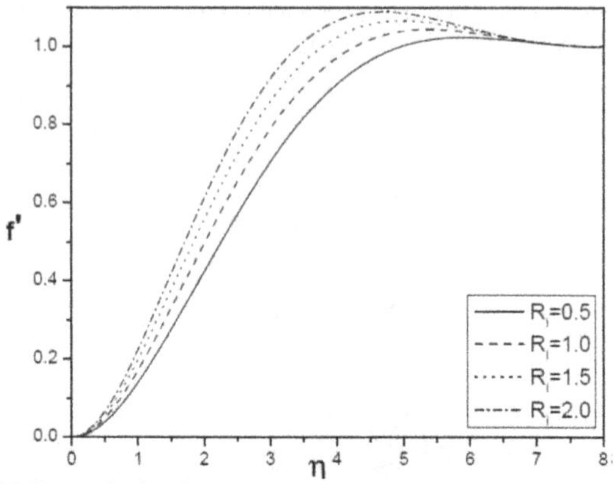

Fig. 2 Effect of mixed convection parameter Ri on velocity when $S = 0.5$, $\varepsilon_1 = 0.1$, $\varepsilon_2 = 0.1$

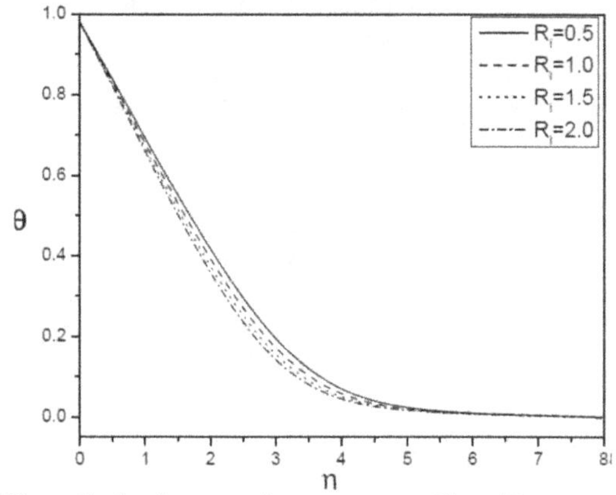

Fig. 3 Effect of mixed convection parameter Ri on Temperature when $S = 0.5$, $\varepsilon_1 = 0.1$, $\varepsilon_2 = 0.1$

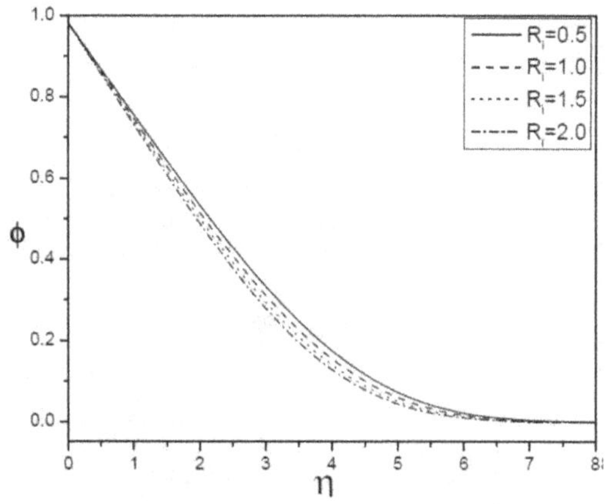

Fig. 4 Effect of mixed convection parameter Ri on Concentration when $S = 0.5$, $\varepsilon_1 = 0.1$, $\varepsilon_2 = 0.1$

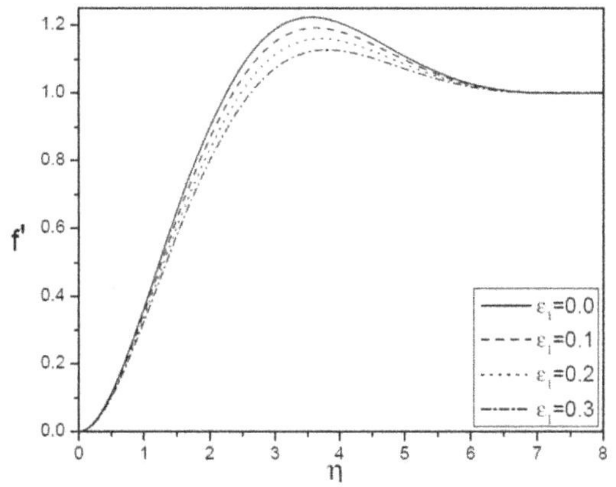

Fig. 5 Effect of thermal stratification parameter ε_1 on velocity when $S = 0.5$, $Ri = 2.0$, $\varepsilon_2 = 0.1$

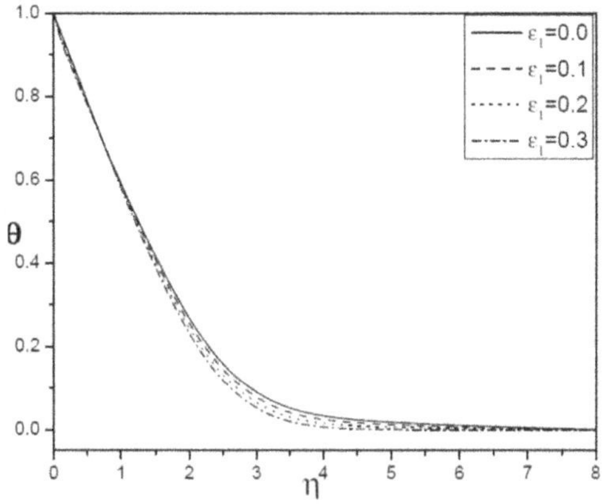

Fig. 6 Effect of thermal stratification parameter ε_1 on velocity when $S = 0.5$, $Ri = 2.0$, $\varepsilon_2 = 0.1$

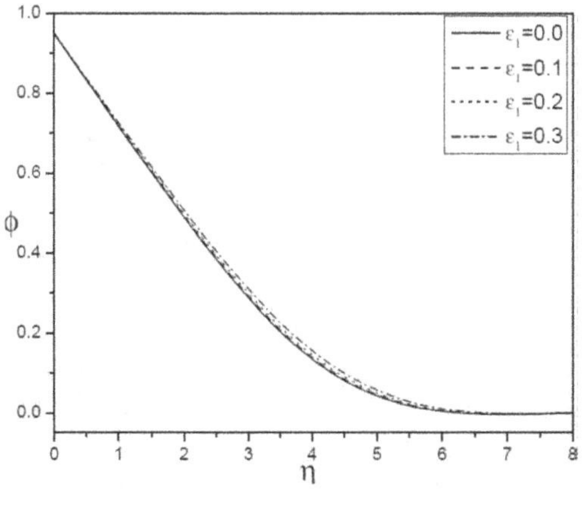

Fig. 7 Effect of thermal stratification parameter ε_1 on velocity when $S = 0.5$, $Ri = 2.0$, $\varepsilon_2 = 0.1$

The effect of stratification parameter ε_1 on non-dimensional velocity is depicted in Fig. 5. It can be seen that as the temperature stratification increases, the velocity decreases. As there is any decrease in thermal stratification, the effective convective potential between the heated plate and the ambient fluid in the medium reduces. Therefore under the boundary layer the velocity reduces with the thermal stratification. Figure 6 shows that the effect of ε_1 on temperature profile for fixed values of *S, Ri* and ε_2. It can be found that increase in ε_1 leads the decrease in dimensionless temperature. The effective temperature difference between the plate and the ambient fluid will decrease under the influence of thermal stratification effect. Hence the temperature reduces as the thermal boundary layer thickened. At fixed values of S = 0.5, Ri=2 and ε_2=0.1, the thermal stratification parameter effect on concentration profile is shown in Fig. 7. It is observed that the dimensionless concentration increases with an increase in ε_1.

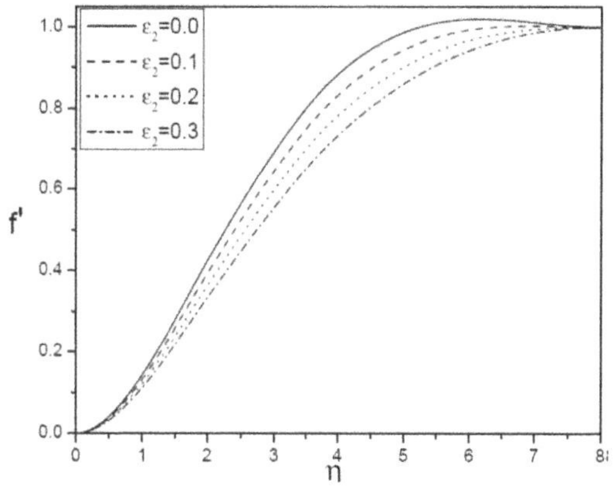

Fig. 8 Effect of solutal stratification parameter ε_2 on velocity when $S = 0.5$, $Ri = 2.0$, $\varepsilon_1 = 0.05$

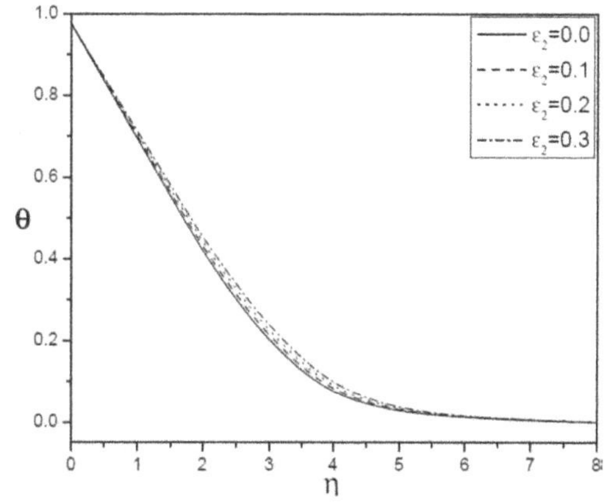

Fig. 9 Effect of solutal stratification parameter ε_2 on velocity when $S = 0.5$, $Ri = 2.0$, $\varepsilon_1 = 0.05$

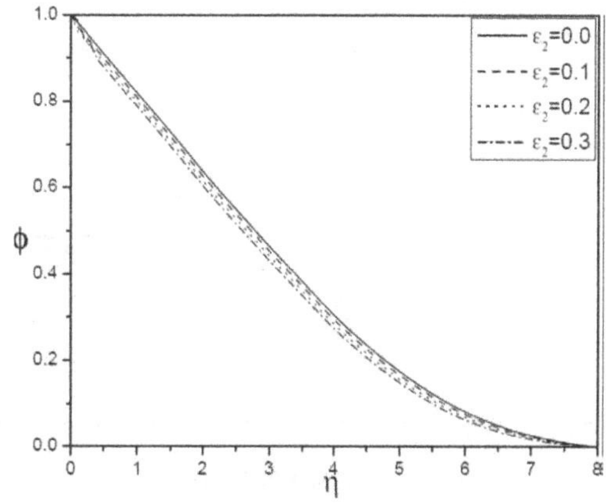

Fig. 10 Effect of solutal stratification parameter ε_2 on velocity when $S = 0.5$, $Ri = 2.0$, $\varepsilon_1 = 0.05$

Figure 8 analyzes the effect of ε_2 on the velocity profile by taking S = 0.5, Ri = 2 and ε_1 =0.05. We see that the dimensionless velocity decreases as an increase in ε_2. Figure 9 represents the ε_2 effect on dimensionless temperature profile at S = 0.5, Ri = 2 and ε_1 =0.05. As the solutal stratification increases, the dimensionless temperature profile increases. The effect of ε_2 on concentration profile can be seen in Fig.

10. It can be seen from the figure that as an increase in ε_2 leads to decrease in concentration of the fluid.

In Figs. 11-13, the profiles of the dimensionless velocity, temperature and concentration is shown with the influence of couple stress parameter S for fixed values of $Ri = 2.0$, ε_1=0.1 and ε_2=0.1. It can be seen that the maximum velocity decreases in amplitude and the location of the maximum velocity moves far away from the wall as an increase in *S*. Since in couple stress fluid the rotational field of the velocity will be amplified. Also, as there is an increase in *S*, the temperature profile increases. It is also observed that the concentration profile increases as an increase in *S*.

The dimensionless heat and mass transfer rates are presented in Table 2 at different values of *Ri*, ε_1 and ε_2 for fixed value of $S = 0.5$. It is observed that increase in ε_1 leads to increase in heat transfer coefficient and decrease in mass transfer coefficient. But in the case of increase in ε_2, the reverse trend is found. Also the effect of *Ri* on Nusselt and Sherwood is shown in the same Table 2 with fixed values of ε_1 and ε_2. It can be seen that there is an increase in nature on heat and mass transfer rates when *Ri* increases. Therefore the mixed convection parameter plays key role in controlling temperature and concentration of the flow.

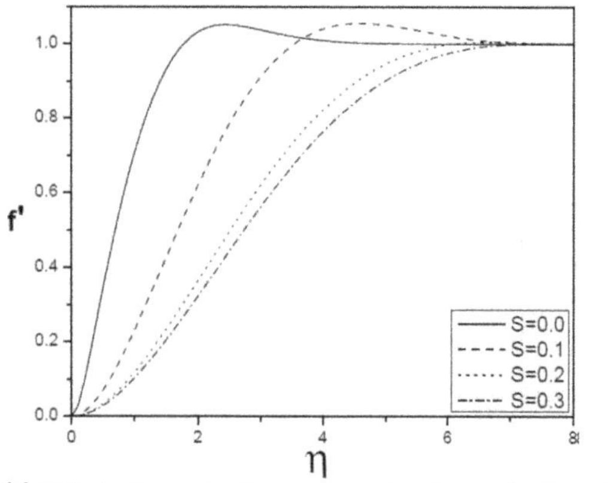

Fig. 11 Effect of couple stress parameter *S* on velocity when $Ri = 2.0$, $\varepsilon_1 = 0.1$, $\varepsilon_2 = 0.1$

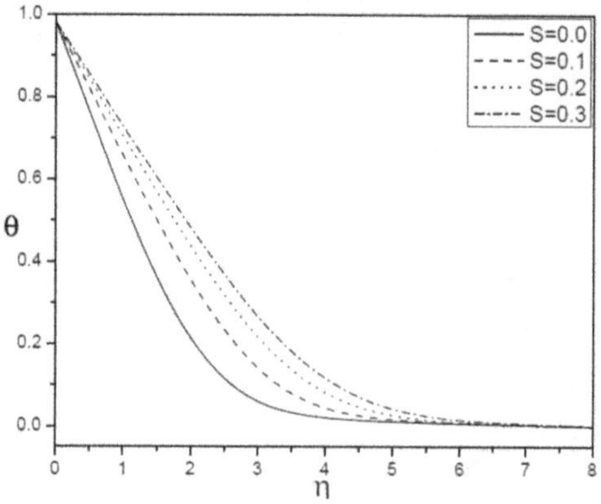

Fig. 12 Effect of couple stress parameter *S* on Temperature at $Ri = 2.0$, $\varepsilon_1 = 0.1$, $\varepsilon_2 = 0.1$

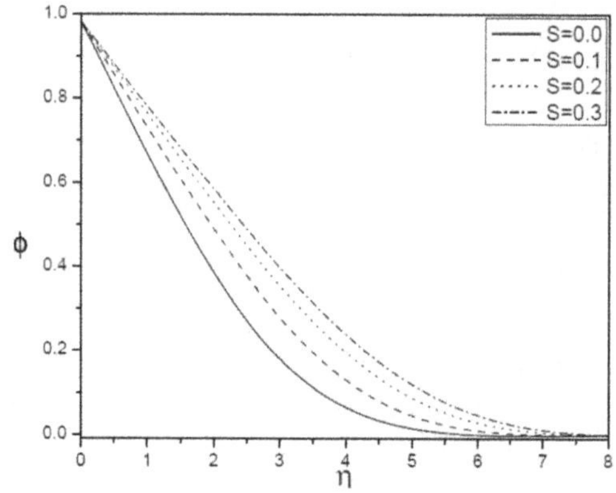

Fig. 13 Effect of couple stress parameter *S* on concentration at $Ri = 2.0$, $\varepsilon_1 = 0.1$, $\varepsilon_2 = 0.1$

Table 2: Variation of non-dimensional heat mass transfer coefficients versus ε1 and ε2 for different values of *Ri* with $S = 0.5$

		$Nu_x Re_x^{-1/2}$				$Sh_x Re_x^{-1/2}$			
ε_1	ε_2	*Ri*=0.5	*Ri*=1	*Ri*=1.5	*Ri*=2.0	*Ri*=0.5	*Ri*=1	*Ri*=1.5	*Ri*=2.0
0	0.1	0.21632	0.21972	0.22295	0.22602	0.18454	0.18684	0.18896	0.19117
0.1	0.1	0.22417	0.22773	0.23111	0.23432	0.18453	0.18682	0.18896	0.19113
0.2	0.1	0.23202	0.23573	0.23926	0.24261	0.18452	0.1868	0.18899	0.1911
0.3	0.1	0.23987	0.24374	0.24741	0.2509	0.18451	0.18678	0.18902	0.19106
0.4	0.1	0.24772	0.25173	0.25596	0.25939	0.1845	0.18676	0.18905	0.19102
0.05	0	0.24389	0.24792	0.25175	0.25544	0.18386	0.18613	0.18831	0.19040
0.05	0.1	0.24379	0.24773	0.25148	0.25512	0.18451	0.18677	0.18895	0.19103
0.05	0.2	0.2437	0.24755	0.25121	0.2548	0.18515	0.18741	0.18958	0.19166
0.05	0.3	0.2436	0.24736	0.25094	0.25449	0.1858	0.18805	0.19021	0.19228
0.05	0.4	0.2435	0.24717	0.25102	0.25417	0.18644	0.18868	0.19085	0.19289

4. CONCLUSIONS

In this present study, mixed convection and boundary layer analysis for couple stress fluid in presence of uniform wall temperature and concentration with thermal and solutal stratification effects is presented. The pseudo-similarity variables are used to transform the governing equations to non-dimensional form. The numerical solution is provided in terms of the profiles of dimensionless velocity, temperature, concentration and heat mass transfer rates. From this study the main findings are:

- Velocity, non-dimensional heat and mass transfer coefficients increases and temperature and concentration profiles are reduces as *Ri* increases.
- As ε_1 increases, the velocity, temperature distributions and non-dimensional mass transfers decreases where as non-dimensional heat transfer coefficient, concentration distribution in boundary layer are increases.
- The dimensionless velocity, concentration distributions and non-dimensional heat transfer coefficients reduces and the temperature distribution, nondimensional mass transfer coefficient increases as an increase in ε_2.
- It is noticed that the presence of couple stresses in the fluid decreases the velocity and increases the temperature and concentration.

NOMENCLATURE

A	Slope of ambient temperature.
B	Slope of ambient concentration.
C	Concentration.
$C_{\infty,0}$	Ambient concentration.
D	Solutal diffusivity.
f	Reduced stream function.
g	Acceleration due to gravity.
Gr	Thermal Grashof number.
k	Thermal conductivity.
L	Length of the plate.
Nu_x	Local Nusselt number.
N	Buoyancy ratio.
Pr	Prandtl number.
q_w	Heat transfer from the plate.
m_w	Mass transfer from the plate.
S	Couple Stress Parameter.
Sc	Schmidt number.
Sh_x	Local Sherwood number.
T	Temperature.
$T_{\infty,0}$	Ambient temperature.
u_∞	Characteristic velocity.
u, v	Velocities in x and y directions.
x, y	Coordinates along and normal to the plate.
α	Thermal diffusivity.
β_T	Thermal expansion coefficient.
β_C	Solutal expansion coefficient.
γ	Spin-gradient viscosity.
η	Pseudo-similarity variable.
η_1	The coupling material constant.
θ	non-Dimensional temperature.
ϕ	non-Dimensional concentration.
μ	Dynamic viscosity.
υ	Kinematic viscosity.
ξ	non-Dimensional stream wise coordinate.
ρ	fluid Density.
ψ	Stream function.
ε_1	Parameter Thermal stratification.
ε_2	Parameter of Solutal stratification.
Subscripts	
w	Wall condition.
∞	Ambient condition.
C	Concentration
T	Temperature
Superscript	
'	Differentiation with respect to η.

REFERENCES

Bejan, A., 1994, *Convection Heat Transfer*, John Wiley.

Cebeci, T. and Bradshaw, P., 1984, *Physical and Computational Aspects of Convective Heat Transfer*, Springer-Verlin, New York.

Hadjesfandiari, A. R., Dargush, G. F., and Hajesfandiari, A., 2013, "Consistent Skew-symmetric Couple Stress Theory for Size-Dependent Creeping Flow," *Journal of Non-Newtonian Fluid Mechanics*,**196**, 83–94. http://DOI10.1016/j.jnnfm.2012.12.012.

Hayat, T., Mustafa, M., Iqbal, Z., and Alsaedi, A., 2013, "Stagnation-Point Flow of Couple Stress Fluid with Melting Heat Transfer," *Appl. Math. Mech. -Engl. Ed*, **34**(2), 167–176. http://10.1007/s10483-013-1661-9.

Jaluria, Y. and Himasekhar, K., 1983, "Buoyancy Induced Two Dimensional Vertical Flows in a Thermally Stratified Environment," *Computer and Fluids*, **11**(1), 39–49. http://10.1016/0045-7930(83)90012-9

Lloyd, J. R. and Sparrow, E. M., 1970, "Combined Free and Forced Convective Flow on Vertical Surfaces," *Int. J. Heat Mass Transfer*, **13**, 434–438. http://10.1016/0017-9310(70)90119-5.

Makinde, O., and Eegunjobi, A., 2013, "Entropy Generation in a Couple Stress Fluid Flow through a Vertical Channel Filled with Saturated Porous Media," *Entropy*, **15**, 4589–4606. http://doi:10.3390/e15114589.

Murthy, P. V. S. N., Srinivasacharya, D., and Krishna, P. V. S. S. S. R., 2004, "Effect of Double Stratification on Free Convection in Darcian Porous Medium," *J. Heat Transfer*, **126**(2), 297–300. http://doi:10.1115/1.1667525.

Muthuraj, R., Srinivas, S., and Selvi, R., 2013, "Heat and Mass Transfer Effects On MHD Flow of a Couple-Stress Fluid in a Horizontal Wavy Channel with Viscous Dissipation and Porous Medium," *Heat Transfer-Asian Research*, **42**(5), 403–421. http://10.1002/htj.21040.

Najeeb, A. K., Mahmood, A., and Asmat, A., 2013, "Approximate Solution of Couple Stress Fluid with Expanding or Contracting Porous Channel," *Engineering Computations*, **30**(3), 399 – 408. http://10.1108/02644401311314358.

Patil, P. and Chamkha, A. J., 2013, "Heat and Mass Transfer from Mixed Convection Flow Of Polar Fluid along A Plate In Porous Media With Chemical Reaction," *Int. J. Numer. Methods Heat Fluid Flow*, **23**(5), 899–926. http://10.1108/HFF-03-2011-0060.

Prandtl, L., 1952, *Essentials of Fluid Dynamics*, London: Blackie.

Somers, E. V., 1956, "Theoretical Considerations of Combined Thermal and Mass Transfer from a Flat Plate," *ASME J. Appl. Mech*, **23**, 295–301.

Srinivasacharya, D. and Kaladhar, K., 2012a, "Mixed Convection Flow of Couple Stress Fluid between Parallel Vertical Plates with Hall and Ion-Slip Effects," *Commun Nonlinear Sci Numer Simulat*, **17**(6), 2447–2462. http://DOI:10.1016/j.cnsns.2011.10.006.

Srinivasacharya, D. and Kaladhar, K., 2012b, "Natural Convection Flow of A Couple Stress Fluid between Two Vertical Parallel Plates with Hall and Ion-slip Effects," *Acta Mech. Sin*, **28**(1), 41–50. http://10.1007/s10409-011-0523-z.

Srinivasacharya, D. and Kaladhar, K., 2013, "Analytical Solution for Hall and Ion-slip Effects on Mixed Convection Flow of Couple Stress Fluid between Parallel Disks," *Math. Comput. Modell*, **57**(9-10), 2494–2509 http://DOI:10.1016/j.mcm.2012.12.036

Srinivasacharya, D., Pranitha, J., and RamReddy, C., 2011, "Magnetic Effect on Free Convection in a Non-Darcy Porous Medium Saturated with Doubly Stratified Power-Law Fluid," *J. of the Braz. Soc. of Mech. Sci. & Eng*, **33**(1), pp. 8–14. http://dx.doi.org/10.1590/S1678-58782011000100002.

Stokes, V. K., 1966, "Couple stresses in fluid," *Physics of fluids*, 1709–1715.

Szewczyk, A. A., 1964, "Combined Forced and Free-Convection Laminar Flow," *J. Heat Transfer*, **C86**(4), 501–507.

Permissions

All chapters in this book were first published in FHMT, by Global Digital Central; hereby published with permission under the Creative Commons Attribution License or equivalent. Every chapter published in this book has been scrutinized by our experts. Their significance has been extensively debated. The topics covered herein carry significant findings which will fuel the growth of the discipline. They may even be implemented as practical applications or may be referred to as a beginning point for another development.

The contributors of this book come from diverse backgrounds, making this book a truly international effort. This book will bring forth new frontiers with its revolutionizing research information and detailed analysis of the nascent developments around the world.

We would like to thank all the contributing authors for lending their expertise to make the book truly unique. They have played a crucial role in the development of this book. Without their invaluable contributions this book wouldn't have been possible. They have made vital efforts to compile up to date information on the varied aspects of this subject to make this book a valuable addition to the collection of many professionals and students.

This book was conceptualized with the vision of imparting up-to-date information and advanced data in this field. To ensure the same, a matchless editorial board was set up. Every individual on the board went through rigorous rounds of assessment to prove their worth. After which they invested a large part of their time researching and compiling the most relevant data for our readers.

The editorial board has been involved in producing this book since its inception. They have spent rigorous hours researching and exploring the diverse topics which have resulted in the successful publishing of this book. They have passed on their knowledge of decades through this book. To expedite this challenging task, the publisher supported the team at every step. A small team of assistant editors was also appointed to further simplify the editing procedure and attain best results for the readers.

Apart from the editorial board, the designing team has also invested a significant amount of their time in understanding the subject and creating the most relevant covers. They scrutinized every image to scout for the most suitable representation of the subject and create an appropriate cover for the book.

The publishing team has been an ardent support to the editorial, designing and production team. Their endless efforts to recruit the best for this project, has resulted in the accomplishment of this book. They are a veteran in the field of academics and their pool of knowledge is as vast as their experience in printing. Their expertise and guidance has proved useful at every step. Their uncompromising quality standards have made this book an exceptional effort. Their encouragement from time to time has been an inspiration for everyone.

The publisher and the editorial board hope that this book will prove to be a valuable piece of knowledge for researchers, students, practitioners and scholars across the globe.

List of Contributors

Keunhan Park
Department of Mechanical, Industrial and Systems Engineering, University of Rhode Island, Kingston, RI 02881, USA

Zhuomin Zhang
G.W. Woodruff School of Mechanical Engineering, Georgia Institute of Technology, Atlanta, GA 30332-0405, USA

Tapas Ray Mahapatra
Department of Mathematics, Visva-Bharati, Santiniketan - 731 235, India

Sumanta Sidui
Department of Mathematics, Ajhapur High School, Burdwan - 713 401, India

Samir Kumar Nandy
Department of Mathematics, A.K.P.C Mahavidyalaya, Hooghly - 712 611, India

B. V. K. Reddy, Matthew Barry, John Li and Minking K. Chyu
Department of Mechanical Engineering and Materials Science, University of Pittsburgh, Pittsburgh, PA 15261, USA

Ravishankar Sathyamurthya, Hyacinth J. Kennadya and T. S. Ravikumar
Department of Mechanical Engineering, Hindustan Institute of Technology and Science, Chennai, Tamil Nadu, 603103, India

P.K. Nagarajan
Department of Mechanical Engineering, S.A. Engineering College, Chennai, Tamil Nadu, India

V. Paulson
Department of Aeronautical Engineering, Hindustan Institute of Technology and Science, Chennai, Tamil Nadu, 603103, India

Amimul Ahsan
Department of Civil Engineering, Green Engineering & Sustainable Technology Lab, Institute of Advanced Technology, University Putra Malaysia, Selangor, Malaysia

Tomoki Hirokawa, Masahiko Murozono and Haruhiko Ohta
Department of Aeronautics and Astronautics, Kyushu University, Fukuoka, 813-0385, Japan

Oleg Kabov
Institute of Thermophysics, Russian Academy of Science, Siberian branch, Novosibirsk, 630090, Russia
Tomsk Polytechnic University, Tomsk, 634050, Russia

Mahyar Kargaran and Mahmood Farzaneh-Gord
Faculty of Mechanical Engineering, Shahrood University of Technology, Shahrood, Iran

Junjie Chen, Longfei Yan and Wenya Song
School of Mechanical and Power Engineering, Henan Polytechnic University, Jiaozuo, Henan, 454000, China

Krishnendu Bhattacharyya
Department of Mathematics, The University of Burdwan, Burdwan-713104, West Bengal, India

Suripeddi Srinivas
Fluid Dynamics Division, School of Advance Sciences, VIT University, Vellore, Tamil Nadu, 632014, India

Akshay Gupta and Ashish Kumar Kandoi
School of Mechanical and Building Sciences, VIT University, Vellore, Tamil Nadu,632014, India

Manimaran Renganathan and Thundil Karuppa Raj Rajagopal
School of Mechanical & Building Sciences, VIT University, Vellore-632014, Tamil Nadu, India

A. Najjaran, Ak. Najjaran, A. Fotoohabadi and A.R. Shiri
Islamic Azad University Branch of Shiraz, Shiraz, Fars, Iran, 7154845589, Iran

Odelu Ojjela and N. Naresh Kumar
Department of Applied Mathematics Defence Institute of Advanced Technology, Deemed University, Pune - 411025, India

Sally M. Smith, Brenton S. Taft and Jacob Moulton
Air Force Research Laboratory Space Vehicles Directorate, Kirtland AFB, NM 87117, USA

Hsiu-Hung Chen and Chung-Lung Chen
Department of Mechanical and Aerospace Engineering, University of Missouri, Columbia, MO 65211

Yuan Zhao
Teledyne Scientific, Thousand Oaks, CA 91360

Machireddy Gnaneswara Reddy
Department of Mathematics, Acharya Nagarjuna University Campus, Ongole - 523 001, Andhra Pradesh, India

Bradley Doleman and Messiha Saad
North Carolina A&T State University, Greensboro, NC, 27411, USA

Dipak Sen, Probir Kumar Bose, Rajsekhar Panua, Ajoy Kumar Das and Pulak Sen
Mechanical Engineering Department, National Institute of Technology, Agartala , Tripura, 799055, India

Greg Ball, John Polansky and Tarik Kaya
Department of Mechanical and Aerospace Engineering, Carleton University, Ottawa, Ontario, K1S 5B6, Canada

Heather Dillon
Dept of Mechanical Engineering, University of Portland, Portland, Oregon, 97203, USA

Ashley Emery and Ann Mescher
Dept of Mechanical Engineering, University of Washington, Seattle, Washington, 98105, USA

Foued Chabane, Noureddine Moummi and Abdelhafid Brima
Mechanical Department, Faculty of Technology, University of Biskra 07000, Algeria
Mechanical Laboratory, Faculty of Technology, University of Biskra 07000, Algeria

Said Benramache
Material Science Departments, Faculty of Science, University of Biskra 07000, Algeria

Biswajit De and Rajsekhar Panua
Mechanical Engineering Department, National Institute of Technology, Agartala, Tripura, 799055, India

Shihong Zhang and Zhihua Wang
School of Environment and Energy Engineering, Beijing University of Civil Engineering and Arch., Beijing, China

Renganathan Manimaran
Thermal and Automotive Research Group, School of Mechanical and Building Sciences, VIT Chennai, Tamilnadu, India

Rajagopal Thundil Karuppa Raj
Energy Division, School of Mechanical and Building Sciences, VIT Vellore, Tamilnadu, India

Shigenao Maruyama, Junnosuke Okajima and Atsuki Komiya
Institute of Fluid Science, Tohoku University, Katahira, Aoba-ku, Sendai 980-8577, Japan

Masud Behnia
School of Aerospace, Mechanical and Mechatronic Engineering, The University of Sydney, NSW 2006,Australia

Masasazumi Chisaki and Takuma Kogawa
School of Engineering, Tohoku University, Aoba 6-6, Aramaki-aza, Aoba-ku, Sendai 980-8579, Japan

Vijayakumar Thulasi and Thundil Karuppa Raj Rajagopal
School of Mechanical and Building Sciences, VIT University, Vellore 632014, India

Biswajit De and Rajsekhar Panua
Mechanical Engineering Department, National Institute of Technology, Agartala, Tripura, 799055, India

Manoj Kr. Triveni, Dipak Sen and RajSekhar Panua
National Institute of Technology, Agartala, Tripura, 799055, India

Etim S. Udoetok
Mechanical Engineering Department, University of Uyo, Uyo, Nigeria

K. Kaladhar
Department of Mathematics, National Institute of Technology Puducherry, Karaikal-609605, India

D. Srinivasacharya
Department of Mathematics, National Institute of Technology, Warangal-506004, India

www.ingramcontent.com/pod-product-compliance
Lightning Source LLC
Chambersburg PA
CBHW080518200326
41458CB00012B/4249
9781632385130